The Fordism of Ford and Modern Management
Volume II

Wherever possible, the articles in these volumes have been reproduced as originally published using facsimile reproduction, inclusive of footnotes and pagination to facilitate ease of reference.

For a list of all Edward Elgar published titles visit our site on the World Wide Web at www.e-elgar.com

The Fordism of Ford and Modern Management

Fordism and Post-Fordism

Volume II

Edited by

Huw Beynon

Director
School of Social Sciences, Cardiff University, UK

and

Theo Nichols

Distinguished Research Professor
School of Social Sciences, Cardiff University, UK

An Elgar Reference Collection
Cheltenham, UK • Northampton, MA, USA

Published by
Edward Elgar Publishing Limited
Glensanda House
Montpellier Parade
Cheltenham
Glos GL50 1UA
UK

Edward Elgar Publishing, Inc.
136 West Street
Suite 202
Northampton
Massachusetts 01060
USA

A catalogue record for this book is available from the British Library

Library of Congress Cataloguing in Publication Data

The Fordism of Ford and modern management : Fordism and post-Fordism / edited by
Huw Beynon and Theo Nichols.
p. cm. — (Elgar mini series) (Elgar reference collection)
Includes bibliographical references and index.
1. Industrial management. 2. Industrial organization. 3. Industrial relations. I. Beynon, Huw. II. Nichols, Theo. III. Series. IV. Series: Elgar reference collection

HD31.F5727 2006
658—dc22 2006042365

ISBN-13: 978 1 85898 948 8 (2 volume set)
ISBN-10: 1 85898 948 5 (2 volume set)

Printed and bound in Great Britain by MPG Books Ltd, Bodmin, Cornwall.

Contents

Acknowledgements

The editors and publishers wish to thank the authors and the following publishers who have kindly given permission for the use of copyright material.

American Sociological Association for article: Steven P. Vallas (1999), 'Rethinking Post-Fordism: The Meaning of Workplace Flexibility', *Sociological Theory*, **17** (1), March, 68–101.

Blackwell Publishing Ltd for articles: Harley Shaiken, Stephen Herzenberg and Sarah Kuhn (1986), 'The Work Process Under More Flexible Production', *Industrial Relations*, **25** (2), Spring, 167–83; Ash Amin (1989), 'Flexible Specialisation and Small Firms in Italy: Myths and Realities', *Antipode*, **21** (1), April, 13–34; James P. Womack (1996), 'The Psychology of Lean Production', *Applied Psychology: An International Review*, **45** (2), 119–22.

Labour and Industry for article: Thomas Bramble (1988), 'The Flexibility Debate: Industrial Relations and New Management Production Practices', *Labour and Industry*, **1** (2), June, 187–209.

Monthly Review Foundation for article: James Rinehart (1999), 'The International Motor Vehicle Program's Lean Production Benchmark: A Critique', *Monthly Review*, **50** (8), January, 19–27.

Anna Pollert for her own article: (1988), 'Dismantling Flexibility', *Capital and Class*, **34**, Spring, 42–75.

Sage Publications, Inc. for article: Bruce Pietrykowski (1999), 'Beyond the Fordist/Post-Fordist Dichotomy: Working through *The Second Industrial Divide*', *Review of Social Economy*, **LVII** (2), June, 177–98.

Sage Publications Ltd for articles: Chris Smith (1989), 'Flexible Specialisation: Automation and Mass Production', *Work, Employment and Society*, **3** (2), June, 203–20; Christian Berggren (1993), 'Lean Production – The End of History?', *Work, Employment and Society*, **7** (2), June, 163–88; George Ritzer (1996), 'The McDonaldization Thesis: Is Expansion Inevitable?', *International Sociology*, **11** (3), 291–308; Ben Dankbaar (1997), 'Lean Production: Denial, Confirmation or Extension of Sociotechnical Systems Design?', *Human Relations*, **50** (5), May, 567–83; Ian Hampson (1999), 'Lean Production and the Toyota Production System – Or, the Case of the Forgotten Production Concepts', *Economic and Industrial Democracy*, **20** (3), 369–91; Andy Danford (2003), 'Workers, Unions and the High Performance Workplace', *Work, Employment and Society*, **17** (3), 569–73.

Taylor and Francis Ltd (http://www.tandf.co.uk/journals) for articles: Karel Williams, Tony Cutler, John Williams and Colin Haslam (1987), 'The End of Mass Production?', *Economy and Society*, **16** (3), August, 405–39; Mike Parker and Jane Slaughter (1990), 'Management-by-Stress: The Team Concept in the US Auto Industry', *Science as Culture*, **8**, 27–58; Paul Hirst and Jonathan Zeitlin (1991), 'Flexible Specialization versus post-Fordism: Theory, Evidence and Policy Implications', *Economy and Society*, **20** (1), February, 1–56.

John Tomaney for his own article: (1990), 'The Reality of Workplace Flexibility', *Capital and Class*, **40**, Spring, 29–60.

Tribune Media Services International and Massachusetts Institute of Technology for article: Michael A. Cusumano (1994), 'The Limits of "Lean"', *Sloan Management Review*, **35** (4), Summer, 27–32.

University of California Press and the Copyright Clearance Center's Rightslink Service for article: Steven P. Vallas and John P. Beck (1996), 'The Transformation of Work Revisited: The Limits of Flexibility in American Manufacturing', *Social Problems*, **43** (3), August, 339–61.

Every effort has been made to trace all the copyright holders but if any have been inadvertently overlooked the publishers will be pleased to make the necessary arrangement at the first opportunity.

In addition the publishers wish to thank the Marshall Library of Economics, University of Cambridge, UK, the Library at the University of Warwick, UK, and the Library of Indiana University at Bloomington, USA, for their assistance in obtaining these articles.

Part I
Flexible Specialisation, the Third Italy and the Wider Debate

[1]

Fergus Murray

The paper discusses the growing importance of the decentralisation of production, as one capitalist response to declining profits and workers' resistance in Italian manufacturing industry. It argues that decentralisation and automation have reduced the traditional strength and quantity of male workers in large factories and have generated new sectors within the industrial working class. The paper ends with the suggestion that the labour movement needs to reshape its organisation and its strategies which erroneously still continue to reflect only the needs of the traditional mass worker.

The decentralisation of production —the decline of the mass-collective worker?

IN THIS PAPER I want to examine one of the changes that have been taking place in the organisation of production and the labour process since the early 1970s, that is, the decentralisation of production. While the geographical dispersal of production is a long established feature of capitalism, in the last ten years decentralisation has undergone a quantitative increase and qualitative change. For example, in Italy large firms have reduced plant size, split-up the production cycle between plants, and increased the putting-out of work to a vast and growing network of small firms, artisan workshops, and domestic outworkers.[1] In Japan large firms using advanced production techniques have insisted that their small supplier firms raise productivity through technological innovation, while moves are underway to link the small firms by computer to the large ones, thereby greatly increasing the control of the large corporations over production. In America and Britain increasingly mobile international capital in high technology small units has been moving into areas of high unemployment, for example in the southern 'sun belt' states of the US and in S. Wales and Scotland in Britain, where careful labour recruitment exploits and exacerbates the segmentation of the labour market and divisions in the working class. And recently a statement in the Soviet press drew attention to decentralisation when it criticised the way in which Russian industrialisation continues to be based on huge factories and proposed a policy for the reduction of plant size and the development of small, flexible, highly specialised and technologically advanced production units. The article cited the example of General Electric which continues to reduce plant size despite the fact that all its 400,000 employees already work in factories of less than 1,500 workers.[2].

There is then a growing body of evidence which challenges the idea that the progressive centralisation and concentration of capital necessarily leads to a physical concentration of production that the small production unit is the remnant of a disappearing traditional, backward sector of production. For generations Marxists have assumed that the tendency of capitalism was to the greater and greater concentration of production, and massification of the proletariat. Indeed there were excellent historical reasons for making this assumption, as the development of both the basic commodity industries of the first industrial revolution and the mass production industries of the post-war boom led to a high concentration of workers in large integrated plants in large industrial towns.[3] Nevertheless the above evidence suggests that the size and location of production cannot be drawn from theoretical premises but rather that they are historically determined, depending on the particular circumstances capitalist production faces in different periods.

This paper draws on empirical material from Italy to show how the use of decentralisation has been intensified and has changed through the introduction of new technology as Italy's dominant firms have sought to restructure production in their struggle against declining profitability. In Italy the combination of automation and decentralisation has been specifically aimed at destroying the power and autonomy of the most militant and cohesive section of the Italian proletariat and this strategy has met with considerable success. This suggests that the political hopes pinned on the mass-collective worker in the seventies need to be carefully reconsidered in the light of decentralisation and the recomposition of the proletariat this implies.

The paper is organised as follows:

The first section examines the determinants of the dominant organisational form of post-war industry, the large factory, and suggests that this form is historically specific, being contingent on the balance of class forces and the technologies available to capital.

Using empirical material the second section attempts to define the different forms of decentralisation in order to bring out the wide variety of different workplaces and workers which decentralisation creates through its physical fragmentation of the labour process.

The third section analyses the way in which the application of information technology in production management not only gives capital a greater potential control over labour in the large factory, but also gives it the possibility of coordinating production and labour exploitation that is increasingly dispersed in small production units, artisan workshops and 'home-factories'.

The last part of the paper suggests that decentralisation has created new divisions in the industrial working class by increasing the number of workers living and working in conditions that greatly differ from those of the mass-collective worker. The transformation of the large factory and the rise of small production units has made collective action considerably more difficult. The paper ends by asking how both old and new divisions can be effectively challenged by the labour movement and the left, with a strategy and organisations that give voice to the different needs and desires of different parts of the proletariat, while also giving them a unity that can overcome divisions rather than exacerbating them.

The large factory: Is it inevitable?

The term 'decentralisation of production' has been used in Italy to describe a number of distinct features of the organisation of production. In general, decentralisation refers to the geographical dispersal and division of production, and particularly to the diffusion and fragmentation of labour. However this can take place in a number of ways:

i) The expulsion of work formerly carried out in large factories to a network of small firms, artisans or domestic outworkers.

ii) The division of large integrated plants into small, specialised production units.

iii) The development of a dense small firm economy in certain regions such as the Veneto and Emilia Romagna in Italy.

In Italy 'decentralisation' has been used to cover all the above developments. In this paper 'decentralisation' is used to refer to the expulsion of production and labour from large factories, either in the form of in-house decentralisation (splitting-up) or inter-firm decentralisation (putting-out) within the domestic economy. This is because the paper focuses on the way large and medium firms in Italy have used decentralisation to reduce costs and increase labour exploitation, rather than on the development of districts of independent small firms that are not directly subordinate to larger firms. The analysis of this latter process has been an important part of the Italian debate on decentralisation (e.g. Brusco, 1982; Paci, 1975; Bagnasco et al, 1978).

DECENTRALISATION

An assumption has prevailed that large corporations operating in such sectors as engineering and electronics will organise production in large factories, in that they will amass large amounts of fixed capital and workers in particular, on any given site. However factory size is not given, and least of all does not necessarily correspond with the size of a firm or corporation's turnover, or their market and financial strength. Rather it is determined by the specific configuration of the conditions for profitable production prevailing in any given period. For example, the integrated car plant developed in rapidly expanding markets, with the balance of class forces intially in capital's favour, which made possible and profitable a particular combination of technology (mechanised flow line production) and labour domination (Taylorism). It was the coincidence of all these factors that made the integrated plant the most profitable form of production organisation in the post-war consumer durables industries. When labour rebelled and markets began to stagnate the 'efficiency' of this form of production was undermined and both capitalists and bourgeois economics discovered 'diseconomies of scale'. The ending of the long wave of expansion, the development of new technologies, and new management techniques have all contributed to change the form of the division of labour and the labour process within the large corporation. Five of the more important factors that influence factory size are the type of product being made, the technologies available, product control, industrial relations and State legislation. I shall consider the role of these factors in turn.

Product Type

Product type is important in determining the degree to which the production cycle for a given product can be divided between separate factories. Industries where there is a high divisibility of the production cycle include aeronautics, machinery, electronics, clothes, shoes, and furniture. In contrast the steel and chemical industries tend to require a large unified production site, although the optimum plant size is not always as large as some people, for example BSC management, think (Manwaring, 1981:72).

One particularly important development that has been taking place in the structure of some products is a process known as modularisation. Although there has been a diversification in the number of models in many ranges of consumer goods, this has been underlain by a standardisation of the major sub-assembled parts of the product. These sub-assembled parts are the basic modules of the product and can be made in different factories and put together at a later date. For example, as argued in Del Monte (1982:154-6) at one time televisions were assembled in a linear manner on a long assembly line. The frame of the television would be put on the line, and individual parts then added to it. In modular production each module is assembled separately, and a much shorter process of final assembly is required. At present modular production is mainly limited to commodities from the electronics sector, but advances in product redesign facilitated by the introduction of micro-electronic components suggest that it will be used elsewhere. (See the example of Fiat later.) If we recall how the bringing together of large numbers of workers on assembly lines in the sixties fuelled workers' spontaneous struggles, modular production, plus the increasing automation of the assembly areas themselves, can serve as important weapons for capital in reducing worker militancy through decentralisation.

Technology

Brusco (1975) argues that Marx's explanation for the concentration of production in large factories was partly based on the necessity of running machines from a central energy source – the steam engine. As steam was replaced by electricity as the principal energy source for industry this particular decentralising tendency was weakened. Initially the expense of electric engines meant that one central engine and a system of transmission shafts and belts were used to drive the different machines. But as electrical technology developed and the price of engines fell, each machine was fitted with its own motor.[4] Other technological changes that affect the product and the organisation of production include shifts in materials, for instance, from steel to plastics, but the most important change that has been taking place in the last decade is the introduction of the microchip into the production of many commodities. While the microchip tends to a lessening of worker control over machines, it is also changing the nature of those machines. Generally there is a trend towards a replacement of electro-mechanical parts with microelectronic components, and from worker control of the machine to the installation of the unit of control in the machine which leads to changes in the production of the product and its associated labour process. Olivetti has been transformed from an engineering multinational to an electrical one over twenty years, and in many engineering firms electrical control systems are now taking over from mechanical ones. This implies a reduction of machine shop work in production. It is also interesting to note that electrical work, such as wiring and the assembly of circuit boards has in some cases proved suitable for putting-out to tiny firms employing semi-skilled women workers – so suitable, according to Wood (1980) in Japan there are an estimated 180,000 domestic outworkers in the electrical components industry alone. Similarly, a firm in Bologna making control units for machine tools did some quite radical experimenting with decentralisation as it shifted from electro-mechanical to electronic control systems. According to the Bologna metalworkers' union (FLM Bologna, 1977:78), with the appearance of micro-electronics in the seventies the firm began to run down its machine shops and progressively intensified putting out which eventually accounted for 60% of production costs. At this time the firm employed about 500 workers directly and over 900 indirectly as outworkers. A couple of years later with the introduction of automation the firm re-centralised production and an estimated 600 outworkers lost their jobs.

There are then techological changes taking place that allow decentralisation and falling factory sizes but it needs to be stressed that these changes don't automatically lead to decentralisation. It is the particular capitalist's use of technology and the conditions of profitability that will determine how the organisation of production changes.

Product Control

The making of many commodities requires huge amounts of co-ordination and control of production and pressure to reduce dead time, stocks, and all types of idle capital has increased markedly since 1974. In a big plant, production is difficult to supervise at every level and the sheer size of the factory and the bureaucracy needed to run it can hide huge amounts of waste. This would suggest that for the capitalist the division of production and management into smaller and more easily controlled units would be a cost effective strategy.[9] The introduction of computer assisted management allows production to be split-up by making the co-ordination of production in different plants considerably easier. General Motor's new 'S' car, for example, is being built in GM's European production network which employs 120,000 workers split-up in 39 plants in 17 countries (*Financial Times*, 28.9.82)

Industrial Relations

The reduction of factory size and relocation of production are contingent upon the extent to which 'unfavourable' industrial relations are an important reason for restructuring in different industries in different countries. Prais (1982) suggests that factories in the UK with over 2,000 workers are 50 times more vulnerable to strikes than those with less than 100 workers, and he goes on to say in his academically refined union bashing tone, that big plants in UK car assembly, steel production, and shipbuilding develop endemic strikes "which impedes the pursuit of efficiency, and leads ultimately to self-destruction".(p.103).

In the late 60's labour militancy in many Italian industries reached levels that directly threatened firm profitability and management undertook a series of strategies designed initially to reduce the disruptiveness of militant workers. One of these strategies, decentralisation, was in part underlain by a management view, typified by the director of a Bologna engineering firm to whom I spoke, which saw a direct correlation between factory size and industrial relations in Italy in the 1970s. This director argued that a significant improvement in industrial relations could be achieved in a factory employing 100 rather than 1000 workers.

This is not to say there is an automatic relationship between industrial relations, labour militancy and factory size. Rather large plants in the post-war boom appear to have created conditions favourable to an intense and often 'unofficial' shop floor struggle that has been very disruptive for capital. It would be wrong therefore to equate the rise of smaller production units with the end of labour militancy on the shop floor. It seems that capitalists expect substantial 'improvements' in industrial relations from smaller scale production units. Clearly this will impose new and real difficulties for the autonomous organisation of workers and the forms it should take in small plants. However, the struggles at Plessey Bathgate and Lee Jeans have shown that these are not unsurmountable.

State Legislation

Central and local state legislation will be important in determining factory size and location in a number of ways. Incentives, grants subsidies, and factories themselves may all be used to persuade firms to set up additional sites, as can be seen by the unco-ordinated efforts of the various regional development agencies in the UK.

Employment legislation, and its implimentation, may also be very influential. In Italy important parts of the Worker's Statute do not apply in firms employing less than 15 workers. And the smaller the plant the more possibility there is of using illegal employment practices, such as the use of child labour, and the evasion of tax and national insurance payments[5].

DECENTRALISATION

Different forms of decentralisation

81

Using empirical material from the Bologna engineering industry, this section examines the two forms of decentralisation that have been used most extensively in Italy by large and medium sized firms. The intention here is to examine the way decentralisation changes the nature of work and workers and the relationships that exist between firms. An analysis of the relationships between firms is important for the left, especially in view of assessing the accuracy and implications of two trends that are supposedly taking place, one is the vertical disintegration of many corporations and the other is the growing wave of support, in Britain especially, for small business from the State and even the banks. On the basis of Macrae's analysis (1982), one would, think that the power of monopoly capital was withering away to open a new golden age for the entrepreneur. However, while it may be true that some corporations are withdrawing from direct control of some production this in no way implies a weakening of their power. Rather, through decentralisation these corporations may maintain a strict control over production while letting the small firm pay the costs and face the risks of production, thereby using decentralisation as a means for reducing and shifting the corporation's risks and losses. In this way corporations maintain their ability to cover fluctuating markets while concentrating on the most profitable areas of production. This of course, does not mean that *all* small firms are subordinate to a particular corporation and many may even find a degree of independence.[6]

Putting-out

Putting-out involves the transfer of work formerly done within a firm to another firm, an artisan workshop or to domestic outworkers. After the initial transfer, putting-out can be used to describe a semi-permanent relationship between firms.

Within the Italian economy putting-out appears to have contributed significantly to the rise of small firms and to the surprising shift that has taken place in industrial employment in the last ten years. In 1971 22.9% of the total industrial workforce were employed in 'mini-firms' of less than 19 employees. By 1978 this figure had risen to 29.4%, an expansion of employment in the 'mini-firms' of 345,000. Furthermore the number of men employed in these firms rose by only 8.3% in this period, whereas the number for women grew by 33.8%. While it is difficult to generalise from such disaggregated data, they do indicate a steady growth of employment in very small production units for which the putting-out and the geographical fragmentation of production have been partly responsible. The period from 1974-8 is particularly interesting as a fall of employment of 52,000 occured in firms of over 500 employees, whereas employment rose by 160,000 in the 'mini-firms' (see Celata, 1980:85)

In the Bologna engineering industry, in the period 1968-80, the number of artisan firms employing between 1-15 employees rose from 6,602 to 9,436, an increase of 42.9% and nearly a third of the Bologna engineering labour force of 88,000 was working in these workshops in 1980 (see FLM Emilia Romagna, 1981:18-19)

The existence of this dense network of artisans workshops and small firms and its expansion due to an initial restructuring of the Bologna engineering industry in the 1950s, has been one of the vital preconditions for the development of putting-out and the increasing division of labour between small firms. As the example that follows suggests, decentralisation has passed through two phases: a first phase between 1968-74 when putting-out was used less out of choice than necessity due to intense shop-floor struggles in the large and medium factories; and a second phase, since 1975, of more systematic use of decentralisation, with the introduction of information technology into production planning and the appearance of numerically controlled machine tools in increasingly specialised artisan shops, accompanined by a gradual reversal of some of labour's gains on the shop floor. In this second phase it is possible to see an implicit shift from the direct control of labour on the shop-floor in the large Taylorised factory to a more articulated and flexible system of the organisation of production where the labour process extends beyond the factory into the artisan workshop. In the artisan workshop the unmediated forces of the market that threaten the artisan's very existence ensure a high degree of 'self-exploitation' often reinforced by the paternalistic despotism of the small entrepreneur.

In the Bologna engineering industry there appear to be three motives for putting-out: to reduce fixed costs to a minimum; to benefit from wage differentials between firms; to maximise the flexibility of the production cycle and of labour exploitation. The nature of putting-out is examined below through its use in a Bologna precision engineering firm.

The strategy of this firm, according to the management, has been to invest in labour and machinery just below the level of minimum expected demand. Any increase of production above this level has been met by putting-out, rather than risking an expansion of the factory or the workforce. However, contrary to management's claims, it is not true that the size of the labour force has always depended upon the level of demand. Until 1969, that is until when the first big strikes occurred, the size of the workforce grew steadily. However, after 1969, although production output rose rapidly for a number of years, the level of employment of production workers and productivity in the firm, actually fell. It therefore appears that a decision was taken to limit employment in the firm as militancy on the shop-floor increased and to cover rising demand by massively raising putting-out. In 1972 46% of production work was put-out of the firm, employing indirectly the equivalent of 570 full-time workers in small firms and workshops, whereas in 1969 only 10% had been put-out. In 1974-5 production fell rapidly, and work put-out dropped to almost nothing, resulting in the loss of approximately 550 jobs. That is, while the level of employment in the firms working for the company went through a massive fluctuation, employment in the company itself was relatively stable. The company putting the work out did not then pay a penny of redundancy money and nor was there any disruptive and socially embarassing struggle over job losses. This illustrates clearly the flexibility putting-out can provide. In this instance the reason for putting-out was not so much the exploitation of wage differentials as the minimisation of costs and conflict over job losses with the union.

However, the same firm does also put-out work for savings on wages, where the outworkers are paid up to 50% less than their counterparts in the factory. The work put-out here is not mechanical work, but wiring and circuit board assembly and involves women working in small firms and sweatshops where they have no legal or union protection.

With the introduction of computer assisted management and with the changes taking place in modular design, the firm has recently overhauled its putting-out system. Formerly, work of a once only basis was put-out to artisan shops the basis of very short lived and verbal agreements. The firm now encourages these artisans, who often employ less than five people, to group themselves together in order to amass the machinery and skills necessary for the production and sub-assembly of modules on a more regular basis. Meanwhile, management has won back some of its former power on the shop-floor with the help of computer aided production and an increase in internal labour mobility. The introduction of the computer has given management an increasingly refined control over the co-ordination of production both within and outside the factory, and putting-out is now used more routinely, while special and rush jobs are done in the factory due to the increased mobility of labour, achieved after six years of almost total rigidity.

Putting-out here has gone from a contingency solution of special problems to a more structured system. Initially flexibility was found in putting-out to artisan workshops to get around rigidity in the factory. Now it is the whole system, factory production and putting-out, that works to give flexibility.

Putting-out in Bologna engineering varies from skilled well-paid work using advanced technology to dirty dangerous and deskilled work. Within this there is a clear division of putting-out based on sexual and racial divisions in the labour market. The skilled workers and artisans are almost exclusively middle aged men, while women, the young, and migrants from the South of Italy and North Africa are concentrated in the dirtiest, most precarious and worst paid work.

The other extensive form of putting-out is to domestic outworkers in industries like clothing, electrical components, and toys. This form of putting-out has received a good deal more attention than putting-out to small firms (e.g. Young, 1981; Rubery and Wilkinson, 1981; Goddard, 1981) and will therefore not be dealt with here.

Another increasingly important type of putting-out is that which takes place across national frontiers where either parts of the production cycle are contracted out or the firm contracts out the production of the finished commodity it already makes, using its own specifications and technology for production in the sub-contracting firm and its marketing network for the sale of the commodity. An example of the former type of international putting-out is cited in Frobel et al (1980:108) and refers to the extensive use the West German textile industry makes of textile firms in Yugoslavia, where firms send out semi-finished products from Germany to be worked up into the final product. And an example of the latter can be found in Del Monte's study (1982) of the electronics industry in Southern Italy, where, again West German firms making televisions, contract out the production of complete sets to medium sized firms around Naples. The firms doing the work use the German firm's know-how and marketing services, not being big enough themselves to break into the world market. They, in turn, put out work to smaller firms in the area. (pp.150-1)

Putting-out then cannot be equated with an archaic and disappearing system of production. Rather it seems to have been reinforced as specific sectors of industry have faced altered condition in the harsh economic and political climate of the seventies as the long-wave of expansion ground to a halt. Therefore it would be mistaken to continue to segment firms in terms of the dualist opposition between large firms using high techology and small firms using outdated technology and traditional production techniques.[7]

Splitting-up production

The second form of decentralisation is the splitting-up of production between factories of the same firm. Clearly firms will relocate factories, and change the organisation of production between them for many and inter-linked reasons. Here, I want to look specifically at splitting-up where it has been strongly motivated by management's desire to make the workers' organisation as hard as possible, and where management has realised the potential dangers involved in concentrating large numbers of workers in large factories located in the large industrial town. While, with the internationalisation of production the fate of the domestic industrial working class is increasingly linked to the fate of the international working class, it is important to understand how the location and structure of the domestic proletariat is changing in a period of restructuring in the national and international economy.[8] Here I will examine some of the ways localised splitting has been used in the Italian economy.

In one of the Bologna engineering firms referred to previously, the upsurge of union militancy in the early seventies was met not only by an increase in putting-out but also by a partial splitting-up of production. While employment was allowed to fall in one factory in the firm, another small factory employing 80 workers was established an hour's drive away in a depressed agricultural region. Although the shop stewards were not slow to make contact with the workers in the new factory it has been difficult to take unified action. The workers at the small plant came from rural areas, do semi-skilled work, and are willing to work 'flexibly', that is they are prepared to change shifts and work over-time so that they can also work their plots of land. In contrast, the workers in the main factory are more skilled, they come from an urban background and are endowed with a militant trade union tradition.

Once the small factory was set up management then tried to put-out work from it into the surrounding area, but found that there were not enough small firms in the area to allow this. However, the tendency to set up 'detached workshops' has been widespread where production permits this. One of the few studies of Fiat's decentralisation of production into Central Italy (Leoni, 1978) has shown how, in its lorry division a mixture of splitting-up and putting-out has been used to maximise the dispersion of the directly and indirectly employed workforce in many very green 'greenfield' sites in a rundown agricultural area.

Another type of splitting-up is when the firm loses a central factory to become an agglomeration of 'detached workshops'. Although this strategy is less common one example from Bologna is striking. In this firm there are three 'major' production sites, three 'minor' ones, a stores site, a research site, and an administrative site spread out in the periphery of Bologna. In all, the firm employs 300 people dispersed in the different sites. Along with this fragmentation of production the firm also practices a high level of putting-out, and is progressively running down its machine shops to concentrate only on assembly, design and marketing activities.

A final example of splitting-up is provided by the electric domestic appliance company belonging to Vittorio Merloni, who is the head of the Italian employers federation, the Confindustria. The firm employs 2,000 workers who work in nine different sites and no factory has substantially more than 200 workers. The

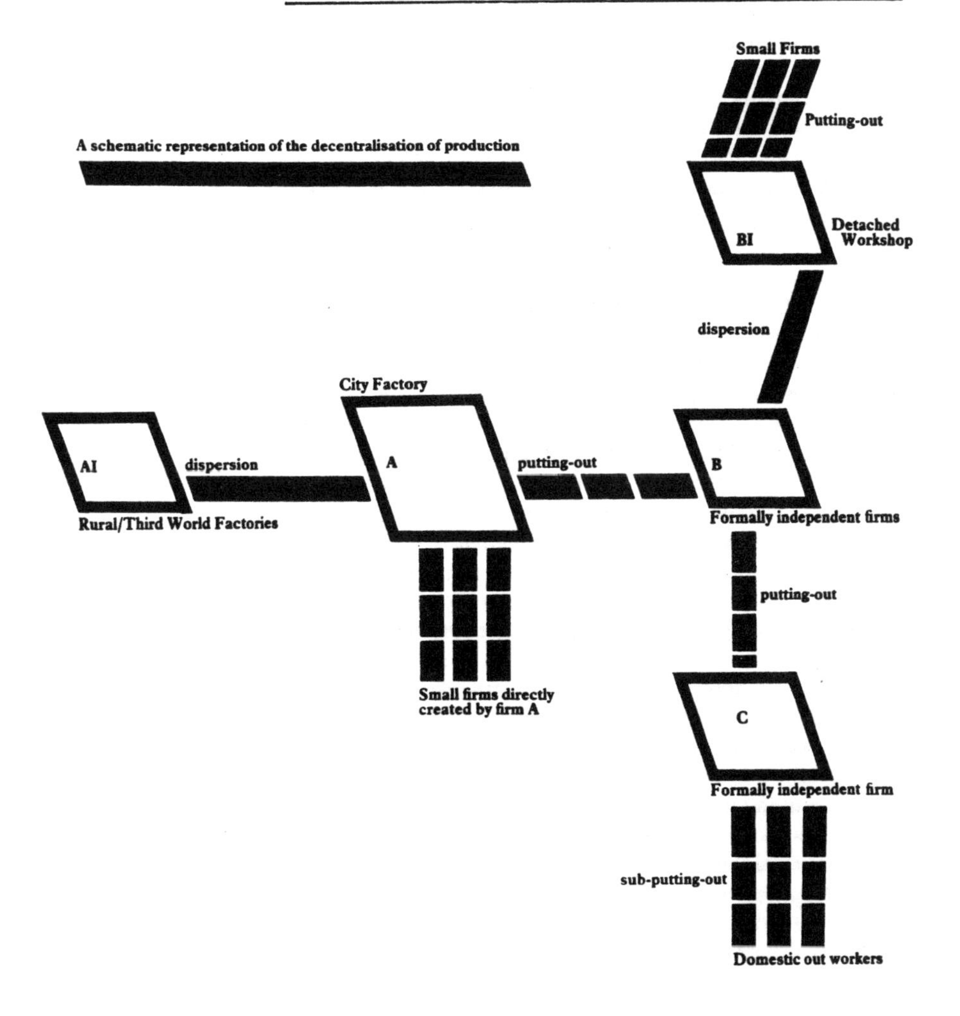

basis of managerial strategy is to take work to the workforce in the depressed agricultural regions of Central Italy, where higher transport costs are easily offset by the 'industrial tranquility' of the environment. One of the Merlonis specifically acknowledges that it is "an advantage to have reduced concentrations of workers, and where possible, to know each worker". And he goes on to explain that the firm has tried to create "a group spirit in and outside the plants" to encourage workers to identify with the firm without losing their roots in the rural community. The idea behind this is to soften and control the traumatising and often radicalising, transition from peasant production to work in a capitalist factory. Meanwhile, to keep things even more 'tranquil' the Merlonis concentrate their efforts on doing pressed steel, assembly and finishing work while the rest of production is put-out to small firms and artisan shops often directly created by the Merlonis, who have paternalistically handed ex-workers the chance to 'go it alone' (see *Lotta Continua*, 22.5.80 and 23.5.80).

By way of ending this section on decentralisation, the diagram below shows how different types of decentralisation could be used by one firm to create a diffused production network, or as some Italians say, a 'diffused factory'.

The Computer in the factory

Within any mode of production the collection, analysis and circulation of information is vital. Within capitalism a particular form of factory production has arisen where one of the functions of the factory is the provision of a structure where information can be collected, co-ordinated and controlled. As communication technology has developed, the emergence of multi-plant and multinational enterprises has been made possible. Although telephones, telex and teletransmitters and the like are in no way determinants of the organisation of production, they have allowed the centralisation of control over capital to increase with the internationalisation and geographical dispersion of production. However, the large factory has remained the basic unit of capitalist production.

The structure of the factory has developed, among other things, to ensure the free flow of information from the bottom of a pyramidal hierarchy to its top, and the free flow of control from the top downwards. Information, and access to it, are the key to formulating and understanding a firm's strategy. For this reason a firm uses a lot of people to collect and transmit information in the factory and this information is carefully guarded. The people who have the greatest amount of information are in a superior position to judge and make decisions, and they will argue that they are 'objectively' correct because of their access to recorded 'knowledge'. In short, access to and the control of information is an instrument of class and sexual power.

In an engineering firm making complex automatic machines there may be as many as 20,000 separate pieces circulating in the factory. For management, this represents big problems and costs. As orders come in and are changed, the production of each piece must be planned and co-ordinated so that the final product is ready on time. Fixed capital and workers must not be allowed to stand idle, detailed plans of machine loadings, stocks and work schedules have to be made and a change in orders, a delay by a supplier, a strike, an overtime ban or a breakdown can all upset these plans. At present many firms incur high management costs to ensure the co-ordination and monitoring of production within the factory. Traditionally this monitoring has been carried out by people writing things on bits of paper, passing them up the hierarchy, amassing them, analysing them and issuing orders based on them. Yet an increasingly flexible production orgnisation is needed to get round worker-imposed rigidity, to ensure the full use of increasingly large amounts of fixed capital and to cut costs 'down to the bone', in the face of the burgeoning contradictions of the system.

The introduction of computer assisted management is a potentially valuable weapon for capital because it can increase management's control over all aspects of production, firstly through the further expropriation of worker's knowledge (mental labour) and secondly, through an 'objectification' of control over labour that ensures the maximum saturation and co-ordination of labour time.

In one Bologna engineering firm there is a computer terminal for every thirteen employees. The terminals are used to both issue orders and to collect, feed back, memorise and co-ordinate information. The course of each part is monitored and information about individual machines and workers, such as work times and 'performance' are constantly recorded. Information from the four basic divisions of the factory, production, marketing, stock control, and planning arrives at the central computer and data base and is recorded and analysed on a day-to-day basis. Information arriving from one department will automatically lead to co-ordination with other departments through the computer's central programme. This gives the management the possibility to foresee where and when bottlenecks will occur, and allows management to experiment with 'dry' production runs on the computer to examine the ways in which potential blockages in production, including strikes, can be overcome through changing production plans in the factory and by increasing or changing plans for putting-out.

I'll now briefly point to three other areas where management benefits from the computer in production. Firstly, idle capital can be reduced to a minimum, whether through a greater control of labour or of stocks, as is achieved by the Japanese 'kanban' (just-in-time) system of stock control. This system uses computers to co-ordinate in-house production and to link its surrounding ring of external suppliers so that stock requirements are calculated on an hourly and not a daily or weekly basis. Production is maintained by

> 'suppliers feeding a wide array of components, in the right order, through the right gate in the assembly complex to reach the line at the right time'.

Secondly, automatic machines and robots can be linked together and run by a central computer, as is beginning to happen in the fully automated flexible manufacturing system. For example, General Electric has recently announced a new computerised system of information control and co-ordination which will enable robots 'to communicate with each other' and link all machines with electrical control into an integrated system, the remote parts of which can be connected by satellite links. (*Financial Times* 30.3.82.) Thirdly, computerised information allows the decentralisation of day-to-day management decisions while centralising strategic control in the hands of a slimmed-down board of directors.[10]

For supervisory staff the introduction of information technology makes their information gathering role potentially obsolete, as the factory hierarchy changes from a function of production command to a more subtle one of political mediation. Fiat has taken this process further and in workshops and offices where now there are no shop stewards,

> 'Fiat takes care of the problem of mediation with its sociologists, its new 'vaseliners' who talk to the workers about their problems'.[11]

For shop-floor and office workers, computers mean stricter control through an impersonal and distant centre, rather than through face-to-face confrontation with the factory hierarchy. Anything a worker does may be recorded by the computer and used against her/him at a later date, while informal breaks won through struggle tend to be formalised and handed out as and when management see fit. And the versatile computer doesn't lose its temper, can also issue orders in Swedish, Finnish, Yugoslavian and Turkish, as the ones used at Volvo do. (See Zollo, 1979; Dina, 1981; Ciborra, 1979)

However, a computer system is only as good as its programme and the degree to which workers are willing to co-operate with management. That is, the potential gains from the introduction of information technology are contingent upon management's ability to erode worker resistance to the technology and prevent new forms of resistance from developing. In one Bologna firm the introduction of terminals on the shop-floor was met by an 'information strike' where the workforce refused to co-operate in the collection of information.

One of the major benefits for capital is that computer assisted management can largely replace the function of the factory hierarchy as an information collecting network. And this in turn opens up the theoretical possibility of changing the organisation of production radically through restructuring. Ferraris (1981) sums up the situation well.

> 'The new technology of the product (modularisation), of production (automation), and information (distributed information and telecommunications) opens up new spaces to the process of decentralisation of work and machines, which advances simultaneously with the concentration of management and control. This permits the overcoming of the historical tendency of the physical concentration of labour and fixed capital as a necessary condition for the centralisation of command and profits.' (p.25)

So far, I have tried to show how the tendency towards decentralisation of production and centralisation of command is taking place. In order to reinforce the argument put forward, I shall cite some Italian examples where it is possible to see this process taking place.

Olivetti

Olivetti's gradual transformation from an engineering group to an electrical one has been speeded up rapidly in the last few years, with the appearance of the dynamic management techniques of C. De Benedetti. Four particular processes can be seen at work:

i) At the financial level, Benedetti has arranged a bewildering series of deals with other international electronics producers which include, Hitachi (marketing), St. Gobain (funds and access to the French market), Data Terminal System (acquisition) and Hermes (take-over of a Swiss typewriter producer).

ii) Within Olivetti's Italian plants there is a move towards automation, using robots and the introduction of computer controlled testing of standardised modules.

iii) Most assembly work is still done manually, but the increasing flexibility needed due to the rapid development and obsolescence of models led management to introduce non-linear assembly in the form of work-islands.

iv) While most assembly work is done in the factory some operations like circuit board assembly and wiring are put out to domestic outworkers in the North of Italy. This process is discussed in Pervia (1980).

Benetton

Benetton is an Italian clothes producer with a turnover of £250 million a year and sells under the names of Jean's West, Mercerie, Sisley, Tomato, OI2, My Market and Benetton. Production and marketing strategies are aimed at achieving two things, the minimisation of costs, the maximisation of flexibility and naturally, profits. This is achieved in the following ways

i) Since the fifties Benetton has increasingly decentralised production. It now directly employs only 1,500 workers and puts work out to over 10,000 workers. The directly employed workers work in small plants of 50-60 employees, where the union is 'absent or impeded'.

ii) In its marketing structure, Benetton has 2000 sales points, but owns none of them. It gives exclusive rights to them. This strategy effectively reduces not only the selling price of the product by cutting the wholesaler out of operation, but it also externalises risks ensuing from fluctuating demand.

iii) Computers are used to keep track of production and sales and to swiftly analyse market trends. Stocks are kept to a minimum of undyed clothes that are dyed when required. (See Ferrigolo, 1980)

Fiat

At Fiat there are four particular things to note:

i) a massive expulsion of labour after the defeat of the 1980 strike

ii) a big move towards automation with the LAM engine assembly plant and the Robogate body plant, both of which are highly flexible robots operated by a centralised computer system.

iii) the introduction of work islands in the LAM system

iv) Fiat's use of decentralisation. This has taken three forms: firstly the export of integrated production units to E. Europe, Turkey and Latin America in the early 1970s; secondly the splitting-up of the integrated cycle and the creation of small specialised plants in the South of Italy, which also began in the early 1970s; and thirdly, the putting-out of work from the Turin plants to local firms, artisans and outworkers.

Following the Japanese model Fiat has recently declared

that in addition to assembly work, it will only produce the suspension systems and technologically important parts of the car in house. All the rest of the work is to be decentralised, although it is unclear what form this decentralisation will take. There has recently been a devastating rationalisation of outside suppliers, with Fiat cutting the number of its suppliers by two-thirds and 'encouraging' the survivors to raise productivity and begin to sub-assemble parts in their own firms. Already 40% of the Ritmo model is sub-assembled outside of Fiat's factories. Vittorio Ghidella, managing director of the car division says,

> 'What we have done is to transfer employment from Fiat to outside companies'[12]

in order to disintegrate vertically as the Japanese have done.

A worker from Fiat's Lingotto pressed steel plant said in 1978 that small is hardly beautiful when you're working in one of the 70 firms with 30-50 employees that make parts of Fiat's decentralised lorry bodywork, where you work Saturdays, and do 10-12 hours overtime each week. He maintained that,

> 'The question of decentralisation and the lack of unity between small and big factories has been the weakest link in the struggles of the past years.' (*Il Manifesto*, 5.10.78)

Fiat's policy then seems to be aimed at automising what can be automised and decentralising as much as possible so that 'decentralisation is the other, almost necessary, face of robotisation and the LAM.' (*Il Manifesto*, 4.4.80)

The decline of the mass-collective worker?

In Italy the increased pace of decentralisation, automation, internationalisation and an eventual frontal attack on the working class were provoked by two principal developments – the emergence of a militant, well organised labour movement and the stagnation of world markets. The heightened shop floor struggles in the large and medium factories threatened the very 'efficiency' of Fordist production techniques, based on the maximum flexibility and total subordination of labour to capital. The strength and combatitivity of the large and medium factory proletariat made impossible a restoration of managerial control through economic recession and increased factory repression, as had happened in 1963-4. Increased competition in world markets and the slump of 1974 made it difficult for firms to pass on the costs imposed on them by labour's gains, while labour rigidity reduced their ability to respond to fluctuations in increasingly unstable markets. As a consequence large firm profit rates fell.

Decentralisation was then grasped on initially as a short-term strategy aimed at evading the labour movement's advances, in that it attempted to compensate high labour costs and low flexibility in the large and medium factories by directly creating or putting work out to small production units, artisans and domestic outworkers, where the influence of the unions was minimal (the small firms in question often being hidden in the submerged economy). However, the longer term aim of decentralisation, automation, and the over-arching control of production by electronic information systems is the destruction of the spontaneous organisation of the mass worker on a collective basis. The dramatic confrontation at Fiat in 1980 hides a strategy which implies much more than a temporary political defeat for the large factory proletariat. Whereas decentralisation was initially a short-term response, its very efficacy has largely precluded a recentralisation of production. Indeed it has been used in conjunction with automation to begin to dismember the large factory proletariat through the increasing division and dispersion of into small plants and into the sweatshop where accumulation is unrestrained by organised labour.

This is not to imply that the mass-collective worker is now politically insignificant. Indeed the power of organised labour based largely on the mass-collective worker is such, that Frobel et al (1980) say,

> 'Any company, almost irrespective of its size, which wishes to survive is now forced to initiate a transnational reorganisation of production.' (p.15)

in order to take advantage of the cheap abundant and well disciplined labour of the underdeveloped countries. Undoubtedly an international reorganisation of capital is taking place but as Graziani argues (1982:34), decentralisation draws attention to the fact that an abundant, potentially cheap and well disciplined labour force is also available within some advanced capitalist countries. In addition, decentralisation reveals how capital gains access to that labour, while at the same time attempting to 'run down' the large factory proletariat, in an effort to restore the competitiveness of mature technology commodities in European markets.

If the aim of decentralisation is ultimately the destruction of the large factory proletariat, its consequence is the recomposition of the industrial working class along new lines and divisions. As we have seen decentralisation takes many forms and to each of these forms correspond different and often new, types of worker. The splitting-up of the production cycle, which is often combined with a restructuring of the labour process creates highly mobile small production units. As Amin (1983) shows, the firm undertaking splitting-up may then search out a particular labour force that embodies the socio-economic characteristics that it considers to be optimal for profitability, taking the fixed capital to the labour force rather than risking its 'contagion' through migration and education in the large industrial town.

Putting-out creates a whole myriad of workers who are seldom immediately visible. In the small firm the labour process and conditions of work vary enormously between firms in the same industry, while the composition of the labour force, its traditions, experience and aspirations largely remain a mystery. An 'apprentice' working in a tiny firm in Turin expresses some of the contradictions that are lived by a small firm worker,

> 'The tiny firm is an inferno, but it is also a hope, and something near to yourself. Yes, but I know . . . that here the work is also being deskilled, but the idea still exists that you can learn a skill here, that they'll teach you something. You're a worker, but at least you can hope to become a good one. Its not really like this deep-down, and everyone knows it, but where do you go if not here? Do you think Fiat's better? The big factory, in a certain sense, scares everyone; these days you only go when you've given up hope. . . . Here they exploit you but you're part of town, your place. You're treated badly, slapped around, but in that place, you see yourself in the work you do.' (*Il Manifesto*, 16.5.80)

Paternalistic relations are common on the shop-floor, with absolute power resting in the hands of the entrepreneur, whereas familial and social ties often link worker and boss outside the factory. In the small firm the relation of labour to capital is often unmediated by unions and labour legislation. It is factory despotism without the large factory and implies the reproduction of the mass, but non-collective, worker at a higher stage of the real subordination of labour to capital where the labour process is fragmented between many small production units, or into the minute division of labour between outworkers and artisans who supervise their own exploitation.

Graziosi (1979) who has done some fine work on restructuring in Italy, makes an important point when he says,

> 'The kernel of the strategy of decentralisation lies in the marginalisation, the increasing precariousness of vast social strata starting with the young, women and the old.' (p.152)

It needs to be stressed that the marginality of these social strata is not economic – since they play a vital role in capitalist acuumulation – but rather it is political and social.[14] The Bologna engineering industry illustrates the complexity of the composition of just one part of the proletariat and the divisions and potential for marginalisation that exist in it are many. In it are found so called 'unskilled' women workers doing assembly work in the submerged economy, N. African men in small foundries, workers in artisan shops supervising numerically controlled machine tools, workers with strong economic and cultural ties with the land working in remote rural factories, plus the workers in the larger factories with their militant uniion tradition and relatively privileged position. It is conceivable that at one time all these workers might have been employed in the same factory and joined by the formal and informal networks and organisations that workers establish, from which their demands and grievances are voiced and from which a collective response is developed. With workers in a firm scattered territorially, socially and culturally, in different conditions of work and often invisible from one another, the problem of uniting a single workforce, let alone the class, is daunting. This raises the question as to whether the shop-floor organisation of unions – in Italy, the factory council and its delegates – can be an effective unifying organisation if it is confined to one factory when the production cycle is being fragmented between plants and firms and domestic outworkers.

The recomposition of the Italian industrial working class is then exacerbating and creating new divisions which are leading to the growth of new sections of the proletariat and to the future weakening of a declining and besieged large factory proletariat. A first conclusion that can be drawn from this is that any faith in a recuperation of the union movement 'in the economic upturn' is fundamentally misplaced and it is sadly ironic, but indicative, that the Fiat workers were beaten when the Italian economy was experiencing a mini-boom. A 'clawback; is made unlikely because the mass-collective worker is being displaced and probably no longer has the strength and cohesion to lead the industrial working class: in future struggles. This does not imply however that the decline of the large factory and the mass-collective worker can be equated with the end of the shop floor or class struggle. Rather the problem is finding the strategy and organisational forms that will allow new and changed members of the proletariat to express their needs and desires and unite with the older sections of the class to fight for common ends.

The Italian experience shows that this is a difficult task and many mistakes have been made. Unions forged out of the struggles of the mass-collective worker have too often tried to impose unsuited strategies and organisations on small firm and diffused workers, while obstructing the creation of organisational forms more suited to their particular circumstances and grievances. This can be seen especially in the failure to form horizontal organisations that link workers in different firms at the local level in Italy, particularly in areas where decentralisation has led to the weakening of informal social and political networks that link workers and collectivise their experiences. In Britain it can be seen by the continuing lack of official support for combine committees. (See Lane, 1982:8)

The Italian labour movement has been quick to recognise that 'diffused' workers exist but for many reasons it has been extremely slow to find out what these workers want from the unions. A consequence of this is that there is a great deal of misunderstanding between the labour movement, which sometimes see the 'diffused' workers as docile, passive and of marginal significance, and the 'diffused' workers themselves, who see the labour movement as being deaf and blind to their grievances and vulnerability.

Britain is not Italy and the mass-collective worker has not dominated the British labour movement to the same extent as in Italy, but this paper has suggested that decentralisation, automation and information technology are particularly effective means for attacking organised labour's power and autonomy, through the expulsion and dispersion of labour from large factories, sites and industrial towns. In Britain, the US and Japanese firms in S. Wales and Scotland are the result of but one type of decentralisation, while the domestic outworkers recently reported to be earning less than £35 a week are another. The textile firm director who 'optimistically' told the *Financial Times* (4.8.82),

> 'I have this vision that St Helens could become the Hong Kong of the North West'

is the voice of a growing submerged and dispersed economy.

The British industrial working class is iteslf being rapidly restructured but the labour movement still largely clings to craft organisations and traditions. Holland (1982) and Lane (1982) have both recently drawn attention to decentralisation in Britain and raised serious doubts about what Lane calls the unions' attempts to,

> 'take themselves by the scruff of the neck and shake themselves into the shape necessary to cope with what is effectively a new environment.' (p.13)

This paper suggests that the reshaping of industry and the working class may accelerate further and faster than has yet been generally realised by the labour movement and the left in Britain. Hopefully the issues are becoming clearer, even if the answers seem to be a long way off.

I gratefully acknowledge the financial support of the SSRC for this research. And many thanks to Ash Amin, Bob Mannings, Donald MacKenzie, Mario Pezzini, Harvie Ramsay and everyone else who read and commented on earlier drafts of this paper.

Notes

(1) For other articles on decentralisation in Italy in English see Amin (1983) Brusco (1982) Goddard (1981) and Mattera (1980).
(2) "'Small is lovely' says Soviet economist" *Financial Times* 9.12.82.
(3) Blair (1972) says, p.113
"Beginning with the new technologies of the Industrial Revolution, the veneration of size has come to take on the character of a mystique, and, like most mystiques, it has come to enjoy an independent life of its own."
(4) See Brusco in FLM Bergamo (1975) p.45-7. Prais (1976) p.52-3 Blair (1972) ch. 5 and 6 and Marx (1976) p.603-4.
(5) see Marx (1976) p.604-5 for a discussion of the Factory Acts and the effect they had on domestic industry.
(6) For a typology of small firms see Brusco and Sabel (1981). They suggest a lot of small firms in Emilia are relatively independent whereas Del Monte (1982) is less optimistic about the position of small firms in the South of Italy (p.125). And many small firms in Japan are 'wholly dependent on a single buyer' Patrick and Rosovosky (1976) p.509-513.
(7) In Japan there has been a 'rather rapid filtering down' in the form of numerically controlled machine tools from big to small firms (Financial Times Survey (1981). Macrae (1982) cites the example of the small Japanese firm where a leased, second hand robot system hammers out components in a 'backshed' workshop.
(8) For work on Britain in this area see Massey and Meegan (1982) Fothergill and Gudgin (1982) and Lane (1982).
(9) 'US Auto makers reshape for world competition', in *Business Week* 21.6.82. See also Griffiths (1982).
(10) See Manacorda (1976). See also the excellent pamphlet produced by the Joint Forum of Combine Committees (1982).
(11) Quote from a union militant in Turin, in *Il Manifesto*, special supplement on Cassa Integrazione, 1982.
(12) Cited in, 'Fiat Follows Japan's Production Road Map' *Business Week* 4.10.82. See also *Sunday Times Business News* 10.10.82 and Amin (1983).
(13) For a discussion of Taylorism, the mass-collective worker and the changing class composition in Italy see Ferraris (1981) Rieser (1981) Santi (1982) and Accornero (1979) and (1981).
(14) This process of marginalisation and division has been aided by left analysis where 'women are seen as marginal workers and hence as marginal trade unionists'. (CSE *Sex and Class* the labour process debate has limited its analysis to those labour processes, that are found in big factories largely employing men. The fact that in Britain, men have largely theorised this labour process, while women have been largely responsible for an analysis of domestic outwork is indicative of the difficulties facing the labour movement and the left. It is vital that left theorists should avoid reproducing the very divisions they are studying.

References

Accornero, A. (1979) 'La classe operaia nella societa' italiana' *Proposte* n.81.

Accornero, A. (1981) 'Sindicato e Rivoluzione Sociale. Il caso Italiano degli anni '70" *Laboratorio Politico n.4.*

Amin, A. (1983) 'Restructuring in Fiat and the Decentralisation of Production into Southern Italy' in, Hudson R and Lewis J,

Dependent Development in Southern Europe, Methuen, London.
Bagnasco, A. Messori, M. Trigilia, C. (1978) *Le Problematiche dello sviluppo Italiano* Feltrinelli, Milan.
Blair, J.M. (1972) *Economic Concentration; Structure, Behaviour and Public Policy* Harcourt Brace Jovanovich, New York.
Brunetta, R. Celata, G. Dalla Chiesa, N. Martinelli, A. (1980) *L'Impresa in Frantumi* Editrice Sindacale Italiana, Rome.
Brusco, S. (1982) 'The Emilian Model; Productive Decentralisation and Social Integration' *Cambridge Journal Of Economics* n.2, June.
Brusco, S. and Sabel, C. (1981) 'Artisan Production and Economic Growth' in Wilkinson (1981).
Celata, G. (1980), 'L'operaio disperso', in Brunetta, R et al.
Ciborra, C. (1979) L'automazione nell 'industria dell 'auto' *Sapere* n.816.
CSE Sex and Class Group (1982) 'Sex and Class' *Capital & Class* n.16.
Del Monte, A. (1982) *Decentramento internazionale e decentramento produttivo Il caso dell'industria elettronica* Loescher, Turin.
Dina, A. (1981) Lotta operaia e il nuovo uso capitalistico delle macchine' *Unita' Proletaria* 3/4 1981.
Ferraris, P. (1981) Taylor in Italia: conflitto e risposta sulla organizzazione del lavoro '*Unita' Proletaria* 3/4.
Ferrigolo, A. (1982) 'Sogno italiano per famiglia veneta' *Il Manifesto* 3.6.82.
Financial Times Survey: Japan the Information Revolution 6.7.81.
FLM Bologna (1975) *Occupazione, Sviluppo Economico, Territorio* SEUSI, Rome.
FLM Emilia Romagnia (1981) *Quaderni di Appunti*, Bologna
FLM Bergamo (1975) *Sindacato e Piccola Impresa* De Donato, Bari.
Fothergill, S. and Gudgin, G. (1982) *Unequal Growth* Heinemann, London.
Frobel, F. Heinrichs, J. Kreye, O. (1980) *The New International Division of Labour* CUP Cambridge.
Goddard, V. (1981) 'The Leather Trade in Naples' *Institute of Development Studies Bulletin* 12,n.3
Graziani, A. (1982) 'La macchina dell'inflazione e la mano invisibile dei padroni.' *Unita' Proletaria* n.1-2, September.
Graziosi, A. (1979) *La Ristrutturazione nelle Grandi Fabbriche 1973-6* Feltrinelli, Milan.
Griffiths, J. (1982) 'Robots March into European Factories' *Financial Times* Survey of the Motor Industry 19.10.82.
Hall, S. (1982) 'A Long Haul' *Marxism Today* November.
Joint Forum of Combine Committees (1982) *The Control of New Technology*.
Lane, T. (1982) 'The Unions: Caught on the Ebb Tide' *Marxism Today* Sept.
Leoni, G. (1978) 'Economia sommersa, ma non troppo' *I Consigli* 57/8.
Macrae, N. (1982) 'Intrapreneurial Now' *Economist* 17.4.82.
Manacorda, P. (1976) *Il Calcolatore del Capitale* Feltrinelli, Milan.
Manwaring, T. (1981) 'Labour Productivity and the Crisis at BSC: Behind the Rhetoric' *Capital & Class* I4.
Marx, K. (1976) *Capital* Vol.I Penguin, Harmondsworth.

Massey, D. and Meegan, R. *The Anatomy of Job Loss* Methuen, London.
Mattera, P. (1980)'Small is not beautiful: decentralized production and the underground economy' *Radical America* October/September 1980.
Paci, M. (1975) 'Crisi, Ristrutturazione e Piccola Impresa' *Inchiesta* October/December 1975.
Paci, M. (1980) *Famiglia e Mercato del Lavoro in un'economia periferica* Angeli, Milan.
Patrick, H. and Rosovsky, H. (eds) (1976) *Asia's New Giant* Brookings Institute, Washington.
Perna, N. (1980) 'L'operaio, punto debole du una macchina altrimenti perfetta' *Quaderni di Fabbrica e Stato* 14.
Prais, S.J. (1976) *The Evolution of Giant Firms in Britain 1909-1970* CUP, Cambridge.
Prais, S.J. (1982) 'Strike frequencies and plant size: a comment on Swedish and UK experiences' *British Journal of Industrial Relations* March, XX, I.
Revelli, M. (1982) 'Defeat at Fiat' *Capital & Class* 16.
Rieser, V (1981) 'Sindacato e Composizione di Classe' *Laboratorio Politico* 4
Rubery, J. and Wilkinson, F. (1981) 'Outwork and Segmented Labour Markets' in Wilkinson (1981).
Santi, P. (1982) 'All'origine della crisi del sindicato' *Quaderni Piacentini*
Wilkinson, F. (1981) *The Dynamics of Labour Market Segmentation* Academic Press, London.
Wood, R.C. (1980) 'Japan's Multitier wage system' *Forbes* August 18th.
Young, K. (ed) *Of Marriage and the Market* CSE Books, London.
Young, K. (1981) 'Domestic Outwork and the Decentralisation of Production' Paper presented to ILO Regional Meeting on Women and Rural Development, Mexico.
Zollo, G. (1979) 'Informatizzazione, Automazione e Forza Operaia' *Unita' Proletaria* 3/4.

[2]

Fergus Murray

Flexible specialisation in the 'Third Italy'

Connecting with earlier debates in *Capital & Class*, this article criticises the model of flexible specialisation, popularised by Sabel, Piore and others. Drawing upon his own extensive research in the Emelia-Romagna district of Italy, Fergus Murray rejects the view that workers stand to benefit, materially and politically, from new flexible labour processes. The paper briefly considers the implications for alternative strategies of economic restructuring.

● In the gloom of Thatcherite de-industrialisation the promise of economic regeneration by way of 'flexible specialisation' has proved seductive to many on the left in Britain. This paper takes issue with the work of Sabel, one of the more optimistic advocates of the healing powers of flexible specialisation (see Sabel, 1982; Sabel & Piore, 1984).

In Sabel's work 'flexible specialisation' denotes a new phase of capitalist production characterised by craft labour, small-scale industry using the latest technology, and diversified world markets and consumer tastes. The main empirical base for this theoretical development is drawn from a brief analysis of changes taking place in small firms in the engineering industry in the Italian region of Emilia-Romagna, which forms part of the so-called 'Third Italy', standing between the congested industrial north and the underdeveloped south of the country (see Brusco, 1982; Brusco & Sabel, 1981).

The 'Third Italy' includes the regions of Emilia-Romagna, The Marches, Tuscany, and Umbria. It is characterised by the presence of small, medium and artisan firms in the engineering, textiles and clothing industries, and areas of extremely rich, and peasant subsistence, agriculture.[1] These regions are sometimes also referred to as the 'Red Belt' because their local and regional administrations have been dominated by the Italian left in the post-war period. Nowhere is this more so than in 'Red Emilia' where the Italian Communist Party (PCI) has reigned supreme since the fall of fascism.

The limited aim of this paper is to return to the Emilia-Romagna of the early 1980s to examine some of Sabel's claims about the 'high technology cottage industry' of flexible specialisation.

Sabel's analysis of flexible specialisation

Sabel tends to see in the Emilian economy the emergence of a post-Fordist production regime based on small-scale, high-technology cottage industry that will hand back to labour some of the creativity of work Fordism has eliminated.[2] Sabel sings the praises of Emilia's small firms because here the strict division between the conception and execution of work, and the minute fragmentation of work typical of Fordism are absent. In short, Sabel considers Emilia's small firms to be 'something more utopian than the present factory system' (1982: 220).

Sabel does mention that it is not every small or tiny firm in Emilia that is a utopia of unalienated labour, although he makes no attempt to quantify the percentage of utopias. However his analysis steadfastly concentrates on Emilia's most 'progressive' firms.

The emergence of these firms is the result of two processes: the diversification of world markets and the strength of Italian workers' struggles. In Italy large firms hamstrung by strong shopfloor worker organisation were unable to respond to changing markets as mass production markets became increasingly diversified. This gave small firms a chance to enter these markets. Decentralisation, through sub-contracting, provided additional work for the small firm sector and some skilled workers left large firms to seek higher wages in the small firm sector.

Small firms were at first heavily dependent on larger ones. Gradually, however, piecemeal innovation made it possible for some firms to work together to produce customised products for world markets. These firms developed the ability to 'create new demand by filling needs that potential customers may only have begun to suspect were there' (1982: 223).

According to Sabel the small firm's success is based on the craft-skilled entrepreneur, the collective elaboration of new products or components within the firm, and the flexible use made of the most modern production technologies. Of particular significance is the need for 'collaboration between different kinds of workers and across levels of official skill hierarchy' and the subsequent blurring that takes place between intellectual and manual work, and the conception and execution of that work' (ibid: 224).

Innovative small firms in Emilia faced two dangers in the early 1980s. Their performance could have been badly affected by either skill shortages or a reduction in their innovative drive. Sabel suggested regional apprenticeship schemes to train-up workers to keep innovation going. He was confident that this would take place because the 'innovative proprietors' in the small firm sector, the trade unions and the regional government were 'intertwined by common political ideas'. The innovative proprietors are loyal to the PCI-dominated local government, says Sabel, because the majority are members of the National Confederation of Artistans, which is closely affiliated to the PCI and the Italian Socialist Party.

Sabel concludes his analysis of Emilia by suggesting that if small firms in the region continue on their present path, 'they will create industrial structures and careers at work that will appear more improbable still against the backdrop of Fordist ideas' (ibid: 230). He assigns a particular importance to the daring and imagination of trade unions, politicians and entrepreneurs in the creation of this improbable future.

Sabel paints an extremely rosy picture of the Emilian economy where work and politics mix to replace multinational corporations and the particular labour processes developed by these firms in the mass production industries. This view is very similar to Brusco's which extols the virtues of the Emilian model where, 'when demand is expanding, anyone accustomed to factory life, and able to work intensively, even if not very skilled, can find work where he or she pleases' (1982: 175).

These observations need to be placed in context, by recognising that the Emilian model continues to exist within the wider Italian economy. The massive companies which dominate the national economy have neither withered away nor been converted into cottage workshops. In fact Fordist and neo-Fordist labour processes are still very much in evidence.

Sabel tends to represent the co-existence of, and connections between, a buoyant small/medium firm sector and a northern multinational-dominated sector, as the former replacing the latter. This is not the case, but the formulation neatly avoids the question of multinational capital and the much more problematic transfer of ownership and control, and labour process transformation, in this sector.

The image that Sabel paints of Emilia is a mixture of resurrected and dignified craft work, where the conflict of capital and labour is largely absent in the small firm. Overarching and protecting the rather fragmented workforce stands the united force of the trade unions, local government and Communist Party.

The next section examines other evidence on flexible specialisation in the Third Italy, concentrating on the province of Bologna, the capital city of Emilia-Romagna.

A post-Fordist economy?

It is doubtful that the selected small firms Sabel extols are representative of a post-Fordist regime of production. Firstly, the firms all have one aspect in common – they produce short batch, often customised or highly specialised, investment goods. Given that short batch produced investment goods have not been subject to Fordist mass production techniques the continued existence of these firms if hardly proof of post-Fordism.

Secondly, Sabel has chosen to highlight one sector of the Bologna engineering industry at the expense of more Fordist ones. In 1976 the investment goods sector (machine tools, automatic machines etc.) accounted for 26 per cent of engineering employment in Bologna, while consumer durables (car parts and motorcycles) accounted for 17 per cent, electronic goods for 13 per cent, and foundry, forging, and heavy metalworking accounted for 14 per cent (FLM Bologna, 1977: 18).[3] The Bologna, and larger Emilian, economy is characterised by the presence of a number of different engineering sectors, Hence it is premature to suggest that the regional economy is dominated by flexible specialisation when this is only one of a number of changes taking place.

I have argued elsewhere that larger Bologna engineering firms in the investment goods sector have successfully subcontracted into the artisan sector to offset labour rigidity and market fluctuations in their own factories. This subcontracting has followed two distinct phases. In the early 1970s subcontracting was a contingency solution to a radical shift of power on the shopfloor following the hot autumn of workers' struggles. In the latter part of the 1970s a more stable form of subcontracting emerged where groups of artisan firms were making complete sub-assemblies. This second stage has taken place alongside large investment in new information and production technologies by Emilia's bigger firms (Murray, 1983, 1984). Furthermore these firms have held onto their market shares, whilst forming a closer bond with multinational capita. This development casts some doubt onto the extent of market diversification and the independence of small capital, said to be characteristic of post-Fordist flexible specialisation.

The subcontracting network of one such firm was carefully analysed by its factory council (shop stewards committee). This investigation revealed that many different types of work were being subcontracted to different types of firms. Much of this work can in no way be described as prized and non-alienating craft labour, and the conditions under which it was carried out were in some cases described as 'tragic' by the factory council (FLM Emilia-Romagna, 1981b).

It is misleading to characterise the artisan sector as one of post-Fordist craft labour. One of its virtues for firms of all sizes is the overall labour flexibility it provides. Racial, gender and skill divisions are essential to the operation of this economic model. The quality craft work that Sabel discovers is work for middle-aged, Emilian men. Semi-skilled assembly work, plastic moulding, and wiring work is carried out by women, while heavy foundry and forging work is done by southern Italian and North African workers.

The position of women in Bologna engineering is little different from that of Britain or other advanced capitalist nations. Roughly a fifth of engineering employees are women, and they are concentrated in the electrical engineering, motor cycle and toy sectors. While 66 per cent of male engineering workers are in the three highest of the six engineering grades, 96 per cent of women workers are in the three lowest grades (Emilia-Romagna, 1981a). Flexible specialisation therefore does not appear to offer very much to women engineering workers.

The geographical fragmentation of distinct phases of a product's labour process works to create maximum wage differentials between different groups of workers. This may well be to the advantage of the skilled male machinist. However, it effectively undermines the radical solidaristic wage bargaining pursued by the Italian engineering workers union, the FLM (Federazione Lavoratori Metalmeccanici) throughout the 1970s. And it leaves the majority of workers in this sector, who do not possess these much-demanded skills, open to the unmediated vagaries of market forces.

Post-Fordist trade union organisation?

Sabel claims that many small firms in Emilia are unionised. His evidence for this assertion is extremely limited. He quotes research which revealed the presence of a shop steward in 25 per cent of engineering firms in the Bologna province employing less than 50 workers (Sabel, 1982: 221), in short a trade union presence in only a quarter of these small firms. However, Sabel goes on to argue that trade union influence is more widespread than shop-floor organisation. There are, for example, regional contracts for workers in the artisan sector (firms employing less than 16 employees). Rather than undertaking an analysis of the existing evidence, or gathering his own, Sabel simply states that the wage rates from these contracts are 'often displayed in the small firms' (ibid: 269). Perhaps recognising the weakness of his evidence, Sabel suggets that the unions in Emilia can utilise their allies in local government to close down small firms where wages and conditions do not comply with the regional contracts.

In the early 1980s the Bologna section of the FLM carried out a census of its engineering workers. The census concludes:

- 88,000 workers are employed in the engineering industries in the province of Bologna.
- 28,000 (32 per cent) of these workers are in artisan firms employing less than 16 employees.

The survey states that unionisation amongst these 28,000 artisan workers is 'scarce' in this the most highly unionised industrial sector. It goes on to estimate that unionisation in the whole of Emilian manufacturing may be less than 50 per cent. And it then argues that the union needs to develop a strategy to organise workers in the small firm and artisan sector (FLM Emilia-Romagna, 1981a: 15-24).

Union organisers in the artisan and small firm sector face an uphill task. One organiser I interviewed in 1982 estimated that union membership amongst the 3-4,000 artisan firm engineering workers in his patch of Bologna was around 200 (6 per cent) and falling, despite the recruitment drive he had organised with shop stewards from larger firms in the area.

Further evidence on unionisation is available from the survey carried out by shop stewards referred to earlier. In 1977 the stewards contacted many of the firms doing subcontract work for their employer. Amongst other things, they discovered that of the 71 firms contacted 40 had no employment contract. In the firms with contracts these solely concerned wages. Hours and conditions were subject to no trade union supervision or control. There was a shop steward in 28 of the 40 firms although this tended to be in the larger firms. Union representation in the artisan sector was particularly low (FLM Emilia-Romagna, 1981b).

The evidence presented here suggests that in the early 1980s there was a scarce to non-existent union presence in the extensive artisan or 'cottage industry' segment of Bologna engineering. This implies that the 32 per cent of engineering workers in this sector stood largely outside Italy's powerful union movement. In reality the Bologna FLM's membership is concentrated in firms employing over 100 workers. These firms account for 70 per cent of union membership (FLM Emilia-Romagna, 1981a: 24; FLM Bologna, 1977: 219).

There is little evidence to suggest that the unions' 'allies' in local government have closed down bad employers. A study of the artisan sector in the early 1980s was sharply critical of Emilia's political leadership because it had failed to make financial support to the artisan sector conditional on improvements in working conditions. Rather, financial assistance has been given in such a way as to 'avoid any choice of which products or sectors to sustain, and omits to enquire into the results of the investment and the quality of the work created' (Lungarella, 1981: 34). It would anyway be extremely difficult to police such a vast number of small firms, many of which exist on the edge of, or completely outside, the formal economy.

In the early 1980s the small, artisan firms in Bologna engineering experienced strong growth resulting in increased employment. Larger firms in the sector have tended to invest heavily in new production technologies and employment in these has stagnated or fallen. Were this trend to continue the engineering union could be seriously weakened.

At a more general level Sabel's cottage industry Utopia spells bad news for trade unions. The problems of contacting, let alone organising, workers in this sector are immense. And it is inconceivable that a union could attempt to organise these workers without the resources and organisational strength it had previously built up in larger workplaces. Furthermore the history of the Bologna FLM shows how it has been union activists in the larger firms who have pushed for the extension of trade union organisation into the small firms (Murray, 1984: Chapter 4). The 'common political ideas' that link trade unions, regional government and small entrepreneurs championed by Sabel hide from view the conflicts between FLM and PCI attitudes and strategies regarding the small firm.

An analysis of the particular nature of trade union representation in Emilia also needs to recognise the historical dominance of the region's trade unions by the Communist Party. This is evidenced in the nature of workers' struggles in Emilia. These were less explosive and more disciplined than those which characterised the actions of workers in the industrial North during Italy's 'hot autumn'. Strike figures suggest Emilia's working class tended to strike for demonstrative rather than contractual ends. These ends were often linked to PCI policy which has dominated the unions in Emilia more than in the North (Daneo, 1972: 21-2). This policy categorically discouraged disruption in the small firm sector (Trigilia, 1981: 122-3).

A further weakening of the large and medium factory base of the engineering union and its once fiercely autonomous factory councils will tend to weaken the organised rank and file in the union. Workers in small firms have less opportunity to elaborate strategy collectively and will tend to be more subject to control from the trade union bureaucracy. This could be seen as a worrying development for trade union democracy and action, particularly as ideological and party differences once more divide and dominate the Italian union movement.

Concluding remarks

The argument of this paper can be summarised in five points. Firstly, the Emilian economy supports a wide variety of engineering sectors, significant parts of which cannot be labelled 'post-Fordist' either because they were never Fordist, or because they are still characterised by variants of a Fordist labour process. Secondly, the artisan sector contains a wide variety of working conditions. Trade union organisation is weak and many workers are exposed to unmediated market forces. This tends to create wide differentials of wages and conditions, which exacerbate gender and racial divisions further. Thirdly, there is little evidence that the regional administration strategically directs the development of, or polices the employment conditions in, the artisan sector. Our fourth point concerns the implications of fragmentation of Emilia's working class. The main effect is to undermine the solidaristic strategies developed by the FLM in the 1970s, and rank-and-file organisation and participation in the union movement. Left affiliated unions may thus once more become the local PCI's policy 'transmission belt' as happened in the 1950s and 1960s. Finally, the Emilian model provides enormous labour flexibility for capital. For a majority of male Emilian skilled machinists, fitters and technicians this offers the possibility of functional labour flexibility between a wide range of manual and conceptual tasks unlikely, but not impossible, in the larger firm. For artisans employing less skilled workers, or operating in labour market sectors where skilled workers are abundant, labour flexibility may tend to take a numerical form. There are few or no restrictions on hiring and firing, lay offs, and conditions of employment. Yet there is evidence that larger firms have subcontracted work out into the artisan sector through the decentralisation of production to take advantage of both functional and numerical labour flexibility. This practice is likely to continue in one form or another. Hence, a vigorous and innovative trade union campaign is still needed to provide a minimum of protection for workers in the artisan sector. Unfortunately in the past the PCI has exerted pressure to contain and drastically curtail the development of such a campaign in order to protect its 'productive middle classes'.

The Emilian model can be seen as one where the development of market forces 'has been more authentically capitalist' than in other parts of Italy (Brusco, 1982: 183). The trade unions and local government have been singularly unsuccessful in controlling or directing these forces. It is perhaps ironically this lack of success in the private sector, rather than daring political choices, which above all accounts for Emilia's enviably booming economy and high levels of employment.

For the vast majority of workers who do not possess the market power of an elite of male machinists, technicians, and designers, a shift towards a fragmented, informal and casual cottage industry spells a return to the worst excesses of industrial capitalism. Fortunately a significant majority of Emilia's engineering workers are not employed in the cottage industry sector.

Whilst it is necessary to recognise that the Emilian model now provides employment and economic growth that are the envy of many of Europe's depressed industrial regions we also need to recognise the particular social, political and economic development of the Emilian model. Here there is not the space to examine this process in detail. However, if a longer historical view is taken it can be argued that the present success of the Emilian economy goes back to the defeats imposed on the region during the first Cold War. The retreat into self-employment, co-operatives, and the poor conditions of the artisan sector was a defensive reaction which followed the dismantling of Emilia's large, and well-organised, industrial bases. It has taken the region's small firms 35 years (over half of which were years of the post-war boom) to develop to a stage where some of them have been able to enter market niches on an independent basis.[4] It is difficult to see how the Emilian model might be transferred to other regional or national economies in the short term.

In the past the Communist Party has to an extent offset the worst consequences of Emilian economic development through the creation of an efficient local welfare state. It has also used its historical popular credit, built up in the Resistance and Cold War, and impressive mass party structures to create an enduring electoral and political hegemony. There are signs that this political dominance is beginning to weaken as the party fails to appeal to young precarious workers of the decentralised economy (see Sechi, 1978: 33) and as its factory base is further undermined by the use of subcontracting and casual labour in the parallel economy. If the left in Britain is looking to create economic regeneration of the manufacturing base through 'Emilianisation' there are a number of points which need to be made.

Firstly, as I have argued, Emilia is not a workers' Utopia. Nor can its economic successes be explained by an economistic analysis combined with an overemphasis of the role of political will and choice. The development of the Emilian model needs to be located in its wider historical, economic and socio-political context.

Secondly, if small firm networks are going to be set up, or intervened in, without strong trade union organisation, the control of these firms on a day-to-day basis begins to look very difficult. In Emilia it is suggested that the hegemonic position of the PCI limits the extent of labour exploitation in the 'cottage industry' sector. I have already argued that this claim is open to some doubt. But in Britain what agency is to play the part of the PCI with its mass party, and molecular and organised presence in the majority of economic institutions and agencies in Emilia?

Thirdly, devoting scarce resources to the organisation of small business networks may replicate the more unpleasant features of the Emilian model. On the other hand attempts to set up co-operatives may be one of the few private sector initiatives local authorities can take short of supplication to internationally mobile companies. But the Emilian experience suggests that it takes many years for a successful and relatively independent small firm sector to develop. The evolutionary stages before independence include dependent low-cost subcontracting to large firms, and the atrocious conditions associated with a low wage, low investment, cottage or sweated industry.

In the early post-war years the artisan sector in Emilia was characterised by economic survival and political resistance rather than prosperity. That is also likely to be the case for Emilian experiments in Britain for a long time to come. It is for local authority Economic Development Units to decide if this is the best use they can make of their resources in private sector initiatives. In some cases I suspect it may be. But if this type of self- and collective-reliance is to be encouraged it might be more fruitful to examine ways of integrating small-scale production and consumption of goods and services into the creation of a limited, and local, parallel economy, rather than attempting to directly compete on the open market with the vastly superior resources of large and international companies.

And finally, large and multinational companies continue to dominate the international economy. They therefore present a central problem for progressive strategies of labour-led economic regeneration. It is unlikely that the local state, however daring it may be in its political choices, can begin to exert control over these firms. Here we need to look to agencies such as central government and local, national and international trade union organisation for the realisation of daring alternative economic strategies.

Notes

1. Many of the industries that characterise the Third Italy like clothing, textiles, and furniture were until relatively recently regarded as being technologically mature, labour intensive, low value-added sectors indicative of economic backwardness. Since that time major technological advances and the emergence of affluent and sophisticated consumer demands in the international economy have brought about a transformation in some of these sectors, which were well placed to respond to these changes.

2. For our purposes here Fordism can be defined as a particular labour process characterised by semi-automatic assembly-line mass production. In this labour process the individual worker has little or no job autonomy and workers collectively are subjected to the uniform movements of a complex machine system (Aglietta, 1979: 118) as typified in the large car assembly plant.

Neo-Fordism is a further development of Fordism. It consists of the automation of areas of the labour process (robotisation of welding, painting, and some assembly), the breaking-up of large integrated plants into geographically scattered specialist units and the overall co-ordination of production through the massive use of information and communication technology.

Generally debates on Fordism have not carefully examined the real spread of Fordist techniques in manufacturing and have, in my opinion, exaggerated its numerical impact on the industrial working class. Even at its height in the 1960s it is unlikely that a majority of industrial workers were subjected to Fordist labour processes.

Present developments in non-Fordist sectors, like capital goods production (plant and machinery etc.) suggest to me a shift towards a 'flexible automation' rather than a 'flexible specialisation'. 'Flexible automation' because emphasis is placed on both the flexibility (rather than specialisation) of production techniques and product mixes and the flexibility of labour to move between job tasks and jobs. At the same time the automation of production and the expulsion of labour is tending to increase (for a more detailed discussion see Murray, 1984).

3. The extent of the influence of multinational capital and large national firms in Emilia has been the subject of some debate. A 1975 FLM study suggested that over 11,000 of the region's engineering workers were employed directly or indirectly by Fiat (FLM Bologna, 1975: 30). Direct Fiat employment in the region in 1977 was 7,280. Amongst Emilia's large engineering firms employing over 500 workers 18 were owned by groups with head offices outside of the region. Of these only three were foreign-owned (FLM Bologna, 1977: 22-3).

4. Levels of industrial injury were considerably higher in Emilia in the mid-1970s than the Italian average. Industrial incomes were marginally lower than the national average, and between 10 and 15 per cent lower than wages in the Northern regions of Piedmont and Lombardy (FLM Bologna, 1977: 26-9).

References

Aglietta, M. (1979) *A Theory of Capitalist Regulation*, London: Verso.

Brusco, S. (1982) 'The Emilian Model: Productive Decentralisation and Social Integration', *Cambridge Journal of Economics*, no. 3.

Brusco, S. & Sabel, C. (1981) 'Artisan Production and Economic Growth', in Wilkinson, F. (ed.), *The Dynamics of Labour Market Segmentation*, London: Academic Press.

Daneo, C. (1972) 'Struttura Economica e Strategia Politica del Dopoguerra', *Note e Rassegne*, nos. 33-4.

FLM Bologna (1975) *Ristrutturazione e Organizzazione del Lavoro*, Rome: SEUSI.

FLM Bologna (1977) *Occupazione Sviluppo Economico Territorio*, Rome: SEUSI.

FLM Emilia-Romagna (1981a) *Quaderni di Appunti*, no. 2, November.

FLM Emilia-Romagna (1981b) *Analisi del Decentramento Produttivo*, Bologna: FLM.

Lungarella, R. (1981) *Investimenti, Occupazione, Governo dell'Artigianato: L'Esperienza dell'Emilia-Romagna*, Bologna: CRESS.

Murray, F. (1983) 'The Decentralisation of Production and the Decline of the Mass-Collective Worker?' *Capital & Class* 19.

Murray, F. (1984) *Industrial Restructuring and Working Class Politics in Post-War Italy*, unpublished PhD Thesis, Bristol University.

Piore, M. & Sabel, C. (1984) *The Second Industrial Divide: Possibilities for Prosperity*, New York: Basic Books.

Sabel, C. *Work and Politics*, Cambridge University Press.

Sechi, S. (1978) 'Il PCI e le Contraddizioni del suo Blocco Sociale in Emilia', *Unitá Proletaria*, no. 3.

Triglia, C. (1981) 'Le Subculture Politiche Territoriali', *Quaderni Feltrinelli*, no. 16.

[3]

Antipode 21:1, 1989, p. 13–34
ISSN 0066 4812

FLEXIBLE SPECIALISATION AND SMALL FIRMS IN ITALY: MYTHS AND REALITIES

ASH AMIN†

Introduction: On Flexible Specialisation

Utopian myths tend to be hinged upon partial truths, obscured realities, discontinuities with the past, and attainable panaceas. Consequently, once in place, they are not easy to dispel. One such myth which is rapidly becoming a new orthodoxy relates to the organisation of work and industry in the advanced countries along the lines of flexible specialisation. This regime is claimed to be radically different from and more humane than the preceding *ancien regime* which came into existence in the 1920s and became defunct in the 1970s. The era which has ended is that of Fordism, loosely distinguished by the mass production of standardised commodities in large metropolitan factories. Its economic strengths are said to have derived from: vertical integration; the detailed division and separation of tasks; machinery dedicated to specific tasks; the full and continuous use of capacity; the rigid codification and hierarchical implementation of work rules; and the gradual erosion of skills, job multiplicity and workers' control in the production process. This organisational principle was best equipped to deal with the demands of mass consumption resulting primarily from the extensive growth in the size and the expendable income of the wage earning class, as well as the provision of a social wage through Keynesian welfarism during the post-war period.

The crisis of the *ancien regime*, provoked by heightened class conflict, cheaper imports from the Newly Industrialising Countries, and technical limits within Fordism to further gains in

†Centre for Urban and Regional Development Studies, The University, Newcastle upon Tyne, NE1 7RU, England

productivity, is understood to be on the wane as a new 'post-Fordist' regime emerges to replace and dominate over the old one. Some of the more sophisticated theorisations of this transition, notably by the exponents of the regulation approach in France, have been more circumspect about claiming that the Fordist crisis has ended. The regulationists have been careful not to describe the emerging regime as something completely new, as is evident from the preference by Aglietta (1979) and others for the term neo-Fordism, instead of post-Fordism. Nor is it assumed that neo-Fordism displays a fixed or unitary empirical pattern in its organisational form. For instance, Lipietz (1987) and Leborgne and Lipietz (1988) refer to many possible strategies and uncertain outcomes among contemporary organisational experimentations, ranging from the combination of decentralised production with worker participation and renewed craft traditions, to more centralised, vertically integrated and ruthless forms of automated Taylorism. Finally, regulationists such as Boyer (1986) and Coriat (1982) are less sanguine about the emancipatory effects of neo-Fordism on the working class, and refer to the new developments in production (especially those associated with changes in the wage relation and new technology) as attempts to extend and reshape rather than relinquish capitalist control and class exploitation.

Other writers, however, who have relied on particular examples of innovation and change to rewrite the socialist agenda for industry and labour into the 1990s have chosen to throw caution to the wind and be more assertive about the consolidation of a very new and less conflictual regime. The work of Hirst and Zeitlin (1988), Murray (1985 and 1988), Piore and Sabel (1983 and 1984), Sabel (1982 and 1988), Scott (1988), and Zeitlin (1988) falls into this emerging tradition that is gaining wide currency among academics. Notwithstanding certain differences in opinion, these 'optimists' tend to agree that through the progressive saturation of markets, growing market instability, and a massive increase in the demand for better quality and more customised products, mass markets are no longer the characteristic feature of demand. This, together with the development of new techniques that eliminate diseconomies of scale and embedded fixed capital costs, has paved the way for a new organisational principle, described as 'flexible specialisation'. As a result the advanced economies in the West can now reassert their threatened hegemony in the international market. Broadly, flexible specialisation has the following characteristics which distinguish it from the *ancien regime*:

i) A reduction of product, production, transaction and management costs and rigidities as a result of the fragmentation of the technical division of labour among interconnected and task specialist units and firms.
ii) A renewed craft tradition which combines paternalistic and participatory practices with flexible work arrangements reliant upon: the use of general purpose machinery; core workers with composite skills and part-time or peripheral workers who are easily expendable; flexible working times and work rhythms; and the 'modular' rather than 'linear' arrangement of work stations. This schema maximises on inventiveness, quality, and adaptability, and minimises on costs associated with the rigid demarcation of tasks, surveillance, co-ordination, re-tooling, task committed machinery, and so on.
iii) The resurgence of regional economies, usually through the build-up of a product specialism by particular areas (eg. textiles in Prato, electronics in Silicon Valley), and the local containment of the various social, economic and institutional mechanisms enabling further growth in independent firms. Each area resembles the Marshallian Industrial District (Becattini, 1987), and acts as an integrated and self-contained economic system which offers its constituent firms an innovative *milieu*, a pool of skilled labour, an inter-firm vicinity which helps to reduce transaction and inventory costs, and ready market opportunities deriving from the interdependence of task specialist firms for goods, services and information.

Flexible specialisation, therefore, is the new organisational principle that best responds to the growth of flexible markets. Furthermore, it introduces new actors into the centre of capitalist development – small firms and previously non-industrialised areas – and it is also morally more defensible than Fordism as it goes some way towards restoring to workers their skills, job satisfaction, involvement in the decision-making process, and control over the pace and flow of work.

Some Critical Questions

This perspective on change is an interesting one, but not without its problems. In addition to the question of diversity and uncertainty raised earlier in discussing the regulation approach, there are other difficulties which challenge the conceptual rigour of the flexible specialisation thesis. One of the cornerstones of the thesis, most fully articulated by Piore and Sabel (1984), is the break-up of mass markets, which has doomed mass production technology and made way for the total or quasi-simultaneous production of many products in small batches. This assertion has been roundly criticised for its lack of attention to the widespread diffusion of batch and craft production during Fordism itself (Williams *et al.* 1987). Doubts about the 'end of mass markets' argument have also been raised by Smith (1988), who marshalls evidence from the food industry to show an even greater focus than before by the major firms on product standardisation and mass production through the internationalisation of brand names. Similarly, Williams *et al.* (1987) draw attention to the persisting importance of scale, capacity raising investment and mass demand even in the so-called 'mature and saturated' markets such as motor vehicles and consumer electronics. This persisting 'Fordism' is explained in terms of the continued expansion of replacement demand (a point also made by Solo, 1985), the domination of ever increasing market shares by the major competitors, and the ability of the market leaders to raise volume and value by introducing new products.

Williams *et al.* (1987) also refer to the many strategies which orthodox mass producers have been able to deploy to cope with greater consumer demand for differentiated products, leading to the production of families of inter-related models which draw upon a very large number of common, standardised and mass produced parts. In other words, there is no reason to assume that the 'old actors' of Fordism cannot continue to assert their dominance with or without flexible specialisation (Blackburn, Coombs and Green, 1985), as it is equally erroneous to equate the greater importance today of product differentiation with the break up of mass markets (Williams *et al.* 1987). Finally, the assertion that flexible specialisation can overcome the market problems (uncertainty, saturation and instability) that challenge Fordist techniques is highly contentious, for as Solo argues, 'no matter how flexible flexible specialisation is, it cannot wriggle out from the descending force of a decline in aggregate demand. . . . Nor would future uncertainty be less debilitating for the flexible specialisation technology than for that of mass production' (1985: 834 – 35).

Through its selectively narrow focus on polyvalent, well-paid, responsible and self-monitoring craft workers for whom employers have found a renewed role, the literature on flexible specialisation (with the exception of Scott, 1986) has also ended up playing down the importance of many real problems that labour flexibility poses for the working class. Firstly, as argued recently by Schoenberger, 'flexible manufacturing, even with increasing levels of output, does not seem likely to absorb the quantity of production workers characteristic of Fordism in its heyday (1988: 259). Schoenberger also draws attention to Bluestone and Harrison's evidence (1986) for growing social polarisation in the US job market, with weaker sections of the workforce facing conditions far inferior to their previous circumstances; a social polarisation which also penalises the low income groups as consumption becomes increasingly geared towards the high income groups (Harvey, 1987).

It is, however, not only the wider social context, but also the less palatable experiments with labour flexibility that are either ignored or ascribed temporary significance by those who argue the case for flexible specialisation. Boyer has recently argued (1987) that there are many forms of flexibilities which are confronting workers in the advanced economies. These include the demand for polyvalence and task mobility, a weakening of the legal and negotiated constraints on the employment contract, the deregulation of wages and wage security, and the possibility of enterprises dispensing with rules restricting their freedom of management as well as social and fiscal contributions for their employees. Drawing upon a range of indicators to measure the evolution of these different forms of flexibility in the OECD countries (an exercise noted for its absence in the flexible specialisation literature), Boyer concludes that 'in these times of crisis, flexibility strategies have entailed, under various euphemisms, the downward adjustment of most of the hitherto established conditions of employment of workers' (1987: 115). Add to this evidence of little job and wage security, poor pay, reduced formal protection of basic rights, and greater work intensification among the peripheral firms and the weakest sections of the labour market caught up in the web of flexible specialisation (Hyman, 1988; Murray, 1987; Pollert, 1988), and the universally enhancing social benefits of the new regime begin to appear to be less and less clear.

Finally, the new orthodoxy also raises some awkward questions in relation to the space in which the new developments in production are being orchestrated. There is the assumption, as described earlier, that spatial agglomeration and the local containment of the functional division of labour best responds to the organisational demands of flexible specialisation. Indeed, Sabel in a recent article (1988) goes so far as presenting the resurgence of regional economies as a *fait accompli*. This perspective underplays the existence of significant differences in the characteristics and origins of various contemporary industrial districts, such as the importance of large retailers and subcontractors as key actors in some areas, small firms networks in other areas, or even government expenditure especially in the defense sector, elsewhere (Amin, Johnson and Storey, 1986; Saxenian, 1985 and 1987; Swyngedouw, 1986).

Secondly, there is a certain uni-directional evolutionism – counter historical in this instance – which pervades the new literature's efforts to demonstrate the return to local production complexes after a period of metropolitan and then internationally fragmented mass production during Fordism. Accurate or not, this approach nonetheless ends up treating the modern local industrial complex as a point of arrival, paying little attention to the forces which are likely to reshape or even threaten their contemporary evolutionary status. Most importantly, however, this approach runs the risk of endorsing an unwarranted localism, if it fails to seriously consider that 'late-Fordism' has witnessed the globalisation and international integration of industry as well as the division of labour, on a massive and extensive scale, through the activities of the multinational corporations and the growth of distance-shrinking new information and communications technologies. It is possible that the 'global network' may express its strengths through some of the most developed local complexes, but it is equally possible that it may come to treat others as a new kind of territorially orchestrated branch-plant with a limited set of functions, or it may disregard them completely. Whatever may be the eventual scenario, the existence of another geography of production which involves functional and inter-firm integration on a global scale (Cooke, 1988), and the determining influence of this process upon the development of local industrial complexes ought not to be disregarded.

On conceptual grounds, therefore, the proclamations about the rise of a new regime of flexible specialisation remain controversial. Myths, however, tend to obscure uncomfortable realities, and select those aspects which support the argument. The present one is no exception to the rule: the 'ideal type' of the model has been defined, and all that remains is to build up a good portfolio of examples to render it fully legitimate. A heterogeneity of examples has been invoked to represent the new industrial landscape. They include the rise of new organisational patterns within new industries such as electronics in the US Sunbelt (Scott and Angel, 1987), and Route 128 on the periphery of Boston (Dorfman, 1983), as well as small firm based high technology industrial complexes such as those at Cambridge (Keeble, 1987), and Montpellier and Sophia Antipolis in France (Perrin, 1986; Planque, 1986).

In particular, however, it is the example of craft firms scattered in the once agricultural regions of central and north-eastern Italy (Emilia Romagna, Toscana, the Marche, Abruzzi and the Veneto), which has become a proxy for flexible specialisation thanks to the eulogisations of Piore and Sable (1984) and selective interpretations of case studies written in or translated into English (Bagnasco, 1981; Becattini, 1978; Brusco, 1982 and 1986; Garofoli, 1984; Lazerson, 1988; Russo, 1985; Solinas, 1982; Trigilia, 1986). Clusters of small, innovative and inter-connected firms producing internationally competitive machinery and machine tools in Bologna and Modena, knitwear in Carpi, textiles in Prato, ceramics in Sassuolo, and furniture or shoes in the Marche have now become the exemplary and feasible standard-bearers of flexible specialisation. In the words of Piore and Sabel,

> The success of the Italian small firms looks like a fortuitous leap forward to a new and desirable form of production. If mass markets are broken up, the capacity to make the largest number of different products at the lowest total price in the shortest total time will prove more important than the ability to turn out one standard product at the lowest possible cost. And the defining characteristics of the small Italian firm nicely meet the general specification for such flexible production: close collaboration between different groups within the firm, between the firm and its neighbours; and, as a corollary to these, general-purpose machines and a broadly skilled workforce (1983: 404).

In view of the popularity and weight which the flexible small firm in Italy appears to be gaining as *the* visible expression of the new regime of production, the remainder of this paper provides a critical assessment of the degree to which the small firm in Italy conforms to its ascribed role.

The Flexible Italian Small Firm

Aggregate Trends

The first macro-economic indicator for renewed small firm activity in the Italian economy during the 1970s – which was a difficult period for the large firms which spearheaded the economic miracle of the 1950s and 1960s – was the publication of the 1981 Industrial Census. This showed that between 1971 and 1981 there was a substantial increase in the number of firms in manufacturing industry (+20.8%) and a more moderate increase in the number of employees (+12.3%). The most dynamic branches of industry were metal goods, engineering and vehicles, in which the increase both in the numbers of firms (+43.3%) and in the number of employees (+23.8%) was higher than the average for manufacturing industry as a whole. In contrast, the heavy industries (metallurgy, non-metallic minerals, chemicals and rubber) displayed none of their employment dynamism of the preceding decade. In the so-called 'traditional' industries (food and drink, textiles, leather and leather goods, footwear, clothing, timber and wooden furniture, paper, printing and publishing, and rubber and plastics), notwithstanding the negative employment growth rate in the food and drinks and textiles industries, the 1970s saw an increase in both the number of firms (+13.8%) and employment (+8.3%). Thus, in contrast to their long-term relative decline in the other advanced economies, the traditional industries in Italy have continued to play a significant role, and in 1981 they still accounted for 46.2 per cent of total manufacturing employment.

More significant than the sectoral changes, however, has been the evolution of the size structure of manufacturing firms in the inter-census period. Small and medium sized firms (less than 99 employees) increased in number by 21 per cent, and by 28.5 per cent in terms of employment between 1971 and 1981. This growth was particularly high in the metal and mechanical engineering, vehicles and most of the 'traditional' industries (with the exception of food and drink, and textiles). Very high growth rates in the number of firms (+61.8%) and employees (+60.0%) occurred in firms of between 10 and 19 employees, and by 1981, small and medium sized firms accounted for 55.5 per cent of the total workforce (as opposed to 48.5% in 1971). The increase in the share of small and medium sized firms was absolute, as it occurred in the context of a 5 per cent *increase* in the number of firms with more than 100 employees between 1971 and 1981. Although the 1981 census overestimated the actual size of small firm growth as a result of its wider coverage in comparison to previous ones (King, 1986), it is commonly accepted that even after allowing for statistical differences, the growth of small firms has been substantial.

These statistics provoked considerable speculation about the origins of the resurgence of the small firm economy. Early conflicts of opinion gradually gave way in the 1980s to a consensus that the *initial* rise of small firms notably in the engineering sectors was due to the extensive use of subcontracting by the medium-sized and large corporations most affected by the 1969–74 wave of intense labour conflict (Brusco, 1986; Fua, 1983; Garofoli, 1984). However, the consolidated *growth* of these firms, especially in the central and north-eastern regions, commonly referred to as the Third Italy, was put down to the ability of the small firms to break out of the subcontracting relationship and sell specialist goods and services in the open market and to other small specialist firms. This was due to their ability to draw upon, within their industrial districts, the continual growth of new specialist firms, cooperative agreements with workers and other firms, low-cost premises, family labour, appropriate new technologies, local banks, designer and entrepreneurial ingenuity arising from the craft tradition, and a variety of local and national incentives such as exemption from 'restrictive' labour legislation for firms employing less than 15 workers.

The growth of small firms in the 'traditional' sectors was put down to the revitalisation of a classical form of small entrepreneurship and self-employment, resulting from job loss and falling opportunities for full-time and permanent employment elsewhere in the economy during the economic crisis which began in the mid 1970s. It is also argued that, in certain semi-rural areas of the 'Third Italy', the latter phenomenon has combined with some of the factors mentioned above in connection with the 'modern' industrial district, to produce clusters of small firms, specialising in a particular industry. This has occurred especially in places which in the past possessed a specialism in the industry concerned (eg. ceramics in Sassuolo, textiles in Prato).

To a very large extent, this perspective on the origins and evolution of the small flexible firm has remained intact during the 1980s owing to the unavailability of comprehensive macroeconomic data since the 1981 Census. However, a very recent paper in English, by Guido Rey (1989) the Director of the Italian National Institute of Statistics (ISTAT), combines the results of hitherto unpublished and published large-scale surveys for 1983 and 1985, to provide some invaluable new insights into the *overall* evolution of small, medium and large firms during the first half of this decade. The information is drawn from ISTAT annual surveys of gross product in firms with more than 20 employees; a survey of a large sample of firms with 10–19 employees carried out by ISTAT in 1983 and 1985; an ISTAT survey on technological innovation among 24,000 firms with more than 20 employees; and the fairly comprehensive annual surveys of the Mediocredito Centrale into the economics and finances of firms with more than 10 employees. Although the surveys are not comprehensive and reliant upon the declarations of entrepreneurs, they provide some very useful and unique insights into certain general trends which case studies are unable to offer.

According to Rey, the data provided by the surveys on gross product in 1983 and 1985 tend to confirm the size and sectoral dynamics which were present during the 1971 – 81 period, with small to medium sized firms (20 – 99 employees) retaining their relative share of total employment, and raising over half their turnover from the traditional industries. What is more interesting, however, is what can be gleaned from the surveys in terms of aggregate trends regarding particular aspects of firm behaviour during the 80s. The technical innovation survey shows that while 90 per cent of firms with more than 500 employees had introduced product, process or organisational innovations in the 1983 – 85 period, the corresponding figure for firms with 20 – 49 employees was lower (63.3%). Paradoxically, however, the poorer level of innovation among smaller firms appears to have had little effect on profitability (ratio of profits to value added), which, according to the Mediocredito Centrale Surveys, *declines with increases in the size of firm*. In 1985, the highest rate of profit in over half the industrial sectors was recorded among firms with 15 – 19 employees, while the lowest was recorded among the largest firms with more than 999 employees. A large profit rate was also evident among firms with 30 – 49 employees. However, one small firm category (the so-called craft firm with less than 15 employees) did experience a clear drop in its overall rate of profit between 1983 and 1985. The first conclusion, then, is that eulogisations of the Italian craft firm should be set in the context of a general decline in the aggregate profitability of such firms, and that while a considerable proportion of small firms have continued to innovate into the 1980s, the response of larger firms has been much better.

Secondly, particular labour circumstances, which tend to be played down in the literature on flexible specialisation, appear to have played a crucial role in helping Italian small firms raise their profit rates in the early 1980s. According to Rey's figures, unit labour costs in 1983 were on average about 40 per cent higher in large firms than in smaller firms, with the greatest differences occurring in food and drink, textiles, footwear and clothing, followed by chemicals, metal and mechanical engineering and vehicles (i.e. the majority of the sectors of strongest small firm growth). The surveys also found that the major differences between the two size categories related to salary levels rather than the payment of social security contributions. Part of the salary difference is put down by Rey to the presence in the larger firms of a higher proportion of management and white-collar workers, who tend to be more expensive than manual workers. However, the rest is explained in terms of differences in the hourly wage rates of manual workers, which, on average, were *50 per cent lower* in the small firms. One cause of this gap is the ability of the unions, which are better represented in the large firms, to negotiate adequate wage increases for all categories of workers, while in the small firms, especially those in the traditional sectors which rely heavily on women and youth workers, the absence of industry wide agreements allows employers to minimise the salary level of the weakest sections of the workforce.

Another interesting statistic revealed by the surveys was that in 1983 manual workers in smaller firms (especially under 50 employees) worked an average of 12% more hours than those in large firms, and that by 1985 this gap had actually increased by 4 per cent between the smallest and the largest firms. Rey's figures, therefore, tend to show that at the aggregate level, if the term 'flexibility' is to be applied to the Italian small firm, it would appear to have more to do with poor wages and long working hours, than with their craft or technical ingenuity.

Two observations, however, corroborate some of the claims of the small firm optimists. The first is a significant level of export penetration by firms with 20 – 99 employees. This is particularly evident in the metal and mechanical engineering, office machinery, vehicles, electrical and electronics, textiles, leather goods, and rubber and plastics industries, where exports account for between 20 – 35 per cent of total turnover. Secondly, the surveys, by eliciting information on the relative weight of intermediate production costs on turnover, tend to confirm the existence of a high and growing level of interdependence between firms. While large firms have tended to maintain a high degree of vertical integration, the smaller ones appear to have embraced specialisation as a common principle, and to rely on sub-contracting or normal market practices for the purchase of a growing proportion of services and intermediate goods. However, flexibility through the division of tasks among firms is not restricted to small firms, but is a strategy which, according to Rey, has also been pursued by medium sized firms (100 – 499 employees) during the 1980s.

Finally, Rey argues that despite the greater profitability of small firms recorded in 1983, large firms have been progressively closing the profit gap. Between 1983 and 1985, there was a 40 per cent decrease in the gap, owing to an enormous increase in the profitability of large firms, precisely in those sectors in which small firms have been the better performers hitherto (textiles, food and clothing, metal goods, engineering and vehicles). Furthermore, owing to greater technological change by larger firms in recent years as well as their success in implementing mass redundancies and curtailing labour resistance, the higher productivity of large firms has also been on the increase. In 1983, output per employee in small firms was 20 per cent lower than in large firms, and by 1985 the gap had increased by a further 20 per cent.

These details suggest that it may be premature to write off, as has been the tendency of the exponents of flexible specialisation, the ability of the large firm in Italy to reassert its economic power through the development of new organisational strategies. Some economists (Camagni, 1986; Camagni and Capello, 1988) have argued that this process of industrial restructuring, involving predominantly the greater uptake and better utilisation of new technology, has already begun to restore to the North-West its traditional economic hegemony. During the latter half of the 60s and the early 70s, industrial productivity (manufacturing GDP per capita) in Lombardia grew at an average annual rate of 3.5 per cent and that in Piemonte at a rate of 4.1 per cent, which was well below the growth rates in the regions of greatest small firm activity – 4.5% in the Veneto; 5.9% in Emilia Romagna; 4.4% in Toscana; and 4.8% in the Marche. During the second half of the 70s, however, industrial productivity in Piemonte rose annually by 4.5 per cent and by 3.9 per cent in Lombardia, well above that in the Veneto (3.5%), Emilia Romagna (2.8%), Toscana (2.9%), and the Marche (3.5%). The study by Camagni and Capello (1988) also shows that by 1981 the regional wage/productivity gap, which worked to the locational advantage of the central and north-eastern regions during the 60s and early 70s because productivity grew much faster than wages in these regions, had virtually reversed in favour of the north-west regions, in which productivity had come to grow faster than wages.

None of the protagonists of flexible specialisation would, of course, claim that the model described earlier in this paper, is intended to offer a description of all small firms in Italy. The model has been derived from a study of particular industrial districts in the Third Italy, and is supposed to describe a new emerging organisational principle which could spread. However, new dogmas, despite the intentions of the originators, often end up as assumed universal truths. More importantly, the evidence above tends to show that it is probably in only very specific circumstances that small firms are capable of leading economic development. Indeed, on the basis of observed aggregate trends, the only generalisations which can be made about Italian firms are firstly, that labour flexibility (which usually amounts to poor wages and long hours) and productive interdependence appear to be the two key relative strengths of the small firm economy; and secondly, that signs of an alternative form of restructuring on the part of large firms are also becoming apparent.

Small Flexible Firms in Industrial Districts – Which Ones?

Flexible specialisation is said to exist as a visible phenomenon in the Third Italy. How widespread, then, is the phenomenon? If defined loosely in terms of the geographical clustering of new and small firms in particular product markets, then many areas in the Third Italy would qualify (Garofoli, 1984). If, on the other hand, a more restricted definition that encompasses some of the key characteristics of the ideal model is adopted, then the picture appears to be significantly less encouraging.

A recent paper in English (Sforzi, 1989) contains the results of a statistical attempt to identify the number of localities in Italy (local labour market areas based on travel-to-work data) which possess the characteristics of a Marshallian industrial district (a locally contained and inter-connected industrial system). In Sforzi's analysis, eligibility is dependent upon the combined presence of the following characteristics, which, in fact, are also those emphasised by the flexible specialisation model: clustering and renewal of small firms specialising in the different phases of a particular industry (a 'system of interacting parts'); a distinct social structure composed of a high level of small entrepreneurs and artisans, skilled workers, working wives, and extended family and youth labour; and proximity between home and work in the labour market. The areas which qualify as industrial districts on the basis of Sforzi's criteria are shown in Figure 1, which confirms their preponderance in the Third Italy regions. The precise location and sectoral characteristics of the industrial districts is provided in Table 1. Two features which stand out immediately are firstly that the innovative engineering districts which lay at the foundation of Piore and Sabel's (1983) work (The Emilian Model) are extremely limited in number. Most of the industrial districts produce fashion wear or furniture. Secondly, the total number of industrial districts, though perhaps high by Western European standards, is relatively contained, even within the Third Italy.

Figure 1 Industrial Districts in Italy

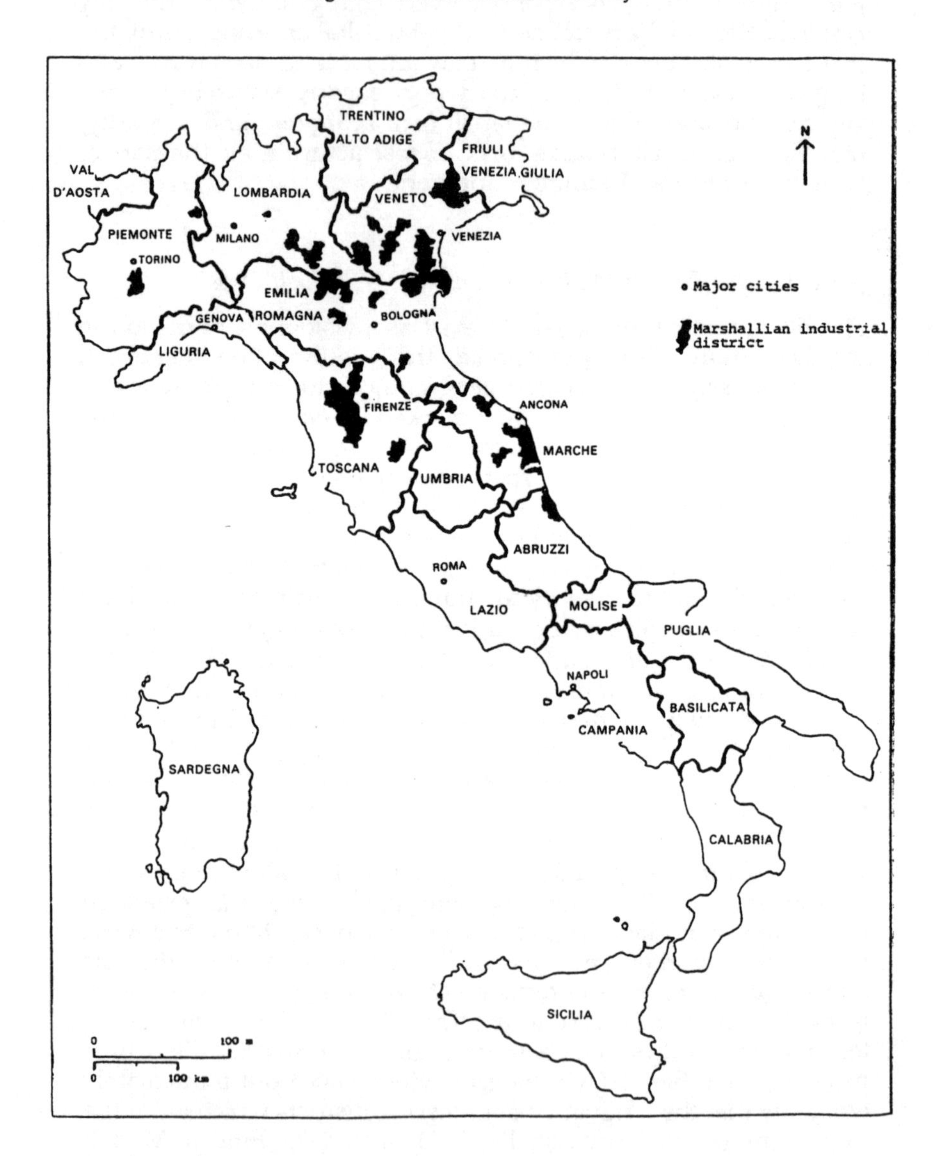

Source: Sforzi (1989)

Table 1 The regional distribution and sectoral specialisation of localities (communes) with industrial districts in Italy

REGION / *SECTOR*	PIEMONTE	LOMBARDIA	EMILIA ROMAGNA	VENETO	FRUILI VENEZIA GIULIA	TOSCANA	MARCHE	ABRUZZI
METAL GOODS	Carmagnolo	Rivarolo Mantovano						
MECHANICAL ENGINEERING		Suzzara	Novellara Cento Copparo	Conegliano				
ELECTRICAL AND ELECTRONIC ENGINEERING			Guastalia					
TEXTILES		Asola Urgnano Quinzano d'Oglio	Carpi			Prato		
CLOTHING	Oleggio	Manerbio Pontevico Verolanuova Ostiano		Noventa Vicentina Piazzola sul Brenta Adria Porto Tolle		Castelfiorentino Empoli	Mondolfo Urbania Corinaldo Filottrano	Rosetto degli Abruzzi
LEATHER TANNING				Arzignano		Santa Croce sull'Arno		
LEATHER GOODS							Tolentino	
FOOTWEAR				San Giovanni Ilarione Piove di Sacco		Lamporecchio Montecatini Terme	Civitanova Marche Fermo Grottazzolina Montefiore dell'Aso Montegranaro Monte San Pietrangeli Torre San Patrizio	
WOODEN FURNITURE		Viadana	Modigliana	Bovolone Cerea Nogara Motta di Livenza Oderno Montagnana	Sacile	Poggibonsi Sinalunga	Saltara	
CERAMIC GOODS			Sassuolo Casalgrande					
TOYS		Canneto sull'Oglio						
MUSICAL INSTRUMENTS							Potenza Picena Recanati	

Source: Sforzi (1989)

The industrial districts listed in Table 1 display a certain superficial similarity insofar as they possess the defining characteristics laid down by Sforzi's model. However, a closer reading of the impressively large body of literature that now exists on the various localities tends to suggest that there are also significant differences between them in terms of the forces which have enabled growth based on small firms. Camagni and Capello's study (1988) draws upon much of this literature to show, for instance, that in Toscana the flexibility generated by the high division of labour among small firms and the region's highly integrated, self-supporting, and co-operative social structure have been two key success factors. In contrast, areas in the Veneto appear to have benefited from low labour costs, the availability of specialised skills as well as part-time labour from agriculture, and the pervasive influence of Christian Democracy and strong Catholic traditions, which have played a crucial role in shaping common social and business goals. The decentralisation of tasks among small firms appears not to be a distinguishing characteristic of the Veneto areas (Belussi, 1988). In the Marche, the areas of small firm development have been able to exploit similar labour circumstances as those in the Veneto, but they have also had to rely more on active local authority participation, rather than consolidated political subcultures or social traditions for the provision of common social and economic resources. In other words, significant differences exist between the industrial districts of the Third Italy and the ones described above refer to only those areas which all produce similar traditional consumer goods!

It may well be true that it is only the older and well-known industrial districts of Toscana and Emilia Romagna (eg. Carpi, Sassuolo, Prato) which bear some resemblance to the ideal model of local economic development through flexible specialisation. The contribution of factors such as the following ones should not be undervalued: specialist and task-integrated small firms producing designer goods for the export market; a competitiveness reliant upon artisanal skills, new technologies and constant product innovation; active support from local authorities and other local institutions; populist and consensus-based local subcultures which permit social collaboration and the exchange of ideas; and the availability of local business support services as well as a name (eg. 'made in Prato') that attracts buyers. To these factors, however, should also be added others which tend to be played down in the literature, such as poor wages and the absence of job security as well as other social benefits (eg. redundancy payments, pension contributions) for above all women and migrant workers. These are conditions which affect the numerous peripheral workers of both the well established knitwear and textile areas (Solinas, 1982) as well as the high-tech engineering industries (Murray, 1983 and 1987) which are to be found even in the most buoyant regions of flexible specialisation such as Emilia Romagna.

The possibility of a departure within the older industrial districts, from growth based upon small firm synergies and the local containment of the division of labour, is another question that the literature which celebrates the Italian experience has failed to address. A recent report (Signorini, 1988) based upon a Bank of Italy survey of technical change among firms of different sizes in Toscana shows a relative slow down in the innovativeness of small firms – a process that may lead to the replacement of the region's traditional mass entrepreneurship by a small number of larger, 'elite' firms. Similarly, Murray (1987), and Tolomelli (1988) refer to the growth in takeover activity within Emilia's engineering industries. For Tolomelli, the growing concentration of ownership is leading to the vertical integration of tasks and functions, as well as the insertion of the local economy into a wider spatial division of labour. These processes could not only alter the basis upon which the Emilian industrial districts have developed, but they could also threaten their survival:

> we see that the traditional organisation of the industrial districts will be overshadowed and replaced by an organisation based on groups of enterprises, with a variety of interconnections and tied to Italian and foreign corporations . . . groups of firms are not only taking over individual firms, but through them can enter the web of relations between firms within the districts and integrate them into their corporate network (Tolomelli, 1988: 8).

Whether the older industrial districts will disappear as did those which characterised the British industrial landscape in the 19th Century, remains an open question. What is certain, however, is that their organisational structure may be changing.

The majority of the remaining industrial districts, especially those in the traditionally less urbanised and less developed regions (Table 1 – Veneto, Marche, Abruzzi) are new and quite different from the older established ones. Many of the areas of recent specialisation in the shoe, clothing and furniture industries appear to be little more than rural clusters of small, family firms producing the same finished or intermediate goods (usually of medium-to-poor quality) for large subcontractors or, on an erratic basis, for wholesalers selling in the open market. Often, these producers have few entrepreneurial or marketing skills. They have little immediate access to technology, specialised business services, finance and intermediate products, because of the lack of institutional support and because the inter-firm division of labour in the area is not sufficiently well developed. They are often in fierce competition with each other as a result of close product similarities as well as severe pressures on price, notably from sub-contractors. In other words, it is only in the loosest possible sense that these areas can be described as an articulated industrial system. The producers possess a restricted degree of freedom in the marketplace, and receive little support from the rest of the system. The firms tend to be isolated, highly dependent on a few buyers, and barely able to improve their market position.

'Flexibility' in the above example, tends to refer to an ability simply to survive, and on the basis of an artisanal capacity to respond to new designs and new market signals, as well as other factors such as self-exploitation and the use of family labour, the evasion of tax and social security contributions, low overhead costs, and the use of cheap female and young workers, especially in the area of unskilled work. These conditions are also typical of the vast numbers of very small and small firms in the traditional industries which, in numerical terms, dominate the Italian small firm economy, especially in the South (see Amin, 1989, for a description of the footwear industry in Naples). They provide poor publicity for the 'ideal' model of flexible specialisation, but they do have a very real and steadfast existence in Italy.

Conclusion

There is little doubt that the viability of Fordist principles (social, technical and organisational) has been severely threatened in the last two decades. There are also clear signs that the search for new economies and new forms of control in production is leading to the consolidation of innovative forms of organisation in many sectors of industry. The introduction of more flexible labour practices, task specialisation and less rigid technological arrangements may well be facilitating product innovation, greater productivity, and reduced transaction costs. The crisis of the *ancien regime* and the emergence of something new seems fairly evident.

What is unclear, however, is whether *one* single and dominant principle is emerging, and what is even more dubious is whether it resembles the model of flexible specialisation most fully theorised by Piore and Sabel (clusters of high-tech cottage industries in industrial districts). The present period is one of turmoil and transition for capitalist industry, with firms experimenting with a range of different strategies in order to restructure. There is little evidence to suggest that one particular solution, especially the one predicated on small firm and craft-based flexible specialisation, is emerging as the dominant solution. Furthermore, the continuities of Fordism and the power of the 'old' economic actors to reassert their dominance and reshape the geography and the social relations of production in ways which work to the disadvantage of local communities and sections of the working class, should not be underestimated.

The empirical foundation for the theory of small firm flexible specialisation also appears to be thin. Evidence from the country which has inspired so much of the theory tends to demonstrate that very few, and only the oldest of the industrial districts come close to resembling the 'ideal type'. Furthermore, the organisational structure of these areas also appears to be changing, as new forces emerge to favour industrial concentration and the integration of the areas into a wider corporate and spatial division of labour. The kind of flexibility which seems to be common among the other industrial districts and the majority of small firms in Italy is not particularly new in historical terms: artisanal skills and adaptability; revenue and expenditure flexibilities typical of the family firm; margins of manoeuvrability in local business transactions resulting from kinship and communal ties; and working conditions and practices which fall outside of the remit of protective labour legislation and trade-union agreements. These conditions are not sufficient for the emergence of an economically buoyant system of flexible specialisation that is capable of guaranteeing healthy profit margins to its firms as well as stable and more rewarding employment to its workers.

It could, of course, be argued that the new industrial districts will in time develop the same characteristics as those of the more successful districts, or that new ones could be nurtured through appropriate public or private sector intervention in places where some of the basic conditions are available. Quite apart from the absurdity of the proposition that the building-blocks and the complex inter-relationships of an industrial *system* can be policy induced, the fact remains that, despite assertions to the contrary by Piore and Sabel (1983), some of the key elements of the successful industrial district are highly place specific. As Bagnasco

(1985) and Trigilia (1986) have so carefully pointed out, traditions such as the family-based and petty bourgeois entrepreneurial culture, community based forms of cooperation and consensus, and municipal mercantilist institutions and structures, are historically sedimented in particular areas, and virtually impossible to transfer to areas where they do not exist.

Acknowledgements

I am grateful to Fergus Murray for his particularly valuable suggestions on an earlier version of this paper, as well as to the two referees of this journal (Russell King and Ron Martin) who commented on the first draft.

References

Aglietta, M. (1979) *A Theory of Capitalist Regulation.* London: New Left Books.

Amin, A. (1989) Specialisation without growth: small footwear firms in an inner-city area of Naples. In Goodman, E. (Ed.) *Small Firms and Industrial Districts in Italy.* London: Routledge, forthcoming.

Amin, A., S. Johnson and **D. Storey** (1986) Small firms and the process of economic development. *Journal of Regional Policy* 4: 493–517.

Bagnasco, A. (1981) Labour market, class structure and regional formations in Italy. *International Journal of Urban and Regional Research* 5: 40–44.

Bagnasco, A. (1985) La costruzione sociale del mercato. *Stato e Mercato* 13: 9–45.

Becattini, G. (1978) The development of light industry in Tuscany: An interpretation. *Economic Notes* 3: 107–23.

Becattini, G. (1987) (Ed.) *Mercato e Forze Locali: Il Distretto Industriale.* Bologna: Il Mulino.

Belussi, F. (1988) 'Innovation diffusion in traditional sectors: an empirical investigation'. Paper presented to the Regional Science Association's European Summer Institute, 17–23 July, Arco (Trento).

Blackburn, P., R. Coombs and **K. Green** (1985) *Technology, Economic Growth and the Labour Process.* London: Macmillan.

Bluestone, B. and **B. Harrison** (1986) 'The great American jobs machine: the proliferation of low wage employment in the US economy'. A study prepared for the Joint Economic Committee of the US Congress, Washington DC.

Boyer, R. (1986) *La Theorie de la Regulation: Une Analyse Critique.* Paris: Editions La Decouverte.

Boyer, R. (1987) Labour flexibilities: many forms, uncertain effects. *Labour and Society* 12, 1: 107–29.

Brusco, S. (1982) The Emilian model: productive decentralisation and social integration. *Cambridge Journal of Economics* 6, 2: 167–184.

Brusco, S. (1986) Small firms and industrial districts: the experience of Italy. In Keeble, D. and F. Weever (Eds.) *New Firms and Regional Development.* London: Croom Helm.

Camagni, R. (1986) New technologies as a response to the crisis in the Italian North-West. In Federwisch, J. and H. G. Zoller (Eds.) *Technologie Nouvelle et Ruptures Regionales.* Paris: Economica.

Camagni, R. and **R. Capello** (1988) Italian success stories of local economic development: theoretical conditions and practical experiences. *Mimeo,* Istituto di Economia Politica, Universita Luigi Bocconi, Milano.

Cooke, P. (1988) Flexible integration, scope economies and strategic alliances: social and spatial mediations. *Environment and Planning D: Society and Space* 6: 281 – 300.

Coriat, B. (1982) Relations industrielles, rapport salarial et regulation: L'inflexion neo-liberale. *Consommation – Revue de Socio-Economie* 3: 31 – 47.

Dorfman, N. S. (1983) Route 128: the development of a regional high technology economy. *Research Policy* 12: 299 – 316.

Fua, G. (1983) Rural industrialisation in later developed countries: the case of northeast and central Italy. *Banca Nazionale del Lavoro* 147: 351 – 377.

Garofoli, G. (1984) Diffuse industrialisation and small firms: the Italian pattern in the 70s. In R. Hudson (Ed.) *Small Firms and Regional Development*. Copenhagen: School of Economics and Business Administration.

Harvey, D. (1987) Flexible accumulation through urbanisation: reflections on 'post-modernisation' in the American city. *Antipode* 19: 260 – 86.

Hirst, P. and **J. Zeitlin** (1988) Crisis, what crisis? *New Statesman* 18 March: 10 – 12.

Hyman, R. (1988) Flexible specialisation: Miracle or myth? In Hyman, R. and W. Streeck (Eds.) *New Technology and Industrial Relations*. Oxford: Blackwell.

Keeble, D. (1987) 'Entrepreneurship, high technology and regional development in the United Kingdom: The case of the Cambridge phenomenon'. *Mimeo,* Department of Geography, University of Cambridge.

King, R. (1986) *An Industrial Geography of Italy*. Beckenham: Croom Helm.

Lazerson, M. (1988) Organisational growth of small firms: An outcome of markets and hierarchies? *American Sociological Review* 53: 330 – 42.

Leborgne, D. and **A. Lipietz** (1988) New technologies, new modes of regulation: some spatial implications. *Environment and Planning D: Society and Space* 6(3): 263 – 80.

Lipietz, A. (1987) *Mirages and Miracles,* London: Verso.

Murray, F. (1983) The decentralisation of production and the decline of the mass collective worker? *Capital and Class* 19: 74 – 99.

Murray, F. (1987) Flexible specialisation in the Third Italy. *Capital and Class* 33: 84 – 95.

Murray, R. (1985) Benetton Britain. *Marxism Today* November: 28 – 32.

Murray, R. (1988) Life after Henry (Ford). *Marxism Today* October: 8 – 13.

Perrin, J. C. (1986) Le phenomene Sophia-Antipolis dans son environment regional. In Aydalot, P. (Ed.) *Milieux Innovateurs en Europe*. Paris: Gremi.

Piore, M. and **C. Sabel** (1983) Italian small business development: lessons for US industrial policy. In Zysman, J. and L. Tyson (Eds.) *American Industry in International Competition: Government Policies and Corporate Strategies*. Ithaca: Cornell University Press.

Piore, M. and **C. Sabel** (1984) *The Second Industrial Divide: Possibilities for Prosperity*. New York: Basic Books.

Planque, B. (1986) La Zone d'Aix-Marseille. In Aydalot, P. (Ed.) *Milieux Innovateurs en Europe*. Paris: Gremi.

Poittier, C. (1988) Local innovation and large firm strategies in Europe. In Aydalot, P. and D. Keeble (Eds.) *High Technology Industry and Innovative Environments*. London: Routledge.

Pollert, A. (1988) Dismantling flexibility. *Capital and Class* 34: 42 – 75.

Rey, G. (1989) Small firms: a profile of their evolution, 1981 – 85. In Goodman, E. (Ed.) *Small Firms and Industrial Districts in Italy*. London: Routledge, forthcoming.

Russo, M. (1985) Technical change and the industrial district: the role of inter-firm relations in the growth and transformation of ceramic tile production in Italy. *Research Policy* 14: 329–343.

Sabel, C. (1982) *Work and Politics.* Cambridge: Cambridge University Press.

Sabel, C. (1988) La riscoperta delle economie regionali. *Meridiana* 3: 13–71.

Saxenian, A. (1985) Silicon valley and Route 128: regional prototypes or historical exceptions? In Castells, M. (Ed.) *High Technology, Space and Society*. Beverly Hills: Sage.

Saxenian, A. (1987) 'The Cheshire Cat's Grin: innovation, regional development and the Cambridge case'. *Mimeo*. Department of Political Science, MIT, Cambridge, Massachussetts.

Schoenberger, E. (1988) From Fordism to flexible accumulation: technology, competitive strategies, and international location. *Environment and Planning D: Society and Space* 6(3): 245–62.

Scott, A. (1986) High technology industry and territorial development: The rise of the Orange County Complex, 1955–1984. *Urban Geography* 7(1): 3–45.

Scott, A. (1988) Flexible accumulation and regional development: the rise of new industrial spaces in North America and Western Europe. *International Journal of Urban and Regional Research* 12: 171–86.

Scott, A. and **D. Angel** (1987) The US Semi-conductor industry: a locational analysis. *Environment and Planning A* 19: 875–912.

Sforzi, F. (1989) The geography of industrial districts in Italy. In Goodman, E. (Ed.) *Small Firms and Industrial Districts in Italy*. London: Routledge, forthcoming.

Signorini, L. F. (1988) 'Innovation in Tuscan firms'. Paper presented to the Regional Science Association's European Summer Institute, 17–23 July, Arco (Trento).

Smith, C. (1988) 'Flexible specialisation, automation and mass production'. Paper presented to Industrial Relations Research Unit Conference on Questions of Restructuring Work and Employment, 12–13 July, University of Warwick.

Solinas, G. (1982) Labour market segmentation and workers' careers: the case of the Italian knitwear industry. *Cambridge Journal of Economics* 6: 331–352.

Solo, R. (1985) Across the Industrial Divide: A review article. *Journal of Economic Issues* 19(3): 829–36.

Swyngedouw, E. (1986) The regional pattern and dynamics of high technology production in France. In ASRDLF and CEEE, *Technologies Nouvelles et Developpement Regional*. Paris, Gremi.

Tolomelli, C. (1988) 'Policies to support innovation processes: Experiences and prospects in Emilia Romagna'. Paper presented to the Regional Science Association's European Summer Institute, 17–23 July, Acro, (Trento).

Trigilia, C. (1986) Small firm development and political subcultures in Italy. *European Sociological Review* 2: 161–75.

Williams, K., T. Culter, J. Williams and **C. Haslam** (1987) The end of mass production? *Economy and Society* 16(3): 405–39.

Zeitlin, J. (1988) The Third Italy: inter-firm co-operation and technological innovation. In Murray, R. (Ed.) *Technology Strategies and Local Economic Intervention.* Nottingham: Spokesman Books.

[4]

Work, Employment & Society, Vol. 3, No. 2, pp. 203–220 June 1989

Abstract: This paper examines the parallels between French and American perspectives on automation as a break with mass production in the 1960s, and French and American debates on Neo-Fordism/Flexible Specialisation as a break with mass production in the 1970s and 1980s. It explores the question of continuity and discontinuity in technological paradigms, and identifies common political themes in the two eras. In particular it notes that American interpretations of French theory, Blauner in the 1960s and Sabel in the 1980s, translate socialist or anti-capitalist transformational elements of these theories into pluralistic and liberal reformist political practice. The importance of the market, the role of craft skills, and the search for social solidarity and community within 'new' capitalist production regimes are recurrent elements in the American literature. French theorists, on the other hand, are more concerned with social agency, continuity of managerial control and capital/labour dynamics within new production forms which are contradictory rather than integrated. The paper examines the genealogy of the concept flexible specialisation, and evidence of its existence in a primary area of mass production, the food industry. It concludes by rejecting both the utility and political message of the flexible specialisation thesis in favour of a more grounded marxist analysis of work restructuring.

FLEXIBLE SPECIALISATION, AUTOMATION AND MASS PRODUCTION

Chris Smith

The term 'flexible specialisation' appeared in the Piore and Sabel book, *The Second Industrial Divide* (1984). The first industrial divide had occurred with the development and diffusion of mass production at the end of the last century, the new divide dated from the late 1960s and represented a break with mass production and consumption. 'Flexibility' refers to labour market and labour process restructuring, to increased versatility in design and the greater adaptability of new technology in production. 'Specialisation' relates to niche or custom marketing, the apparent 'end of Fordism', mass production and production standardisation. Hence the concept unites changes in production and consumption. Piore's background in labour market economics, especially dualism and segmentation theory, is evident in the concept and theory of the book and has been discussed in detail by Pollert (1988). Sabel's contribution has not been properly documented, and yet he would seem to have both

Chris Smith is a Lecturer in Industrial Relations, Aston University. He is author of *Technical Workers: Class, Labour and Trade Unionism* (1987), co-author of *White Collar Workers, Trade Unions and Class* (1986) (with Peter Armstrong, Bob Carter and Theo Nichols) and *Innovations in Work Organisation: The Cadbury Experience* (forthcoming, with John Child and Michael Rowlinson). His current research interests are in white collar union mergers, business policy in the transnational food industry and the comparative social position of technical workers.

developed and diffused the concept of flexible specialisation to audiences in industrial relations (Katz and Sabel 1985), economic history (Sabel and Zeitlin 1985) as well as local state initiatives in acting as consultant to the former GLC.

In *Work and Politics* (Sabel 1982) the term 'flexible specialisation' is not used. Instead the typology of Fordism, defined as 'the efficient production of one thing' (p.210) using unskilled labour and dedicated equipment, is represented as undergoing reform into 'neo-fordism' (p.209), the 'flexible assembly line' (p.212), 'flexible automation' (p.211) or 'other elastic forms of fordism' (p.217). These rubbery refinements express the partial modification of the production system in terms of an extension of product ranges, but substantially within the same framework of tight managerial control, deskilling and mechanisation. By contrast Sabel suggests that a new production typology, 'high technology cottage industry' (p.220), has been developing which combines craft forms of production with advanced technology and regional political supports, and offers a genuine alternative to mass production in all its variants. It is however, a very particular alternative, represented by the politics and industrial structure of artisanal firms in the Emilia-Romagna and Tuscany regions of Italy, and not one to be generalised without definite political pre-conditions.

The distance between qualified Fordism and 'high technology cottage industry' narrows in the Piore and Sabel book, as the term 'flexible specialisation' is constructed to embrace both handicraft production supported by local states and mass production sectors undergoing restructuring, batch industries using new technologies, as well as the construction and fashion goods sector. In Sabel and Katz (1985) centralised and concentrated centres of Fordism like autos are represented as moving towards 'flexible specialisation' as workers' skills become more 'flexible' and products less standardised.

It is the aim of this paper to explore the transition from a narrow alternative to mass production in artisanal work, to the broad alternative of flexible specialisation. To explore this through its parallels with the earlier debates on the transitional technological paradigm of automation, which was variously perceived as constituting a continuity with mass production, mechanisation and increasing managerial control, or as a discontinuity or radical break from it.

Automation: the French Debate

The flexible specialisation thesis argues that discontinuity with mass production characterises the current period. The automation thesis in the 1960s was similarly heralded by some as discontinuity with various elements of mass production. The French tradition in the *sociologie du travail* was concerned with the social identity of new developments against existing and older production regimes. While Sabel (1982), Piore and Sabel (1984), Sabel and Zeitlin (1985) distance themselves from elements of technological determinism evident in French and American writing, they nevertheless reproduce the tensions between their all embracing typologies – craft and mass production, Fordism, flexible specialisation – and the contingency of history.

In the post war period French writers, especially Marxist ones, examined the consequences of automation for the division of labour, worker alienation, the class structure and class struggle earlier than in other countries. Because France was, in the words of Friedmann (1955: 247) 'the country of diversity', with millions of small farmers and strong handicraft forms coexisting with advanced industry, debates about the social and political impact of modernising technologies occurred earlier and developed conceptual models which structured later debates (Gallie 1978; Rose 1979).

The questions posed by French writing on stages in the evolution of skills (Touraine), production types (Mallet) or later 'regimes of accumulation' (Aglietta) concern the extent to which different technological systems are continuous or discontinuous with each other. Did mass production, Fordism or the dominance of the assembly line lay the foundation for a more advanced but equally narrow division of labour in automation? Was automation a simple extension of earlier patterns of mechanisation, or a qualitative break?

Friedmann (1955) held both positions at different periods of his life. At certain times automation was interpreted as creating a 'new artisan class' and treated as a positive development, because 'the worker . . . ceases to play a part of an appendage to a machine, bent to its rhythm, and his functions of control may be intelligent ones in which interest in the job is reinstated' (Friedmann 1955: 187). At other times such developments are a poor substitute for the moral influence and community of craft control, and automation remains closer to mechanisation and mass production: 'each wave of mechanisation and rationalisation reduces [the moral influence of work] as skill is fragmented into repetitive operations demanding less and less attention and technical knowledge' (Rose 1979: 36–37).

For Mallet automation represented discontinuity, which while not offering the conditions of worker autonomy in craft regulation, did produce forms of work and industrial action that enhanced the potential for workers' control which mass production weakened. Mallet saw, like Friedmann, a novel agency, the new working class of technical and polyvalent labour emerging out of automation. The working class in process industries and those using electronic technologies was considered to be more integrated into capitalism, hold polyvalent or all-round skills, and an identification with the company that was absent in mass and craft production. Automation undermined piecework and created group or collective payment systems; it generated industry-specific skills, not occupational ones, and increased security of employment because the growing capital intensity of work demanded labour market stability. Automation reversed the tendency in mass production of job fragmentation, deskilling, alienation and the subordination of the worker to the rhythm or routine task discipline. Skills were not tied to occupational elites, but to industry-specific work groups and learnt on the job.

Mallet's interest in automation was chiefly through the perspective of agency and change. The working class in the older industries could not 'formulate a positive alternative to neo-capitalist society . . . It is only the social groups most integrated into the most advanced processes of civilisation which are in a position to formulate the manifold forms of alienation and envisage superior forms of development' (Mallet 1975a: 14). Unlike Blauner and Touraine, Mallet saw the integration of the worker in automated production as contradictory, 'producing the objective possibilities for the development of generalised self-management of production' while being constrained by 'techno-bureaucratic' management hierarchies, short-term profitability and private ownership (Mallet 1975b: 82). Mallet was only interested in the new conditions of class struggle that automation supported.

For Touraine mass production, standardisation and deskilling were also necessary steps or stages in the development of automation, which represented both a continuity with mass production and a break from it. His stagist model of technology – craft production (Phase A), mass production (Phase B) and automation (Phase C) – was not an evolutionary schema of deskilling and degradation as in Friedmann. Touraine in his studies of skill changes at Renault argued, like Ford had done, that mass production opened-up shop floor promotion by extending job hierarchies and training possibilities to unskilled workers, opportunities that craft regulation had blocked. However, the standardisation of products and production developed under mass production set necessary preconditions for automation. In terms of autonomy, craft workers possess technical skill individually and defend it within the craft community against economic incursions by capital. Under 'automation or complete mechanisation the worker no longer actively intervenes in the manufacturing process. He superintends, he records, he controls. His job can no longer be defined as a certain relationship between man and materials, tools and machines, but rather by a certain role in the total production picture . . . For the workers, the work possesses a direction and value which depends entirely on social factors, a situation which is the opposite to that which craftsmanship promotes' (Touraine 1972: 56–7).

The paradox of automation was that it provided a stable technical background against which to design varied organisational structures and social relations. For Touraine there were two totalities in the relationship to work under automation: the job, which is strictly a social environment, because skills have been absorbed from the worker and placed in machinery and organisational structure; and workers' attitude towards work. These totalities are based upon two false assumptions. Firstly that automation somehow 'solves' the technical problems of production and all that remains is the subjective question of attitudes. And secondly that work is only composed of technology and labour, when product markets, management structure and style also intervene to shape work organisation.

For Touraine the social choice or contingent element of automation meant that 'all narrow managerial ideologies ... and the present division between technicians and workers' could be overcome with the right attitudes to work (Touraine 1972: 60). Such attitudes are shaped by education and management policies, not by class struggle, and Touraine is close to Human Relations orthodoxy in this voluntarist approach. Close to Blauner, as we shall see, but different from Mallet.

Automation: Blauner and American Voluntarism

Friedmann began to utilise Marx's concept of alienation in 1956 as a way of studying the wider historical significance of change in work (Rose 1979: 41). By the time this is taken up by Blauner in the early 1960s the term had been 'operationalised' and transformed from a concept of political economy to one of socio-psychology – a rejection of Marxism in favour of a subjective social science (Eldridge 1973). There are parallels with Sabel here. Blauner, aware of the French literature on automation, not surprisingly found most sympathy with Touraine, and the belief in managerial reform of work through technological change and the assistance of the social scientist. He shares with Piore and Sabel a belief in the compatibility of workers' and managers' interests, hence in the conclusion of his book he calls for: 'joint research by industrial and social scientists into new methods of ... industrial design and job analysis, orientated to the goals of worker freedom and dignity, as well as the traditional criteria of profit and efficiency' (Blauner 1964: 185).

Blauner's thesis is well known and can be briefly stated. It concerns community and the social integration of workers at work, and relates these to different technological systems: craft, machine minding, assembly line work and process production. Social solidarity is high in craft work, but declines in mass production only to begin to be reversed again in process production and the spread of automation. The familiar elements of French writing are all present in Blauner, although he stresses the importance of production diversity and criticises the attention given to mass production which he claimed accounted for 'no more than five percent of the entire workforce'. Secondly, while generally reproducing the technological determinism and stagist models and belief in the 'long-run trend towards increased mechanisation' (Blauner 1964: 8), following Bright (1958), Blauner is more sensitive to the effect of the product market on technological trends. He gives one illustration of technological reversal; in the car firm of A.O. Smith there was a shift from automated back to hand manufacture when car frames ceased to be standardised. But he does not interpret product market changes as necessitating re-skilling of workers, for whom 'a black two door coupe differs little from a two-tone station wagon in overall structure, basic components or method of production' (Blauner 1964: 90). It is the stability of demand, and not the 'vagaries of consumer spending' that provide the economic supports for job security and high wages for workers in chemicals. Hence, the market conditions of increased worker integration and community are built on demand stability, not diversity as in the flexible specialisation thesis.

Blauner through an orientation to community, worker's social integration and anomie rather than alienation, found a potential resolution in the commodity status of labour in particular technological systems. The connections and continuities between craft and process production in his now infamous 'U' curve of 'worker satisfaction' are in marked contrast to radical French theories of automation. Whereas Mallet and Gorz (1967) looked to the confidence created by increased worker autonomy in process production for widening the contradictions of the social relations of production and class struggle, Blauner saw in automation a potential resolution of these contradictions. There is in both sets of writings illusions in technology, and a tendency to treat it in a self-sufficient or deterministic way, but there are strikingly different political interpretations attached to the same technological paradigm of automation. There are in these political transformations similarities to Sabel's treatment of 1970s paradigms such as Neo-Fordism as I will now discuss.

Neo-Fordism: French and American Interpretations

The Regulation School: Markets, Production and Discontinuity

Expansion of the product market was, for Friedmann, Touraine and Mallet, a necessary precondition for the establishment of mass production and automation. It was never doubted that such expansion should occur in sector after sector. Their theories, with the exception of Friedmann, were largely premised on conditions of economic growth and a concern with the full employment struggles between capital and labour. They were also production not market centred, counterposing the emergence of planning from below, increasing specialisation of production and supports for self-management against the authoritarianism of management and irrationality of capitalist markets. The market was always despotic, arbitrary and anarchic, whereas the increasingly qualified labour inputs into production, the needs of automation and 'technicisation' of labour were all signs of latent social order and rationality.

Writers in France in the 1970s who continued to typologise or periodise the capitalist labour process not only rejected the technological determinism of the earlier writers, they did so within conditions of economic crisis, growing world competition in product markets and unemployment in labour markets. Writers of the 'regulation school' looked for explanations of crisis through the social character of production, just as the earlier tradition had examined affluence through this particular lens. However, the optimistic search for transformation agencies – the new working class for neo-capitalism – is replaced by the narrower and more pessimistic concern with how the system is able to regulate itself.

The 'regulation school' of French socialist economists argued that capitalism is most successful at managing its contradictions when there is synergy between particular forms of technology, the labour process, consumption patterns and social institutions. Such eras of synergy, called 'regimes of accumulation' because they primarily relate to production and productivity, are divided into different periods characterised by the particular arrangements between production, consumption and managerial control. These are Taylorism, Fordism and Neo-Fordism. Without detailing the character of each of these, for my purposes the significant element of their analysis is the crisis and continuity between regimes of accumulation.

Neo-Fordism for Palloix (1976) who coined the phrase, was a 'purely formal attempt to abolish the collective worker'. Although offering the 'appearance' of autonomy, Neo-Fordism is essentially about 'setting-up absolute despotism in the coordination of the labour processes based on automation. Industrial job recomposition and enrichment seem to be . . . only an adaption of labour processes in mass production (Taylorism and Fordism) to new conditions of control of labour, to new conditions of reproduction of the domination of capitals in relation to the conditions for the production of the surplus-product, and constitute a new capitalist practice: Neo-Fordism' (Palloix 1976: 65). For Coriat (1980) Neo-Fordism arose out of internal problems of Fordism, in particular, the discontent with conventional assembly lines of young workers with high expectations and 'new constraints on the realisation of value' stemming from the growth of product market variability. Neo-Fordism retains the principles of Fordism, especially the essential discipline of the assembly line, non-marketability of jobs and routine tasks, but introduces 'segmentation of lines into distinct spaces, each supplied with its own stock of components and tools; and the introduction of team or group working, rather than the Fordist principle of one man/one position/one job. This is "controlled autonomy" which leaves the rhythm of work outside workers' control' (Coriat 1980: 35–6). Neo-Fordism is best represented in Japanese automobile assembly lines where the organisational principles of Fordism remain, but 'under conditions in which managerial prerogatives are largely unlimited' (Dohse, Jurgens and Malsch 1984: 35). Managerial control is transferred to the work group, which is autonomous only to the extent that '"autonomy" becomes a tool of self-discipline' and this has nothing to do with '"control" of the production process' (Coriat 1980: 40).

Neo-Fordism, like automation in the 1960s, concerns the themes of worker integration into the firm, team work and the blending of mass production within new technological conditions. It represents a continuity with managerial control. Participation, through Quality Circles, for example, is designed not to enhance worker control, but managerial coercion, as surveys of workers experience of QCs by Japanese trade unions demonstrate (Tokunaga 1983). Such 'participation' occurs 'in a controlled context in which the topics, goals and forms of articulation are practically limited to company interests' (Dohse *et al.* 1984: 37). Neo-Fordism for Palloix, Aglietta (1979), Coriat and the more critical accounts of Japanese car plants, is a capitalist solution to a central problem of scientific management, namely 'how to utilize knowledge of employees for the purpose of rationalisation' (Dohse *et al.* 1984: 37). It is a continuity of capitalist managerialism, with no 'benefit' for labour. A 'new practice' within the constraints of capitalism and as such can only be analysed through the contradictions of capitalist social relations.

Sabel and Neo-Fordism

Sabel's (1982) discussion of Neo-Fordism lacks the global context of French writers, is disengaged from a Marxist political economy and, in keeping with voluntarism and contingency of Blauner and Touraine, addresses the strategic choices it offers corporate management rather than labour or the objective 'regulation' of capital. In particular the introduction of new technologies such as CAD and NC create the possibility both of extending managerial control, but also enhancing workers' skills. The strategic direction of technology and conditions of Neo-Fordism are shaped by management action, not labour pressure and this can be for the good of all 'if management could step back from their work, as only a few of them can do, they would see that the present reorganisation reinterprets fordism as much as it perfects it' (Sabel 1982: 209).

On the one hand Sabel follows Coriat in acknowledging that Neo-Fordism is management utilising the 'innovative technologies and organisation devices to increase flexibility of production while holding to a minimum and sharply circumscribing discretion at the workplace' (Sabel 1982: 211). On the other he interjects a sense of workers benefitting from this process through notions of 'flexible automation' or 'more flexible neo-fordism' providing management seize these opportunities for 'flexibility'. Just as Blauner put alienation through the mangle of American liberalism and arrived at a situation in which the key strategic agents at work were managers and their job designs, so there is in Sabel's discussion of Neo-Fordism a similar pluralism. This is, however, most strikingly represented by the creation of the new typology, flexible specialisation, which blends employers' and workers' interests in a prescription for a shared future of industrial growth.

From Neo-Fordism to Flexible Specialisation

The dualism between craft production and Neo-Fordism evident in Sabel's *Work and Politics* is reconciled in *The Second Industrial Divide* through the concept of 'flexible specialisation'. By identifying the engineering, industrial relations and production restructuring of monopoly corporations not as continuity with managerialism and problems of labour control, but rather a potential – always a potential – discontinuity which could shift the American corporate economy towards a new era of growth through flexible specialisation, Sabel has abandoned the earlier dualism.

He has left behind the romanticism of craft in Emilia-Romagna, and embraced the restructuring of Boeing, General Electric, GM and Fords as indicative of a new era of corporate community and solidarity. Piore and Sabel caution that 'without the unifying vision of craft production' such reforms, which are largely inspired by Japanese competition, remain internally contradictory, which 'might serve as a point of crystallization of an economy of flexible specialisation; but then again it might not' (Piore and Sabel 1984: 283). However such vacillation is less prominent in later articles, for example in Katz and Sabel (1985) and Piore (1986), who argue that labour institutions represent the main obstacle blocking 'progress' towards the new corporate community. Particularly significant in these articles is the glorification of Japanese and German corporatism, Katz and Sabel (1985) leaving unintegrated the more critical accounts of the history and practice of Japanese industrial relations, in their concern to celebrate the apparent end of Fordism and arrival of flexible specialisation.

What then are the claims of the thesis? What elements of the earlier debate on technological paradigms exist in the current typologies? And what evidence is there to support the case for 'flexible specialisation'?

Fundamental to the thesis is the assumption that changes are occurring in the product markets of monopoly capitalist sectors, and that these changes offer a potential benefit to capital, labour and consumers – more flexibility in the use of resources, more skilled work and greater choice. 'Instead of producing a standard car by means of highly specialized resources – workers with narrowly defined jobs and dedicated machines – the tendency is to produce specialized goods by means of general purpose resources – broadly skilled workers using capital equipment that can make various models' (Katz and Sabel 1985: 298). The claim is that the break-up of the mass product market, and arrival of new 'flexible' computer technologies (Piore and Sabel 1984: 261) will improve the skills of workers.

Piore and Sabel suggest that the product market exercises the dominant influence on technology, and not qualified labour, self-sufficient technical ends or capital-labour dynamics. Computer technology in the office and on the shop floor is supposedly shaped by the prevailing trends in product markets. It therefore follows that if mass markets are breaking-up, and it is assumed that they are despite no evidence for this being advanced by the authors, then technology will respond accordingly. 'Had the mass market of the 1950s and '60s endured in the 1970s, computer technology would have mirrored the rigidity of mass production' (Piore and Sabel 1984: 261). Despite this demand-centred perspective on technology, Piore and Sabel are in addition careful not completely to rubbish the belief in an 'immanent logic of technological progress' and notions such as 'technological trajectories' and 'technological paradigms' are used throughout the book. These imply the idea of an independent determination to technology which possesses fixed courses, direction and linear tracks, and links their writing to the earlier stagist models of technology. They may reverse the weight given to the role of product market by Blauner, but they are close to the work of another American, Bright (1958) who stressed the limiting role product markets have on the application of technology. They are therefore firmly within earlier American approaches to automation.

The market, competition and corporatism are all central themes. Much of the theoretical justification for the necessity of competition and innovation that underlies their concern with 'industrial prosperity' stems from the political economy of craft and Proudhon. Rose (1987) in reviewing their book notes their 'stress on the moral as well as the social and economic superiority of craft culture'. Clearly this echoes Friedmann, and the debates in French writing between Mallet's support for Proudhon against Marxist critics (Rose 1979: 66). However the return to Proudhon in the 1960s was shaped by an attempt to draw parallels with the independent, anti-capitalist spirit of the craft workshop and the potential for similar forces affecting the shape of work in new process industries. In other words, Proudhon's view on work and workers' control was central to this revival. In contrast Sabel and Zeitlin (1985: 143–53) take from Proudhon the legitimacy of the market, beneficial effects of capitalist competition, and necessity of collaboration between labour and capital. And in Piore and Sabel (1984: 5) Proudhon is similarly used to show that 'economic success depended as much on cooperation as on competition'. An altogether different reading from the 1960s.

The role of the market, and the link between markets and work organisation that lie at the centre of the flexible specialisation thesis require empirical examination. This I shall do through an assessment of the current corporate restructuring of a major area of mass production, the UK food industry. Sabel and Zeitlin (1985) characterise Britain as an intermediate case between national industrial types – the US embodying the ideal type mass production economy and France the ideal type specialist, high-quality goods economy. I have already suggested that the diversity of production systems in France encouraged an early and sustained theoretical and empirical examination by French writers into the consequences of modernising ideologies and technologies. Conceptualising changes in mass production sectors as Neo-Fordism has been the dominant response to the product and labour market conditions Piore and Sabel see as creating a new state of workplace corporatism, namely flexible specialisation. By taking the example of a mass production industry in the intermediate, not extreme case, it may be possible to assess whether current employment and work organisation strategies are, as Piore and Sabel suggest, pushing towards flexible specialisation, or as French writers would argue, towards Neo-Fordism.

Work Restructuring in the Mass Production Food Industry

The connection between shifts in demand and flexible production is made most strongly in relation to consumer goods sectors because it is here where Piore and Sabel's sovereign consumer is most influential. The 'individualisation' of consumption patterns is the main lever against Fordism. Mallet, Gorz and even Coriat's young rebels in the production process of the 1960s and 1970s, are, in the language of the 1980s market place, transformed into radical consumers, who by their good 'taste' are restructuring workers' lives in capitalist labour processes. Sabel warns us of the purchasing power of yuppie shoppers: 'do not forget all those fashion, health and quality conscious consumers who, quite independently of foreign competition, are unsettling the manufacture of everything, from shirts to bread' (Sabel 1982: 212). His case is echoed in statements from leading mass food manufacturers, such as Sir Adrian Cadbury, who in a speech to French industrialists in 1982 said that he expected 'people to adopt a more individual life-style in years ahead and to be less ready to accept the offerings of the mass market. The problem for us as manufacturers will be to meet these individual needs without losing the advantages of long production runs' (Smith, Child and Rowlinson 1989). The question we have to ask is, what evidence is there of a break-up of mass production in consumption areas? And further, what connection, if any, does change in the structure of the product market have on the labour process and workers' skills?

Sabel (1982) discusses the case of bread in *Work and Politics*. He suggests that there have been two responses to 'fluctuations in demand taste', one is large firms subcontracting to a 'blender who aggregates demand from different speciality bakeries, buys in bulk, blends the base and ships a cartload of the desired assortment to the large bakery' who then bake on conventional assembly lines. The other is a return to 'artisanal methods of production' (Sabel 1982: 218–219). We have a Neo-Fordist and craft response to the same problem, and Sabel implies that the former is only short term expediency: 'how many different types of bread can the large bakery make on its assembly lines [and] what happens if the blender sets up his own business?' (Sabel 1982: 219). Heavy on rhetoric, and thin on evidence, does the demise of white sliced bread, cornflakes, strawberry jam and the rest of our processed fare really threaten mass production which began with and has been dominant in foods?

In a forthcoming book, I have examined product and work organisation changes in a major oligopolistic segment of the food industry, chocolate confectionary (Smith, Child and Rowlinson, 1989). We found the evidence of differential demand contradictory. In chocolate, biscuits, cheese, cakes, sweets, and jam market segments there has been a reduction in the variety of products and concentration on a limited number of core brands. This has been a deliberate marketing policy – chiefly guided by concentration in British and European retailing – to leave low-volume products and promote core brands produced intensively and advertised globally. A strategy of product and company globalisation, not fragmentation and decentralisation of products or ownership. For all Sir Adrian Cadbury's worry, Cadburys has been warmly embracing Fordism, reducing the number of products by over half in the last ten years and simultaneously internationalising their market (Smith 1987; Child and Smith 1987).

It should be stressed that greater product variety in the past, or as it exists in specialist confectionary producers like Thorntons, has always been met by employers using batch production and the 'flexible' and cheap hand skills of women workers, who have never been socialised through the craft route taken by bakers. The gender construction of skills in these mass production sectors is essential to understanding employer strategies and control, but Sabel appears blind to this and selects an example from one of the few segments of food processing where craft skills exist. In most food sectors women on assembly lines and semi-skilled men in manufacturing is the typical division of labour, and craft skills are absent or confined to maintenance work.

In bread, prepared milk and yoghurt, ready-to-eat cereals, snack food, cooked meats and margarine there has been an expansion of product 'variety' and 'quality', although both terms should be used cautiously. However, there is no evidence from case study or survey data on cooked meats (Liff 1986), snack foods and cereals (Leach and Shutt 1983) of such variety breaking-up assembly lines and increasing skills. Neither is there evidence, outside of prepared or cook-chill foods, of the growth of petty bourgeois enterprises using craft skills fulfilling the 'broadening' of consumer tastes. Own-labelling by retail stores has been the major factor in the differentiation of demand, and this has been achieved by multiple retailer buying abroad, creating dependency relationships with small and medium sized firms who are tied into the product market of the large retailer or giant firms producing branded goods and 'own label' for the large retailers. The vision of the 'blender' setting up shop to satisfy discerning consumers fits neither dominant manufacturing nor retailing practices and trends.

The globalisation of tastes is the central aim of mass food producers, and the market share of the top few companies in all the major food segments indicates stability. Spreading processed foods to developed and developing countries is achieved at the expense of indigenous products and tastes, just as increasing choice of exotic fruits and vegetables through the concentrated buying power of food retailers is accomplished creating cash-crop dependency and destroying local product variety in supplying countries. Globalisation reinforces mass production principles (Burbach and Flynn 1980; Leopold 1985; Sorj and Wilkinson 1985).

It is not clear that the concern with labour 'flexibility' in mass manufacturers or mass retailers has anything to do with increasing or decreasing product variety. In the Cadbury case we found efforts directed at restructuring work organisation and undermining job demarcations occurring with the movement towards fewer products. Flexibility had little to do with extending skills or worker satisfaction, and a lot to do with undermining the remaining areas of craft control in maintenance, extending managerial authority over labour mobility and ensuring greater utilisation of expensive capital equipment (Smith *et al.* 1989).

Household, handicraft and industrialised food systems have long co-existed, and Sabel is wrong to imply that there will be a 'return' to one system at the expense of the others. Hegemony shifted from agriculture to manufacturing with the rise of mass markets and branded goods, but is today under 'threat' from the concentrated buying power of retailers who are emerging as the dominant player in the food chain. Handicraft was fragmented and weak under mass process hegemony and remains in a dependent position under retail hegemony. This does not mean it will 'disappear', but there is no empirical evidence of any renaissance. Craft bakers have co-existed alongside mass factory systems, as have butchers, fishmongers and greengrocers alongside concentrated or industrialised alternatives. In Britain in the post-war period, however, these independent 'handicrafts' have all declined: butchers from 41,799 to 21,488 between 1950 and 1979; fishmongers and poulterers from 9,511 to 2,725; greengrocers and fruiterers from 43,948 to 14,380; and bread and flour confectioners from 24,181 to 13,210 (Smith *et al.* 1989). The downward spiral is continuing with increasing concentration of ownership in retailing. The 'return' to brown bread has not reversed this trend as the Neo-Fordist solutions have largely been the ones adopted, which has expanded own-label brown bread sales through concentrated retailers or hot bread chains, not through independent bakers. Sabel's example of bread is ill chosen. Not only does it not stand up to empirical testing, but it reveals his male-centred definition of 'craft' which is atypical in the food industry where it is women who are the mass workers and not men.

Conclusion

The theory of flexible specialisation has re-kindled an older debate about the technological typologies which were developed in different forms and political contexts in the 1960s. A central problem in the debate on automation was a tendency to dislocate the particular conditions of production from their social form. Abstract pronouncements on the fixed character of capitalist social relations are inadequate, but creating theories on the appearance of 'particular configurations of hardware and social organisation' (Harvey 1982: 133) without attention to their contradictory nature and particularity is not a satisfactory approach to a theoretical understanding of capitalist diversity.

> The concrete forms of technology, organisation and authority can vary greatly from one place to another, from one firm to another, as long as such variations do not challenge the accumulation process. There are evidently, more ways to make a profit than there are to skin a cat. And if the value productivity of labour can be better secured by some reasonable level of worker autonomy, then so be it.' (Harvey 1982: 116)

Technological determinism stems from the isolation of one factor, technological state or organisational form, albeit a feature of labour (its skill structure) or of fixed capital (the versatility or dedicated nature of machinery) from this valorisation process. The flexible specialisation thesis is premised on technological determinism not simply because, like the American debate on automation, hardware was given unusual powers to shape work organisation, but rather that its promoters have invested particular combinations of labour and capital with qualities to reconcile capitalist social contradictions. They have also invested particular production or technological systems with a completeness or totality, which does not mirror the ongoing production diversity within capitalism as a whole.

Unlike certain strands of the French debate on automation, whose *discontinuity* was judged as new source of opposition to *weaken* accumulation, and challenge the hierarchy, authority and capitalist control at work, todays debate on flexible specialisation searches for new ways of *enhancing* accumulation. This reflects the politics of the period, but also the theoretical traditions of American pluralism. Piore and Sabel accept 'competition' of the market as a permanent and positive element of 'industrial society'. Their main policy objective is to find that combination of hardware and labour that can be competitive in product markets and simultaneously minimise social antagonism and low trust in production. At times this is found in the 'solidarity and communitarianism' of craft (Piore and Sabel 1984: 278); at others in the autonomy of the petty bourgeois, as in Sabel's vision of the self-employed baker; and at other times in the high trust and corporate culture of their version of Japanese working practices (Katz and Sabel 1985: 298). In all production settings their concerns, like those of Blauner, are the search for industrial community, an end to alienation and class struggle within the constraints of the capitalist market place.

The case for flexible specialisation has been created by a false polarisation between craft and mass production which initially leaves out of account well-established industrial classifications of technology into unit/small batch, mass and process types (Woodward 1958). By erecting a dichotomy between extremes, craft and mass, the continuing presence of non-mass technologies and small firms, can be presented as 'among the most discordant facts about the mass prodcution economy' (Sabel and Zeitlin 1985: 137–8). But this is only anomalous to a vision of capitalism as a mass production paradigm, and not as an economic system committed to profitability and capital accumulation, where the existence of diversity, whether in mass and batch production or large and small capital, is normal.

Industrial diversity is endemic to capitalist competition and reflects the separation of capitals, and the costs attached to innovation and inertia. Harvey (1982: 120) lists nine possible strategies to deal with heightened competition between capitals, which included lowering wage rates; increasing intensity of an existing production system; new investment; economising on constant capital inputs; developing more efficient 'factor' combinations; changing the social organisation of production – job structures, chains of command – in search of more efficient management; appeals to workers to cooperate and work harder; new marketing strategies; and changes in the location of production.

'Through one, or any combination of these responses, individual capitalists can hope to preserve or improve their competitive position'. Each separate capital has the 'possibility to alter his own production process so that it becomes more efficient than the social average'. Therefore production diversity or concentration of capitalists within dominant typologies, does not negate the 'imperative to revolutionise the productive forces by whatever means of whatever sort' (Harvey 1982: 120). Diversity does not undermine capitalist market relations. The imperatives of the market place are a given to Piore and Sabel, but their continuing capitalist nature and therefore the continuing social cleavage between capital and labour are not. The latter can be reconciled if the appropriate combination of inputs and outputs is chosen.

Part of the aim of this paper is to recall the earlier debate on technological typologies, and observe the common and divergent readings of the impact of automation in the workplace. In all versions the principles or idea of a technological type – automation, mass production, flexible specialisation – has an inner logic, a set of defining characteristics and appropriate labour/capital relations, which in the case of automation, were found not to exist as hypothesized (Gallie 1978; Rose 1979).

In relation to the evidence of flexible specialisation, I have suggested in the case of the British food industry that a coexistence of various production systems characterise most food segments, but that mass production dominates. Artisanal systems are not undergoing a renaissance, but continuing decline. The factory bakery may have moved to the high street, but it has done so through the multiple form with concentrated companies like Don Millers or 'hot' bread supermarket outlets. In this form the division of labour and gendering of occupations remain the same as in the factory, the main difference being that trade union organisation in the factory is strong, but weak or non-existent in the high street. But perhaps, as Piore and Sabel (1984: 278) suggest, the privilege of so called 'craft production' is preferable to mass production 'regardless of the place accorded to unions'.

British mass producers, unlike their American counterparts, have never fully embraced Fordism, but have rather persisted with a hybrid type of mass/batch production better suited to Britain's more diverse product markets. Such product market diversity has always been absorbed by female workers' flexibility, rather than male craft labour. The dominant trends amongst mass producers appears to be towards cutting products and mass producing on a national and global scale a limited number of core brands. This represents in product and production terms a movement towards classical Fordism. Work and employment restructuring has been about employers increasing their utilisation of their capital and labour, not enhancing skills and worker satisfaction (Child and Smith 1987). In other words embracing the qualities of Neo-Fordism described earlier by French writers. Flexible specialisation is nowhere in sight (Smith *et al.* 1989).

Attacking the theory of flexible specialisation on the basis of mustering evidence of continuity within existing structures is, however, unlikely to upset or unseat those committed to the paradigm. Rose (1979) shows that most surveys of the impact of automation in France were hesitant, qualified or even negative of its hypothesized qualitative break with mass production. Yet such studies did little to reduce the strength of the theory. Case studies into the 'existence' of flexible specilisation will therefore be unlikely to undermine the typology. They are more likely to help diffuse the concept to wider academic circles. It is for this reason that I have attempted to question the novelty of the theory in terms of a longer debate on technological change. In examining the parallels with earlier debates on the 'end of mass production', this paper has hopefully cautioned against the utility of explaining capitalist restructuring through the constraints of technological paradigms which, of logical necessity, positively value one typology over another and conceive of capitalism through the technical rather than social relations of production.

A more grounded theoretical approach to change in work is required. This should not begin inside a mystifying technological framework, but rather upon the historical interaction between product and labour markets, management strategy, technical change and the labour process in the context of the dominant trends in capitalist restructuring. Kelly's (1985) studies on the relationship between product markets and work organisation, or historical sector studies of work restructuring developed by the Work Organisation Research Centre at Aston University, offer theoretically informed starting points. In a period of major and diverse structural change within a context of heightened capitalist competition, technological typologies only obscure developments and encourage utopian prognostications, rather than much needed sensitive and detailed case studies of change and continuity in the capitalist labour process.

Acknowledgements

The revised version of this paper was prepared for a conference at the Industrial Relations Research Unit, Warwick University, *Questions of Restructuring Work and Employment*, 12–13 July 1988. I am grateful for the comments on the paper by participants at that Conference. I would also like to express my appreciation for the referees' comments and those of the Editor of this journal.

References

Aglietta, M. (1979) *A Theory of Capitalist Regulation*, London: New Left Books.
Blauner, R. (1964) *Alienation and Freedom*, Chicago: University of Chicago Press.
Burbach, R. and Flynn, P. (1980) *Agribusiness in the Americas*, New York: Monthly Review Press.
Bright, J.R. (1958) *Automation and Management*, Boston: University of Harvard Press.
Child, J. and Smith, C. 'The Context and Process of Organizational Transformation

– Cadbury Limited in its Sector'. *Journal of Management Studies*, 24, 6, 565–594.
Coriat, B. (1980) 'The Restructuring of the Assembly Line: A New Economy of Time and Control', *Capital and Class*, 11, 34–43.
Dohse, K., Jurgens, U. and Malsch, T. (1984) 'From "Fordism" to "Toyotaism"? The Social Organization of the Labour Process in the Japanese Automobile Industry', Berlin: International Institute for Comparative Social Research, Labour Policy, *Working Paper*, April, 1–42.
Eldridge, J.E.T. (1973) *Sociology and Industrial Life*, London: Nelson.
Friedmann, G. (1955) *Industrial Society*, New York: The Free Press.
Gallie, D. (1978) *In Search of the New Working Class*, Cambridge: Cambridge University Press.
Gorz, A. (1967) *Strategy for Labor: A Radical Proposal*, Boston: Beacon Press.
Harvey, D. (1982) *The Limits to Capital*, Oxford: Basil Blackwell.
Katz, H.C. and Sabel, C.F. (1985) 'Industrial Relations and Industrial Adjustment in the Car Industry', *Industrial Relations*, 24, 2, 295–315.
Kelly, J. (1985) 'Management's Redesign of Work: Labour Process, Labour Markets and Product Markets' in D. Knights, H. Willmott and D. Collinson, *Job Redesign*, Aldershot: Gower.
Leach, B. and Shutt, J. (1983) 'Chips and Crisps: The Impact of New Technology on Food Processing Jobs in Greater Manchester'. Paper to British Association Conference, *New Technology and the Future of Work*, Brighton, August.
Leopold, M. (1985) 'The Transnational Food Companies and Their Global Strategies', *International Social Science Journal*, 105, XXXVII, 3, 315–329.
Liff, S. (1985) 'Women Factory Workers – What Could Socially Useful Production Mean for Them?' in Collective Design/Projects, *Very Nice Work If You Can Get It*, Nottingham: Spokesman.
Mallet, S. (1975a) *The New Working Class*, Nottingham: Spokesman.
Mallet, S. (1975b) *Essays on the New Working Class*, St. Louis: Telos Press.
Palloix, C. (1976) 'The Labour Process: from Fordism to neo-Fordism', In Conference of Socialist Economists (eds.) *The Labour Process and Class Strategies*, London: Stage 1, 46–67.
Piore, M.J. and Sabel, C.F. (1984) *The Second Industrial Divide*, New York: Basic Books.
Piore, M.J. (1986) 'Perspectives on Labor Market Flexibility', *Industrial Relations*, 25, 2, 146–166.
Pollert, A. (1988) 'Dismantling Flexibility', *Capital and Class*, 34, Spring, 42–75.
Rose, M. (1979) *Servants of Post Industrial Power*, London: Macmillan.
Rose, M. (1987) Review of M.J. Piore and C.F. Sabel (1984) *The Second Industrial Divide: Possibilities for Prosperity*. In *European Sociological Review*, 3, 1.
Sabel, C.F. (1982) *Work and Politics*, Cambridge: Cambridge University Press.
Sabel, C.F. and Zeitlin, J. (1985) 'Historical Alternatives to Mass Production: Politics, Markets and Technology in Nineteenth Century Industrialisation', *Past and Present*, 108, 133–176.
Smith, C. (1987) 'Cadburys and the Management of Consent'. Work Organisation Research Centre, Working Paper Series, Aston University. No. 27, February, 1–71.
Smith, C., Child, J. and Rowlinson, M. (1989) *Innovation in Work Organisation: Cadbury Ltd 1900–1985*, Cambridge: Cambridge University Press (forthcoming).
Sorj, B. and Wilkinson, J. (1985) 'Modern Food Technology: Industrialising Nature', *International Social Science Journal*, 105, XXXVII, 3, 301–313.
Tokunaga, S. (1983) 'A Marxist Interpretation of Japanese Industrial Relations, with Special Reference to Large Employers' in T. Shirai (ed) *Contemporary Industrial Relations in Japan*, Columbia: University of Columbia Press.

Touraine, A. (1972) 'An Historical Theory in the Evolution of Industrial Skills' in L.E. Davis and J.C. Taylor (eds.) *Design of Jobs*, Harmondsworth: Penguin.
Woodward, J. (1958) *Management and Technology*, London: HMSO.

Organisation Studies and Applied Psychology
University of Aston Business School
Aston Triangle
Birmingham B4 7ET

[5]

John Tomaney

The reality of workplace flexibility

Evidence from Japan, West Germany and the UK shows that there is no generalisable trend toward techno-logy driven flexible craft work. Insofar as it exists, workplace flexibility does not guarantee the benefits for labour implied by some of its proponents. Rather flexibility strategies have been used to weaken worker constraints on management pre-rogatives and to intensify work.

● This paper takes a look at current changes in the organisation of work and production in the manufacturing sectors of the advanced capitalist countries. Specifically it examines the detailed, empirical evidence for change and asks what some of the implications of these might be for workers.

Among sections of the Left in Britain there is a belief that rapid technological change is transforming work (and society as a whole) in new and unprecedented ways. A good example of this view is contained in the Communist Party document *Manifesto for New Times*, which states:

> At the industrial heart of the new times will be production based on a shift to information technology and microelectronics. New technology allows more intensive automation and its extension from large to smaller companies, pulling together the shopfloor and the office, the design loft and the showroom. It allows production to be both more flexible, automated and integrated . . . Work is being reorganised around new technology. Traditional demarcation lines between blue and white collar

worker, skilled and unskilled are being torn down in the wake of massive redundancies in manufacturing. In future, work in manufacturing will be about flexible team working within much smaller, more skilled workforces (Communist Party, 1989: 6–7).

This paper seeks to examine the evidence for such claims and to test the assumptions of the theory which underlies them. The CP position, through the work of Murray (1985, 1988), has been much influenced by research which suggests that industrial production in capitalist societies is being transformed in the direction of 'flexible specialisation' and is associated with the work of Charles Sabel and Michael Piore (Sabel, 1982, Piore and Sabel, 1984) and to a lesser extent with that of Horst Kern and Michael Schumann (1984, 1987, 1989).

In the section below, the flexible specialisation thesis is outlined, and in the following sections the empirical evidence for the kinds of changes suggested by the thesis is assessed through an examination of work and production organisation in Japan, West Germany and the UK. A final section attempts to draw some theoretical and political conclusions.

Flexible specialisation in the workplace: Theory and evidence

Theory

Piore and Sabel (1984) ascribe current changes in production and work organisation to limits inherent in the post-war model of industrial development. This 'paradigm' is referred to as 'mass production' which is seen as superseding a 19th century 'craft' paradigm. The mass production paradigm was concerned with the production of standardised commodities for stable mass markets. It is in the disintegration of these mass markets that the crisis of mass production is located. The new market segmentation and volatility places a new imperative on enterprises to move toward a more 'flexible' system of production which can cope with rapidly changing demands. This means that whereas work under the mass production paradigm was characterised by an intense division of labour, the separation of conception and execution, the substitution of unskilled labour for skilled labour and special purpose for universal machines; the quest for specialisation prompts a more flexible organisation based on collaboration between

Workplace flexibility

31

designers and reskilled craftworkers to make a wide variety of goods with general purpose machines. In the first instance, these tendencies are identified in areas containing autonomous 'industrial districts' – loose alliances of small firms of which the Third Italy is the most cited example (Sabel, 1982: 220–27; Piore and Sabel, 1984: 226–29). The tendency toward flexible specialisation is also identified within some 'mass production' industries, such as motor vehicles (e.g. Katz and Sabel, 1985); and at the level of the national economy in the case of Japan (Piore and Sabel, 1984: 156–62). The competitive success of these areas and enterprises is linked to the presence of tendencies toward 'upskilling' and the reintegration of tasks. Hence, it is implied, the tendency toward *flexible specialisation* is largely progressive in its implications for workers.

While operating at a different theoretical level, Kern and Schumann (1984, 1987, 1989) arrive at conclusions which are remarkably convergent with those of Piore and Sabel. Kern and Schumann are not engaged in constructing a theory of the transition from one mode of industrial development to another but rather report empirically observed trends from their studies in one country, West Germany. They argue that the capitalist pursuit of higher productivity of human labour with the aim of improved capital utilisation remains unchanged. However, the means by which this was achieved is altered. They maintain that whereas under previous forms of production, labour was viewed solely as a cost of production and obstacle to efficiency, enterprises within the core sectors of the West German economy now regard labour more strategically. They identify labour processes in industries such as cars, machine tools and chemicals where the reintegration of tasks is occurring with a consequent demand for highly qualified and highly skilled labour. In some cases they identify the emergence of teamworking. They associate such developments largely (though not exclusively) with the presence of new technologies. In a manner similar to Piore and Sabel they argue:

> Higher productivity cannot be attained under the present conditions without a more considerate, 'enlightened' treatment of labour – that is something that capital too must learn (Kern and Schumann, 1987: 162).

They go so far as to argue:

> *The end of the division of labour*: that is what development in an important part of industrial production could lead to under the influence of the new production concepts (Kern and Schumann, 1987: 163, emphasis added).

According to Hirst and Zeitlin (1988, 1989) the notion of flexible specialisation has much usefulness in explaining the relative competitive performance of UK manufacturing performance. They argue that despite the improved productivity witnessed in UK manufacturing since the early 1980s, Britain still performs relatively badly in world markets (as is evidenced by a whole range of economic indicators). This poor performance is attributed to the persistence of mass production methods and mentalities. They suggest:

> Most accounts of the competitive success of countries like Japan and West Germany assume that they are simply more efficient in using these [mass production] methods. But a closer look at manufacturing in these countries suggests that they have responded to the changed international environment by an alternative which *reverses* the principles of mass production: flexible specialisation involves the combination of general-purpose capital equipment and skilled, adaptable workers to produce a wide and changing range of semi-customised goods (1989: 168).

In policy terms, therefore, the aim appears to be to encourage the development of the types of flexible production processes and institutional supports which are said to characterise the successful enterprises and social formations present in Germany, Japan and so on (see also Murray, 1985). In relation to the main concern of this paper, the nature of the 'new' workplace flexibility, a number of important issues are raised. Given that the principles of production and work organisation in countries such as West Germany and Japan are to be emulated, it is worth subjecting the principles under which they operate to some critical analysis. Also, given the political implications of Piore and Sabel and Kern and Schumann for the strategies of workers and their organisations, it is important to ask what are the actual implications of these forms of work organisation. In an attempt to address these

Workplace flexibility

33

questions and before turning to an examination of the UK evidence, a critical analysis of the nature of workplace flexibility in Japan and West Germany is advanced.

Evidence

Flexible specialisation in Japan: a worker's nirvana?

Piore and Sabel's account of production and work organisation in Japan is unremarkable in so far as it propagates what is now a fairly orthodox view about the nature of capital-labour relations in that country. The implication is that Japanese work organisation is essentially progressive and largely beneficial for workers. Piore and Sabel contend that following the defeat of Japanese militarism in 1945, an alliance of the US occupiers, the reconstituted Japanese state and the *zaibatsu* (enterprise confederations) combined to prevent the left-wing trade unions from forming a powerful national labour movement. Instead the US:

> allowed a system of flexible shopfloor control to emerge through collective bargaining in export-oriented plants (Piore and Sabel, 1984: 160).

They argue that the *zaibatsu* system of inter-firm organisation was based on a particular form of production and work organisation where the distinction between the conception and execution of tasks was 'deliberately blurred' through teamworking and where foremen were regarded as the work teams' 'representatives' to management. In turn these developments were facilitated by the institutionalisation of long term employment guarantees and 'extensive job rotation to familiarize workers with the context of their work and to increase their flexibility' (Piore and Sabel, 1984: 161). They conclude:

> Finally, once the workforce was multiskilled, used to consulting with management, and committed to the firm, it was easy to introduce, in the 1960s, 'quality circles', in which labor-management teams improved the phases of production for which they were responsible (ibid.).

Thus by the 1970s Japanese production methods are seen as exemplifying 'the re-emergence of the craft paradigm amidst the crisis' (Piore and Sabel 1984: 205–8).

This interpretation of the Japanese 'economic miracle' can be contested on a number of grounds. These are mainly: its selective use of evidence; its reliance on solely management sources (especially US management consultants); and its misrepresentation of key events in Japanese labour history. At best Piore and Sabel offer a partial view of the real nature of work in Japan, but one which is now widely held in the west. Attempts to 'sell' Japanese work methods to unions (often as prelude to inward investment) has led to attempts to present these work methods as involving inherently higher levels of skill and workforce participation (e.g. Wickens, 1987).

Sayer's (1987) attempt to broaden the discussion of the just-in time (JIT) system is a useful starting point in analysing the reality behind the myths. Sayer argues that JIT as a method of industrial organisation should be seen as a search for time economies in the circulation of capital and new ways of extracting surplus value 'based on a sophisticated method of learning by doing' (1987: 52). The now familiar aspects of JIT such as reduced set up times, the use of small, simple machines (instead of large, complex ones), careful plant layouts and the reduction of buffer stocks are all seen as contributing to a more rapid throughput and greater capital utilisation. Reductions in buffer stocks reveal waste in the production process – wasted labour, materials and imbalances in the line – and lead to a new emphasis on total quality control. As such it is not technology itself which is significant, but the flexibility and intensity with which it is utilised.

According to Sayer, worker participation is essential to this process. Under Japanese production systems, for instance, lines are constructed piecemeal utilising workers' knowledge. There is a tendency under Japanese systems toward multiskilling. Wickens reports an example of this process of continuous improvement (known as *kaizen*) at Nissan's Sunderland plant:

> The Nissan body construction shop is now considerably different from the original layout and the vast majority of these changes have been thought up and implemented by the people working in the area (1987: 46).

Like Piore and Sabel, Sayer emphasises the role of worker participation in production planning and like them he appears to perceive this automatically as evidence of worker approval of the system and 'consent'. But does participation indicate

Workplace flexibility

35

evidence of a consensus between capital and labour? Moreover, what is the real nature of multiskilling under these methods?

It is possible to go along with the description of JIT and Japanese work methods as involving an element of worker participation without seeing these as having fundamentally beneficial consequences for workers. More critical accounts see 'cooperation' as part of the system of exploitation which operates in Japanese manufacturing enterprises. For instance, according to Turnbull, JIT is essentially:

> a highly developed form of work intensification which belies any notion of job enrichment through teamworking, flexibility and job rotation claimed by the many proponents of JIT (in fact, job rotation, teamworking, flexibility and the like are the very tools of work intensification under the JIT system) (1987: 8).

Turnbull sees the JIT system as an organic whole in which the securing of time economies in the circuit of capital and increasing the productivity of labour are inextricably linked. The link lies in the fact that by increasing the rate of throughput of the factory the total productivity of the plant can be raised. The reduction of buffers and total quality control are the first step on the road to JIT:

> but JIT aims to eliminate *all* costs surrounding the production process . . . of equal importance are the unproductive and wasteful elements of the worker's labour such as waiting time, downtime and excessive set up times (Turnbull, 1987: 9).

Turnbull argues that JIT does alter several of the principles of the classic assembly line by moving to 'single unit' production. Under this system worker's move between several machines in conjunction with work moving through the factory. As well as saving on direct labour, this process is more efficient as a means of maintaining the productive flow than simply speeding a conventional line which often only led to the problems of bottlenecks. However, the emphasis is on *multiple-machine minding* and as such Turnbull's account casts a new light on notions of multiskilling as inherent in Japanese work methods. The chief process here is one of intensification.

'Participation' can, therefore, be seen as coerced rather than a voluntary consensus. Total quality control and quality circles ensnare workers themselves into this system of intensification. In practice, and contrary to a widespread myth, quality circles are less concerned with product than with process innovations, principally the elimination of wasteful activities. According to Abernathy *et al* (1983):

> One of the principal thrusts of quality circles in Japan is to achieve a full 60 minutes of work each hour by each worker (quoted by Dohse *et al.*, 1985: 128).

At Nissan's UK manufacturing facility in Tyne and Wear, a system of 'neighbourhood checks' are used to enforce total quality control. In essence these involve groups down the line reporting on poor work performance by groups upstream (Garrahan *et al.*, 1989). In Japan, line foremen and even quality circle members can receive training in method study techniques.

The elimination of buffers (described earlier) and the principle of making visible all waste, especially wasteful labour, places extraordinary pressure on workers to comply in the rationalisation production. As Abernathy *et al.* (1983) put it:

> The determination to make all problems visible is not an unmixed blessing. It offers the hope of thoroughly efficient operations by substantially raising the social costs and consequences of failure. Reducing inventory levels places increasing pressure on managers and workers alike to remove whatever problems remain . . . by ratcheting up the level of stress at which the workforce is expected to perform . . . it is wonderful how a little fear and danger can clear the mind (quoted in Shaiken *et al.*, 1986: 176).

In addition, Japanese managers aim for total time flexibility (in Japan this can mean unlimited unpaid overtime) and functional flexibility (the will and capacity to undertake a wide variety of tasks) on the part of workers in order to cope with unforeseen. For instance, Komatsu, the Japanese earth moving equipment manufacturer at Birtley, Tyne and Wear, included in its single union agreement with the AEU provision for:

> Complete labour flexibility, interchangeability and mobility . . . in order to maximise productivity and correct imbalances in the production flow (quoted in IDS, 1988: 24).

Workplace flexibility

37

In sum, Japanese methods of work organisation can be seen ***primarily*** as a means of work intensification based upon the elimination of wasted time in production. The intensity, and even brutality, of this system has been described by Satoshi Kamata (1980) in his 'insider's account' of life in a Toyota plant in the 1970s. Flexibility in this account means the ability to move uncomplainingly from one deskilled task to another at a moment's notice and to meet production shortfalls through an open ended commitment to exhausting levels of overtime.

Such an analysis leads Dohse *et al.* (1985) to characterise these methods as variants of Taylorism or Fordism (see also Garrahan *et al.*, 1989). The US management consultant Schonberger goes so far as to say that 'the Japanese out-Taylor us all' (quoted by Dohse *et al.*, 1985: 127). For Dohse *et al.*, (1985) Japanese management techniques are a solution to the capitalist problem of workers' refusal to place their knowledge of production in the service of rationalisation.

Notwithstanding these arguments Wilkinson and Oliver plausibly suggest that, in theory:

> while JIT heightens the visibility of worker behaviour and increases the internal substitutability of labour, its high pervasiveness, high immediacy and low external substitutability means that overall the power capacity of workers will be enhanced (1989: 52).

The refusal to work overtime or to be flexible about tea breaks or task mobility could have ramifications throughout the production process. Yet this appears not to occur in Japan. Wilkinson and Oliver attribute this to 'goal homogeneity' on the part of capital and labour in Japan. This is achieved through ensuring worker dependence to the company through workplace welfare provision and the system of lifetime employment (*nenko*). According to Wilkinson and Oliver:

> The result, at least since the 1950s, has been relatively few strikes or other industrial action, and a dedicated, loyal and flexible workforce – exactly the conditions necessary for a JIT production system to operate (Wilkinson and Oliver, 1989: 55).

However, the recourse to explanations which emphasise worker loyalty and dedication, although common in writing on Japan, seem particularly unsatisfying. Ultimately such explanations rely on ethnocentric accounts of the Japanese system. The key factor glossed over by Wilkinson and Oliver and, significantly by Piore and Sabel is the absence of independent trade unions in Japan. It is this absence which allows the complete functional flexibility necessary for Japanese production methods to operate. The failure to address the historic defeats inflicted on militant Japanese workers organisation in the period before and after the Second World War and the widespread implantation of company unionism (Moore, 1983; Cusumano, 1984; Ichiyo, 1984; Dohse *et al.*, 1985, Jurgens, 1989) leaves their account suffering from a serious deficiency. The 'harmonious industrial relations', which they describe, were won by capital in struggle against labour. It is the consequences of these defeats with which the Japanese working class live today.

Workplace flexibility in West Germany: the end of the division of labour?

The thesis that current trends in workplace restructuring indicate the beginning of the end of the division of labour in core manufacturing industries is based upon the premise that while the aim of capitalist rationalisation remains unchanged – i.e., an attempt to raise labour productivity with the aim of improved capital utilisation – the means by which this end is achieved is changing (Kern and Schumann, 1984, 1987, 1989; Lane, 1988). The thesis maintains that under 'mass production', workers are regarded as obstacles to production to be substituted by capital as far as possible. With the emergence of more flexible forms of production technology a new attitude to labour is becoming dominant, based upon the active participation of labour in production. High levels of automation (through robotisation, etc.) are seen as eliminating unskilled work and changing the content of the labour which remains. As Kern and Schumann put it:

> The new worker is a sort of scout – sensitive to breakdown with quick reactions and the ability to improvise and take preventative action (1987: 162–3).

This is leading to the acquisition of new, more cognitive skills and higher levels of training on the part of shopfloor

Workplace flexibility

39

workers and blurring of blue and white collar functions. Therefore, a new imperative to co-operate is placed on both capital and labour. Also increased skill levels and more fulfilling job design are seen as *necessary* for the high levels productivity made possible by the new technology.

The evidence of Kern and Schumann is drawn from three principal sectors: motor vehicles, machine tools and chemicals. Outside of these sectors the evidence for flexibility is seen as less clear cut and in contrast to Piore and Sabel they acknowledge the existence of growing social division within the labour market. Nevertheless the political implications of this thesis are clear:

> German management have adopted a consistent strategy of flexible specialisation which, on balance, constitutes a progressive move from the point of view of labour (Lane, 1988: 167).

The West German car industry is seen as the exemplar of these trends. Within Volkswagen for example, a trend toward task integration and a holistic conception of work has been identified.

Production jobs, it is argued, have been merged with some maintenance and quality control tasks and a general increase in skill levels is observed. Shopfloor work is seen as evolving toward a new 'production mechanic' grade. In general, jobs are seen as 'enriched', even if workers do complain about increased levels of stress. The most developed form of task integration and holistic conceptions of work is 'semi-autonomous' teamworking which heralds the end of the division of labour. Similar arguments have been advanced for other parts of the car industry such as Audi (Heizmann, 1984) and in the machine tool sector skilled labour is apparently no longer seen as a 'necessary evil' but as a 'positive planning concept' (Kern and Schumann, 1987).

The case of the car industry is particularly instructive because in much of the flexible specialisation literature it is seen as a harbinger, prefiguring changes in the wider economy. In this respect the work of Kern and Schumann forms part of new consensus which sees such developments as offering solutions to the widespread problems of labour unrest, sabotage, absenteeism and alienation which characterised assembly line work in the past (e.g. Katz and Sabel, 1985, Kochan,

Katz and McKersie, 1986, on the United States). Such interpretations, however, need to be treated with extreme caution.

However, a broad range of studies of the widespread introduction of new technology, particularly in West German and Austrian car plants, do not support the idea that such changes are accompanied by significant upskilling. Windolf's (1984) case studies of three plants

> departmental boundaries in the sense of a division of labour and areas of responsibility. The team concept has this as a starting point. The concept leads to production teams, that means that all those directly participating are in 'one boat' and all can do all of the work tasks within the area of responsibility of the team – quality, production, volume and capacity utilisation. The joining together of previously separated individual areas of responsibility in a team opens up the possibilities for reducing the sum of time lost. Or, expressed differently *to achieve the highest possible ratio of time worked within the working time of the individual employee.* The ideal leads automatically to the formation of a team based on different areas of responsibility (quoted in Jurgens *et al.*, 1988: 269, emphasis added. The essence of the description is conveyed in Figure 1).

Such developments complement innovations such as preventative maintenance and machine monitoring which improve asset utilisation. Moreover, in the management conception, 'semi-autonomous group working' can still be controlled by the application of tight centrally imposed time standards.

In an early, but farsighted study of group working at Renault, Coriat (1980) discovered a similar process occurring at various plants. According to management figures staggering productivity gains were made by applying traditional forms of rationalisation to the activities of groups rather than individual workers. For instance, at Renault's Choisy le roi engine plant, on the introduction of 'modular assembly' productivity was almost doubled.

Like the West German studies, Coriat's paper illustrates that by assigning tasks on a group basis and expanding the roles and responsibilities therein, Renault were able to achieve a reduction in idle time and the conversion of time thus saved into productive time: principally by assigning non-directly productive tasks to production workers. This led to a reduction in transfer and waiting times.

Coriat argues that the pace of work is still ultimately determined by management not by workers as the optimistic accounts would have it. Jurgens *et al.*, (1988) draw a similar conclusion in the case of GM's experiments described above. They note that despite the apparent transfer of responsibilities to groups, management is not giving up its prerogatives of control. At best the new system can be described only as partial 'autonomy' and is controlled by tight time standards and is increasingly subject to electronic monitoring. It represents intensification because, invariably, it means an increase in the number of productive actions in the course of the working day.

West German experiments share some common characteristics with (and owe some inspiration to) developments occurring in Sweden since the mid 1970s. However, the Swedish evidence seemingly is less conclusive about the emergence of new production concepts. According to Christian Berggren (1989) the Swedish evidence (from the automotive industry) points to two problems with the Kern and Schumann thesis. Firstly:

> In their book *Ende der Arbeitsteilung* they devote a major section to the technical and organisational changes in the automotive industry. Here their favourite instances of new production concepts are the robotized body shops with their novel skilled jobs, such as *Strassenfuhrer* and *Anlagenfuhrer* (monitor of complex equipment). Such positions can be found also in the Swedish body shops. But, it is also quite easy to find counter evidence, instances where robotization of body processing has generated more restricted and fragmented work. Thus, the consequences of automation in one instance may not be generalizable to others (Berggren, 1989: 173).

Secondly, argues Berggren, Kern and Schumann have little to say about changes occurring in more labour intensive parts of car plants such as final assembly.

The Swedish example is instructive because European interest in alternative forms of work organisation could be said to have their origins in Volvo's experimental Kalmar plant in the 1970s. There, the assembly line was 'replaced' by a system of automatically guided vehicles (AGV). The assembly process was segmented with groups of workers building sections of the car in bays, in a method known as 'dock assembly' (see Lindholm and Norstedt 1975 for a contemporary account). Although cited as an example of the humanisation of work (e.g. Aguren *et al.*, 1984), the Kalmar experiment was a very limited departure from the orthodox line: the AGV system operated in a highly centralised manner similar to a conventional line. In addition, no organisational changes were introduced to support the development of teams and the level of worker discretion remained unchanged. According to Berggren:

> At specified predetermined times the carriers started moving, monitored by a central computer, regardless of whether or not assembly workers were finished (thus constituting in practice an indexing line, an intermittently running line) . . . in practice the same pace was maintained throughout the factory (1989: 181. On the significance of central monitoring see also Auer, 1985, Gronblad, 1987).

Berggren argues that a tendency toward expanded job cycles means that up to the mid-1980s the trend was toward 'flexible Taylorism'. For instance, Saab's Trollhatten plant operates on the basis of 'mini-lines'. This involves the segmentation of traditional lines, with each section separated by buffers. Mechanical pacing is retained within sections. Cycle times are only 2–3 minutes. The line is less susceptible to disruption (because of the buffers) and there is reduced final adjustment and repair work (because inspection is carried out in the buffers at the end of each section). Some job rotation is possible but the new line balancing system means that work intensity increases. Throughout the system central pacing is maintained.

Berggren suggests, however, that particularly within the commercial vehicles sector, alternatives to the assembly line are emerging which could be described as new production concepts. At Volvo's LB truck plant non-mechanical work flows have been introduced and are structured for group work with assembly at stationary vehicles. Completed work is transported by carriers which are controlled by the work groups themselves. Cycle times are 30–45 minutes. The groups themselves are responsible for tasks and training, daily planning, detailed line balancing and choosing group leaders. There is job rotation which includes the position of group leader. Despite these improvements, however, workers at the plant still feel the work is intensive and monotonous.

The most advanced Swedish experiments clearly do represent a departure from the principles of Taylorism but the extent to which these methods will prove economically viable from management's viewpoint (and hence generalisable) must remain a matter of conjecture. What is significant about the Swedish situation is the active role of the Swedish metalworkers union in developing a powerful critique of and detailed alternatives to Taylorist work methods.

What kind of flexibility?

The previous sections have challenged the idea advanced by some researchers that the trend toward 'flexible specialisation' has broadly progressive implications for labour. The association of new technology, work flexibility and job enhancement permeates the flexible specialisation thesis in all of its variants. While it would be churlish to deny that job enhancement may have occurred in some instances, there is little evidence that this represents a general trend.

Many researchers seem to take evidence of job expansion as evidence of reskilling, but simply attaching more responsibilities to production jobs does not equal reskilling. A survey of changes in technology and work organisation in North American car plants, conducted for the Canadian Auto Workers Union was less sanguine about the claims made for job enhancement than are some researchers. (D. Robertson, J. Wareham)

It is significant that many of the most advanced Swedish experiments retain centralised forms of pacing and control, which involve microelectronic monitoring rather than the traditional conveyor belt. This is an important point. It seems likely that newly emerging forms of capitalist work relations can involve some 'requalification' on the part of workers (although, as I have suggested, we should be careful about our use of such terms). However, 'requalification' is often at the cost of intensification of the work process which translates into increased stress levels on the shopfloor. This is a consequence of many contemporary changes precisely because the expansion of tasks and the introduction of group working are concerned centrally with the reduction of down-time, waiting-time and other dead periods – the so-called 'balance delay' problem – which are said to have characterised production organised around the principle of the assembly line and its variants (see Aglietta, 1979; also Berggren, 1989: 178).

Moreover, the notion that technological change necessitates and is bringing into existence a new social bargain based on the greater involvement and participation of the workforce in production needs to be challenged. While it does appear that in some instances West German, and more particularly Swedish workers, have gained some benefits from the process of change, these benefits have been *won* by labour and are not simply the consequence of technical change as some researchers imply. In West Germany and Sweden where workers, arguably, have been successful in winning a share in improved productivity for themselves, it must be acknowledged, traditions of co-determination and institutions of corporatism (i.e. the centrality of labour in political life) have not been subject to the same levels of political attack as have been the case in, say, the UK and US. (In the Japanese case effective trade unions, of course, are non existent.) Indeed, in the West German case unions have been concerned with attacking the basis of the bosses' flexibility scenario through the campaign for a reduction in working hours (e.g. Wainwright, 1987).

Workplace flexibility in the UK

The purpose of this section is to examine recent changes in the organisation of work and production in the UK. Within the broader context of this paper this is important because in various ways the experiences of countries such as Japan and concepts such as flexible specialisation have been used to explain the performance of the UK economy.

For instance, there is a burgeoning literature on the 'Japanisation' of British industry (e.g. *Industrial Relations Journal*, 1988; Oliver and Wilkinson, 1988). Similarly, attempts have been made to examine the UK experience in terms of the debate about the transition to flexible specialisation (Hirst and Zeitlin, 1988, 1989; Lane, 1988). For Hirst and Zeitlin, for example, the relatively poor performance of UK manufacturing is due to the failure to introduce and utilise the technology of flexible specialisation and to develop the

Workplace flexibility

45

political institutions which foster and give rise to it (see also Murray, 1985, 1988). Hirst and Zeitlin, however, offer no analysis of the rapid improvement of UK productivity which has occurred since 1979. A problem arises here because they are unable to address evidence which suggests that qualitative changes in the organisation of production took place in the UK in the 1980s and to establish its relation to the changes occurring in other capitalist countries. It is the aim of this section to address the concrete evidence for such change and to assess any implications.

Despite the controversy which surrounds Thatcherite policy toward manufacturing it is possible to make two firm statements about contemporary workplace change. First is that an undeniable increase in productivity has occurred. Output per worker in manufacturing rose by over 42 per cent between 1979 and 1988. Secondly, a consensus is emerging that the origins of this productivity do not lie in increased levels of investment but in greater levels of labour productivity (for example: Bean and Symons 1989; **The Economist**, 20.5.89; OECD, 1988; for a brief but useful review of the debate see Wolf, 1988). This position is summarised in a recent survey of the UK economy by the OECD:

> Stronger labour productivity growth has not been linked to capital investment, which, in fact, has remained lower relative to GDP than in other recovery periods. Rather it seems linked to changes in work organisation with inflexible and outdated job demarcation giving way to more rational job allocation. This would indicate that a large part of the observed growth rates in the 1980s are in fact successive level changes as opposed to underlying growth rates (OECD, 1988: 79).

Interpretations of the flexibility issue in the UK have been dominated by the work of Atkinson which suggests a widespread move toward a flexible firm of a multi-skilled core of workers surrounded by a periphery of unskilled, temporary and part-time workers (e.g. Atkinson, 1987). While the model of the flexible firm has proved influential in policy and journalistic circles, Pollert (1988) has demonstrated convincingly that the evidence for the changes described by Atkinson is limited. However, as Pollert suggests but does not elaborate upon, a rejection of this model does not prelude the possibility that real changes are occurring although in reality the changes in workplace organisation appear to be rather less dramatic than some accounts would have us believe.

For instance according to one survey of *major* flexibility agreements:

> Myths about the advance in flexible working practices have misled many people into thinking that fundamental changes have taken place on a very wide scale. On the contrary, the fundamental changes are extremely narrowly concentrated in particular corners or, very occasionally, particular sectors of industry. And even where these changes have been taking place, they can be incomplete, halting or superficial (IDS, 1986: 4).

According to IDS, it is the erosion of demarcations – not 'multiskilling' – which is the key issue for managers. Such a development may seem unimpressive in relation to strategies pursued by competitor countries such as Japan. However, such moves share some characteristics with these strategies. For instance, to the extent that the IDS survey was able to identify the principles underlying changes in work practices in UK manufacturing, the main motive appeared to be the elimination of idle time in production (downtime, waiting time and so on) and increased rates of machine utilisation.

These principles are reflected in these comments of the Director-General of the Engineering Employers Federation (EEF):

> We need to make maximum use of plant and machinery by eliminating restrictive practices, by having full flexibility between and within trades and occupations and between supervisor and supervised . . . [and] . . . in order to make better use of plant and equipment our member companies need to be able to adopt flexible working times when required (quoted in Wickens, 1987: 41).

In the engineering industry these requirements have led to pressure for *flexibility between crafts*. However, there is no evidence that a 'multi skilled craftworker' is emerging. Negotiations toward the more limited end of a measure of flexibility between trades have been fraught with difficulty. National level talks between the EEF and the CSEU collapsed in 1987.

Workplace flexibility

47

The result has been a highly uneven pattern of local agreements. Weakened unions and the uncompetitive position of British capital in a period of recession has led to a number of attempts to introduce a measure of flexibility between crafts. For example, the shipbuilding industry has seen agreements which have led the erosion of demarcations eroded at the likes of Swan Hunter (**Financial Times**, 27.6.88), Tyne Shiprepair (IDS, 1988a) as well as the recent far-reaching agreement at Harland and Wolff (**Financial Times**, 16.8.88).

One of the main principles underlying these agreements is highlighted in the most advanced agreements. At Babcock Power, Renfrew – an engineering company making boilers – the aim of introducing flexibility between crafts was 'to eliminate non-productive waiting time from direct labour' (IDS, 1986: 4). However, the Babcock management stress that for the most part a worker will use 'his primary skill' (ibid.). Nevertheless waiting time can be reduced when an electrician or a mechanic removes a guard from a machine before repairing, say, a motor, rather than waiting for a fitter. To require more than this from workers would require a deal of training which many firms are not prepared to countenance. (The failings of UK management in this regard are usefully examined in **Financial Times**, 4.9.87. For a view of the very different West German system see **Financial Times**, 25.8.88).

The findings of IDS are supported by those of Cross who reports the views of managers in fifty companies which had attempted to deal with the demarcation issue. He concludes that:

> the major factor causing the blurring [of demarcations – JT] is economic pressure to reduce lost production through breakdowns, labour costs, and as a result of changes in the technology of the machines (1985: 69).

IDS (1986) reports that one Yorkshire manufacturing firm introduced a series of measures to increase labour flexibility as a means to reducing machine downtime. A new pay structure yields payments on hours saved in the changeover and repair of machines. As a result, downtime was reduced from around 5–6% to 2.5% with maintenance engineers usually achieving the maximum available bonus.

While the progress toward 'craft flexibility' has been slow there is a more discernible trend toward introducing some measure of *flexibility into production work*. In part this can be related to altered work requirements as a result of technical change: for instance a higher level of automation might mean that job cycles are lengthened to allow multi-machine minding of the kind common in Japanese plants. However, while developments such as semi-autonomous group working at Borg Warner's South Wales car component plant (**Financial Times**, 21.4.88; Eaton, 1988), or modular batch production at Hewlett-Packard's Bristol plant (**Financial Times**, 21.11.88) and high levels of functional flexibility at Pirelli's automated wire plant at Aberdare (**The Guardian**, 4.4.88), are widely publicised, there is no evidence that they represent the norm in UK industry.

Flexibility in production work can draw on the most basic Japanese methods. This appears to represent the most transferable aspect of Japanese work methods as Wickens account of Nissan in north-east England makes clear. There the approach is to expand all jobs 'as much as possible'. For instance, production workers become responsible for cleaning and maintaining their own work area which has the effect of tightening individual work practices:

> If the man on the line is responsible for keeping it clean, he will be less inclined to make it dirty in the first place (Wickens, 1987: 45).

The significance of this is that heightened effort levels are increasingly built into the system. One Ford worker described the effects of the 'After Japan' strategy at Dagenham:

> Flexibility means that every 102 seconds a car comes by, and not only do you have to screw something into the car, but in between you have to tidy up, check your tools, repair things and check you've got enough parts. You do not have a single job any more. If there is no work on the line they move you to where there is work. You are working the whole time (quoted in the **Financial Times**, 8.2.88).

Horizontal job enlargement appears to be the most commonly reported form of workplace 'flexibility'. It is notable that this much is conceded by Atkinson and Meager (1987). The attachment of minor maintenance tasks and certain quality control functions to production workers, on the other hand, is invariably presented as enriching jobs by proponents of flexibility such as Wickens (1987). However, for workers such 'enrichment' often appears as more of the same routine tasks. Turnbull quotes one West Midlands factory worker thus:

> The jobs are just the same as before, you just do more of them. And there's no big deal to assigning quality control to direct operators – you just stick the components under a feeler gauge several times a day to check things are going OK (1987: 12).

The advantage for the firm is highlighted by Coriat's (1980) study of Renault's Le Mans plant in the 1970s where self-regulation led to a reduction in rectification work and the abolition of certain categories of maintenance work. This translated into savings in production time.

In the UK context, it appears, the principles of scientific management underlie the introduction of information technology, not emancipatory craftwork. Scarbrough and Moran (1985) report that while the Austin Rover (now Rover Group) Metro line at Longbridge is seen as an example of high automation, significant changes have occurred in more labour intensive parts of the plant. In the power train area technological change has occurred in the form of a Machine Monitoring System (MMS) 'an electronic information gathering system' (Scarbrough and Moran, 1985: 209, see also Scarbrough, 1986). One production manager summarised the function of the MMS as 'to look for the idle buggers' (ibid.).

A similar aim can be seen behind the system of remote control and monitoring developed and introduced by British Coal. The system is hierarchical and centralised (Chandler, 1978) and designed to increase capital utilisation (Steel, 1988). This end is partly achieved by monitoring the activity of face equipment and, by implication, face workers. The aim is to remove delays from the coal getting process. One manager described the (theoretical) advantages of the FIDO (Face Information Digested On-line) system:

> The system improves communications. The controller knows immediately the face is stopped; he does not have to rely on a message from underground. Using the status display, and his own knowledge and experience, he can often interpret what is happening on the face. When he

contacts the face his questions are direct and to the point. People underground are aware of the resources that are available to the controller. They realise he will not be fobbed off with imprecise information, hence their replies to his questions are more accurate (Cleary, 1981: 288).

The aim of such technology is the greater utilisation of existing mining machinery as BC management have made clear:

> We must no longer accept our coalfaces operating at perhaps only a third of their potential on each shift. We must increase the proportion of time when the face equipment and the whole infrastructure which serves it, is actually working at full stretch (Sir Robert Haslam quoted in **Financial Times**, 16.4.87).

Whether such systems operate in the ways hoped for by management remains an open question. However, management intentions are revealed as essentially Taylorist, despite the presence of high technology. In sum, the evidence on the UK situation suggests that a number of discernible and significant changes are occurring at the point of production in manufacturing. Hirst and Zeitlin (1989), as proponents of the flexible specialisation thesis have little to say about the nature of the recent UK productivity performance. The UK experience is defined only in negative terms; that is, as the absence of 'flexible specialisation'. In fact, the piecemeal and halting experiments which characterise workplace flexibility do cohere around the problematic of how to heighten levels of capital utilisation and the conversion of higher levels of work time into directly productive activity.

Conclusions

The foregoing sections have been concerned with establishing two points which have serious consequences for the flexible specialisation thesis. Firstly, it is argued that within the workplace there is no generalisable trend toward the technologically driven flexible craft work suggested by the flexible specialisation theorists. Secondly, it is suggested that workplace flexibility, even in its most advanced forms does not guarantee the benefits for labour implied by its proponents. This concluding section attempts to offer some elements of an alternative theorisation and to address some political implications for the labour movement.

The term flexible specialisation has been used to cover such a wide range of experiments and changes in technology, work organisation and political institutions that any precision it might once have had has been lost. The previous sections showed that capitalist enterprises are following many different strategies and that assigning a determining role in such developments to reprogrammable manufacturing technologies is problematic to say the least. One central contention of the exponents of flexible specialisation is that the imperative to which it gives rise is undermining the basis of Taylorist and Fordist methods of production organisation. However, the evidence offered in this paper tends to support a different conclusion.

I suggested above that the factor which united the disparate efforts to restructure workplace relations was the attempt to raise the rate of capital utilisation through a reintegration of work tasks. This is particularly important where investments in new, high cost flexible technologies are concerned. However, as the UK example appears to suggest, the restructuring of work practices within the context of the existing capital stock can give rise to significant improvements in productivity.

The pursuit of such ends is not new. The pursuit of increased labour control and closer management of time in production is a central capitalist problematic. The rise of the factory system (Marglin, 1974), the emergence of factory discipline (Thompson, 1967) and the transition from manufacture to machinofacture (Marx, 1974) are examples of shifting capitalist strategies in the pursuit of these ends (see Aglietta, 1979). In the twentieth century the growth of capital intensity in manufacturing spurred Taylor's original efforts to formulate a means of scientific management, principally as a means to increase plant utilisation and productivity through a heightened control over labour (Sohn Rethel, 1978). For Taylor the use of time and control studies and the assessment of unit times allowed management this new control. This in turn allowed the intense fragmentation of work tasks (based on the separation of conception and execution) and the subsequent speeding of productive operations (Taylor, 1947; Braverman, 1974). Increasingly, and particularly within certain key sectors of manufacturing, such principles came to be embodied within the physical plant of the 'Fordist' assembly line (e.g. Sohn Rethel, 1978; Aglietta, 1979).

The crisis of this system of work and production organisation lay in the heightened levels of class conflict which occurred at the end of the 1960s and in the 1970s, particularly in those industries where such principles were most developed (e.g. Aglietta, 1979; Bosquet, 1977; Coriat, 1984; Holloway, 1987; Negri, 1989). The fragmented work process based around the assembly line was particularly difficult to coordinate and intensify. From management's viewpoint, these problems were exacerbated by the forms of labour organisation adopted by workers in order to retain control over effort levels and the pace of the line (e.g. seniority rules, demarcation lines). The attack on these principles was necessary in order to overcome the productivity crisis. Flexible work practices ranging from job expansion to teamworking can be seen, therefore, as means to reduce the lost time arising from job demarcation and hierarchical division. It was the response to this real and observable crisis of control and productivity which stimulated the present reorganisation of work practices, not the 'thousand imponderables' and technological autonomy described by Piore and Sabel (1984). What is striking about current attempts to restructure the relations of production in new ways is the presence of strong continuities with the past as the examples above showed. The focus on new forms of labour control through electronic surveillance, forms of neo-Taylorism – via Japanese work methods, or through subjecting group working to time study methods – represent new developments of existing tendencies. The re-integration of some tasks or the introduction of controlled group working can be seen as efforts to reduce the 'porosity' of the working day. Moreover, the existence of relatively inexpensive micro-electronic systems means that the work system and productive process can be subject to a higher degree of management regulation. Invariably at the heart of the changes is an intensification of the work process.

The pursuit of these concerns have been as characterised as the search for 'a new economy of time and control' in production (Coriat, 1980; 1984). What can also be seen is that within this broad problematic a wide range of strategies can be discerned. These invariably involve the attempt by management to introduce elements of 'flexibility' (task mobility) and can even, within tight limits, involve a degree of work group autonomy. This is the grain of truth in the flexible specialisation thesis.

However, there is no evidence of a trend toward a computer assisted craft worker. What is clear is that there are many forms of flexibility confronting workers in the advanced capitalist countries. These include the demand for some polyvalence and task mobility, the weakening of constraints on the employment contract, the deregulation of wages and wage security, and the abandonment of rules restricting management prerogatives. As Boyer points out, far from these leading to emancipatory forms of work organisation, most indicators which measure the development of different forms of flexibility in the OECD countries suggests the opposite:

> In these times of crisis, flexibility strategies have entailed, under various euphemisms, the downward adjustment of hitherto established conditions of employment of workers (Boyer, 1987: 115;).

In short, there is no evidence that the tendencies of the past one hundred years are being reversed. Rather what we witness is the extension and redevelopment of existing forms of labour control and efficiency maximisation.

The description of present restructuring as leading to and somehow requiring a new more equal partnership between capital and labour is central to the flexible specialisation thesis. Clearly in a period of working class retreat such a message can have a powerful appeal to new realists. The rhetoric of flexible specialisation (and its 'post-Fordist' variant) have been used by managements, union bureaucracies and London magazines to urge workforces to accept the reorganisation of work practices or inward investment from firms associated with 'flexible' forms of production. One political expression of flexible specialisation is 'business unionism' wherein the reality and diversity of workplace change is lost behind a barrage of exhortation directed at workforces and requiring them to be more 'co-operative' and 'flexible' (Communist Party, 1989).

The dangers in such a strategy are rapidly becoming apparent (see Foster and Woolfson, 1989). To the extent that workers in core industries accept the message of the new realists and adopt intensification and flexibility as guarantors of job security based on ensuring high productivity and competitiveness, this is leading to the peripheralisation of those groups excluded from this new corporatist agreement. However, this peripheralisation can be used to undermine the centrality of core groups.

Ironically, perhaps, this process is highlighted in the recent practices of those countries said to be exemplars of the trend toward flexible specialisation. The decision of Bosch to locate its new car component plant in South Wales is a case in point. A key motivation behind the choice of South Wales is the favourable wage rates prevailing there in comparison to West Germany. In 1987 the average employment cost for a German worker in the component industry was DM33 per hour compared to DM18 per hour in the UK. Consequently, it is estimated by Bosch that it will produce alternators for 15–20% less in Wales than it could West Germany (**Financial Times**, 18.4.89).

The level of mobility possessed by capital makes the West German high wage/high productivity coalition look particularly fragile. The new deal between GM and the West German unions which is seen as having far reaching implications for labour practices in that country, illustrates how flexibility is being used to undermine rather than enhance the position of workers in industries such as cars. The deal on working hours for the intensive operation of plant and equipment was already in place in GM's low wage plants in Zaragoza and Antwerp. IG Metall were unable to resist this encroachment on working conditions in the name of flexibility (**Financial Times**, 8.3.89). The pursuit of low cost, flexible labour also lay behind Ford's decision to relocate production of its Sierra from Dagenham to Genk (**Financial Times**, 19.1.89, 30.1.89). Rather than develop an internationally based strategy to undermine Ford's attempts to play one group of workers off against another, the response of TGWU leaders such as Jack Adams was to urge workers to attain higher levels of productivity in order to persuade Ford to reverse the decision (**Financial Times**, 27.1.89). The point is a simple one: that flexibility is being established on capital's terms. Flexibility, therefore is part of the problem, not the answer.

The significance of the flexible specialisation thesis lies in its attempt to dispense with class struggle – both as a cause of the crisis and as a response to it. As the preceding paragraphs showed, attempts to establish firm-based co-operative industrial relations and new institutions of corporatism are irrelevant so long as the power of multinational corporations to relocate at will and to negotiate flexibility from a position of historic strength remain unchallenged. The political programme associated with the flexible specialisation thesis offers no attempt to undermine the centres of capitalist decision making. In reality, where we can point to positive outcomes to restructuring such as in some of the Swedish or West German experiments the key explanatory variable has been the presence of a strong, well informed and articulate trade union movement which often in the face of tough odds has been able to negotiate benefits for its members (see Wainwright, 1987). What is required are not prescriptive outlines of capital's new agenda but a rigorous analysis of the diversity of contemporary capitalist strategies in the present crisis as a prelude to the construction of a socialist alternative. A strong and independent trade union movement is a pre-requisite of this.

I would like to acknowledge the comments and advice of Ash Amin and Ken Ducatel in the preparation of this paper.

Figure showing reduction in idle time through team concepts at GM's Aspern plant (Vienna)

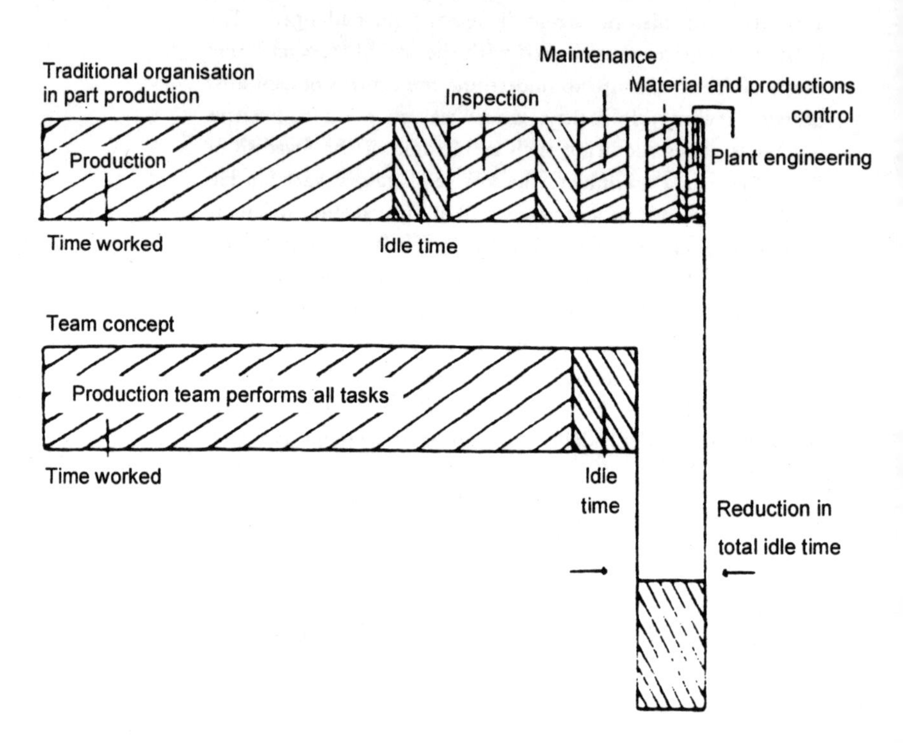

Source: V. Haas, 'Team-konzept: Mitarbeiter planen und betreiben ihr Arbeitssystem'. Paper presented to the IAO-Arbeitsagung 'Weltsbewerbs fahige Arbeitssyteme' 22–23 November, 1983. Bobingen. Cited in Ulrich Jurgens, Knuth Dohse, Thomas Malsch, 'New production concepts in West German car plants' in S. Tolliday and J. Zeitlin (eds) **The Automobile Industry and its Workers** London, Polity Press, 1988.

Bibliography

Abernathy, W., K. Clark, A. Kantrow, (1983) *Industrial Renaissance: producing a competitive future for America* New York, Basic Books.

ACAS (1988) 'Labour flexibility in Britain: the 1987 ACAS survey' *Occasional paper 41*, London,,Arbitration, Conciliation and Advice Service.

Aglietta, M. (1979) *A theory of capitalist regulation: the US experience*, London, NLB

Aguren, S., C. Bredbacka, R. Hansson, K. Ihregren, K. Karlsson (1984) *Volvo, Kalmar Revisited: Ten years of experience*, Stockholm, Efficiency and Participation Development Council (SAF, LO, PTK).

Atkinson, J. (1987) Flexibility or fragmentation? The United Kingdom labour market in the eighties' *Labour and Society* 12, 1.

Atkinson, J., N. Meager (1987) 'Is flexibility just a flash in the pan?' *Personnel Management*, September.

Auer, P. (1985) 'Industrial relations, work organisation and new technology: the Volvo case' Discussion paper IIM/LMP 85–10, Berlin: Wissenschaftszentrum.

Bean, C., J. Symons (1989) 'Ten years of Mrs T' Centre for Labour Economics, London, London School of Economics.

Berggren, C. (1989) '"New production concepts' in final assembly – the Swedish experience' in S. Wood (ed.) *The Transformation of Work?*, London, Unwin Hyman.

Bosquet, M. *Capitalism in Crisis and Everyday Life*, Brighton, Harvester.

Boyer, R. (1987) 'Labour flexibilities: many forms, uncertain effects', *Labour and Society*, 12, 1.

Braverman, H. (1974) *Labor and Monopoly Capital*, New York, Monthly Review Press.

Chandler, K. (1978) 'MINOS – a system for central control at collieries' *2nd International Conference on Centralised Control Systems*, London, Institution of Electrical Engineers.

Cleary, J. (1981) 'FIDO at Bold colliery', *Mining Engineer*, November.

Communist Party (1989) *Manifesto for new times*, London, CPGB.

Coriat, B. (1980) 'The restructuring of the assembly line: a new economy of time and control' *Capital and Class*, 11.

Coriat, B. (1981) 'L'atelier Fordien automatise' *Non!*, 10, November–December.

Coriat, B. (1984) 'Labor and capital in the crisis: France 1966–82' in M. Kesselman (ed.) *The French workers' movement: economic crisis and political change*, Boston, Ma, Allen and Unwin.

Coriat, B. (1987) 'Information technology, productivity and job content', Paper to BRIE meeting on comparative production, Berkeley, September 11–13.

Cross, M. (1985) *Towards the flexible craftsman*, London, Technical Change Centre.

Cusumano, M. (1984) *The Japanese automobile industry*, Cambridge Ma, Harvard University Press.

Dohse, K., U. Jurgens, T. Malsch (1985) 'From Fordism to Toyotism? The social organisation of the labor process in the Japanese automobile industry' *Politics and Society*, 14, 2.

Eaton, J. (1987) 'The flexible firm in Wales: a consideration of its impact on trade unions', *Welsh Economic Review*, 1, 2.

Economist (20.5.89) 'The end of the beginning', (Special Report).

Foster, J., C. Woolfson (1989) 'Post-fordism and business unionism', *New Left Review*, 174.

Garrahan, P., P. Stewart (1989) 'Post-Fordism, Japanisation and the local economy: Nissan in the north-east', Conference of Socialist Economists, Sheffield Polytechnic, 7–9 July.

Gronblad, S. (1987) 'Production supervision of automatic assembly line at the Volvo Skovde plant', *Electronics in Measurement, Automation and Control*, April.

Heizmann, J. (1983) 'Work structuring in automated manufacturing systems exemplified by the use of industrial robots for body assembly' in T. Martin (ed.) *Design of work in automated manufacturing systems*, (IFAC seminar Karlsruhe, FRG) Oxford: Pergamon.

Hirst, B., J. Zeitlin (1988) 'Crisis, what crisis?' *New Statesman* 18 March.

Hirst, B., J. Zeitlin (1989) 'Flexible specialisation and the competitive failure of UK manufacturing', *Political Quarterly*, 60 (2).

Holloway, J. (1987) 'The red rose of Nissan', *Capital and Class*, 32.

Ichiyo, M. (1984) 'Class struggle on the shopfloor – the Japanese case, 1945–84' *Ampo, Japan Asia Quarterly Review*, 16, 3.

IDS (1984) 'Craft flexibility', *IDS Study 322*, London, Incomes Data Services Ltd.

IDS (1985) 'Improving productivity', *IDS Study 331*, London, Incomes Data Services Ltd.

IDS (1986) 'Flexibility at work' *IDS Study 360*, London, Incomes Data Services Ltd.

IDS (1988a) 'Flexible working' *IDS Study 407*, London, Incomes Data Services Ltd.

IDS (1988b) 'Teamworking' *IDS Study 419* London, Incomes Data Services Ltd.

Industrial Relations Journal, (1987) Special issue on Japanization, 19, 1.

Jurgens, U., K. Dohse, T. Malsch (1988) 'New production and employment concepts in West German car plants' in S. Tolliday and J. Zeitlin (eds) *The automobile industry and its workers*, Oxford, Polity Press.

Jurgens, U. (1989) 'The transfer of Japanese management concepts in the international automobile industry' in S. Wood (ed.) *The Transformation of Work?*, London, Unwin Hyman.

Kamata, S. (1980) *Japan in the Passing Lane*, London, Counterpoint.

Katz, H., C. Sabel (1985) 'Industrial relations and industrial adjustment in the car industry', *Industrial relations*, 24, 3.

Kern, H., M. Schumann (1984) *Das ende der Arbeitsteilung?*, Munich, CH Beck.

Kern, H., M. Schumann (1987) 'Limits of the division of labour: new production concepts in West German industry', *Economic and Industrial Democracy*, 8.

Kern, H., M. Schumann (1989) 'New concepts of production in German plants' in P. Katzenstein (ed.) *Industry and Politics in West Germany: Towards the Third Republic*, Ithaca, NY, Cornell University Press.

Kochan, T., H. Katz, R. McKersie (1986) *The transformation of American industrial relations*, New York, Basic Books.

Lane, C. (1988) 'Industrial change in Europe: the pursuit of flexible specialisation in Britain and West Germany', *Work, Employment and Society*, 2, 2.

Lindholm, R., J-P Norstedt (1975) *The Volvo Report*, Stockholm, SAF (Swedish Employers Confederation).

Malsch, T., K. Dohse, U. Jurgens (1984) Industrial robots in the automobile industry. A leap toward 'automated fordism'? IIVG/dp 84–222 Berlin, Wissenscahftszentrum.

Marglin, S. (1974) 'What do bosses do?' in A. Gorz (ed.) *The Division of Labour*, Brighton, Harvester.

Marx, K. (1974) *Capital*, Vol 1, London, Lawrence and Wishart.

Moore, J. (1983) *Japanese workers and the struggle for power 1945–47*, Madison, Wi, University of Wisconsin Press.

Murray, R. (1985) 'Bennetton Britain', *Marxism Today*, November.

Murray, R. (1988) 'Life after Henry (Ford)', *Marxism Today*, October.

Negri, A. (1989) *Revolution Retrieved: Selected Writings*, London, Red Notes.

OECD (1988) *Economic survey: United Kingdom*, Paris, OECD.

Oliver, N., B. Wilkinson (1988) *The Japanization of British Industry*, Oxford, Basil Blackwell.

Piore, M., C. Sabel (1984) *The second industrial divide*, New York: Basic Books.

Pollert, A. (1988) 'Dismantling flexibility' *Capital and Class*, 34.

Robertson, D., J. Wareham (1987) *Technological change in the auto industry: CAW technology project*, Willowdale, Ontario, Canadian Auto Workers Union.

Sabel, C. (1982) *Work and politics*, Cambridge, Cambridge University Press.

Sayer, A. (1985) 'New developments in manufacturing: the just in time system', *Capital and Class*,30.

Scarbrough, H. (1986) 'The politics of technological change at British Leyland' in O Jacobi *et al.* (eds) *Technological Change, Rationalisation and Industrial Relations*, Beckenham, Croom Helm.

Scarbrough, H., P. Moran (1985) 'How new tech won at Longbridge' *New Society*, 7 February.

Schonberger, R. (1982) *Japanese manufacturing techniques: nine hidden lessons in simplicity*, New York, Free Press.

Shaiken, H., S. Herzenberg, S. Kuhn (1986) 'The work process under more flexible production', *Industrial Relations*, 25, 2.

Sohn-Rethel, A. (1978) *Intellectual and manual labour: a critique of epistemology*, Atlantic Highlands, NJ, Humanities Press.

Steel, W. (1988) 'The development of management information systems for running machines at full potential', *Colliery Guardian*, August.

Taylor, F. (1947) *Scientific management*, New York, Harper.

Thompson, E.P. (1967) 'Time, work-discipline and industrial capitalism', *Past and Present*, 38.

Turnbull, P. (1987) 'The limits to Japanisation – just in time, labour relations and the UK automotive industry', *New Technology, Work and Employment*, 3, 1.

Wainwright, H. (1987) 'The friendly mask of flexibility', *New Statesman*, 17 December.

Wickens, P. (1987) *The Road to Nissan*, London, MacMillan.

Wilkinson, B., N. Oliver (1989) 'Power, control and Kanban', *Journal of Management Studies*, 26, 1.

Windolf, P. (1984) 'Industrial robots in West German industry', *Politics and Society*, 14, 4.

Wolf, M. 'Is there a British miracle', *Financial Times*, 16.6.88.

[6]

HARLEY SHAIKEN,
STEPHEN HERZENBERG,
and SARAH KUHN*

The Work Process Under More Flexible Production

RAPID DECLINES in the domestic and international market share of U.S. manufacturing firms over the past few years have prompted growing concern about the future of the U.S. manufacturing base. In response to these declines, domestic producers have undertaken a major restructuring of production by introducing programmable technology, changing shop-floor work organization, and adopting new labor relations strategies. This paper examines the implications of this restructuring for the role of workers in production.

One interpretation of these implications, the "flexible specialization" thesis of Michael Piore and Charles Sabel (1984) suggests, in part, that work restructuring and the use of programmable technology should increase the importance of shop-floor skill in production.[1] Piore and Sabel offer three reasons why successful competition in today's increasingly fragmented markets requires that firms combine programmable technology with a broadly skilled workforce. First, a firm cannot afford repeated trials to perfect each production run in small-batch production and therefore workers play a critical role in debugging programs or intervening when production goes awry. Second, skilled worker knowledge of production is essential to both process and product innovation. Third, workers require broader skills to master new responsibilities when firms repeatedly change product lines. Work organization and labor relations, according to Piore and Sabel, must evolve so that workers acquire

*The authors are, respectively, Associate Professor, Department of Communications, University of California-San Diego, Graduate Student, Department of Economics, M.I.T.; and Graduate Student, Department of Urban Studies and Planning, M.I.T

[1]Piore and Sabel's broader thesis suggests that, in today's economic environment, firms increasingly produce specialized goods for market niches, rather than making mass production goods in large volumes. We take issue with only the implications of this for workers in production. Piore (1985) presents a brief elaboration of this thesis, Piore and Sabel (1984) a more extended version. Reich (1983) develops a related hypothesis in which the importance of skill in manufacturing is regarded as less exclusively tied to small volume, specialized production (see especially pp. 128-130).

INDUSTRIAL RELATIONS, Vol. 25, No. 2 (Spring 1986). ©1986 by the Regents of the University of California.
0019/8676/86/525/167/$10 00

and then use the broad skills necessary to compete successfully in flexibly specialized markets.

We present here an alternative interpretation of the implications of the restructuring of production. As part of their response to competitive pressures, U.S. corporations are seeking to use technology and shop-floor reorganization to remove the constraints on managerial authority vested in worker skill and autonomy, work rules, and strong, independent unions. While managers clearly recognize the need for the flexibility to change the product mix or retool rapidly in uncertain markets, they often pursue strategies which fail to take advantage of the complementarity between programmable technology and skilled shop-floor workers. Instead, U.S. managers apply computers in a way which centralizes control of production and attempts to reduce the unpredictability associated with worker autonomy. Firms also introduce new approaches to shop-floor management primarily to remove restrictions on managers' control of labor deployment and of workers' time on the job. From the individual worker's point of view, the current shop-floor reorganization often intensifies work and reduces autonomy on the job.

The Empirical Evidence

Whether the flexible specialization argument or the control thesis better characterizes the consequences of the reorganization of production for workers and for the importance of shop-floor skill is an empirical question. We explore these consequences using material drawn from four case studies conducted for the Office of Technology Assessment (OTA) in the spring and summer of 1983 (Shaiken, 1984b) and from a subsequent (unpublished) study. The bulk of our primary evidence concerns the introduction of new technology, although we tentatively extend our analysis to organizational and labor relations issues by supplementing our own findings with related research.

The OTA case studies examined 10 firms and 13 work sites spanning an important segment of U.S. manufacturing: two plants and an engineering section of an aerospace company; two plants of an agricultural implements firm; one auto assembly plant; and seven small metalworking shops. The companies were selected because they are leading users of programmable automation. They include a number of firms operating in environments highly suited to flexible specialization, although they were not selected explicitly to test Piore and Sabel's hypothesized connection between market structure and the role of workers. The companies produce parts in batch sizes ranging from 1-1,000 at the small metalworking shops, to small and medium volumes at the aerospace and farm equipment firms, to hundreds of thousands at the auto plant. The farm equipment and aerospace firms are the most successful

U.S. firms in these competitive, increasingly fragmented markets, and five of the seven small shops had maintained or increased employment levels recently despite difficult market conditions in their industry.[2] At the large firms, we spoke extensively with production and engineering managers from vice-president to shop-floor supervisors, and at the small shops we spoke with several of the top managers, often including the owner. At each site, we also interviewed hourly workers and union officials. In total, we spoke with 26 staff managers, 28 shop-floor supervisors, and 94 hourly workers in open-ended interviews lasting from 15 minutes to several hours.

The Impact of Programmable Technology

Programmable technology offers impressive technical and economic benefits. Many of these benefits stem from the inherent flexibility of the technology. This flexibility allows firms to make design changes easily, retool quickly from one product to another, and make a variety of parts with the same equipment. It enables firms to lower the volume break-even point and adjust rapidly to changing markets. Our cases suggest, however, that cutting volumes from hundreds of thousands to tens of thousands does not result in what remains high-volume production taking on the character of batch production. Even in small-batch production, there is little sign of the new craft worker described by Piore and Sabel.

Implementation of numerical control. Numerically controlled machine tools (NC) are critical technologies in the aerospace, agricultural equipment, and small job shops we studied. Examining NC in practice illustrates the way managers deploy flexible technologies in settings where shop-floor skill has traditionally been of great importance.[3] In this section, we look at the use of NC in the seven small metalworking shops. Prototype and small-batch production in these shops are representative of the application of NC to skill-intensive machining. Table 1 details the characteristics of these shops.

In conventional small-batch machining, skilled shop-floor workers play a pivotal role. To understand the effects of technological change on the importance of that role requires a brief look at the machining process itself. With conventional machine tools, the machinists' skill translates a blueprint into a

[2]Two major unions represent workers at the aerospace, agricultural implement, and auto firms. The small shops are unorganized. The small shops and auto assembly plant are in the Northeast, the agricultural equipment plants in the Midwest, and the aerospace firm on the West Coast. Other details on the firms and the nature of their markets can be found in Shaiken (1984b).

[3]Whether or not the dynamics of NC operate in other settings where skilled workers play a critical role is an important research issue. Cases in Wilkinson (1983) and Shaiken (1984a, especially Chapter 5) suggest that they do.

TABLE 1

COMPANY/PLANT

	ALPHA	BETA	GAMMA	DELTA	EPSILON	ZETA	ETA
Total employees	75	19	74	16	10	200	48
Employees on shop floor	60	16	60	15	6	130	40
Annual sales	$8.2M	$900,000	$4.5M	$600,000	$300,000	$25M	$2.75M
NC and CNC machine tools	21 NC and CNC	2 CNC lathes 1 CNC miller 16 NC millers	3 CNC punch presses 1 CNC laser cutter 5 CNC press brakes	6 CNC	4 CNC millers 1 CNC lathe	12 CNC lathes 2 CNC vertical millers 30 NC machines	21 NC and CNC machines; more than half of these are CNC; 2 CNC in prototype
Year company founded	1969	1940	1973	1972	1974	1945	1942
Year first NC or CNC machine tool purchased	1974	1966	1976	1976	1979	1957	ca. 1966
Principal client industries	military aircraft medical	varied: electronics, hydraulics, etc.	mostly electronics	electronics aircraft	aircraft medical	aircraft, both military & commercial	electronics aircraft
Programming system	Digital "APT"	Genesis "Encode"	Webber "Prompt"	Webber "Prompt"	Bridgeport "Easy Cam"	Digital "APT"	General Numeric "Numeridex"
Age of programming system	3–4 years	6 years	3 years	6 months	2 months	11 years	1 year
Lot size range[a]	10–150	25–1,000	10–2,500	100–5,000	50–1,000	1–1,000	1–1,000
Average or typical lot size[a]	50	250–500	100	250–500	250	100	50
Employment level over time	steady growth	stable for 10 years; before that steady growth	steady growth	fluctuates; down from a peak of 19 in 1980	growth, recent layoffs	cyclical; twice as many employees in late 1960s; constant for last 7 years	stable
Size of shop in square feet	23,000	10,000	30,000	8,000	2,500	66,000	28,000

[a]Lot size figures are rough estimates only.

finished part. After planning the cuts, the machinist makes fixtures to hold the part during machining, selects the cutting tool, and determines the speeds (how fast the piece or tool rotates) and feeds (how fast the piece or tool moves along the axes of the machine). During each cut, the machinist observes the vibration, color, or sound of the part or tool and, if necessary, modifies the original cutting strategy.

With computer numerical control (CNC), a computer controls the way a machine tool cuts metal.[4] Since all operations are preprogrammed, CNC guides a machine from cut to cut far more rapidly than manual control, thereby reducing the time required to make a part. In addition, CNC makes possible the machining of more complex parts than conventional equipment because the computer is able to guide the cutting tool through complex arcs and angles that no machinist could duplicate. The ability of CNC to reproduce complex motions also reduces the need for intricate fixtures to orient parts at special angles to the cutter.

Like conventional machining, CNC still requires a combination of conceptual skill and metalcutting experience. It leaves open the question of who will write the program for which this skill and experience are necessary. At one extreme, a full-time parts programmer sits in front of a video screen off the shop floor and determines how a part will be made. At the other extreme, a machinist programs a minicomputer located at the machine.[5] Where between the two possible extremes responsibility for programming lies reflects a managerial choice.

The locus of programming control. There are sometimes compelling technical reasons for initial programming to be done off the shop floor. Devising long, complex programs often requires intricate calculations and takes several days. It may make more sense to assign such calculations to specialized programmers, rather than to provide space and tools for each machinist to work them out. In other cases, however, it is more efficient for machinists to write programs—particularly for simpler parts. Operators at the machines are also especially

[4]The earliest NC machines used a punched tape to guide the cutting tool. The minicomputer inside CNC machines facilitates the integration of machinists into production planning by creating the possibility of programming or editing right at the machine. The machines we looked at in the small shops were all CNC. For this reason, although much of what we say applies with some qualification to tape machines, we restrict discussion in the text to CNC.

[5]What programming by machinists implies about their role in production planning depends on the particular CNC technology being used, as well as on the complexity of the part. With the latest generation of CNC machine tools, especially lathes (which cut more standardized parts), programming at the machine has become highly automated. The machine operator enters dimensions and the material being cut for a given class of parts displayed on a screen. The computer than generates a program. We note later in the text the two cases in our sample where programming by the machinst implied little about the location of planning responsibility. With these exceptions, the shops we studied employed CNC machines and cut parts for which programming requires considerable craft skills.

well situated for debugging flawed programs. In one shop we visited, for example, a program called for making a heavy cut across a block of aluminum—an operation that generates considerable heat—and then boring two holes a precise distance apart. When the machine carried out the steps in this order, the distance between the holes decreased as the aluminum cooled. The machinist corrected the problem by editing the program so that the holes were drilled first.

Given the flexibility to concentrate programming on or off the shop floor, how is CNC applied in practice? The logic of the flexible specialization thesis suggests that, in batch production of the small volumes typical in the job shops, programming (or at least editing) would remain in the hands of skilled shop-floor workers. The premium placed on quality and fast delivery would require that technology and shop-floor skill be used in a complementary fashion. The evidence from our case studies supports a different conclusion, however.

Admittedly, the small shops organized production on CNC machines in diverse ways. In one shop with both prototype and small-batch production operations under one roof, prototype machinists programmed their own machines. In a second shop, a machinist did the highly automated programming on one lathe, while three mills were programmed remotely. In a sheet metal shop, where programming tends to be less complex than in metalcutting, some workers rotated through the programming department located off the shop floor. In the remaining four shops, programmers other than machinists wrote all programs.

As is standard industry practice, in all the shops machinists were allowed to change feeds and speeds in order to deal with variability in metal hardness and types of cuts as well as with programming errors. The machinists made these adjustments, however, by using a manual override available on most stand-alone NC tools, not by editing a program. In one shop we visited, several lathe operators were running their machines at 30 per cent of the programmed feed rate. This new feed rate was clearly near the maximum for this workpiece since the cut was generating a blue chip, which indicates a great deal of heat in a roughing cut. Without the override dial adjusted downward, the metal cutting operation would have quickly burned out the tools or scarred the workpiece.

While the override dial gives the machinist an important measure of control over the operation, it is not the same as editing the program. The overall execution of the job—order of operations, cutter path, etc.—remains outside the machinist's control. According to all managers and former skilled machinists interviewed, operator intervention on NC does not utilize worker skill as much or grant workers as central a role in production as conventional equipment in prototype and varied small-batch production. The intervention at the machine tool that does take place also falls considerably short of the skill levels that would be necessary with greater emphasis on worker programming. In the shops we visited, CNC thus transferred the bulk of the detailed planning of work away from the machinist.

Our case study findings are broadly consistent with the survey results of Donald Hicks (1983). Only 21 per cent of the 1,172 shops (all employing less than 250 workers) Hicks surveyed reported that machine operators were most often responsible for writing NC programs. At 52 per cent of the shops, full or part-time programmers had prime responsibility, while at 19 per cent shop supervisors most frequently wrote instructions for the machines.[6]

Centralized programming: advantages to management. Interviews with shop managers reveal that they centralize programming for a complex set of interrelated technical, economic, and social reasons. The owners saw centralized programming as a means of improving quality or efficiency. While they readily acknowledged that operator intervention is essential and generally agreed that skilled workers operate NC more effectively, owners interviewed feared that routinely allowing machinists to program or edit would, by reducing managerial ability to supervise and coordinate work, threaten shop performance. They worried that decentralized programming might result in less capable machinists editing programs badly, in having different versions of programs for the same part, or in erosion of management's ability to schedule. How much arguments equating centralized programming with efficiency are valid and how much they are rationalizations of owners' preferences for control based on other reasons is difficult to establish. None of the shops determined the division of programming and editing responsibility by systematically varying it and then monitoring performance.

One owner explained his preference for centralized programming in the following terms: "I believe in having control over every program and every part because there's a lot involved. . . . We basically have to have quality control of the programming. . . . You don't want everybody doing their own thing." (Shaiken, 1984a, Case 1, p. 24) Owners also acknowledged centralizing programming operations because they favored transferring responsibility to

[6]Hicks does not explicitly distinguish between simple programming, manual data input, and more complex programming. (On manual data input machines, generally used to cut simpler classes of parts, an operator enters part dimensions by punching a series of buttons at the machine tool.) If he had, the percentage of machine operators responsible for writing programs which require craft skills would probably fall. Noble (1979, 1984) and Wilkinson (1983) reached case study conclusions similar to ours based on interviews in the U.S. and England, respectively. Case studies conducted by Hartmann *et al*. (1983) find that programming remains on the shop floor far more frequently in West Germany than in Britain.

office programmers rather than leaving it with machinists on the shop floor.[7] Since programmers are not responsible for actually running the machines, they have little incentive to use programming to slow the pace of production. Some owners' memory of their dependence on skilled workers before CNC also contributed to their preference for limiting the role of skilled machinists in production planning. One owner complained:

> Five, six years ago, we were very dependent on skilled labor, to the point where I spent half my life on my hands and knees begging somebody to stay and do something. . . . Machinists tend to be prima donnas. This is one of the motivations for bringing in NC equipment. It reduces our dependence on skilled labor. (Shaiken, 1984b, Case l, p. 24)

Flexible manufacturing systems. At the agricultural implement producer, we examined the introduction and use of flexible manufacturing systems (FMS), often referred to as the prototype for the automatic factory. Using computer-controlled materials-handling equipment, these systems link together CNC machines. Prior to FMS, firms produced small batches of parts on individual CNC machines; larger volumes were manufactured on transfer lines—chains of mechanically linked machines dedicated to the production of a single part. FMS combines the flexibility of individual CNC machines with throughput approaching that of transfer lines.

Managers at the agricultural equipment firm referred repeatedly to the advantages of flexibility in interviews about the FMS. The company purchased an FMS to produce transmission cases and clutch housings for a new line of heavy-duty tractors. As a result of the system's flexibility, the company was able to make significant design changes late in the product development cycle. This production flexibility, however, did not translate into an expanded role for production workers, as the flexible specialization thesis would suggest. Compared with stand-alone NC equipment, in fact, an FMS takes the elimination of the workers' direct role in production one step further. Operator responsibilities at the agricultural equipment firm—and on other FMS lines (see Gerwin and Tavondeau, 1981; Blumberg and Gerwin, 1981)—bear a striking resemblance to those of machine operators we have observed staffing dedicated mass-production machining transfer lines. Operators in both settings inspect parts, change tools at periodic intervals or when a tool breaks, intervene when problems occur, and clean the area. FMS operator jobs bear little resemblance to those of skilled machinists on conventional machine tools.

[7]The managerial, vendor, engineering, and engineering society literature is replete with references to the way numerical control can help managers remove planning responsibility from machinists. For example, the officers of the Numerical Control Society said in 1981 that NC has put important decisions "in the hands of manufacturing and professional personnel rather than machine operators." (Jenkins, 1981, p. 185) For additional examples see Shaiken (1984a) and Noble (1984).

Interviews at the plant indicate that managers perceived the ability to limit operator control of the work pace as an important advantage of the FMS compared to stand-alone NC, another technical option considered.[8] The project manager spoke of the advantages of automatic control of unloading and reloading in the FMS: "You don't have people you're relying on. Once the computerized system gets the part, it doesn't wait for a guy who is drinking a cup of coffee." (Shaiken, 1984b, Case 2, p. 42)

Changes in Inventory Management

The technological changes discussed above have frequently been accompanied in recent years by the introduction of just-in-time inventory management.[9] The just-in-time inventory approach seeks to minimize in-process inventory between machining or assembly operations.[10] Parts are passed down the line "just-in-time" to be used, rather than accumulated in large banks between operations in what is sometimes facetiously referred to as "just-in-case" production. In the past, managers viewed in-process inventory as a good way of minimizing disruptions when a machine goes down. With a buffer stock available between two operations, production can keep going on all but the disabled machine. To derive optimal inventory levels, this benefit would be weighed against the combined costs of holding parts on the shop floor and of the extra floor space needed to make room for buffer stock.

At the plants we visited and in the literature (see Abernathy *et al.*, 1983; Schonberger, 1982; Cusumano, 1985), analysts and managers view minimizing in-process inventory as the centerpiece of a new focus on quality. With little inventory, if a machine makes bad parts, it should be discovered more quickly down the line. Inventory is a place where problems can hide. One manager likened traditional inventory management to filling a lake that has rocks—machinery or quality problems—lying at the bottom. Just-in-time removes

---

[8]While an FMS automates materials handling, computer monitoring can control the work pace by supervising stand-alone NC feed rates. An information system at the aircraft firm that monitored production at 66 CNC machines illustrates this use of monitoring. The system, nicknamed "the spy in the sky" by workers, signals to supervisors whenever feed rates drop below 80 per cent of the programmed rate. It also signals and records the time machines stand waiting for parts. In this way, it reduces the control of work pace and parts transfer possessed by operators on unmonitored CNC (see Shaiken, 1984b, Case 3). A management information system at the agricultural equipment firm reduced operator autonomy and control of the work pace in another way. By tracking the amount of time it took workers to perform jobs, the system enabled supervisors to detect when workers discovered short cuts. By tracking the progress of parts through the plant, the system helped identify when workers were hiding "banks" of processed parts. In the past, workers used short cuts and banks to maintain their piece-rate bonus pay in the event of downtime (see Shaiken, 1984b, Case 2).

[9]More than 70 per cent of 245 North American auto suppliers surveyed recently reported some implementation of just-in-time by early 1984 (see Cole, 1985).

[10]Many just-in-time systems also aim to reduce the inventory of purchased parts held at a plant, requiring that suppliers deliver parts in small batches. We focus in the text on in-process inventory because of its implications for workers.

the water so that you can see the rocks. In practice, the purely technical advantages of just-in-time only partly explain its introduction. An equally important managerial rationale for just-in-time stems from its ability to regulate the flow of subassembly or batch production work. In traditional batch or subassembly production, work stations produce a batch in a fixed period of time, or a given quota of parts in a day. Holding in-process inventory equal to a few hours' production decouples the pace of each station from further operations. By minimizing inventory and configuring operations in series, however, managers create assembly line pacing.[11]

A further social motivation for just-in-time appears to be the pressure it puts on workers responsible for repairing machinery. If repair workers don't bring machines back up quickly, one breakdown can end up bringing a series of machines to a halt. Abernathy, Clark, and Kantrow (1983, p. 76), missionaries for the transformation of American approaches to production management, describe the pressure just-in-time places on workers with maintenance responsibility:

> This determination to make all problems visible is not an unmixed blessing. It offers the hope of thoroughly efficient operations by substantially raising the social costs and consequences of failure. Reducing inventory levels places increasing pressure on managers and workers alike to solve whatever problems remain—that is, it directs energy and initiative where they belong by ratcheting up the level of stress at which the workforce is expected to perform. . . . As rough-hewn characters have long been wont to observe, it is wonderful how a little bit of fear and danger can clear the mind.

In the study conducted for the OTA, we observed the consequences of just-in-time for workers at an assembly plant of a major automaker. The company introduced a robotized welding system into its frame welding shop in 1980. Simultaneously, the subassembly area was reorganized by arranging operations in series and limiting inventory between stations to about 15 minutes' output. In subassembly, production workers place two or more parts in automatic welding machines, push a button to start the machine, then place the joined piece on a gravity feed or monorail to the next operation. Before the introduction of the robot welding system, the banks of parts stored beside each machine gave workers some control over the pace of their work: to break the monotony of the day, they could push ahead quickly, pile up a lot of parts, and then have some free time later on. Without banks, however,

---

[11]Richard Schonberger (1982) devotes a chapter of his book on Japanese management techniques to descriptions of how just-in-time helps reproduce the rhythm of the assembly line in various settings. He also quotes (p. 88) the former plant manager of a Kawasaki facility in Nebraska describing the vision guiding the manager's application of just-in-time at the batch production plant. "I envision the entire plant as a series of stations on the assembly line, whether physically there or not " See also Cole (1985).

workers are tied together into a line and paced within narrow limits by the robots to which they feed parts.

All 11 subassembly production workers interviewed preferred subassembly work before the elimination of banks because the workers controlled their own work pace. According to one, "It's not how much you put out, but how much it puts out. Because when it puts out, you put out." (Shaiken, 1984b, Case 4, p. 42) Frequent downtime on the highly interconnected system periodically halted operations, but workers uniformly found these interruptions disconcerting. Unlike rests earned by banking, these stops were outside their control. One worker explained, "What you're doing is you're going and stopping, going and going. You know what that does to your insides?" (Shaiken, 1984b, Case 4, p. 42)

The decision to minimize subassembly inventory, the highly integrated nature of the entire computer-controlled frame welding shop, and the financial difficulties of the firm also placed enormous pressure on the tradesmen and supervisors responsible for maintaining the area. The three maintenance supervisors interviewed regarded stress as the most salient feature of their job.[12] In the 33 months between the time the system began operation and our visit to the plant, eight first-line supervisors had either resigned, transferred, or been demoted, a turnover rate of 150 per cent. "This has been the hardest three years of my life," one general foreman declared, "There isn't any relaxation. . . . I've walked out of here and sat in my car, unable to move, getting myself together." (Shaiken, 1984b, Case 4, p. 25)

Changes in Work Organization and Labor Relations

The computerization of manufacturing has been accompanied by important innovations in work organization, including the elimination of narrow job classifications and other work rules, and the spread of quality of working life or quality circle programs (Kochan, Katz, and Mower, 1985; Katz, 1985a).[13] The flexible specialization thesis views fragmenting markets as the most recent and important impetus behind these changes (Piore and Sabel, 1984). Broad job classifications and worker participation create in

[12]The hourly maintenance workers that repaired the system also complained of pressure. They spoke of stress far less than their supervisors, however, perhaps because, given the shortage of available staff with a good electronics background, the hourly maintenance workers had both good job security and leverage in disputes on the shop floor.

[13]The remaining sections of this paper are based primarily on an unpublished study in which Shaiken interviewed 23 staff managers in a division of a major auto firm. He also interviewed 21 floor supervisors and 60 hourly employees on mass production machining lines in two auto manufacturing plants. The details of findings drawn from this study have been disguised slightly, at the company's request. We emphasize that we have only studied work organization issues in mass production settings and rely on other sources in generalizing tentatively to other sectors.

workers the general skills necessary for smaller batch, more rapidly evolving production. Katz and Sabel (1985, pp. 298, 299) suggest that this logic drives recent changes in shop-floor management even in the auto industry, paradigm of mass production:

> Instead of producing a standard car. . . [using] workers with narrowly defined jobs and dedicated machines—the tendency is to produce specialized goods. . . [with] broadly skilled workers using capital equipment that can make various models. . . . This has required firms to reorganize industrial relations to encourage workers to acquire and deploy the constantly expanding range of skills needed to meet unforeseen production difficulties. . . . Broad job classifications and informal work rules are essential to deployment of skills once acquired.

In our view, however, this misinterprets the impact of shop-floor reorganization on workers' jobs. Managers employ these methods not only to reduce their volume break-even point and achieve the high quality levels now demanded by consumers, but also to redefine the level of work effort and the customary level of active cooperation on the job. For example, at the mass production plants we visited, reducing staff levels was a significantly more important managerial goal than was promoting worker adaptability. The central labor relations approaches—quality of working life programs and repeated managerial emphasis on the pressure of competition—complement attempts to intensify work by encouraging workers to internalize the company goals of efficiency and quality.

The issue of job classifications. In the last few years, recession and competitive pressures have led to consolidation of job classifications and other work rule modifications in much of unionized manufacturing. Managers object to classifications primarily because they believe narrow job definitions increase staffing levels. With narrow classifications, managers cannot reassign skilled or production workers if there is no work which falls within their job definition. Thus, for example, plants must hire enough electricians to limit downtime on days when electrical problems accumulate. Managers cannot, however, reassign electricians to pipefitting work when most electrical systems are operating smoothly. Likewise, if inspectors, machine operators, stockhandlers, and sweepers staff a machining line, a foreman cannot require a machine operator to clean, stockhandle, or inspect, even if the operator's machine either runs smoothly or is down. Traditionally, strict lines of demarcation betwen skilled and production work also prohibit production workers from performing minor maintenance. Even if they know how to replace a faulty limit switch on their machines, production workers must still wait for a skilled electrician to come and install the new one, sometimes extending machine downtime as well as increasing staffing levels.

An incident at one plant Shaiken visited underscores that the classifications issue is primarily a battle over the intensity of work. In 1982, management sought to cut the number of workers on a new machining line 30 per cent below traditional levels, both by merging cleaning and inspection work into the machine operator classification and by reducing the number of machine operators. The production manager at the plant spoke explicitly of gaining eight hours' work for eight hours' pay.

When the line began operation, substandard quality led management to discipline eight workers in a two-month period for poor workmanship, far more than typically disciplined in the machining area. The machine operators on the line complained that the workload made it impossible to check parts adequately. The production manager regarded workers' complaints as a consequence of past management practices rather than physical inability to do the job. Workers, he felt, had become accustomed to working only four or five hours a day due to management sloppiness over the last 20 years.

The incident ended after four months, when management added back about half the number of workers it had originally cut. Ten of eleven machine operators we interviewed still regarded the workload as heavy after the addition of five workers. One machine operator called the workload three times that of ten years earlier, and 50 per cent higher than the workload three years ago.[14]

Worker reaction. This intensification of work might be expected to spark a reaction from workers and unions that would threaten attempts to improve performance. The opposite, however, has been the case. Many unions granted wage, benefit, and work rule concessions in the 1982 bargaining round and a number took steps towards greater labor-management cooperation (see Katz, 1985b).

The most important reason for limited worker reaction almost certainly is the state of the economy. In the auto industry, for example, employment of production workers (SIC code 371) dropped from 800,800 in December 1978 to 487,700 in January 1983 (Katz, 1985a). Moreover, the threat of plant closings continues to loom in the auto and other heavy manufacturing industries

[14]Managers' attempts to consolidate classifications may have similar roots in other settings. Many of the details in an article in *Business Week* (May 16, 1983) support the view that work rule changes often reflect efforts to cut staffing. Charles Jones, director of labor relations at B. F. Goodrich, is quoted as saying, "We're talking about doing the same amount of work with fewer people or more work with the same people" (p. 103).

Cusumano (1985) and Ohno (1982) describe Toyota's effort to maintain broad classifications and related production management initiatives in the early fifties in terms that unambiguously reflect their connection to the intensity of work. Broad classifications are linked to the fragmented nature of the postwar Japanese auto market only in that they facilitate cutting workers and maintaining work intensity when the volumes of different models fluctuate.

as saturated domestic demand, the growth of outsourcing, rising investment in newly industrializing countries like Mexico and South Korea, and continued erosion of U.S.-based multinationals' market share are expected to further reduce U.S. employment (for the auto industry, see Altshuler *et al.*, 1984).

GM, Ford, and other major unionized manufacturing firms have sought more active worker cooperation and have tried to insulate this cooperation from the immediate economic environment through two closely related labor relations strategies: constant reference to the pressure of competition—what we call the "ideology of competition"; and the development of worker participation and labor-management cooperation more generally.

The ideology of competition seeks to persuade workers of a logical connection between higher productivity and job security. As Katz (1985a) notes, management initiatives on the shop floor often center around efforts to convince workers and their unions that a correspondence between work rule concessions and job security does exist. Company newsletters, "information sharing" meetings between managers and workers, informal contact on the shop floor, and participation groups themselves transmit the message at the plants we visited and at those described in the literature. Robert Guest's (1982) description of an Employee Involvement (EI) program at Ford's Sharonville, Ohio transmission plant illustrates how employee involvement programs, broadly defined, reiterate the importance of competitiveness. Early in the program "sessions were held by top management personnel and the union leadership on the challenges of Japanese competition and the need to involve employees in improving quality" (pp. 45-46).[15] Worker participation programs may also integrate workers into the company in more subtle ways. They encourage workers to adopt a managerial perspective when production problems arise on the shop floor and to gain personal satisfaction from contributing ideas that promote efficiency.[16]

Conclusions

These cases illustrate the repeated use of technology and work reorganization to increase managerial control of production at the plants we studied. It is difficult, however, to generalize from a small sample of case studies. Further research might well discover greater variation in the way

[15]Guest notes that the existence of the Japanese challenge was re-emphasized on several other occasions. Apparently the employees got the message. At one point, an EI group wrote, in a letter to the other members of the plant community, "We care and we know you care. In the present crisis, our personal and our company attitudes must change for the better. . . . Many of you know where improvement can be made, we would welcome your help. We trust in your determination to keep the Sharonville Transmission Plant in operation and we need your help to achieve this goal." (Guest, 1982)

[16]Parker (1985) and Rinehart (1985) develop similar arguments, the former using examples from the auto industry, the latter drawing from a case study of a Canadian GM plant.

firms deploy programmable technology in other countries, for example, or in nonunion firms with a tradition of good labor relations. At the very least, however, the absence of new craft forms in successful companies that are already advanced users of programmable technology in batch production indicates that there is no inherent dynamic in this direction.

The cases we examined have disturbing implications for the quality of work in U.S. manufacturing. Managers in the plants we studied introduced new technology guided by a vision of the automatic factory, or continuous process plant, not nineteenth century craft production. They attempted to remove planning responsibility and autonomy from the shop floor more often than they tried to combine flexible technology with broadly skilled workers.

Managers interviewed believed that, within limits, greater managerial control increases firm responsiveness to the market as well as improving product quality, productivity, and other performance measures. Greater managerial control, in their view, implies better coordination and more efficient utilization of firm resources. It also enables professionals—programmers, managers, and engineers—in whose ability and loyalty managers have the most faith, to make more production decisions.

The flexible specialization thesis makes two fundamental errors in its claims about the implications of lower volume production for workers. First, it incorrectly assumes that in small-batch settings it is necessary to broaden the knowledge of all shop-floor workers in order to respond to market fluctuations. In our sample, only in prototype work in one machine shop—where new parts are produced in volumes of one—did the type of production lead management to grant skilled machinists the central role in innovation and debugging that Piore and Sabel regard as generally necessary for competitive success in batch production in rapidly changing markets. Knowledge attainable only on the shop floor remains essential in all batch settings, of course, to design new products and plan and implement their production. In all settings other than prototype job shop production, however, managerial or other white-collar employees with some shop-floor experience contributed this knowledge. In sum, in the cases we studied, market structure had very limited implications for work organization.

Second, in higher volume settings, the flexible specialization thesis misinterprets the assignment of a wider range of unskilled or semi-skilled jobs to production workers as a reflection of the growing importance of shop-floor skill. In many cases, management simply groups together more semi-skilled tasks to increase the intensity of work.

Ironically, the managerially dictated restructuring of production is making batch production work more like mass- than like craft-production. First, with the introduction of NC and then FMS, all-around machinists in batch settings change from skilled craft workers to operators with occasional responsibility for editing programs written off the shop floor and, in turn, to monitors of a process for which they have no planning responsibility. The job of machinist in batch production comes to resemble that of machine operator on dedicated mass production transfer lines. Second, by reducing worker control over the pace of semi-skilled batch production work, just-in-time inventory and, to a lesser extent, computer monitoring, extend assembly line pacing beyond mass production. Third, by increasing the cost of downtime in batch production, just-in-time brings the pressure of assembly line maintenance work to lower volume settings.

The vision—however welcome—of a more broadly skilled workforce emerging from work restructuring and the increased use of programmable technology holds more promise in theory than in practice in the U.S.

References

Abernathy, William J., Kim B. Clark, and Alan M. Kantrow. *Industrial Renaissance: Producing a Competitive Future for America*. New York: Basic Books, 1983.

Altshuler, Alan *et al*. *The Future of the Automobile: The Report of MIT's International Automobile Program*. Cambridge, MA: MIT Press, 1984.

Blumberg, Melvin and Donald Gerwin. "Coping with Advanced Manufacturing Technology." Paper presented to a conference on "Quality of Working Life and the '80's." Toronto, August 30-September 3, 1981.

Business Week. "A Work Revolution in U.S. Industry," May 16, 1983.

Cole, Robert E. "Target Information for Competitive Performance," *Harvard Business Review*, LXIII (May-June, 1985), 100-109.

Cusumano, Michael A. *The Japanese Automobile Industry: Technology and Management at Nissan and Toyota*. Cambridge, MA: Harvard University Press, 1985.

Gerwin, Donald and Jean Claude Tavondeau. "Uncertainty and the Innovation Process for Computer Integrated Systems: Four Case Studies." Unpublished mimeo, March 1981.

Guest, Robert H. "The Sharonville Story." In Robert Zager and Michael P. Rosow, eds., *The Innovative Organization: Productivity Programs in Action*. New York: Pergamon Press, 1982, pp. 44-62.

Hartmann, Gert, Ian Nicholas, Arndt Sorge, and Malcolm Warner. "Computerized Machine Tools, Manpower Consequences, and Skill Utilization: A Study of British and West German Manufacturing Firms," *British Journal of Industrial Relations*, XXI (July, 1983), 221-231.

Hicks, Donald A. *Technology Succession and Industrial Renewal in the U.S. Metalworking Industry*. University of Texas, Dallas: Center for Policy Studies, June 1983.

Jenkins, Lamont J. "Getting More Out of NC," *American Machinist* (October, 1981), 185.

Katz, Harry C. *Shifting Gears: Changing Labor Relations in the U.S. Automobile Industry*. Cambridge, MA: MIT Press, 1985a.

———. "Collective Bargaining in the 1982 Bargaining Round." In Thomas A. Kochan, ed., *Challenges and Choices Facing the American Labor Movement*. Cambridge, MA: MIT Press, 1985b, pp. 213-226.

——— and Charles F. Sabel. "Industrial Relations & Industrial Adjustment in the Car Industry," *Industrial Relations*, XXIV (Fall, 1985), 295-315.

Kochan, Thomas A., Harry C. Katz, and Nancy R. Mower. *Worker Participation and American Unions: Threat or Opportunity*. Kalamazoo, MI: Upjohn Institute, 1984.

Noble, David F. *Forces of Production*. New York: Knopf, 1984.

_________ . "Social Choice in Machine Design: The Case of Automatically Controlled Machine Tools, and a Challenge to Labor," *Politics and Society,* VIII (1978), 313-347.

Ohno, Taiichi. "How the Toyota Production System Was Created," *Japanese Economic Studies,* X (Summer, 1982), 83-101.

Parker, Mike. *Inside the Circle: A Union Guide to QWL*. Boston, MA: South End Press, 1985.

Piore, Michael J. "Computer Technologies, Market Structure, and Strategic Union Choices." In Thomas A. Kochan, ed., *Challenges and Choices Facing American Labor*. Cambridge, MA: MIT Press, 1985, pp. 193-204.

_________ and Charles F. Sabel. *The Second Industrial Divide: Possibilities for Prosperity*. New York: Basic Books, 1984.

Reich, Robert B. *The Next American Frontier*. New York: Times Books, 1983.

Rinehart, James. "Appropriating Workers' Knowledge: Quality Control Circles at a General Motors Plant," *Studies in Political Economy,* No. 14 (Summer, 1984), 75-97.

Schonberger, Richard J. *Japanese Manufacturing Techniques: Nine Lessons in Simplicity*. New York: Free Press, 1982.

Shaiken, Harley. *Work Transformed: Automation and Labor in the Computer Age*. New York: Holt, Rinehart, and Winston, 1984a.

_________ . "Case Studies on the Introduction of Programmable Automation in Manufacturing." Volume II, Part A of *Computerized Manufacturing Automation: Employment, Education, and the Workplace*. Washington, D.C.: Office of Technology Assessment, June 1984b.

Wilkinson, Barry. *The Shopfloor Politics of New Technology*. London: Heinemann Educational Books, 1983.

[7]

REVIEW OF SOCIAL ECONOMY

Beyond the Fordist/Post-Fordist Dichotomy: Working Through *The Second Industrial Divide*

Bruce Pietrykowski
University of Michigan-Dearborn

Abstract The publication of *The Second Industrial Divide* helped to initiate a sustained inquiry into the transformation of work under industrial capitalism in the late twentieth century. The argument that the breakdown of Fordist mass production ushered in a new production paradigm in the shape of flexible systems of work organization is reexamined. The dominant role of high-volume mass production and its craft-based counterpart can continue to coexist well into the future. Nevertheless, current income and employment trends appear to disadvantage the traditional blue-collar Fordist worker and industrial unions. The cause of these trends may not, however, be directly linked to skills associated with computer technology. Finally, the type of flexibility most closely associated with the work of Piore and Sabel—flexible specialization—is discussed. It is argued that flexible specialization within industrial districts that (a) foster the development of socially informed economic action and (b) constrain competitive behavior may form the basis for the creation of different employment opportunities that challenge the dominant logic of capitalist development through which flexible employment strategies are used in tandem with corporate downsizing and increased managerial control.

Keywords: Fordism, flexible specialization, industrial district, work organization, employment, skill

I. INTRODUCTION: THE IMPACT OF *THE SECOND INDUSTRIAL DIVIDE*

Published in 1984, *The Second Industrial Divide* helped to initiate an exploration of contemporary changes in the fundamental nature of capitalist labor markets and industrial structure. By focusing on the evolution from mass production to systems of flexible specialization, Piore and Sabel asserted that a different model of capitalist accumulation was apparent in small industrial districts throughout Europe and the United States. By attending to difference, thereby highlighting key

Review of Social Economy Vol LVII No. 2 June 1999 ISSN 0034 6764 aspects of social and cultural history that form the backdrop of economic activity, and by focusing on the institutional framework within which large-scale economic growth takes place, Piore and Sabel provided a reference point from which to assess the changing postwar economic landscape.

The Second Industrial Divide was not the only work on these themes to emerge around the early 1980s. This period initiated a research agenda that has continued to occupy sociologists, economists, political scientists, cultural studies scholars, and geographers for well over a decade. For example, Aglietta's *A Theory of Capitalist Regulation* (1979, English edition), Gordon, Edwards, and Reich's *Segmented Work Divided Workers* (1982), Bowles, Gordon, and Weisskopf's *Beyond the Waste Land* (1983) and Kochan, Katz, and McKersie's *The Transformation of American Industrial Relations* (1986) each argue for a more institutionally contextualized understanding of the capitalist production process, placing particular emphasis on the capital–labor relation. Later in the 1980s, David Harvey's masterful work, *The Condition of Postmodernity* (1989) linked the debates over Fordism, post-Fordism, and flexibility to the broader methodological critique of modernity entwined within these debates. Harvey questioned whether one could or should reject modernism in favor of a postmodern perspective that focused primarily on the production of symbol and spectacle in economic and social life. He did not, however, reject wholesale the insights obtained from adopting a postmodern view. The deconstruction of labor markets, the hyper-segmentation and recombination of tasks, and the compression of time and space made possible through technological change are all tangible manifestations of advanced capitalist development.

> We thus approach a central paradox: the less important the spatial barriers, the greater the sensitivity of capital to the variations of place within space, and the greater the incentive for places to be differentiated in ways attractive to capital. The result has been the production of fragmentation, insecurity, and ephemeral uneven development within a highly unified global space economy of capital flows. The historic tension within capitalism between centralization and decentralization is now being worked out in new ways.
>
> (Harvey 1989: 295–296)

I make use of Harvey's perspective to argue against adopting an analytical framework in which one must choose between Fordist mass production or Post-Fordist flexible specialization. This is not to deny that the nature and structure of work has indeed changed over the course of the last two decades. The task is to better understand the nature, structure and depth of change and to assess the various frameworks available from which to gain some perspective on the future of work.

In this contribution, I will critically reexamine the thesis framing *The Second Industrial Divide*. This thesis holds that the economic system has reached a branching point in its path of development: the choice involves the attempt, on the one hand, to reconstruct the institutional supports for a reinvigorated system of large-scale global mass production or to adopt, on the other hand, a system of production and distribution premised upon the small-scale associations of producers creating a network of complementary and collectively regulated enterprise units and long-term production relations. This analysis will involve a rereading of *The Second Industrial Divide* that deconstructs the authors' thesis. This rereading eschews the dualist framework wherein economic development is path dependent and locked into a technological and institutional trajectory after a branching point is reached and a new technological/institutional "fix" is chosen. Indeed, I argue that Piore and Sabel themselves seem to doubt that this framework is tenable. Throughout the book, reference is made to the *interdependence* between mass production and its other, flexible specialization. I wish to deconstruct their thesis in order to suggest what a more complex and contingent relationship between mass production and flexible specialization means for the future of work.

II. BASIC THESIS: FROM MASS PRODUCTION TO FLEXIBLE SPECIALIZATION

A. Mass Production and the Era of Fordist Hegemony

The Ford Motor Company in the 1910s and 1920s is often cited as the primary example of large-scale mass production. The impetus to large profits was the production of a high volume of standardized output. When this output is produced using dedicated capital equipment unit costs fall as output rises. The labor requirements for work at Ford were relatively modest. However, the highly routinized labor process was sensitive to high levels of employee turnover. Training costs were relatively low when measured on a per worker basis. But, given the large scale of production and the concomitantly large workforce, high-volume turnover from a work force that could not or would not endure the monotony, noise, and rigid system of labor control, threatened the high profits upon which the production process was predicted (Meyer 1981: 84).

The "five dollar day" was a solution, albeit a temporary one (Meyer 1981: 196–197), to the problem of labor turnover. In 1914 the five dollar day became an enormously powerful symbol. It secured Henry Ford's place in the pantheon of industrial giants. It also elevated Ford, the man and the company, in the eyes of the public. Even though recession and increased competition forced Ford to scale back the real value of the wage (Meyer 1981: 167)[1] the Ford high wage policy had the effect of demonstrating the possibility inherent in combining mass production with mass consumption (Piore and Sabel 1984: 61). Mass production provided the goods that workers could now afford with their expanded earnings. This is the foundation of *Fordist* mass production. It is this link between mass consumption and mass production that ushered in the industrial era that Piore and Sabel seem to suggest we are poised to move beyond. Yet, a rereading of the mass production story reveals:

> Mass production . . . has always necessitated its mirror image: craft production. During the high noon of mass production, craft production was used by firms operating in markets too narrow and fluctuating to repay the specialized use of resources of mass production.
>
> (Piore and Sabel 1984: 206)

Indeed, there were instances in which centralized mass production coexisted with both decentralized mass production and small batch production of the flexible specialization type. For example, Ford's massive Highland Park and Rouge plants coexisted with Ford plants that produced parts via small-scale mass production and small batch, craft production from the beginning of the 1920s into the 1940s (Pietrykowski 1995b).[2]

The actual constitution of a Fordist regime, as opposed to a pocket of Fordism at Ford Motor, required more than the beneficence of enlightened capitalists, however. The industrial union movement, in the form of the UAW, achieved for workers at GM and Chrysler what Ford initiated. Indeed, Ford was the last of the big U.S. auto makers to come to terms with the union. The Fordist social contract moved beyond autos and into the durable goods manufacturing sector and beyond through collective bargaining agreements and the implementation and expansion of government income maintenance programs. So, the pillars of Fordist mass consumption included key sectors of capital, unions, and the state.[3]

---

1 It should be noted that the high-wage policy of Ford did have a longer-lasting impact in areas at a distance from urban Detroit. At Ford's village industry plants the wages for men and women were very similar throughout the 1920s and into the 1930s and village industry wages were higher than the wages of workers in the auto industry in general and in similar parts plants in Michigan (Pietrykowski 1995a; 1995b).

2 See Scranton (1997) for a history of U.S. custom and small-batch producers and the industrial districts they helped to create and sustain into the 1920s. The increased attention paid to organizational and technological difference within capitalism does not deny the important role played by the modern corporation in structuring economic life (Harrison 1994; Chandler *et al.* 1997). Rather, it is meant to point out the heterogeneity between industrial producers and the variety of organizational structures within a given corporate enterprise.

3 Bowles *et al.* (1983) refer to this as the capital–labor accord. The linkage between productivity growth and real wage gains is a cornerstone of the system of Fordist mass production.

BEYOND THE FORDIST/POST-FORDIST DICHOTOMY

The breakdown of Fordist mass production and the structural transformation of the postwar capitalist system is frequently attributed to supply-side shocks, the collapse of Bretton Woods, the productivity slowdown, heightened labor conflict, and the growing popularity of state policies premised upon macroeconomic austerity and free-market ideology, that undermined the very institutions constituting the basis of support for the Fordist regime. In line with this framework, Piore and Sabel argue that shortages in labor, wheat, and oil drove prices up. This effectively set the stage for a decrease in demand due both to deficient levels of aggregate demand and increased uncertainty over transaction costs. This confusion over the composition of demand and the trajectory of input prices led to a breakdown in the mass production system. No longer able to predict the future demand for their product lines and no longer certain about the future costs of resource inputs necessary to sustain long production runs, capitalists could no longer undertake continued investment in fixed-cost special-purpose machinery dedicated to the production of a single product or component type (Piore and Sabel 1984: 183).

In addition to this confluence of special historical events (labor unrest, oil price shocks and wheat shortages), Piore and Sabel cite several historical trends that added to the further weakening of the mass production system. For instance, the saturation of the market in consumer durables in the advanced industrial countries led to a search for new markets. Since this was a rational choice for all domestic producers it had the unintended effect of increasing the level of international trade and intra-industry global competition in the mass production sector. The attempt to increase individual firm market share had the effect of increasing competitive behavior over a wider geographic area. Barring trade restrictions imposed by the home country on foreign imports, this also allowed for the erosion of domestic markets through increased domestic market competition from foreign rivals (1984: 184–187). Added to this was the strategic development of export-led growth sectors in developing countries in Asia and Latin America (1984: 188-189). These historical developments reinforced the competitive pressure on mass production sectors and contributed to the search for alternative technological capabilities and organizational structures.[4]

4 Note that Piore and Sabel's account of the demise of mass production has been criticized for lacking an explicit explanation of the forces leading to a more generalized crisis of capitalist accumulation while their description of the flexible specialization paradigm has been faulted for leaving untheorized the process by which new industrial districts of small batch producers are established and how they respond to technological change (Walker 1995). This is an important and serious critique. However, I wish to focus on the work structures and the configuration of labor markets within an existing system of flexible specialization. I acknowledge that there are unexamined issues relating to the maintenance and sustenance of accumulation under a system of flexible specialization. The case studies provided by Piore and Sabel (1984) and Sabel and Zeitlin

Recalling that Fordism required both mass production and mass consumption, Piore and Sabel briefly entertain the notion that changes in consumer tastes, as well as increased demand for variety and novelty, were determining factors reinforcing the saturation of consumer durable markets. They begin to sketch out an analysis in which consumer tastes and preferences are endogenous to the Fordist system, thereby rejecting the view that consumption behavior is a preexisting, natural urge. In the end, however, they abandon this explanation (1984: 191) and fail to make reference to the institutional supports necessary for the creation of a culture of consumption (Fine 1995). Yet the changing structure of consumer markets is a characteristic feature of the breakdown of Fordism. Advertising, of course, played a role in the rise of mass consumption but other mechanisms were devised or refurbished throughout the twentieth century to acclimate the working classes to the idea of shopping as a productive, culturally legitimate leisure-time activity. The growth in retail employment throughout the post-war period and the increased financial resources devoted to the entertainment and "image production industry" (Harvey 1989: 290) highlight an important dimension of the new economic times: the relative decline of mass production employment and rising employment share devoted to retail trade and services. The growing importance of consumption as an arena for both work and leisure time activity is an important feature of the new worlds of work opening up in retail and services (Pietrykowski 1994; Pfleeger 1996). The effect is to blur the line between production and consumption activity.

With respect to the structure of work in the mass production system, the Fordist labor market was premised upon the relative growth in the demand for unskilled and semi-skilled labor. In autos, requisite skills included manual dexterity, physical strength, hand-eye coordination and the ability to comprehend directions. Frequently, a set of additional cultural skills were also an asset on the shop floor. Such "skills" included deference to authority, high tolerance for rules and bureaucratic structure, conformity, personal autonomy, competitiveness, and belief in the legitimacy of managerial decision-making (Taylor 1967; Braverman 1974; Gintis 1976; Meyer 1981; Lewchuk 1993). Indeed, it may well be the case that the forces underlying changes in labor relations and labor market structures have as much to do with the reconfiguration of these cultural attributes of work as they do with the changing nature of skills as traditionally defined (Applebaum and Batt 1994; Hudson 1997).

(1985; 1997) provide a starting point, however. Also, the (Marshallian) industrial district model, within which flexible production systems operate, is predicated upon economies of agglomeration and localization and, implicitly, regional investment and consumption multipliers (Marshall 1927; 1982: 225–227; Scott 1988).

B. Flexible Specialization and the Image of Labor After Fordism

Although much has been written about the arrival of a post-industrial, Post-Fordist, lean, high-performance production system premised on flexibility, advanced information and communication technology, and global networks, I want to focus on the image of flexible specialization put forth by Piore and Sabel. By arguing that Piore and Sabel advance an image of flexible specialization that is at odds with proponents of lean production or high-performance production, I want to emphasize the institutional prerequisites that need to be established along with a set of technological processes and plant-level production relationships. By offering a rereading of *The Second Industrial Divide*, I also want to recast the main thesis along lines that differ from the dualistic framework suggested by the title. Reference to an industrial *divide* has the unfortunate rhetorical effect of imposing a deterministic framework on what is a decidedly contingent and context-specific analysis. Both flexible specialization and mass production can coexist for long periods of time. Furthermore, elements of flexible specialization can exist as adjuncts to mass production or can function autonomously.

First, attention must be paid to the type of flexibility being advocated. Flexible specialization, as used by Piore and Sabel, refers primarily to *functional* flexibility, not *numerical* flexibility (Atkinson 1987; Wood 1989; Boyer 1988; Biewener 1997; Smith 1997). The difference is that functional flexibility refers to the characteristics of jobs while numerical flexibility refers to the structure and distribution of jobs within the firm. Numerical flexibility is indeed the dark side of flexible production (Harrison 1994; Gordon 1996). The flexibility associated with many of the new agile manufacturing systems makes many workers expendable during periods of slack product demand. The reliance on

a) a core group of workers;
b) a secondary group of workers employed by subcontractors;
c) a subsidiary group of "regular" part-time workers; and
d) a tertiary group of temporary workers

increases the firm's labor market flexibility (Harvey 1989). Numerical flexibility also carries the advantage to the firm of reducing the total labor compensation bill through the restriction of health insurance and other fringe benefits to the core work force. However, numerical flexibility can seriously inhibit the development of long-term employment/transaction relations, trust and interdependence that appear to be highly valued in team-oriented workplaces employing multiskilled workers.

Therefore, in advocating for flexibility it is important to distinguish between the flexible utilization of labor as a disposable input and the flexible utilization of workers as skilled and valued members of an organization. The future of work under a regime of complete numerical flexibility would exacerbate income inequality and economic insecurity for many of those in the secondary and tertiary segments of the labor market. The available evidence that flexibility is currently being used to downsize and discipline the labor force suggests that the rhetoric of flexibility and high-performance has been seized upon to promote a more divided and demoralized working class (Applebaum and Batt 1994; Gordon 1996; Biewener 1997). Numerical flexibility can be achieved by restructuring employment relationships. It can also be accomplished through the development of new work processes, like lean production, in which work tasks are redistributed and workers are made responsible for multiple tasks. If unaccompanied by changes in technology requiring increased education and skill, lean production can simply increase the ability to assign workers to more unskilled or semi-skilled tasks (Rinehart *et al.* 1997). In his description of lean production in the auto industry, Mike Parker concludes that, "The 'skills' required in performing several related jobs of very short duration are manual dexterity, physical stamina, and the ability to follow instructions precisely" (1993: 271). This description applies equally well to workers employed in Ford's Rouge plant in the 1930s and 1940s. This usage of flexibility can be contrasted with an alternative set of meanings (and practices) of flexibility. Doing so highlights some of the conflicts inherent in the new paradigm of flexible work processes.

Flexible specialization, as opposed to numerical flexibility, combines techniques of craft production at several small batch production shops through a series of networks of association and mutual support. The *sine qua non* of flexible specialization of this type is the short production run characteristic of batch production processes. The increasing fragmentation of markets, the creation and development of niche markets and the increasing use of point-of-purchase sales data—in which information about the style, color and size of goods sold is electronically relayed to the production centers of the corporation—that enable just-in-time delivery of finished products to retailers also allows for the profitable implementation of small batch production technologies. The textile production regions of Italy and metal working districts in Germany are cited frequently as models of industrial districts in which flexible specialization techniques flourish. These districts evoke earlier eras in which regions of prosperity arose in Europe and in the United States and were able to sustain a good life for resident workers better than their more laissez-faire counterparts (Sabel and Zeitlin 1985, 1997).

The key characteristics of the industrial district—the ensemble of firms utilizing flexible specialization techniques together with the social institutions that sustain the production process—are detailed in Table 1. For Piore and Sabel, labor practices inside an industrial district are governed by limits on competitive

TABLE 1: Key Features of Industrial Districts

- a cluster of mainly small and medium spatially concentrated and sectorally specialized enterprises;
- a strong, relatively homogenuous, cultural and social background linking the economic agents and creating a common and widely accepted behavioral code, sometimes explicit but often implicit;
- an intense set of backward, forward, horizontal, and labor market linkages, based both on market and non-market exchanges of goods, services, information, and people;
- a network of public and private local institutions supporting the economic agents in the clusters.

Source: Rabellotti 1997: 175

behavior and by the flexible arrangement of resources toward the production of specialized outputs. Resources are employed in the production of a historically and culturally appropriate set of goods and services. In other words, the place of a firm within the community of producers helps to determine the scope of output (Piore and Sabel: 269; Piore 1995).

Several key questions must be raised about the viability of such a system of small batch production. A primary question to ask is to what extent does the system of flexible specialization actually require a reconstruction of worker skills. What type of skills are being altered and do new skill requirements vary by occupational segment? Is the local or regional economy capable of developing the institutional infrastructure to sustain an industrial district predicated on flexible production? Finally, is the fate of industrial unionism inextricably tied to the future of mass production or can labor unions create a presence for themselves on behalf of workers employed in industrial districts? It is to those concerns that I now wish to turn.

III. GENERAL SKILL AND EMPLOYMENT TRENDS: THE AMBIGUOUS IMPACT OF COMPUTER TECHNOLOGY ON JOBS

How have the level and distribution of skills changed since the publication of *The Second Industrial Divide*? Throughout the 1980s and into the 1990s the discussion has heated up over the implications of the rise of a new knowledge and information-based economy predicated, in large part, upon the application of computer technology. Indeed, Piore and Sabel argue that while flexible production processes do not require computer technology, the (re-)emergence of industrial districts is propelled by industrial innovation that benefits directly through the application of computer technology. The degree to which Piore and Sabel link flexibility with computer technology is certainly problematic.[5] One need only refer to Braverman (1974), Edwards (1979), Greenbaum (1979), and Shaiken (1986) to recall that computer technology has been deployed within a particular social and economic setting, complete with its own limitations and structural constraints, in which technical control and class power exerts a strong influence on the design and use of machinery.

Nevertheless, the prevailing view seems to be that skills are being upgraded and jobs reapportioned away from individuals with low levels of education and skill. For example, the secretary-general of the OECD has recently issued a call for the development of policies to guide the development of knowledge-based industries and industrial networks of firms:

> Producing goods and services with high value-added is at the core of improving economic performance and international competitiveness. The fastest rates of growth in output are being recorded by high-technology manufacturing sectors, such as computers and aerospace, and knowledge-based services, finance and communications for example. . . . Investments in R&D, computer software and the like have substantially grown in importance. Increasing intangible investment—which is difficult to measure—in the upgrading of the skills and competencies of workers has become a major issue for enterprises and government. Employment prospects are good for highly skilled workers, less promising for the unskilled.
>
> (Paye 1996: 4)

In the case of the United States, the claim that increased utilization of computer technology shifted the skill distribution of jobs in favor of technical/professional jobs and away from production and low-skilled jobs is linked to a neoclassical narrative in which the demand for high-skilled white collar workers is growing while the demand for unskilled or semi-skilled production jobs is declining. In addition, access to computer technology has caused a substitution of capital in place of production labor (Berman *et al.* 1994). The result is an increasingly bifurcated distribution of employment and income. However, recent analyses by Howell (1995) and Howell *et al.* (1998) are skeptical of the claim that computer technology has caused the shift in employment and income. Indeed, much of the change occurred prior to the widespread introduction of computers. "Put

5 For example, Piore and Sabel maintain, "Whereas most machines have an independent structure to which the user must conform, the fascination of the computer—as documented in the ethnographic studies—is that the user can adapt it to his or her own purposes and habits of thought." To ignore the limits of such flexibility and the mechanisms of control that can be attached to or hard-wired into computer technology is to forget that the use to which technology is put is a subject of contest and the manifestation of power in the production process. See, for instance, Zuboff's (1988) discussion of the panoptic role played by computer technology. Even within a system of flexible specialization, relations of power, control and conflict over the structure of work, the work process and the role and pace of technological change will not disappear.

differently, it is not that low skilled labor is any less necessary; it is simply now more highly concentrated in foreign low-wage regions and paid more poorly in the domestic (formerly) high-wage region" (Howell *et al.* 1998: 22). If this is the case, the view of a new high-tech, computer-driven information economy presents at best a partial and at worst a seriously flawed explanation of trends in the demand for skilled workers.

A clearer picture emerges when we examine skilling trends at a more disaggregated level. Two studies highlight the uneven effect of new technology on labor market segments and the job opportunities of various group of workers. First, Gittleman and Howell (1985), study changing employment and skill levels by labor market segment. By clustering jobs into segments (independent primary private; independent primary public; subordinate primary white collar; subordinate primary blue collar; secondary blue collar and secondary service) based on skills (general education, specific vocational preparation, people skills, motor skills), earnings, working conditions, employment status and institutional setting, they find that the employment share has been rising in the independent primary sector of the labor market from 1973–1990. Furthermore, employment has been falling in the traditional blue-collar segments and little change was noted in the subordinate primary white collar (e.g. nurses, typists, bank tellers) and secondary service (e.g. retail, childcare workers, teachers' aids) segments. By disaggregating segments they find that the most prominent employment share losses are among blue-collar workers. These are the workers most closely associated with the Fordist system of mass production.

Does this mean that flexible specialization is a means to offset the deteriorating economic position of blue-collar workers? Or, is a flexible production strategy prone to accelerate the erosion of employment in the traditional mass production sector? Flexible specialization is a response to the declining importance of mass production as an organizational logic within the United States and Europe.[6] Elements of flexibility—functional flexibility—are currently being adopted in manufacturing industries. Cappelli (1993), studying job titles matched to actual job evaluations for ninety-four manufacturing jobs in ninety-three establishments, finds that production jobs in manufacturing have undergone a general

6 This is not to deny that, within a global setting, Fordism and flexible specialization often coexist. For example, production facilities in the South may produce goods for mass markets in the North while firms in the North fabricate specialty goods using flexible specialization techniques of production. However, I would argue that the relationship is much more fluid and complex than this example would suggest. For instance, high-tech capital-intensive "lean" automobile production technology is being introduced in Mexico (Shaiken 1994; Carrillo 1995) and flexible specialization districts comprising small autonomous production units are to be found in Third World countries (Pederson *et al.* 1994).

upward trend in skill requirements. Osterman (1995), using survey data from 875 respondents asked to comment on the job characteristics involved by those workers directly engaged in making the product or providing the service, finds that skill levels increased more for professional/technical workers than for blue-collar workers. Interestingly, Osterman reports that "While for both groupings shifts in technology and heightened use of computers are important, they are relatively less so for blue-collar workers" (1995: 132). Furthermore, he discovers that, "general behavioral changes are more important for blue-collar workers" (1995: 132). When asked about the hiring criteria for workers involved in these jobs, Osterman comments, "It is quite striking, however, that skill is relatively much more important for professional/technical employees while behavioral traits are more important for blue-collar workers" (1995: 133).[7]

As these studies suggest, the effect of flexible work organization strategies on blue-collar skills is far from clear. Despite the lack of clarity about the determinants, the evidence appears to suggest that income inequality and employment opportunities tend to favor the independent primary professional sector and disadvantage the traditional blue-collar segment of the work force. This trend preceded the large-scale adoption of computer technology. Although firm-level survey data seem to indicate that computer-related skills are being required in professional-sector jobs it is not entirely clear how those new skills are constituted. On the other hand, the skills involved in traditional production work in manufacturing seem to be skills relating to new socialization schemes (like team work, employee involvement and flexible work groups) shaping the labor process.

For these new social skills to be effective, the production organization must be able to generate adequate levels of trust. Communication between workers and between labor and management needs to be able to provide workers with the ability to raise claims as to the truth, sincerity and legitimacy of management statements, rules and regulations. Whether these prerequisites are capable of being created and successfully reproduced over time within the capitalist organization of production is a questionable proposition at best (Babson 1995b; Graham 1995). Conflict and contestation between capital and labor cannot be easily

7 This may well reflect the respondents' assumption that professional/technical workers already possess the requisite behavioral traits necessary to "do their job." Yet, even this caveat suggests that worker behavior, attitudes and patterns of socialization characteristic of routinized mass production labor are undergoing a reassessment. The choice should not be cast in terms of either management-imposed flexibility or mass production since both systems are far from ideal. Writing about the experiences of Linden, New Jersey autoworkers during a period of plant-level restructuring and downsizing, Ruth Milkman states that, "it is crucial to understand their [workers'] contempt for the dinosaurs of the old industrial regime and their eagerness for something better. To take their perspective seriously is to abandon any effort to restore the world of mass production industry as it existed in the past—which is probably impossible in any case" (1997: 19).

overcome even with new information technologies and computer programs that forever seem to hold the promise of banishing alienating labor. However, flexible production within an industrial district may provide an opportunity to imagine worlds of labor outside of the direct sphere of capitalist production.

IV. THE FATE OF FLEX SPEC AND INDUSTRIAL DISTRICTS AND THE FUTURE OF WORK

A. The Promise of Flexible Specialization: Fragmenting the Discourse About Successors to Fordism

The terms "flexible specialization" and "Post-Fordist production" have been used to describe a wide array of production and employment relationships. These are illustrated in Table 2.

TABLE 2: Dimensions of Flexible Specialization

Production Relationships
- craft-based, capital-intensive
- craft-based, labor-intensive
- sweated labor

Primary Distribution Relationships
- exchange governed by cooperation, alliance, interdependence, familial relations
- market exchange (market)
- internal bureaucratic exchange (hierarchy)

Internal Organization
- part of a federated system of producers
- owned subsidiary of larger company
- unit of a multi-unit corporation

Spatial Orientation, Localization Economies, and Commitment to Community (adapted from Markusen 1996)
- Marshallian Industrial District—high level of internal trade in intermediate goods and services; high level of consumption of locally produced goods and services; high level of commitment to community and commitment to employer/employee.
- Hub-and-Spoke—high-level of internal trade in intermediate goods and services; greater commitment to firm or employer than to community.
- Satellite Platform—little intra-regional trade and low level of commitment to community; often high level of commitment to employer due to the parent company's monopsony power.

From Table 2 we can see that there are many possible types of flexible production systems and employment relationships capable of being created. The first entry within each subcategory best illustrates the prototypical flexible specialization system discussed extensively by Piore and Sabel. Note that there are a specific set of spatial, organizational and distributional requirements for such a system to function effectively. Seen in this way, flexible specialization production processes are actually localized economies premised upon an invigorated and sustainable set of institutional and cultural characteristics that seem at odds with the cultural conditions most favorable to the success of capitalist economic development.

Following, Gibson-Graham's (1996) work examining the spaces and fissures within capitalism available for organizing non-capitalist ways of life, I suggest that flexible specialization processes that encourage and depend upon trust, community, and solidarity be developed in those areas ravaged by and abandoned by the logic of Fordist development—the inner cities and crumbling industrial suburbs of the United States. It is in these spaces where community economic development programs, focusing on technical training, reskilling, and community-building could bring together the key constituents of flexible specialization processes. In doing so, one needs to recognize that post-Fordism can be subject to more than one reading. For example, post-Fordism is usually taken as an example of the further manifestation of the flexibility and fluidity of capitalism. Indeed, some proponents of lean manufacturing and high-performance work organization seem to view post-Fordism as the new organizational fashion worn by this year's chic competitive, profit-driven, efficient capitalist industry. Local development, employment and training initiatives might be inclined to adopt this approach as a means to both align with local business interests and lessen their community's dependence on the location decisions of large employers.

> To activists long dismayed by the destruction of traditional industry, post-Fordism in general and flexible specialization in particular offered an inspiring model of industrial regeneration. Rather than a return to the prosperity of the past, this body of theory promised a new world in which small enterprises could thrive and workers could realize their human capacities instead of emulating machines. . . . By promoting locally based post-Fordist development strategies, they could make their communities less vulnerable to multinational firms that milk and close profitable plants. By fostering "modernization" among small and medium-sized firms, they could increase their chances of survival in a competitive global economy.
>
> (Gibson-Graham 1996: 163)

A second reading of post-Fordist flexible specialization foregrounds the role of industrial districts as helping to constitute a set of economic and social practices that allow for the development of differently capitalist or explicitly non-capitalist organization of work settings. I say "differently capitalist" to distinguish the range of organizational and distributional forms that firms can take within the industrial district model. The dependence of industrial districts on community may well introduce an added, and perhaps destabilizing, dimension to the capitalist production process. For example, Gibson-Graham suggests that community allocation of individual firm surplus could be used to promote new enterprises, some of them less closely tethered to the capitalist imperative to accumulate. "Such a community could increase the presence of noncapitalist economic activity and generate a discourse of the value of class diversity for economic sustainability" (Gibson-Graham 1996: 167). For example, the communal organization of production and consumption appears highly compatible with industrial districts in which trust, cooperation and sharing of resources and market opportunities play a paramount role in sustaining the local and regional economy (Hirst and Zeitlin 1989; Sabel 1989). Gibson-Graham argues that "by virtue of their emphasis on teamwork, shared responsibility, and participation, post-Fordist forms of production may foster the conditions for the emergence of communal class processes in some industrial settings" (1996: 171). This marks an important application of flexible production. It is important in that it helps to create spaces of difference and to promote a policy of economic development and employment that realizes the hegemonic role played by market forces but does not succumb to a belief that hegemony means capitulation to blind market forces beyond our control. In an essay tracing out the possible linkages between flexible production and pragmatic local economic development strategies, Piore argues, "The most striking point of conflict between the [neo-]liberal vision and the new reality is the network organisations and the way in which they depend, not upon autonomous economic actors, but upon the interaction of individual actors embedded in a social network" (1995: 85). The industrial district model provides a vision of a world of work that is linked to a form of community life that promotes an alternative to self-interested behavior and instrumental forms of human interaction.

This is not to argue that the social networks of the industrial districts of the past always present us with worthy models (Sabel and Zeitlin 1985, 1997). There are significant lacunae in the flexible specialization model of small-scale, cooperative, craft-based economic development. The craft-based model offers us a romanticized view of the patriarchal family upon which craft production is centered.[8] The male bias of Piore and Sabel's thesis is evident in the treatment of the existence of sweated labor in new industrial districts in textiles (Piore and Sabel 1984: 264; Harrison 1994: 98) and the role of the family as the primary economic unit in the Third Italy. Rather than building upon patriarchal systems of economic

8 See Arestis and Paliginis (1995) for a discussion of how the economic status of women is neglected in both the Regulation School and the flexible specialization approaches.

organization, I suggest the exploration of ways of extending the basic institutional requirements of flexibility in directions which enhance the employment prospects and neighborhood quality of inner-city and suburban residents within the local and regional economy.

B. Threatened Institutions and Elusive Goals in a Post-Fordist World: Mass Production, Unions, and Retail/Service Employment

This is not to deny that flexible specialization of the type highlighted in this paper is a partial and fragmentary economic arrangement. Piore and Sabel argue that a world economy structured around flexible specialization is more liable to economic fluctuations and employment volatility (Piore and Sabel 1984: 276). Full employment may, they claim, be unattainable within this scenario (Sabel 1987: 51). On the other hand, to the extent that full employment of labor resources through work-sharing within an industrial district can occur, this suggests that cooperative employment relations, enforced through community norms, union bargaining and the enactment and acceptance of new economic rituals can counter the competitive logic governing the capitalist enterprise. So, while full employment "policies" can be established within a district, the macroeconomic logic of full employment remains as much of a problem under a system of post-Fordism as it was during the crisis of Fordism.

Furthermore, I contend that the flexible specialization model of industrial districts cannot, by themselves, determine the trajectory of economic growth. I argue that flexible specialization not be viewed as the new essentialist category upon which to inscribe the developmental logic of capitalist economic growth. Rather, flexible labor arrangements and organizational design practices need to be scrupulously analyzed in relation to the institutional setting in which they are to be enacted. Increased flexibility is not a panacea. Flexibility can be deployed to promote a decentralized form of Fordism, to complement corporate downsizing strategies or to further the process of labor market segmentation by creating islands of secure, high-skilled and well paying jobs surrounded by a sea of contingent, overworked, low-wage and partially skilled or unskilled workers.

The fact remains that employment continues to shift away from manufacturing and employment opportunities for the traditional blue-collar worker continue to erode. The mass production sector is contracting, but it is not evaporating. Fordist production processes remain and they continue to be a key locus of strength for organized labor. For example, a generational shift is currently taking place in the U.S. auto industry as a large number of workers are beginning to retire and a cohort of new workers is hired into the auto assembly and parts divisions. The plants into which these new hires are sent encompass à broad spectrum of production technologies. Some plants, for instance, use high-tech computer-controlled machinery within an assembly-line environment that bears a striking resemblance to the labor process in the auto plants of the 1950s. The particular organizational fix that plant managers and company engineers devise in order to combine flexibility with Fordism is, in many ways, peculiar to each plant. Piore and Sabel argue that the particular institutional prerequisites for the successful establishment of flexible specialization make cross-national comparisons difficult.[9] I would also suggest that what appears to be an extended phase of exploration by mass producers to find the proper degree of integration between the rigidity of mass production and the fungibility of flexible specialization actually represents a process of "customizing" the organization to fit the particular political and economic institutions in place at the plant level.[10]

The shift away from manufacturing threatens the ability of workers to organize and join established labor unions.[11] The role of unions under flexible labor processes is unclear at best. Sabel (1987) suggests that unions take the offensive and devise strategies to capitalize on flexibility by helping to reduce the transaction costs associated with the reliance on flexible labor markets and by controlling the pace and process of technological change on the shop floor. Wells (1996) maintains that unions need to both expand their horizontal influence (organizing across nation-state boundaries, for example) and to facilitate more dynamic forms of communication and decentralized decision-making so as to include rank and file members in helping to coordinate innovations at the level of the work task itself. Additional union strategies to organize workers at small and medium-sized batch production plants into regional councils to pool training and share resources represent examples that connect plant-level issues to local economic development needs (Friedman 1997; Rogers 1997). These efforts would respond to flexible arrangements in plants where a union already exists. They do not address the need to organize the new growth segments of the labor market in retail and service-sector employment.

---

9 Interestingly, Kern and Pichierri (1995) suggest that differences in the social scientific methods dominant in Italy and Germany may help to explain part of the difference in case studies of the dynamics of manufacturing technology in each of those countries.

10 This helps, in part, to explain the diverse results emanating from case studies of lean production in the auto industry (Womack *et al.* 1990; Berggren 1992; Babson 1995a; Deyo 1996; Rinehart *et al.* 1997). This view is echoed by Kochan *et al.*: "Companies are not simply choosing between a Fordist (or neo-Fordist) mass production approach and a Toyota-inspired lean production system. A much broader range of alternatives arise as a result of the interactions of production systems, employment relations systems, the competitive strategies, and managerial traditions of different firms and the roles played by governments and unions" (1997: 323).

11 Similarly, the spatial dimension of industrial and organizational restructuring suggests that the paradigm of industrial union organizing within national boundaries must give way to transnational labor organizing and collective bargaining (Moody 1997, especially Chapter 10).

The extent to which flexible employment practices have penetrated the service and retail sectors appears to be quite limited. In the 1990s some firms in the retail sector sought to increase efficiencies through increased training and responsibility. Furthermore, some evidence exists that national chains are decentralizing buying authority in order to improve responsiveness to regional tastes (Bailey and Bernhardt 1997). Based on their case studies of firms in the retail sector, Bailey and Bernhardt find that "In general, the changes that we observed have created somewhat more interesting and varied jobs in which retail personnel have more discretion, engage in a wider variety of tasks, and have more complex interactions with customers" (1997: 191). Nevertheless, this reskilling was not clearly associated with a marked increase in wages. "Probably the most consistent and troubling finding from these case studies is that despite the workplace innovations we observed, there were few changes in terms of wages and benefits for the sales' staff" (Bailey and Bernhardt 1997: 191). While the average age of retail workers is quite low, it is becoming more likely that retail employment may be a permanent home for some of these young people. The role for unions to adopt an organizing strategy geared to occupation rather than industry or firm makes good sense in the case of retail and service employment (Cobble 1991).

The mass production worker is not gone, but mass production work is itself being redefined and supplemented with new labor processes and forms of labor control that modify traditional relationships between capital and labor. Unions that arose and came to power during an era of hegemonic Fordism are destined to continue to decline in economic power and political influence unless they adopt new organizing strategies and develop new relationships with local communities. The industrial union movement, predicated as it was, on the ability to organize an increasingly homogeneous and deskilled work force, confronts a more diffuse, differently skilled, differently gendered, increasingly service-oriented and spatially decentralized working class than it had become accustomed to representing. To the extent that flexible specialization within industrial districts represents one of the future worlds of work, the boundaries between union organizing, community organizing, tenant organizing, community banking and economic development need to be breached and border crossing skills need to be effectively developed and nurtured.

V. CONCLUSION

Despite the numerous changes confronting workers at the end of the twentieth century, the current era of industrial restructuring and reorganization remains tied to the goal of profitability. The result of such restructuring suggests that there will be many future worlds of work. From labor's perspective, increased attention to difference and to the heterogeneity of workplace experiences is crucial in order to better understand and resist changes that undermine worker bargaining power and threaten to worsen working conditions extant under both mass production and flexible specialization.

I want to suggest that *The Second Industrial Divide* presents a partial attempt to understand the current constellation of technological and sociocultural forces driving the economic transformation of working life for many workers in the global economy. To read the thesis espoused by Piore and Sabel as a blueprint for *the* correct path toward economic transformation or as a flawed analysis rooted in a dualistic framework is to neglect the subtle complexity of economic structures that Piore and Sabel seek to elucidate. In particular, Piore and Sabel argue that the adoption and successful implementation of flexible systems of production is dependent on the particular set of national political policies, history of state–society relations, civil societal traditions, and cultural frameworks for understanding and accepting communal modes of economic provisioning. The nature of work is changing in ways workers and their unions often feel powerless to control. For the foreseeable future and for most people, income and consumption standards will remain tied to access to the capitalist labor market. The bifurcation of the wage and skill distribution for jobs suggests that access does not guarantee a satisfactory way of life for millions of workers. Workers remain doubly free—free from control over capital resources and free to participate in the capitalist market economy. However, by opening up new spaces in order to encourage alternative organizational structures outside of the traditional capitalist labor market, we can help to reenvisage an economy where difference is important and where there is more of a future for workers to help to redefine the meaning and nature of work in the twenty-first century.

REFERENCES

Aglietta, M. (1979) *A Theory of Capitalist Regulation*, London: NLB.

Applebaum, E. and Batt, R. (1994) *The New American Workplace: Transforming Work Systems in the United States*, Ithaca, NY: ILR Press.

Arestis, P. and Palignis, E. (1995) "Fordism, Post-Fordism and Gender," *Economie Appliquée*, 48(1): 89–108.

Atkinson, J. (1987) "Flexibility or Fragmentation? The United Kingdom Labour Market in the Eighties," *Labour and Society*, 12(1): 87–105.

Babson, S. (1995a) *Lean Work: Empowerment and Exploitation in the Global Auto Industry*, Detroit: Wayne State University Press.

——— (1995b) "Lean Production and Labor: Empowerment and Exploitation," in S. Babson (ed.) *Lean Work: Empowerment and Exploitation in the Global Auto Industry*, Detroit: Wayne State University Press: 1–37.

Bailey, T. R. and Bernhardt, A. D. (1997) "In Search of the High-Road in a Low-Wage

Industry," *Politics & Society*, 25(2): 179–201.
Berggren, C. (1992) *Alternatives to Lean Production: Work Organization in the Swedish Auto Industry*, Ithaca, NY: ILR Press.
Berman, E., Bound, J., and Griliches, Z. (1994) "Changes in the Demand for Skilled Labor Within U.S. Manufacturing: Evidence from the Annual Survey of Manufacturers," *Quarterly Journal of Economics*, 109 (May): 367–397.
Biewener, J. (1997) "Downsizing and the New American Workplace," *Review of Radical Political Economics*, 29(4): 1–22.
Bowles, S., Gordon, D. M., and Weisskopf, T. E. (1983) *Beyond the Wasteland*, Garden City, NY: Anchor Books.
Boyer, R. (1988) *The Search for Labor Market Flexibility*, Oxford: Clarendon Press.
Braverman, H. (1974) *Labor and Monopoly Capital*, New York: Monthly Review Press.
Cappelli, P. (1993) "Are Skill Requirements Rising? Evidence from Production and Clerical Jobs," *Industrial and Labor Relations Review*, 46(3): 515–530.
Carrillo, V. Jorge (1995) "Flexible Production in the Auto Sector: Industrial Reorganization at Ford-Mexico," *World Development*, 23(1): 87–101.
Chandler, A., Amatori, F., and Hikino, T. (1997) *Big Business and the Wealth of Nations*, Cambridge: Cambridge University Press.
Cobble, D. (1991) "Organizing the Postindustrial Workforce: Lessons from the History of Waitress Unionism," *Industrial and Labor Relations Review*, 44(3): 419–436.
Deyo, F. C. (1996) *Social Reconstruction of the World Automobile Industry*, London: Macmillan.
Edwards, R. (1979) *Contested Terrain: The Transformation of the Workplace in the Twentieth Century*, New York: Basic Books.
Fine, B. (1995) "From Political Economy to Consumption," in D. Miller (ed.) *Acknowledging Consumption: A Review of New Studies*, London: Routledge: 127–163.
Friedman, S. (1997) "We'll Take the High Road: Unions and Economic Development," *WorkingUSA*, 1(4): 58–66.
Gibson-Graham, J. K. (1996) *The End of Capitalism (as We Knew It)*, Cambridge: Blackwell.
Gintis, H. (1976) "The Nature of Labor Exchange and the Theory of Capitalist Production," *Review of Radical Political Economics*, 8 (Summer): 36–54.
Gittleman, M. and Howell, D. (1995) "Changes in the Structure and Quality of Jobs in the United States: Effects by Race and Gender, 1973–1990," *Industrial and Labor Relations Review*, 48(3): 420–440.
Gordon, D. M. (1996) *Fat and Mean*, New York: Free Press.
Gordon, D. M., Edwards, R., and Reich, M. (1982) *Segmented Work, Divided Workers*, Cambridge: Cambridge University Press.
Graham, L. (1995) "Subaru-Isuzu: Worker Response in a Nonunion Japanese Transplant," in S. Babson (ed.) *Lean Work: Empowerment and Exploitation in the Global Auto Industry*, Detroit: Wayne State University Press: 199–206.
Greenbaum, J. (1979) *In the Name of Efficiency*, Philadelphia: Temple University Press.
Harrison, B. (1994) *Lean and Mean: The Changing Landscape of Corporate Power in the Age of Flexibility*, New York: Basic Books.
Harvey, D. (1989) *The Condition of Postmodernity*, Oxford: Basil Blackwell.
Hirst, P. and Zeitlin, J. (1989) "Flexible Specialisation and the Competitive Failure of UK Manufacturing," *Political Quarterly*, 60 (2): 164–178.

Howell, D. (1995) "Collapsing Wages and Rising Inequality: Has Computerization Shifted the Demand for Skills?" *Challenge*, 38(1): 27–35.

Howell, D., Duncan, M., and Harrison, B. (1998) "Low Wages in the U.S. and High Unemployment in Europe: A Critical Assessment of the Conventional Wisdom," Center for Economic Policy Analysis, Working Paper No. 5, New York: New School for Social Research.

Hudson, Ray (1997) "Regional Features: Industrial Restructuring, New High Volume Production Concepts and Spatial Development Strategies in the New Europe," *Regional Studies*, 31(5): 467–478.

Kern, Horst and Pichierri, A. (1995) "New Production concepts in Germany and Italy," in P. Cressey and B. Jones (eds.) *Work and Employment in Europe: A New Convergence?* London: Routledge: 143–158.

Kochan, T. A., Katz, H. C., and McKersie, R. B. (1986) *The Transformation of American Industrial Relations*, New York: Basic Books.

Kochan, T. A., Lansbury, R. D., and MacDuffie, J. P. (1997) *After Lean Production: Evolving Employment Practices in the World Auto Industry*, Ithaca, NY: ILR Press, Cornell University.

Lewchuk, W. (1993) "Men and Monotony: Fraternalism as a Managerial Strategy at the Ford Motor Company," *Journal of Economic History*, 53(4): 824–856.

Markusen, A. (1997) "Sticky Places in Slippery Space: A Typology of Industrial Districts," *Economic Geography*, 72(3): 293–313.

Marshall, A. (1927) *Industry and Trade*, London: Macmillan.

Marshall, A. (1982) *Principles of Economics*, eighth edition, Philadelphia: Porcupine Press.

Meyer, S. (1981) *The Five Dollar Day: Labor Management and Social Control in the Ford Motor Company, 1908–1921*, Albany, NY: State University of New York Press.

Milkman, R. (1997) *Farewell to the Factory: Auto Workers in the Late Twentieth Century*, Berkeley: University of California Press.

Moody, K. (1997) *Workers in a Lean World: Unions in the International Economy*, London: Verso.

Parker, M. (1993) "Industrial Relations and Shop-Floor Reality: The 'Team Concept' in the Auto Industry," in N. Lichtenstein and J.H. Howell (eds.) *Industrial Democracy in America: The Ambiguous Promise*, New York: Cambridge University Press: 249–273.

Paye, J. (1996) "Policies for a Knowledge-Based Economy, *OECD Observer* 200 (June/July): 4–5.

Pederson, P. O., Sverrisson, A., and Van Dijk, M. P. (1994) *Flexible Specialization: The Dynamics of Small-Scale Industries in the South*, London: Intermediate Technology Publications.

Pfleeger, J. (1996) "U.S. Consumers: Which Jobs are They Creating?" *Monthly Labor Review*, 119(6): 7–17.

Pietrykowski, B. (1994) "Consuming Culture: Postmodernism, Post-Fordism and Economics," *Rethinking Marxism*, 7 (1): 62–80.

——— (1995a) "Gendered Employment in the U.S. Auto Industry: A Case Study of the Ford Motor Co. Phoenix Plant, 1922-1940," *Review of Radical Political Economics* 27(3): 39–48.

——— (1995b) "Fordism at Ford: Spatial Decentralization and Labor Segmentation at the Ford Motor Company, 1920–1950," *Economic Geography* 71(4): 383–401.

Piore, M. (1995) "Local Development on the Progressive Political Agenda," in C. Crouch and D. Marquand (eds.) *Reinventing Collective Action*, Oxford: Blackwell: 79–87.
Piore, M. J. and Sabel, C. F. (1984) *The Second Industrial Divide: Possibilities for Prosperity*, New York: Basic Books.
Osterman, P. (1995) "Skill, Training, and Work Organization in American Establishments," *Industrial Relations*, 34(2): 125–146.
Rinehart, J., Huxley, C., and Robertson, D. (1997) *Just Another Car Factory?* Ithaca, NY: ILR Press, Cornell University.
Rabellotti, R. (1997) *External Economies and Cooperation in Industrial Districts: A Comparison of Italy and Mexico*, London: Macmillan.
Rogers, J. (1997) "The Folks Who Brought You the Weekend," *WorkingUSA*, 1(4): 11–20.
Sabel, C. (1987) "A Fighting Chance: Structural Changes and New Labor Strategies," *International Journal of Political Economy* 17 (3): 26–56.
——— (1989) "Flexible Specialisation and the Re-emergence of Regional Economies," in P. Hirst and J. Zeitlin (eds.) *Reversing Industrial Decline: Industrial Structure and Policy in Britain and Her Competitors*, Oxford: Berg: 17–70.
Sabel, C. and Zeitlin, J. (1985) "Historical Alternatives to Mass Production: Politics, Markets and Technology in Nineteenth-Century Industrialization," *Past and Present*, 108: 133–176.
——— (1997) *World of Possibilities: Flexibility and Mass Production in Western Civilization*, New York: Cambridge University Press.
Scott, A. J. (1988) "Flexible Production systems and Regional Development: The Rise of New Industrial Spaces in North America and Western Europe," *International Review of Urban and Regional Research*, 12 (2): 171–185.
Scranton, P. (1997) *Endless Novelty: Specialty Production and American Industrialization, 1865–1925*, Princeton: Princeton University Press.
Shaiken, H. (1986) *Work Transformed: Automation and Labor in the Computer Age*, Lexington, MA: D.C. Heath.
——— (1994) "Advanced Manufacturing and Mexico: A New International Division of Labor?" *Latin American Research Review*, 29 (2): 39–71.
Smith, V. (1997) "New Forms of Work Organization," *Annual Review of Sociology*, 23: 315–339.
Taylor, F. W. (1967) *The Principles of Scientific Management*, New York: Norton.
Walker, Richard. (1995) "Regulation and Flexible Specialization as Theories of Capitalist Development," in D.C. Perry and H. Liggett (eds.) *Spatial Practices*, Thousand Oaks, CA: Sage: 167–208.
Wells, D. (1996) "New Dimensions for Labor in a Post-Fordist World," in W.C. Green and E.J. Yanarella (eds.) *North American Auto Unions in Crisis*, Albany, NY: SUNY Press: 191–207.
Womack, J. P., Jones D. T., and Roos, D. (1990) *The Machine That Changed the World: The Story of Lean Production*, New York: Harper Collins.
Wood, S. J. (1989) *The Transformation of Work?*, London: Unwin Hyman.
Zuboff, S. (1988) *In the Age of the Smart Machine: The Future of Work and Power*, New York: Basic Books.

[8]

The Transformation of Work Revisited: The Limits of Flexibility in American Manufacturing*

STEVEN P. VALLAS, *Georgia Institute of Technology*

JOHN P. BECK, *Michigan State University*

Students of work processes have increasingly debated the likely emergence of a flexible, "post-Fordist" pattern of work organization within advanced capitalist manufacturing industries. This paper uses qualitative methods to address the debate over post-Fordism, hoping to bring to light certain neglected aspects of shopfloor relations that bear upon the transformation of work. The study focuses on the changes that have occurred at four manufacturing plants in the pulp and paper industry located in diverse regions of the United States. Despite abundant opportunities for the redesign of manual workers' jobs, these mills provide only partial support for post-Fordist theory: The skills demanded of manual workers have indeed risen, as post-Fordists expect, but we find little indication of any expansion of craft discretion, nor of any nascent synthesis of mental and manual labor. Instead, we find trends toward the increasing centrality of process engineers within the labor process, a heightening of the barrier between expert and non-expert labor, and the symbolic devaluation of craft knowledge — in short, strong indications that Fordist principles of work organization have remained quite prevalent. The paper concludes by discussing the social and organizational factors that explain the tenacity of Fordism and that seem likely to obstruct the pursuit of workplace flexibility within U.S. manufacturing.

Introduction

In recent years social scientists concerned with the nature of work have increasingly spoken of an emerging "post-Fordist" pattern of work organization within the industrial capitalist nations. Advocates of this view typically argue that as large-scale shifts have occurred in both product markets and process technologies, large corporations have begun to shed their traditional reliance on centralized bureaucracy and standardized tasks, adopting a new set of work arrangements that answers to many names: "flexible specialization" (Piore and Sabel 1984; Sabel 1991), "the post-hierarchical workplace" (Zuboff 1988), and the "learning organization" (Senker 1992), to cite a few. Regardless of the precise formulation, the general argument has been that the organizational models appropriate to mechanized, mass production processes can no longer suffice in a technologically advanced, post-industrial economy.[1]

* This is a revised version of a paper presented at the 1994 Meetings of the American Sociological Association. The data collected for this paper were gathered with the support of the NSF/ASA Small Grant Program/Fund for the Advancement of the Discipline and the Georgia Tech Foundation. We wish to express our gratitude to both. We especially wish to thank the managers, engineers, and workers whose participation made our research possible; to acknowledge the helpful comments and suggestions made by Laurie Dopkins, Daniel Cornfield, William Form, Jeffrey Leiter, Bob McMath, Roger Penn, Daniel Glenday, Daniel Kleinman, Mark A. Shields, Kathy Michaelis, and several *Social Problems* reviewers. We are, of course, responsible for any remaining flaws. Direct correspondence to Vallas, School of History, Technology, and Society, Georgia Institute of Technology, Atlanta, GA 30332-0345.

1. Note that the present analysis seeks mainly to address the "post-Fordist" literature on flexible specialization, and cannot discuss the very different uses of the flexibility concept at the hands of managerial and industrial relations analysts. For treatments of the latter debates, which have fastened on the need for changes in work rules, pay levels, and employment relationships as a means of reducing labor market rigidities, see Piore 1986; Wood 1989; and Penn,

Such bold assertions have succeeded in winning the ears of managerial personnel, public policy analysts, trade union officials, and others concerned with the changing structure of work. Yet despite its sweeping claims of the obsolescence of mass production techniques, post-Fordist theory remains afflicted by abiding ambiguities and conflicting formulations. Disagreements persist as to which factors seem to drive organizational change, with some theorists stressing factors exogenous to the work organization (such as volatile product demand), while others fasten on factors endogenous to the workplace (mainly, the dynamics of technological change). There is also significant uncertainty regarding the coordinates of the post-Fordist work organization: Some theorists anticipate a rehabilitation of the craft tradition within new economic contexts (Piore and Sabel 1984), while others speak of entirely "new production concepts" that synthesize mental and manual skills (Kern and Schumann 1992). Moreover, few accounts have shown how large, bureaucratic firms can shed their traditionally rigid modes of operation and embrace the new post-Fordist arrangements (see Kelley 1990). Mindful of these and other difficulties we will discuss further, many commentators have remained sharply critical of the degree to which post-Fordist theory can provide an empirically useful guide to workplace transformation in the advanced capitalist world today (Hyman 1989; Penn and Sleightholme 1995; Taplin 1995).

This paper addresses the continuing debate over post-Fordism by exploring the transformation of work at four large manufacturing plants in the U.S. pulp and paper industry, all of which are owned by the same multinational corporation. The study's goal is two-fold: first, to contribute to the larger task of assessing the empirical validity of post-Fordist claims; and second, to open up for discussion aspects of workplace change that advocates of post-Fordism have tended to neglect. Our research strategy is to focus upon a small number of establishments in one industry, in effect trading breadth of analysis for greater depth (see Wilkinson 1983; Child 1972; Child et al. 1984; Thomas 1994). Although this paper involves merely one dimension of a larger research project, it does point toward an important aspect of the current work restructuring that has gone largely unnoticed — the often-conflictual relations between technical experts and manual workers — and that calls for important qualifications within post-Fordist theory as typically construed.[2]

We begin by briefly sketching the coordinates of flexibility theory and describing the research strategy. We then describe the labor process that has historically characterized the pulp and paper industry, together with the key organizational and technological changes that have recently occurred in this branch of production. After analyzing the implications of these changes for both the structure and the culture of mill life, we offer some tentative conclusions regarding the bearing of our research on workplace change in U.S. manufacturing more broadly.

Post-Fordism Re-Examined

While sociologists of work have long studied the relationship between organizations and technologies, the area has gained greater prominence in recent years with the spread of what many refer to as a new scientific-technological revolution, often compared in magnitude to the first industrial revolution unleashed in late 18th-century England. Most analysts agree that massive changes are under way in the organization of manufacturing, but sharp differences remain as to which facets of work are likely to be recast, precisely how the structure of work (especially the management function) is likely to change, and why.

Lilja, and Scattergood 1992. For discussion of the nature and breadth of workplace transformation more generally, see Osterman 1994.

2. Three other lines of inquiry our larger project will address, but that cannot be pursued here, concern the perpetuation of gender segregation among hourly and professional occupations, hourly workers' images of the firm, and the relation between union organization and managerial prerogatives.

Adler (1992a) usefully distinguishes four generations of post-war thinking about the relation between technological change and the structure of work. The first, "upgrading" approach is associated in the United States with the work of Blauner (1964), and in England with that of Joan Woodward (1958). Reflecting broader theories of modernization and the logic of industrial society, this generation of analysts concluded that the automation of manufacturing tended to free workers from highly standardized tasks and to give them a fuller, more integrated view of the work process, thus easing capitalism's endemic problems of alienation and industrial conflict. The second, "deskilling" approach of Braverman (1974) and his followers contested these upgrading claims and in some respects stood them on their head. For this second generation of scholars, the pressures of capital accumulation relentlessly forced employers to simplify the labor of skilled manual and "mental" occupations, using new technologies to place production knowledge into the hands of managerial employees.

The conflict between these two perspectives until recently has occupied much of the existing research, as analysts have sought to adjudicate their competing claims (see Spenner 1983, 1990; Vallas 1990, 1993; Smith 1994b). Yet by the early 1980s, a third generation of theorists had explicitly renounced the search for a single dominant tendency in the evolution of work, arguing instead that "the quest for general trends about the development of skill levels, or general conclusions about the impact of technologies, is likely to be in vain and misleading" (Wood 1989:4). This third, "contingent" view has more modestly sought to identify the social conditions that help account for the varied consequences of technological change (e.g., Cornfield 1987; Kelley 1990; Child et al. 1984; Barley 1986; Form et al. 1988).

Frustrated with this contingent view of technological change, a loose assemblage of theorists (including Adler himself) has sought to articulate the elements of a fourth approach. Despite variations in its precise formulation, proponents of this last school of thought broadly agree on the contemporary emergence and spread of a "post-Taylorist" or post-Fordist model of work (e.g., Sabel and Zeitlin 1985; Piore and Sabel 1984; Sabel 1991; Hirschhorn 1984; Kern and Schumann,1992). Their claims can quickly be recited. During an earlier period of capitalism, machine designs were largely consonant with rigid, standardized job structures that accorded workers little responsibility and demanded of them few skills. Organizational structures under this regime reflected a sharp division between managers and the "managed," as the former sought to rule by command. Steady growth in mass consumption, stabilized by Keynesian economic policies, undergirded the Fordist paradigm. Yet now, dramatic shifts in process technologies and consumer markets have combined to generate a crisis of mass production throughout the advanced capitalist nations. To begin with, information technologies have transformed the structure of production in at least two respects: The spread of microprocessor technologies has made small-batch production more economically feasible than before, while at the same time requiring the use of greater analytic or "intellective" skills for the best use of new machines. Equally important, consumption patterns are more subject to rapid and volatile change, inducing firms to favor more flexible, general-purpose production systems over highly specialized (and therefore rigid) ones. What transpires, say post-Fordists, is a heightened level of skills required of manual workers; more generally, an expansion of craft discretion, presaging a synthesis of "mental" and manual functions within the automated plant; and a broader shift from bureaucratic "control" to organizational "commitment" as the principle that undergirds the new structure of work (R. Walton 1986; Zuboff 1988). Ironically, some theorists contend, the "the project of liberated, fulfilling work, originally interpreted as an *anti*-capitalist project," is now "likely to be staged by capitalist management itself" (Kern and Schumann 1992: 111).[3]

3. Note how this view — that management itself assumes the emancipatory functions once attributed to the socialist project — echoes the argument made decades earlier by Kerr and his colleagues (1960).

Owing to the audacity of its claims and to its ability to engage issues relevant to managerial and public policy officials, post-Fordist theory has attracted a growing audience among managers, trade union officials, and academics in many advanced capitalist societies. Unlike previous approaches toward workplace change, the theory is attuned to shifts in the nature of product markets, the economics of capital equipment, and the institutional and spatial contexts in which production takes place. Moreover, because it lends itself to the generation of new organizational models, it has sustained an ongoing dialogue with the practitioners of workplace reform. These are no small contributions. Yet despite its strengths, the theory has remained vulnerable to sharp criticism and debate. Some critics have chided post-Fordists for their exclusive focus on technologies and markets and their consequent tendency to ignore the conflicting interests and world views that exist within work organizations. The latter influences, it is held, often lend managerial strategies a complex and contradictory nature (Wood 1989; Smith 1994a), heighten the problem of organizational inertia (cf. Hannan and Freeman 1984), and limit the probability of paradigmatic shifts in the structure of work (Gahan 1991; Hyman 1989; Taplin 1995). Other critics, especially those sympathetic with contingency perspectives, have suggested that post-Fordist theory ignores important sources of variation in the outcome of work restructuring, not only within particular societies (see Kelley 1990 and Form et al. 1988) but between them as well (Lane 1988; Maurice, Sellier, and Silvester 1984; Hartmann et al. 1983; Taplin and Winterton 1995). Especially important for our purposes here is that much of the literature on flexibility has been framed at the level of regional or national economic structures, and therefore at a great distance from the workplace realities the theory purports to describe.[4]

The Study

Seeking to draw on Wilkinson's (1983) "action" theory of workplace change and, more recently, Thomas's (1994) "power-process" perspective, we have sought to explore the ways in which the fabric of shopfloor relations bears on the outcome of workplace change in U.S. manufacturing (cf. Jones 1989). By shopfloor relations we mean to include the normative orientations that occupational groups sustain among their ranks, the language they use to describe the production process, and the patterns of conflict and negotiation they bring to bear upon technological change — aspects of work that are not easily studied using quantitative techniques. To anticipate our discussion: While all of the paper mills in our study have taken steps to heighten workers' commitment to the firm's goals, in none of the mills do we find an expansion of craft responsibilities, nor any emerging synthesis of mental and manual functions. Instead, the transformation of work has tended to accord technical experts — formally educated process engineers — a position of greater centrality on the shopfloor, both organizationally and culturally, than before. If our findings prove relevant to continuous-process industries more generally, as we suspect, then advocates of flexibility theory will need to address a potential obstacle to workplace reform — the danger that an increasing reliance upon advanced technology will also favor a technocratic language, logic, and method borne by professional engineers, placing major limits on the development of flexible patterns of work organization.

We conducted an intensive study of a single branch of production: the pulp and paper industry. This branch of production has historically made widespread use of continuous-process work methods, which have figured so prominently in the debate over the evolution of industrial work that one flexibility theorist has termed them "the paradigmatic settings of post-industrial manufacturing" (Hirschhorn 1984:99; cf. Blauner 1964; Nichols and Beynon

4. Some theorists — notably, Zuboff (1988) and Hirschhorn (1984) and Hirschhorn and Mokray (1992) — have grounded their research at the workplace level. Yet in these cases, the evidence in favor of a post-hierarchical turn seems rather more equivocal than flexibility theory suggests.

1977; Halle 1984; and Zuboff 1988). Because the industry relies heavily on workers employed as control room operators called "machine tenders" — highly skilled jobs at the center of production control systems — it provides an especially good terrain on which to apply Kern and Schumann's (1992) expectations about the role of the "systems controller" (a position they believe epitomizes the emerging synthesis of theoretical and practical knowledge).

Equally important, the major corporations in this industry have witnessed a massive wave of organizational and technological changes, beginning in the late 1970s and early '80s, which has only now begun to recede. In less than a single decade, microelectronic controls and mill-wide information systems have transformed this terrain from a traditionally organized craft industry into a major outpost of automated manufacturing. Moreover, as the leading corporations in this industry have embraced the "quality movement" in general, and Total Quality Management (TQM) approaches in particular, new organizational principles have rippled throughout the industry, with many suppliers and contractors now formally constrained to demonstrate their application of TQM principles. Because the industry has been marked by the rapid adoption of these and other process innovations, then, it provides an especially opportune site for research on the transformation of work in U.S. industry (cf. Penn and Scattergood 1988; Penn, Lilja, and Scattergood 1992).

The evidence we present has been gathered from four pulp and paper mills located in different regions of the United States, all of which have been acquired in the last 10 years by the same multinational corporation with a reputation for its relatively cooperative labor relations policies.[5] While not a leader in the implementation of team systems, the firm has sought to incorporate such principles into its labor management approach, providing each mill with resources and support toward this end. Our focus on a relatively progressive, forward-looking firm should provide a research context that is relatively favorable to the newer and more flexible production concepts that are so commonly discussed in the literature.

The selection of mills for inclusion in the study was guided by two considerations: the desire to include a broadly representative mix of plants that vary in their size, age, product mix, and locale; and our interest in plants that have adopted the most recent generation of information technologies and process controls. During the early stages of our fieldwork at a large southeastern mill, we encountered job training seminars that were part of the company's plan to introduce new mill-wide information systems at several of its plants. Inasmuch as we were particularly eager to explore workplace relations within technologically advanced settings, we sought and received research access at three other plants that had recently introduced the new systems (although in a slightly different form). The four mills we ultimately selected are broadly representative of the company's production facilities and employ methods that are parallel to the pulp and paper establishments we have toured within the industry at large. The smallest of the mills in our research employs 700 workers in all positions, while the largest employs in excess of 2,000. Two of the mills are in the Southeast and two are in the Northwest. Two are located near urban areas, while two are in outlying rural areas. The oldest of the four mills was built in 1882 and has been repeatedly expanded and modernized; the other three (including both Southern mills) are more modern establishments, built in the late 1950s. All four mills produce a mixture of consumer products such as tissue and paper towels, combined with commercial products such as photocopying paper and coated paper for magazines. Hourly workers at all four mills are unionized, as is the great majority of workers in the industry.

5. In contrast to pulp and paper firms that have adopted a confrontational labor strategy, the company studied here has instead followed what it calls "the high road" in its dealings with workers and their unions, recognizing the need for labor organization and a cooperative relationship with union representatives.

The remarks reported here are based on two periods of study conducted between 1992 and 1995. An initial phase of the research grew out of an exploratory study of work commitment and job satisfaction at one of the mills; this pilot phase of the research ultimately included semistructured interviews with a strategic sample of 50 managers and workers. An additional wave of data collection was then conducted by the first author, who conducted approximately 200 hours of fieldwork and an additional 65 open-ended interviews with roughly equal proportions of process engineers, manual workers, and plant managers. During both waves of research, the authors were granted free rein to sit in with production crews during their shifts, observing their routine activities and work culture while participating in their ongoing conversations. Interviews were conducted under conditions deemed least inhibiting to respondents: for manual workers, in their control rooms, as the rhythm of production allowed; for engineers and managers, in their offices. Finally, for purposes of comparison, we jointly conducted a small number of interviews and observations at three older mills in New England and the South, to understand the nature of the traditional methods of production that predominated during the manual era of production. In many respects these traditional mills served as living museums, showcasing craft-based tricks of the trade that have all but faded from human memory.

The Case of Pulp and Paper

The Labor Process Before Automation

The pulp and paper mills we studied are massive industrial complexes that daily ingest tons of logs, wood chips, water, and chemicals at one end and spew out truckloads of packaged paper products at the other. The process begins when trucks (sometimes railroad cars or river barges) deliver wood and chemicals to the mill. The wood is mechanically debarked and reduced to chips that are then automatically fed into the mill's digesters. These are vertical towers that use either chemical compounds ("white liquor") or mechanical grinding stones to produce liquid pulp, called brown stock. Workers in chemical processing areas are responsible for producing white liquor, other caustic agents, and bleaching compounds used in the pulping stages of production. Other workers oversee the bleaching, washing, and refining of the brown stock and add any necessary dyes or additives before furnishing the pulp to the appropriate paper machines — huge mechanical complexes in their own right. At the wet end of a paper machine, the pulp moves first to a head box, which evenly disperses it onto flat screens called "wires," which press and heat the stock to reduce its moisture content, until it rolls out as a continuous paper sheet at the dry end of the machine. Workers routinely oversee the accumulation of paper onto large, 10-foot-wide reels that often weigh more than 20 tons each. At this point, the process enters a number of final stages, which may include the coating, supercalendering (receiving a glossy finish), and eventual conversion into a shippable, packaged product by being cut into smaller rolls or sheets. Up until the dry end of the paper machine, the process is essentially a chemical production process. At the dry end, however, where the paper is formed, the process becomes a mechanical one, with more immediate contact between workers, product, and machines.

Each of the production areas we studied was closely dependent upon the others in the mill. As workers at a paper machine know only too well, slight variations in the bleaching and digesting process "upsteam" from their work can and often do have huge effects on the quality of their output and even on the intensity of their work effort: slight variations in fiber length, for example may result in frequent "breaks" on the paper machine, forcing workers to engage in arduous and sometimes dangerous efforts to rethread the machines and bring production back on line.[6] Conversely, variations in the behavior of the paper machines can ripple upward, affecting work in bleaching and digesting areas. Frequently, we watched failures on a paper machine force workers in the bleaching and brown stock area to slow down the chemical reactions they controlled, only to have to speed them back up again once their co-workers brought the paper machines back on line. These changes are difficult and sometimes dangerous to control and can result in major spills and wastage of costly stock or white liquor.

Such tight coupling among different production areas requires frequent communication among workers in different areas of the mill who must learn to anticipate the decisions made by their co-workers in other departments. Despite the spread of portable phones, digital pagers, and radio intercoms, communication across production areas, shifts, and organizational ranks is often difficult. Seemingly minor changes in production methods — for example, adjustments in the specifications for particular grades of paper — are often not effectively transmitted. Intensifying this problem among workers in the mill is the fact that seniority rules and promotion sequences are typically intra-departmental. On the one hand, this rewards workers for the slow and patient accumulation of knowledge concerning their area's processes and machines. But on the other hand it has important social effects: It isolates hourly workers from their counterparts in other production areas and spawns departmental allegiances and identities that limit broader forms of cohesion among workers. Despite claims made by Blauner (1964) and some flexibility theorists, the organization of the production process provides little opportunity for workers to glimpse the production process as a whole.

Until the early 1980s, the consoles of most control rooms were equipped with pneumatic controls that provided workers with only a limited set of readings and process control capabilities. Typically, control rooms also contained panel boards — essentially, wall-mounted maps of each production area located above the pneumatic consoles and equipped with flashing indicators that displayed the status of valves, pumps, and tanks out on the shopfloor. Given the paucity of information these process controls provided, workers had to rely on direct physical or sensate means of collecting information about the process, much as Zuboff (1988) has described. Thus the workers we interviewed recalled spitting tobacco juice onto the winder (to gauge its take-up speed), and using wooden sticks to bang on the logs of the finished product (listening for signs of variation in their product's weight). Several recalled having to run their hands over the stock at the dry end of the mill (too much dust or static electricity told workers that their output was too dry). In pulping areas, workers would often take samples of bleached pulp directly from their dryers, judging the fiber length and acidity of the stock on the basis of its look, feel, and even taste. Reflecting the primacy of sensory knowledge during the manual era, respondents often could not describe older forms of production knowledge without referring to one or more of the human senses: Thus workers and supervisors routinely recalled the workers' need to have a good "feel for the machinery," "an eye for the process," and a "nose for trouble."

To accumulate the repertoire of manual skills they needed to perform their jobs, operators had to serve long years of a *de facto* apprenticeship, moving from jobs as fifth or fourth hands to backtender, and (perhaps) even to the most senior job on a paper machine, the machine tender. During the years needed to move through this career progression, workers amassed a stock of knowledge that became an important form of intellectual property, qualifying them for promotion into more rewarding positions and giving them a critical source of

6. One sometimes hears workers on paper machines say that "you can't turn chicken shit into chicken salad" — a not-so-subtle rebuke of their counterparts in the pulping area. Rivalry between workers in these two production areas has been common in the past, with paper machine workers representing themselves as the more highly skilled group. This will be discussed more fully.

power in relation to their supervisors and fellow workers (Halle 1984; Kusterer 1978). One worker we interviewed recalled a machine tender who stopped work whenever his fourth hand was nearby, explaining that such guarding of work knowledge "was his personal job security program." Even today, some workers carefully record vital information gained while making particular grades of paper, rarely offering to share the content of their "black books" with others in the mill.[7]

Remaking the Labor Process

This portrait of paper making would be familiar to virtually all the workers in the mills we studied, but increasingly as an historical representation. For since the early 1980s, craft control over the production process has experienced a profound transformation, brought about by three distinct but interrelated changes in both the technology and the organization of production: Chronologically, these include, 1) the introduction of Distributed Control Systems (DCS) —automated process controls that interpose symbolic, computer-mediated representations between machine tenders and the production process; 2) the adoption of Total Quality Management principles, which have dramatically altered decision-making methods and procedures within production areas; and 3) the introduction of a mill-wide information system that supplements the automated process controls with an array of analytic and communications tools. In the following pages we briefly describe these changes, and then analyze their effects on the structure and culture of shopfloor life.

Distributed Control Systems. Beginning in the late 1970s, the company began to follow an economic strategy based on the acquisition of undervalued capital assets. By the early 1980s it had accumulated significant numbers of older mills and began to modernize many of its key production facilities with an eye toward achieving greater stability in its operations and appreciable reductions in crew sizes. With these ends in mind, top management elected to introduce new, automated process controls — especially within the larger mills producing consumer goods. The result has involved massive changes in operators' working lives.

Especially in their most advanced incarnations, Distributed Control Systems employ computer terminals to provide visual depictions of the production process, using diagrams, graphs, and maps to represent the functioning of myriad pipes, valves, and pumps operators must oversee. In some respects, it is as if the old panel boards had sprung to electronic life. By using computer keyboards and touchscreen monitors, workers can peer into and adjust the most remote details of the process. Patiently moving through dozens of computerized screens of graphical maps and data, workers track key process variables — the acidity, fiber length, and brightness of the pulp, or the thickness, weight, and tensile strength of the paper — monitoring even minor changes during the course of their shifts. It is true that workers can sometimes put their systems on "cascade" —an automatic state of functioning in which the process controls become self-adjusting, and change in one variable triggers appropriate actions in other process variables. Yet even when workers leave the process on cascade, they must continue to track the state of their operations, lest the machine make inappropriate changes with potentially costly or even disastrous results.

The coming of DCS has granted operators access to a wealth of production data, but it has also placed workers at much greater distance from the process they control. Time on the floor checking the process has been replaced with time in the control room tending DCS controls and video monitors showing key bottlenecks in the process outside. Now, workers directly engage the machines only when they are down — e.g., during scheduled maintenance or when a "break" occurs in the flow of paper coming off the machines. In the latter

7. The term "black book" is commonly used among workers and supervisors, for whom it denotes a sign of commitment to one's craft. For a discussion of similar practices among machinists, see Shaiken 1984.

case, workers must hurriedly intervene, feverishly cooperating to bring production back up. During "normal" periods of their working day, however, workers are now largely isolated from the material objects of their labor and work mainly with symbolic representations of the production process: a development that Zuboff (1988:58-96) has appropriately termed the increasing "abstraction" of industrial work.

Most workers recall having great difficulty in adapting to the new work methods. The new processes intimidated many workers, especially those with less education, and left them fearful of the consequences of a personal failure to master them. In one mill where trade union consciousness is fairly well developed, the introduction of DCS in the mid-1980s prompted workers to slash the tires of an engineer's car and to ostracize the technical implementation team. Workers remain wary even now, checking and double-checking the truth of computer-mediated data — as one worker put it, "to make sure the machine ain't lyin' to you." Workers often experience an enduring tension between two conflicting dispositions toward the machines: *hope* that they are correct (reflecting the workers' concern for the product, and their wish to avoid hard physical labor), and *fear* that the machines (and therefore a process that had relied on their personal skills) have grown beyond their control. Most find that the capacities of the new machines — their ability to enhance the system's stability — gain for the whole system (and therefore for management) a greater control at the expense of individual workers who must accept an end to their craft knowledge in the process. Many workers seem to have reconciled themselves to the loss of individual creativity and discretion in their jobs by focusing on the material benefits of the new technologies, such as the reduction of physical labor. This trade-off is perfectly expressed in one worker's comment that "I'd rather be bored to death than worked to death."

Total Quality Management. As top management began to introduce DCS technology into the company's larger mills, it also set about rethinking its organizational strategies. After having experimented with other innovations, by the mid-1980s it elected to incorporate many of the principles of Total Quality Management into its operations (for discussion see Appelbaum and Batt 1994; Hill 1991; Walton 1986). Viewing TQM against the background of previous management efforts at reform, many of which were both superficial and short-lived, production workers have commonly adopted a cynical view of TQM as the latest "program of the month." Yet TQM seems to comprise more than just another managerial fad. Unlike earlier forms of workplace reform, TQM is directly tied to the technical methods and procedures of the work process itself (Hill 1991). Indeed, subsequent developments in process technology that we will discuss have begun to build TQM principles into their very design.

One of the key elements of TQM is its emphasis on Statistical Process Control (SPC), a heavily quantitative system of interpreting fluctuations in production outcomes that uses probability theory to distinguish between random and systematic variations. Production areas, led by engineers, have established target values (called "centerlines") for each critical process variable and have defined confidence intervals at given distances above and below these centerlines. Managers have directed the workers on each shift to take scheduled readings of key measures, plot the datapoints on control charts, and inspect the resulting patterns of deviation from the centerline value. Workers are trained to alter key process variables only when the observed pattern violates a rule, indicating that a "special" (or non-random) cause of variation has upset the equilibrium of the production process. Management has posted the rules governing centerlining on the walls of most control rooms, and workers who defy the rules must justify their actions. Although workers often comply in a ritualistic fashion (completing control charts only at the end of their shifts, rather than as an active tool during the work itself), the effect of centerlining and SPC has been to standardize production methods, removing a considerable degree of autonomy from workers' hands (Klein 1994).

Most of the managers we interviewed seemed quite aware of the potentially alienating effects that SPC can have and are often at pains to remind workers that they drew on workers' knowledge when defining centerline values. Moreover, in training sessions many technical staff have tried to translate quantitative concepts into more colloquial language that will be less off-putting from the workers' point of view. Thus one engineer liked to use a highway metaphor to explain centerlining and confidence intervals:

> Imagine you're riding down the highway on a motorcycle. If you get too far from the center stripe, you're gonna wind up in the ditch, right?

Regardless of the rhetorical devices managers employ, the reality is that centerlining has constrained workers' customary methods of process control. One departmental manager who had previously been a process engineer observed:

> it's tightened output up around the standard deviation, but it's eliminated a lot of individuality. Workers used to set things the way they liked to, with lots of variability. That's gone now.

In addition to the introduction of SPC methods, the shift toward Total Quality has also involved an increasing reliance on team systems of management, which the company hopes will foster greater cooperation across departments and levels in the organizational hierarchy. Often composed of both workers and supervisory personnel, teams vary widely in the breadth of their mission. One relatively pedestrian example concerned an *ad hoc* team composed of maintenance craftworkers, whom management had charged with designing a system for the storage of technical manuals and documents that are vital to expeditious machine repair. A somewhat more substantively important case was that of a small group of senior craftworkers who were invited to design a training and certification program for highly skilled operators. Regardless of the particular task that teams have been assigned, their effects have been fairly limited: Decisions to authorize the work of each team remain in the hands of departmental managers; the hierarchical chain of command has remained in place; and production standards continue to distinguish between the performance of each bureaucratic unit. In short, despite an overlay of cross-functional teams, the logic of bureaucratic hierarchy has been left largely intact.

Information systems. A last and most recent set of changes began during the late 1980s, when process engineers in the company's technical center developed a generation of information systems that have been introduced into several key plants, including all four of the mills in our study. The most important of these is the Mill-wide Information Network on Economics (MINE).[8] Initially intended as a tool for middle managers to control production costs, it was designed for use on minicomputer systems and incorporated few user-friendly features. Since then, the company's development team has rewritten the system's software, incorporating a graphical user-interface and PC compatibility to make it more accessible to non-technical employees. MINE does not have operational functions: The task of controlling production remains the province of the DCS equipment. Instead, MINE provides a set of communications and analytic functions that even the most advanced DCS controls lack. The new information system provides for on-line bulletin boards, e-mail systems, and shift reports, all of which can now be accessed from any work station in the mill. Moreover, MINE enables workers to "see" into remote production areas — even into areas at mills elsewhere in the country — thus providing information that may have bearing on their own operations. The intent of the designers has clearly been to break down the pattern of departmental isolation that characterizes even the most advanced DCS-equipped production area. In so doing,

8. We have altered the system's real acronym in order to protect the company's identity.

MINE provides a broader, more inclusive overview of the mills' operations, rendering the process more transparent, both to workers and to managers, than was possible before.[9]

MINE also provides an array of programs that support the analysis of data on the causes of downtime, variations in quality, and fluctuations in production costs, most of which make explicit use of SPC terminology. In theory, operators who want to understand why significant deviations from certain centerline values have occurred in the past can access MINE's database and construct bivariate bar charts that show the proportion of such deviations attributable to each cause. Although many mill managers expressed the hope that not only technical personnel but also manual workers would share in the use of MINE, participation has been almost entirely limited to managerial and engineering personnel, for reasons that will be discussed.

The Changing Nature of Shopfloor Life

The changes just described make clear that there has indeed been ample opportunity for the transcendence of Fordist work structures. The profusion of information about the work process, the spread of team systems, and the availability of analytic tools like MINE render post-Fordist work structures more conceivable than ever before. The question to be addressed is how these technological and organizational developments have combined to reconfigure the structure and culture of the shopfloor. The answer that emerges from our research centers on three decisive shifts: the increasing centrality of process engineers within the production process; the redefinition of what constitutes legitimate work knowledge; and the growing standardization of decisions made by non-expert workers, in accordance with quantitative expertise. We discuss these shifts in turn.

Process engineers. The first and most obvious change in shopfloor life is the increasingly decisive role played by process engineers in the day-to-day operations of the mill. The simplest manifestation of this trend is the disproportionate growth in the number of process engineers now directly involved in each mill's production process.[10] This growth partly is due to the hiring of engineers into newly created positions such as that of the "shift engineer" — technical employees assigned to work alongside production crews more closely than before. It also stems from the creation of new technical positions that center on the monitoring of production standards in various departments. A final and perhaps most important source of growth in the ranks of the process engineers has stemmed from their increasing representation in supervisory and second-line management jobs that had traditionally been filled from the ranks of hourly personnel. The experience of one of the mills is representative of this latter development. As recently as 1986, 48 percent of all first- and second-level supervisors at this mill held B.S. degrees in an engineering field; only nine years later this proportion had increased by more than a third, rising to 62 percent. The superintendent of this mill has since declared that *all* supervisory openings will be filled with applicants with engineering degrees. When we inquired into this decision, the mill manager articulated an organizational strategy predicated on the expansion of technically qualified expertise. To hammer his point home he held up an organizational chart with the names of all salaried personnel. The names of degree-holding engineers had been marked with a yellow highlighter, indicating each department's relative strength at a glance.

9. There is a clear resemblance between MINE and the information systems that Zuboff (1988:255-284) describes, as well as the broader phenomenon of panopticism originally described by Foucault (1979:195-231). This will be discussed further.

10. The company has not made personnel statistics available to us. The following data have been pieced together with the cooperation of local mill managers and through corroboration with multiple informants.

The increasingly prominent role played by process engineers has had important effects on the structure of opportunity within the mills, as many manual employees have experienced a narrowing of their promotion prospects. For one thing, supervisors who have come up through the ranks no longer enjoy reasonable chances of promotion into jobs as second-line managers. Because these supervisors find themselves stuck in their present positions, and because technical education is increasingly required for even first-line supervisory jobs, hourly workers too find their opportunities reduced. These changes have erected or (in some cases) solidified a credential barrier between expert and non-expert labor (see Burris 1993).

Legitimate work knowledge. As technically trained personnel have grown more prominent, more subtle changes have occurred in the social context in which manual workers are employed. Especially during day shifts, when salaried employees are most strongly represented, the numbers of engineers sometimes begins to approach those of the hourly personnel, especially when grade changes are introduced. This has brought engineers and hourly workers into much closer contact, as Blauner (1964) anticipated some decades ago; the results, however, have been at odds with what Blauner expected.

Bolstered by their increasing prominence in the mill and by the spread of sensors and other instruments that displace manual functions, engineers have set in motion a process that might be termed an "epistemological revolution" within mill life: an inversion in the criteria that define legitimate knowledge of the work process. Implied here is an overturning of traditional craft methods for generating knowledge about production and their replacement by a newer set based on scientific and engineering discourse — a process that has symbolically devalued craft expertise. In lunch-room discussions or meetings now, one sometimes hears engineers portray workers' knowledge in derisive terms: as either amateurish ("sandlot baseball") or else as a form of superstition, ("black magic" or "voodoo"). The general thrust of such portrayals is that craft knowledge is indicative of a backward, pre-scientific approach toward work that is steeped in the dogma of tradition. As one process engineer told us:

> It drives me *crazy* when operators say you can't control the whole process with the computers. They'll stand there and scrape the stock with their thumbnail, and say they can tell me more about the stock than the $40 million Accuray nuclear instruments we just installed! They're just feeling threatened by us, like all their secrets are being taken away, and they don't like that at all.

This engineer is convinced of the inherent superiority of the new equipment; he rejects the argument that craft knowledge might perceive things that sophisticated measurement instruments cannot detect. In this respect, his views are shared by many of his fellow process engineers. Another engineer said:

> I get frustrated whenever people talk about the "art" of papermaking . . . From my standpoint, it's a lot better if I put another control loop and some calibrated instrumentation on a paper machine, and just put out a memo telling the operators to leave the system on cascade, not to touch it. I'd rather have things that way than depend on a 50-year-old man filling out control charts and applying complex rules by himself.

The dominant view, as expressed here, is that nothing makes it harder to achieve consistency in the quality of output than workers who feel they must change production values on the basis of their own experience. This is precisely why control loops are often installed: to overcome what we have termed the "surplus creativity" of the operators.

In most of the production areas we studied, this shift in mill epistemology unfolded gradually. Yet, in a few cases, the process found expression in a single defining event, as occurred in the brown-stock area of one mill in the South. The following incident was related by a widely respected machine tender whose experience is especially noteworthy, for he has been extremely accommodating toward management during the restructuring of mill operations.

I used to put my hands in and take out a handful of brown stock, squeeze it, tell you whether it was good quality Kraft [pulp], good pine stock or what. I'd tell you about fiber lengths. But then they started bringing in these sensors and machines and whatnot. I still used to put my hands in the dryer, see what I could tell. One day when I did, I said to myself, "that's not right, that's not good quality Kraft." So I told my superintendent. They got the lab testers to work and got the engineers down here. But the lab tests came out OK. So the question was, who they gonna go with? This went on for some time, until finally, this was two years ago, my superintendent said, "J.W., we're gonna go with the lab tests." See, they didn't trust my way of knowin' anymore.

It later turned out that J.W. was right: He had detected an impurity in the white liquor used to digest the wood that had gone unnoticed by the laboratory tests. In this case the result was not serious, but the event was significant in J.W.'s eyes; for to him it symbolized the end of an era. Asked how he felt, he said simply, "I don't put my hands in the brown stock anymore." At this point, he knew that the language of engineering had gained hegemony over its craft equivalent.[11]

Despite these shifts in the definition of legitimate production knowledge, one sometimes hears evidence of a dissenting world view among engineers that breaks with the newly dominant epistemology and seeks to acknowledge the legitimacy of craft knowledge. Such engineers sometimes become defenders of traditional forms of working knowledge within highly automated control rooms. Consider for example the words of one process engineer in his early 20s, born and raised near the Southern mill at which he now worked:

I've been told to idiot-proof things, to lock people out. I've been told to put in [automatic feedback] loops that locked people out. But I'm not comfortable with people sitting back in their chairs, waiting for an alarm.

The same sentiments were put even more clearly by an engineer in one of the Northwestern mills, who explicitly challenged his peers' emphasis on electronic sensors and computer-mediated process controls.

Some of our engineers think everything can be characterized in technical terms, and that if it can't, then it doesn't make sense. They just aren't able to listen to an hourly person talk about paper making. I mean, sometimes I think we get hooked on control rooms. It helps from the process point of view. It's quiet, so we can talk and use the computers to play "what if" games. But it takes you away from the process, isolates you. Out there, you can smell it, hear it, taste it. We have lots of sensors, but there's a lot of things that we don't have sensors for, and some of them can be extremely important for keeping the process on line.

Engineers who subscribe to this latter, more worker-centered view are in the minority. Yet they play an important role in shopfloor relations, often serving to maintain the fabric of trust among occupational groups that would otherwise be at odds.[12]

11. Michel Foucault speaks of "subjugated knowledges," by which he means "a whole set of knowledges that have been disqualified as inadequate to their task [or as] "lying beneath the required level of cognition or scientificity" (1977:82). Referring to much the same phenomenon, Bourdieu (1984:23, 387) speaks of "uncertified cultural capital," by which he means knowledge that has been denied the official stamp of legitimacy, and therefore enjoys limited currency. (See also Bourdieu and Passeron 1977:42). Their arguments regarding knowledge, power, and social inequality have major implications for theory and research on skill. This point must, however, be addressed in another context.

12. It is interesting to speculate on the sources of variation in engineers' conceptions of manual workers' skills. Although our data are far from conclusive, we suspect that a key factor has to do with the engineer's relation to the production locale. That is, engineers who favor placing constraints on manual workers' discretion are often recruited within national labor markets, whether within the firm or beyond. As "cosmopolitans," they often have little in common with the local culture of the mill to which they are assigned, and workers sometimes describe them as merely "writing their resumes" rather than setting down local roots. Engineers who are indigenous to an area ("locals"), by contrast, are more likely to share normative assumptions with their subordinates and to enjoy a "solidarity of place" with them (Newby 1975:157). On the distinction between locals and cosmopolitans, see Gouldner 1957, 1958.

Standardization of decisions. A further way in which shopfloor life has been reconfigured centers on the ways in which analytic functions and decision-making powers have been distributed. Recall that post-Fordist theory expects the process of work restructuring to reallocate a portion of these tasks downward, blurring or even transcending the traditional division between mental and manual labor. We find little evidence of such a trend. Instead, our research indicates that the dominant tendency has involved a pattern of *tightened* constraints upon manual workers' judgment rather than the "*relaxation* of constraints" that flexibility theory foresees (Hirschhorn 1984).

The point is clearly manifest in the workings of Statistical Process Control. In essence, SPC involves the effort to define expected values for key process variables and to formulate detailed rules that govern how workers respond to deviations from centerline values. A key issue here, which managers and engineers have rarely considered in any depth, concerns the process through which centerline values themselves are established. When we asked managers and engineers to explain how particular centerlines were chosen, many referred somewhat vaguely to "collective experimentation" or "inherited wisdom" — as if the target values were a consensual product of shopfloor history. Occasionally, managers would sense a contradiction here, even stopping in mid-sentence to reformulate their thoughts, as in the case of one quality engineer who oversaw tissue production. After he spoke somewhat critically of top management's effort to direct his *own* behavior, we asked him whether he thought that workers ever resented the imposition of centerlines on *their* routines:

> Yeah. And I can understand that. It's the same thing as we're being told [by headquarters] what starch we should use. *We're* telling *them* [hourly workers]... [Pauses] We're not . . . I mean, it can come across that way, that we're telling them how we want them to run the machine, and to a certain point that's true. But actually, if we do our job right, we should be able to explain to them that we're *not* telling them how to run the machines, but rather, this is where we *want* them to run it, and if they can't run it there, they need to understand why they can't and be able to explain that.

Displeased with the imposition of rigid, centralized patterns of authority on his own work situation, this man was reluctant to acknowledge participating in the standardization of craftworkers' jobs. Other respondents spoke more bluntly. Asked if the practice of centerlining made operators feel they were being told how to run their machines, a young engineer replied:

> Sure. That's natural. And you *are* telling them how to run. You're saying, in the collective opinion of the people who are good at running this process, it runs best at X. Therefore, we're gonna run it at X. When we deviate from X, it'll be under experimental, controlled conditions, and *we'll* determine if the deviation is in fact better than the old one. If it is better, we're gonna move to that one. So you *are* in fact telling them how to run.

This statement accords quite closely with our own field observations: The definition of centerline values has typically been defined as a technical question best left to the judgement of formally trained engineers.

There were of course dissenting voices among managers and engineers. Asked whether workers might play a more active role in establishing parameters for the process, some engineers felt that it would be both possible and desirable. Said one business manager with long experience as a mill engineer:

> We just don't . . . We tend to think only technically trained people can understand SPC. We're missing the boat in the formation of centerlines.

Other engineers acknowledge that centerlining was "largely a management program," expressing their regret at this turn of events. But again, such dissent was exceptional. Most engineers felt that workers had sufficient powers of consultation, and that any efforts to involve workers more fully would only magnify the problem of inconsistency and instability (surplus creativity) they had long sought to transcend. A high-ranking engineer in the company's technical center who himself oversaw the design of the new information systems said:

> We've got to get over *to* the boring stage. *We've got to make workers' jobs more boring.* We have to go from the chaos we have now to the stability that makes us money. I don't know how to deal with that [the social consequences of boredom]. Teams, or whatever. But this is where we have to go.

In this view, workers' discretion is equivalent to "chaos." Only a standardized work regime, achieved via the coupling of SPC and automation, can deliver the stability the company needs.

Discussion

It is clear that the structure of work in these mills has been evolving in a direction that departs in certain important respects from the expectations of post-Fordist theory. It is true that, with respect purely to skill requirements, manual workers have indeed encountered a rising set of skill demands, as they have had to learn the use of DCS equipment and to cope with the wealth of process information it makes available. In addition, workers have had to learn when to accept computer-provided directions, and when to intervene. Yet even as workers' *skill* requirements have apparently increased, we find little evidence of any expansion of craft *discretion* or *autonomy*, or any imminent synthesis of mental and manual labor, as post-Fordists predict. Before we can address the larger implications of this finding, however, the question must be asked: Why has the hierarchical, Fordist structure of work proved so tenacious throughout the mills in our research?

Advocates of post-Fordist theory might point to the external market environments of these plants, and suggest that the structure of product demand has limited the development of alternative work structures. The argument here is that since much of the demand for pulp and paper products involves mass consumer markets for towel and tissue products — commodity items most economically produced through standardized work structures — the ingredients necessary for the cultivation of flexibility are largely absent from this case. While this thesis is plausible, we are compelled to reject it on a number of empirical grounds.

To begin, all of the mills in our study are complex, multi-product mills supplying heterogeneous markets that range from mass consumer products (towel and tissue, napkins) to smaller, specialty niches (unbleached coffee filters, specialty coated papers). There was no difference in work organization across mills or production areas oriented toward one or another product type. Smaller, specialty product lines whose paper machines undergo more frequent grade changes were no more disposed to adopt flexible work methods than their counterparts supplying larger, commodity markets. Indeed, of the company's two major product divisions — consumer products and communications paper (which includes such items as coated paper) — support for self-directed teams was appreciably stronger in the former division, which is more fully oriented toward mass production for commodity markets. We therefore doubt that the structure of product markets can suffice to explain the limits of flexibility in these mills.

While our observations are somewhat tentative, we believe that the limits of flexibility can be traced to two alternative influences. First, and at a macrosocial level of analysis, is the nature of the wider culture and society, whose institutional structure confers relatively little legitimacy on craft knowledge and provides relatively few resources for vocational training, certification, or recruitment (Lane 1988). Although we have not interviewed corporate managers, indirect evidence (such as directives expressing their desire to "catch up" with rival firms' technical strengths) indicates that corporate managers view the engineering composition of each mill's personnel as a symbol of its modernity. A second set of influences operates on the shopfloor itself, involving the distribution of resources among engineers and manual workers, affecting which groups are able to benefit from the restructuring of work. The following remarks are confined to the latter, organizational influences.

Recall that manual workers are typically embedded within intra-departmental seniority systems and progression ladders that anchor them within particular production areas. This system, codified in collective bargaining agreements, serves important functions for the company and its workers alike: It encourages workers to accumulate specialized knowledge of their immediate production locale, while it establishes a system of job security that protects the position of skilled workers in particular. At the same time, however, it perpetuates a pattern of local identification and occupational rivalry that often constrains workers' capacity to view the production process as a whole. Thus, one worker on a paper machine had been employed at the same Northwestern mill for more than 30 years, but he had never set foot in the pulping area that furnished his machines with its raw materials — a situation that leads pulp workers to comment that "it's like they think their furnish arrives through a magic pipe in the wall." Moreover, the use of computerized process-controls has meant a shift toward fewer control rooms and sharp reductions in crew sizes, expanding the production areas that workers must oversee, increasing the pressure they feel to keep production on-line. Workers thus find themselves even more tightly confined within their traditionally defined job duties than before. Finally, although many workers are critical of formal, theoretically based knowledge and expertise, which they sometimes view as a tool of self-aggrandizement, they typically lack more than the rudiments of formal education and often seem intimidated by the cultural capital their superiors can wield (Bourdieu and Passeron 1977; Lamont and Lareau 1988). All these factors combine to limit craftworkers' capacity to compete for larger, more autonomous roles within the mills.

It is important to point out that despite the new-found dominance that engineering discourse enjoys, manual workers have nonetheless preserved a residue of informal, experiential knowledge that remains vital to production. As one worker on a paper machine admitted, "the knowledge stays, but in the cracks. We like to keep it hidden" (cf. Halle 1984; cf. Hodson 1991a, b). The question thus arises: If workers' knowledge has in effect been driven underground, can they not use this hidden knowledge as a strategic resource with which to weaken the hegemony of the engineers? At times, the workers we have studied did indeed employ their knowledge as a weapon, typically to retaliate against assaults upon their own dignity. In one representative case, workers suffered a series of insults at the hands of the same engineer. Led by their machine tender, they withheld their knowledge during a production change, rendering the engineer helpless as pulp eventually began to spill out (as one worker recalled), "looking like oatmeal all over the shopfloor." Workers refused to intervene until the engineer had retreated to his office, enabling them publicly to reaffirm their own competence.

Such instances prompt two observations. First and most obvious is the fact that open resistance like this is quite rare, in no small part because it represents a double-edged sword: it involves costly and often laborious disruptions in production, and it can damage workers' own performance records. Second and more important is the fact that even when such resistance occurs, it is almost always aimed at the *"abuse"* of engineering authority — as in the above example, an assault upon their dignity — leaving untouched the engineers' normal claim to superior expertise. We believe that such acts of resistance represent part of a broader process of informal negotiation, through which workers impose certain limits on engineers' interpersonal practices. Such tactics enable workers to exact a modicum of respect from their superiors — in workers' terms, ensuring that engineers "learn to play ball"— but in ways that rarely if ever challenge the hegemony of engineering knowledge itself.

Process engineers, for their part, are far better situated to benefit from the process of workplace change. Although they are also divided by their engineering specialties and their attendant levels of prestige — electrical engineers are "top dog," while civil engineers are merely "the concrete guys"— these differences seem to have little material effect. For the most part, process engineers constitute a relatively cohesive occupational group whose members spontaneously feel a shared sense of mission in relation to the production process. One major reason, we surmise, lies in the structure of mobility established for technical personnel within all of the mills. In contrast to the situation of hourly employees, engineers are expected to progress through a number of distinct positions within far-flung areas of the mill. (A newly-hired engineer might be assigned to a technical group in a given department, to gain hands-on experience in technical support. After a few years, he or she commonly moves into a job as a technical specialist on a paper machine, and then into a position as a first-line supervisor in production. He or she would than be poised to assume a position with substantial authority within an engineering department.) Such mobility patterns make possible the accumulation of knowledge regarding the production process as a whole, which management increasingly values. These patterns also equip engineers with shared organizational experiences, enabling them to sustain a common sense of purpose in relation to the mill as a whole. Coupled with the generalized authority conferred on technical expertise by the wider culture and organizational environment and the traditional presumption that analytic functions are properly lodged in the engineers' hands, these factors quite naturally position engineers to reap the benefits of workplace change (cf. Wilkinson 1983; Child et al. 1984).

Conclusion

This paper has explored the transformation of work under conditions that should be relatively favorable to the flexibility thesis. These mills all employ continuous-process methods of production that make widespread use of programmable control systems and sophisticated information technologies. All four mills are owned by a large multinational corporation with a reputation as an innovative, forward-looking firm. Opportunities for the adoption of flexible, post-Fordist structures are therefore present in abundance. Yet our research has uncovered only partial support for the flexibility thesis. Indeed, the thrust of this paper has drawn attention to a set of obstacles to flexibility — centering on the power of engineering knowledge and its symbolic devaluation of skilled manual labor — that post-Fordist theory has ignored.

We do indeed find some halting developments that lead beyond the rigid bureaucratic structure of industrial organization. For one, management has sought to cultivate workers' normative commitment to the mill, especially through the articulation of cross-functional teams in all of the mills. Moreover, the evidence does indeed suggest that the complexities of new process technologies brought about an increasing level of skill requirements among manual employees, who must learn to master an increasing array of data generated by the computerized machine. This latter point effectively refutes Braverman's (1974:224) characterization of continuous-process work as involving little more complexity than "learning to tell time." Finally, a number of managers have expressed the hope that new methods of skill deployment could take root and would empower production workers as full citizens within the labor process. Yet these tendencies have been overwhelmed by a different set of influences that have imposed a rigid, hierarchical pattern founded on the language of technical expertise. As we have seen, process engineers play an increasingly salient role within mill operations, multiplying in relative numbers and importance, and by implication placing limits on manual workers' career opportunities. The definition of legitimate knowledge at work has shifted in ways that increasingly favor scientific discourse over local or experiential knowledge, rendering craft skill a frequent object of derision among salaried personnel. Finally, as management has introduced Statistical Process Control and other elements of Total Quality Management, work methods have been subject to increasing levels of standardization, as centerline values have come to be defined on the basis of formal expertise. These developments, it seems clear, are not easily reconciled with the concept of post-Fordism. Indeed, there may be some virtue in speaking of the prevalence of important *neo*-Fordist tendencies in these mills, or of a neo-Taylorist search for the "one best way" to run the production process.

To say this is by no means to suggest that a simple, uniform managerial regime has begun to unfold within the mills. To the contrary: The evidence identifies a set of work relations that are fraught with contradictions. On the one hand, mill managers want operators to maintain high levels of commitment to the quality of output, and thus to scan their process controls with the utmost patience. On the other hand, mill managers have embraced a conception of work that places growing limits on craftworkers' traditional discretion and demands that production settings be defined in accordance with the logic of technical expertise. Thus even as they take some halting steps in the direction of greater flexibility, these mills are pulled even more decisively in the direction of a technocratic operational regime.[13]

These findings hold a number of theoretical and practical implications. Perhaps most obviously, our research suggests that flexibility theorists will need to pay closer attention to the micro-political effects that occur when plants place more emphasis upon technical expertise. The neglect of this aspect of work organization in the literature seems to stem from an abiding assumption that scientific and technical knowledge comprises a largely neutral or objective force. It was precisely this assumption that led Blauner (1964) to predict that the increasing frequency of contact between engineers and control room operators would contribute to the integration of plant life, as manual workers derived symbolic rewards by working alongside high-status technical personnel. Yet our research indicates that such contact may have very different effects. Rather than contributing to social *integration*, it may instead prompt organizational *conflict* between rival groups, who are in effect the bearers of competing rationalities. The theoretical point here bears emphasis: Forms of production knowledge are not necessarily innocent; instead, they inevitably harbor certain normative or political assumptions about the organization of work and the distribution of valued responsibilities (see Foucault 1979). In more practical terms, the danger here is that the more emphasis that organizations place on the orthodoxy of technical expertise, the less space there may remain for the cultivation of truly flexible structures of work.

A second point concerns the variables that have been included in workplace flexibility debates. Much of the literature on post-Fordism has implicitly privileged economic structures (e.g., patterns of demand and inter-firm linkages) and technical innovations (skill requirements), viewing these as the prime movers of workplace change. Such a perspective advances our thinking in some respects — notably, by acknowledging the linkage between the organization of work and its market environment. Our research finds that increasing competition and volatility within product markets have reverberated throughout the work process, bringing about an effort to attain greater quality and inter-departmental cooperation than ever before. Yet our research also indicates that an exclusive focus on markets and technologies runs the risk of neglecting other, less tangible but often decisive aspects of shopfloor life, involving formally and informally sanctioned conceptions of skill, knowledge, and expertise. For under circumstances in which craft knowledge has been defined as a sign of backwardness, and definitions of officially sanctioned knowledge shift to the advantage of engineering discourse, flexible work structures seem unlikely to take root *regardless* of the market or technological conditions that obtain.

These observations serve to raise an important question about the bearing of our research on developments within U.S. manufacturing more broadly. It is important to acknowledge the hazards of generalizing from this case study to the pulp and paper industry

13. David Noble (1984) found a similar set of contradictions in his study of the G.E. machine-tool plant at Lynn, Massachusetts.

more broadly, let alone to U.S. manufacturing processes writ large. Moreover, one can easily point to numerous efforts at work restructuring that have gone much further in the direction of flexibility than in the mills studied here (for example, see Adler 1992b; Kanter 1988; Thomas 1994; and the discussion in Appelbaum and Batt 1994). Nonetheless, a number of considerations lead us to expect that "successful" adoption of flexible work practices is in many respects an exceptional phenomenon in the United States; and that the neo-Taylorist tendencies we have observed are likely to predominate within U.S. manufacturing industries.

For one thing, much of the cross-national research on flexibility has suggested that the capacity of manufacturing establishments to transcend the Fordist model varies significantly across societies: Nations that have sustained a strong craft tradition, nourished by educational, industrial relations, and labor market structures that secure the material and symbolic value of skilled manual labor, are especially likely to embrace more flexible models of production (Hartmann et al. 1983; Lane 1988; Lash and Urry 1987; Dankbaar 1988; cf. Piore and Sabel 1984; Jaikumar 1986). By contrast, "countries without a pervasive craft ethos" are likely to adopt "a hybrid strategy, combining a changed market orientation with a high-tech version of production along the old Taylorist lines" (Lane 1988:163). Arguably, efforts to embrace flexible specialization in the United States will tend to yield precisely the sort of "hybrid," contradictory pattern that we have found here. If so — and more directly comparative research, including American workplaces, is needed on this point — the patterns we have identified may better represent the "normal" outcome of workplace restructuring in U.S. manufacturing than the much-celebrated instances of successful flexibility.

Yet if the structural and cultural conditions of workplace change do tend to reproduce strongly hierarchical models of work in the United States, the emergence of neo-Fordist regimes is by no means preordained. Indeed, research that identifies the obstacles to the achievement of flexibility may contribute to the furtherance of organizational change. Comparative research that documents the existence of divergent patterns of skill deployment even under similar economic and industrial conditions may serve to "denaturalize" the division of labor within U.S. plants, stimulating dialogue and debate concerning the advantages of alternative work structures to those established here. Mindful of the tenacity of Fordism in the United States, we nonetheless offer three practical suggestions that might conceivably lead in the direction of greater flexibility.

First, the mills in our study have all applied Statistical Process Control and centerlining practices, which have led to a quantitative-formulaic search for the "one best way" of defining production parameters. The engineers and managers we interviewed largely viewed the firm's highly centralized application of SPC as a simple fact of technical life. The alternative, we heard on numerous occasions, was disorder and instability —the "chaos" to which one designer referred. Yet clearly the application of SPC is more complex than this, with alternatives that few managers seem to have explored. Decisions about centerline values might easily emerge from production teams, on the basis of the collective knowledge and experience of manual employees (cf. Adler 1992b). Reallocation of technical functions downward in the production process, embedding decisions more fully within a system of teams, might well have favorable effects on mill operations (e.g., by reducing downtime or production costs). It might also enable craftworkers to bring the value of their knowledge more prominently into view.

Second, we have found some indications that process engineers who were indigenous to the production locale were more likely to share family, community, and cultural ties with the workers they oversaw, and thus felt a greater sense of allegiance with their subordinates. They seemed more likely to share the same domain assumptions as their manual co-workers, to speak the same language, and to valorize craft expertise. If this pattern holds true then one further way to foster flexible work structures might emphasize the cultivation of indigenous technical expertise. Recruitment strategies that focus on local (and by implication, non-elite) sources of technical knowledge might thereby provide an important precondition for the legitimation of craft expertise.

Finally, our research suggests that the opportunity structures established within the mills constitute a significant impediment to workplace flexibility in at least two respects. First, by reserving supervisory jobs for degreed engineers, mill managers send the implicit signal that manual workers are not competent to wield positions of authority. And second, because of the current seniority arrangements, manual workers tend to remain embedded within job progression sequences that restrict their mobility within the mill and "localize" their knowledge of mill operations. (As one engineer told us, "when you're stuck on the wet end of a paper machine, you don't learn much about the process as a whole.") The danger is that as the need for inter-departmental knowledge grows, manual workers will be left ever further behind. The question, then, is whether alternative forms of mobility within the firm — e.g., formal training programs that lead to craft certification with widened job-bidding rights — can be devised that might broaden workers' knowledge and opportunities without seeming to threaten their job security arrangements. It remains to be seen whether these and other reforms might begin to pry open the technocratic iron cage that decades of U.S. manufacturing have so efficiently forged.

References

Adler, Paul

1988 "Automation, skill and the future of capitalism." Berkeley Journal of Sociology 33:1-36.

1992a "Introduction." In Technology and the Future of Work, ed. Paul Adler. New York: Oxford.

1992b "The 'learning bureaucracy': New United Motor Manufacturing, Inc." Research in Organizational Behavior 15:111-194.

Appelbaum, Eileen, and Rosemary Batt

1994 The New American Workplace: Transforming Work Systems in the United States. Ithaca, N.Y.: ILR Press.

Attewell, Paul

1992 "Skill and occupational changes in U.S. manufacturing." In Technology and the Future of Work, ed. Paul Adler, 46-88. New York: Oxford University Press.

Barley, Steven

1986 "Technology as an occasion for structuring." Administrative Science Quarterly 31:78-108.

Blauner, Robert

1964 Alienation and Freedom: The Factory Worker and His Job. Chicago: University of Chicago Press.

Block, Fred

1990 Post-Industrial Possibilities: A Critique of Economic Discourse. Berkeley: University of California Press.

Bourdieu, Pierre

1984 Distinction: A Social Critique of the Judgement of Taste. Cambridge, Mass.: Harvard University Press.

Bourdieu, Pierre, and Jean-Claude Passeron

1977 Reproduction: In Education, Society and Culture. Thousand Oaks, Calif.: Sage.

Braverman, Harry

1974 Labor and Monopoly Capital: The Degradation of Work in the Twentieth Century. New York: Monthly Review.

Burris, Beverly

1993 Technocracy at Work. Albany, N.Y.: SUNY Press.

Child, John
1972 "Organizational structure, environment and performance: The role of strategic choice." Sociology 6:2-22.

Child, John, R. Loveridge, J. Harvey, and A. Spencer
1984 "Microelectronics and the quality of employment in services." In New Technology and the Future of Work and Skills, ed. P. Marstrand, 163-190. London: Pinter.

Cornfield, Daniel
1987 Workers, Managers and Technological Change: Emerging Patterns of Labor Relations. New York: Plenum.

Dankbaar, Ben
1988 "New production concepts, management strategies and the quality of work." Work, Employment, and Society 2:25-50.

Form, William, Robert L. Kaufman, Toby Parcel, and Michael Wallace
1988 "The impact of technology on work organization and work outcomes: A conceptual framework and research agenda." In Industries, Firms and Jobs: Sociological and Economic Approaches, eds. G. Farkas and P. England, 303-330. New York: Plenum.

Foucault, Michel
1977 Power/Knowledge: Selected Interviews and Other Writings, 1972-1977. New York: Plenum.
1979 Discipline and Punish: The Birth of the Prison. New York: Vintage.

Gahan, Peter
1991 "Forward to the past? The case of 'new production concepts'." Journal of Industrial Relations 33:155-177.

Gouldner, Alvin
1957 "Cosmopolitans and locals: Toward an analysis of latent social roles — I." Administrative Science Quarterly 2:281-306.
1958 "Cosmopolitans and locals: Toward an analysis of latent social roles — II." Administrative Science Quarterly 2:444-480.

Halle, David
1984 America's Working Man. Chicago: University of Chicago Press.

Hannan, M.T., and J.H. Freeman
1984 "Structural inertia and organizational change." American Sociological Review 49:149-164.

Hartmann, G., I. Nicholas, A. Sorge, and M. Warner
1983 "Computerized machine tools, manpower consequences and skill utilization: A study of British and West German manufacturing firms." British Journal of Industrial Relations 21:221-231.

Hill, Stephen J.
1991 "Why quality circles failed, but total quality management might succeed." British Journal of Industrial Relations 29:4.

Hirschhorn, Larry
1984 Beyond Mechanization. Cambridge, Mass.: MIT Press.

Hirschhorn, Larry, and Joan Mokray
1992 "Automation and competency requirements in manufacturing: A case study." In Technology and the Future of Work, ed. Paul Adler, 15-45. New York: Oxford.

Hodson, Randy
1991a "The active worker: Compliance and autonomy at the workplace." Journal of Contemporary Ethnography 20:47-78.
1991b "Workplace behaviors: Good soldiers, smooth operators, and saboteurs." Work and Occupations 18:271-290.

Hyman, Richard
1988 "Flexible specialization: Miracle or myth?" In New Technology and Industrial Relations, eds. R. Hyman and W. Streeck, 48-60. Oxford: Basil Blackwell.

Jaikumar, R.
1986 "Postindustrial manufacturing." Harvard Business Review 64:69-76.

Jones, Bryn
1989 "When certainty fails: Inside the factory of the future." In The Transformation of Work? Skill Flexibility and the Labour Process, ed. S. Wood. London: Unwin and Hyman.

Kanter, Rosabeth Moss
1988 "When a thousand flowers bloom: Structural, collective and social conditions for innovation in organization." Research in Organizational Behavior 10:169-211.

Kelley, Maryellen
1986 "Programmable automation and the skill question: A reinterpretation of the cross-national evidence." Human Systems Management 6:223-241.
1990 "New process technology, job design and work organization: A contingency model." American Sociological Review 55:191-208.

Kern, Horst, and Michael Schumann
1992 "New concepts of production and the emergence of the systems controller." In Technology and the Future of Work, ed. Paul Adler, 111-148. New York: Oxford.

Kerr, Clark et al.
1960 Industrialism and Industrial Man. Cambridge, Mass.: Harvard University Press.

Klein, Janice
1994 "The paradox of quality management: Commitment, ownership, and control." In The Post-Bureaucratic Organization: New Perspectives on Organizational Change, eds. C. Heckscher and A. Donnellon, 178-194. Thousand Oaks Calif.: Sage.

Kusterer, Kenneth
1978 Know-How on the Job: The Important Working Knowledge of Unskilled Workers. Boulder, Colo.: Westview.

Lamont, Michele, and Annette Lareau
1988 "Cultural capital: Allusions, gaps and glissandos in recent theoretical developments." Sociological Theory 6:153-168.

Lane, Christel
1988 "Industrial change in Europe: The pursuit of flexible specialization in Britain and West Germany." Work, Employment and Society 2:141-168.

Lash, Scott, and John Urry
1987 The End of Organized Capitalism. Madison, Wis.: University of Wisconsin Press.

Maurice, M., F. Sellier, and J.J. Silvester
1984 "Rules, contexts and actors: Observations on a comparison between France and Germany." British Journal of Industrial Relations 3:346-364.

Newby, Howard
1975 "The deferential dialectic." Comparative Studies in Society and History 17:139-164.

Nichols, T., and H. Beynon
1977 Working for Capitalism. London: RKP.

Noble, David
1984 Forces of Production: A Social History of Industrial Automation. New York: Knopf.

Osterman, Paul
1994 "How common is workplace transformation and who adopts it?" Industrial and Labor Relations Review 47:173-188.

Penn, Roger, and Hilda Scattergood
1988 "Continuities and change in skilled work: A comparison of five paper manufacturing plants in the UK, Australia and the USA." British Journal of Sociology 39:69-81.

Penn, Roger, Kari Lilja, and Hilda Scattergood
1992 "Flexibility and employment patterns in the contemporary paper industry: A comparative analysis of mills in Britain and Finland." Industrial Relations Journal 23:214-223.

Penn, Roger, and David Sleightholme
1995 "Skilled work in contemporary Europe: A journey into the dark." In Industrial Transformation in Europe: Process and Contexts, eds. E.J. Dittrich, G. Schmitdt, and R. Whitley, 187-202. London: Sage.

Piore, Michael
1986 "Perspectives on labor market flexibility." Industrial Relations 25:146-166.

Piore, Michael, and Charles F. Sabel
1984 The Second Industrial Divide: Possibilities for Prosperity. New York: Basic.
Sabel, Charles
1991 "Moebius-strip organizations and open labor markets: Consequences of the reintegration of conception and execution in a volatile economy." In Social Theory for a Changing Society, eds. P. Bourdieu and J. Coleman 23-53. New York: Russell Sage.
Sabel, Charles, and Jonathan Zeitlin
1985 "Historical alternatives to mass production: Politics, markets and technology in nineteenth century industrialization." Past and Present 108:133-176.
Senker, Peter
1992 "Automation and work in Great Britain." In Technology and the Future of Work, ed. Paul Adler, 89-110. New York: Oxford University Press.
Shaiken, Harley
1984 Work Transformed: Automation and Labor in the Computer Age. New York: Holt, Rinehart.
Smith, Vicki
1994a Institutionalizing flexibility in a service firm: Multiple contingencies and hidden hierarchies." Work and Occupations 21:284-307.
1994b "Braverman's legacy: The labor process tradition at 20." Work and Occupations 21:403-421.
Spenner, Kenneth
1983 "Deciphering Prometheus: Temporal change in the skill level of work." American Sociological Review 48:824-837.
1990 "Skill: Meanings, methods and measures." Work and Occupations 17:399-421.
Taplin, I.
1995 "Flexible production, rigid jobs: Lessons from the clothing industry." Work and Occupations 22:412-438.
Taplin, Ian, and J. Winterton
1995 "New clothes from old techniques: Restructuring and flexibility in the U.S. and U.K. clothing industries." Industrial and Corporate Change 4:615-638.
Thomas, Robert J.
1994 What Machines Can't Do: Politics and Technology in the Industrial Enterprise. Berkeley: University of California Press.
Vallas, Steven
1990 "The concept of skill: A critical review." Work and Occupations 17:379-398.
1993 Power in the Workplace: The Politics of Production at AT&T. Albany: State University of New York Press.
Walton, Mary
1986 The Deming Management Method. New York: Perigee.
Walton, R.E.
1986 "From control to commitment in the workplace." Harvard Business Review 63:77-84.
Wilkinson, Barry
1983 The Shopfloor Politics of New Technology. London: Gower.
Wood, Stephen
1989 "Introduction." In The Transformation of Work? Skill, Flexibility and the Labour Process, ed. Stephen Wood. London: Unwin Hyman.
Woodward, Joan
1958 Management and Technology. London: HMSO.
Zuboff, Shoshana
1988 In the Age of the Smart Machine. New York: Basic.

[9]

Rethinking Post-Fordism: The Meaning of Workplace Flexibility*

STEVEN P. VALLAS
Georgia Institute of Technology

Social scientists increasingly claim that work structures based on the mass production or "Fordist" paradigm have grown obsolete, giving way to a more flexible, "post-Fordist" structure of work. These claims have been much disputed, however, giving rise to a sharply polarized debate over the outcome of workplace restructuring. I seek to reorient the debate by subjecting the post-Fordist approach to theoretical and empirical critique. Several theoretical weaknesses internal to the post-Fordist approach are identified, including its uncertain handling of "power" and "efficiency" as factors that shape work organizations; its failure to acknowledge multiple responses to the crisis of Fordism, several of which seem at odds with the post-Fordist paradigm; and its tendency to neglect the resurgence of economic dualism and disparity within organizations and industries. Review of the empirical literature suggests that, despite scattered support for the post-Fordist approach, important anomalies exist (such as the growing authority of "mental" over manual labor) that post-Fordism seems powerless to explain. In spite of its ample contributions, post-Fordist theory provides a seriously distorted guide to the nature of workplace change in the United States. Two alternative perspectives toward the restructuring of work organizations are sketched—neoinstitutionalist and "flexible accumulation" models—which seem likely to inspire more fruitful lines of research on the disparate patterns currently unfolding within American work organizations.

INTRODUCTION

Since the early 1980s, the concept of "flexibility" has occupied an increasingly central place in social scientific and managerial thinking about work, as analysts have come to view traditionally bureaucratic patterns of workplace hierarchy as all but obsolete. There seems little consensus on the reasons that underlie this putative shift, with theorists variously stressing the rise of global competition, changing patterns of consumer tastes, and the demands of new information technologies. Relatively constant, however, is the belief that contemporary capitalism is undergoing an "epochal redefinition of markets, technologies, and industrial hierarchies" (Sabel 1982:231) leading toward "the displacement of mass production . . . as the dominant technological paradigm of the late twentieth century" (Hirst and Zeitlin 1991:36). Put differently, post-Fordists contend that "the project of liberated, fulfilling work, originally interpreted as an *anti*-capitalist project" is now "likely to be staged by capitalist management itself" (Kern and Schumann 1992:111, emphasis added). Inscribed on the banners of a growing intellectual movement rooted in many

*Direct correspondence to the author at School of History, Technology, and Society, Georgia Institute of Technology, Atlanta GA 30332-0345; email: STEVEN.VALLAS@HTS.GATECH.EDU. Originally presented at the 92nd Annual Meeting of the American Sociological Association, Toronto, 1997. Among those whom I wish to thank for their helpful comments on this paper are Michael Allen, Robert Althauser, Beverly Burris, Craig Calhoun, William Form, George Gonos, Heidi Gottfried, Randy Hodson, Arne Kalleberg, Ian Taplin, and two anonymous reviewers for *Sociological Theory*.

Sociological Theory 17:1 March 1999

countries and disciplines, post-Fordist thinking has provided the elements of a new theoretical orthodoxy that has begun to rival in importance the older Taylorist movement it seeks to eclipse.

As flexibility theorists have gained the ears of powerful decision makers in business, labor, and policy arenas, these would seem to be heady, even revolutionary times for students of workplace change. Yet enthusiasm for the new logic of production has by no means been universal. Rather, a gnawing skepticism has persisted regarding both the nature and the magnitude of the vaunted post-Fordist turn. Some critics contend that post-Fordist theory has imposed a set of simplistic dichotomies on the history of work organizations, without regard for the distortions that are likely to ensue (e.g., see Hyman 1988; Wood 1989). Other critics have taken issue with the theory's one-sided attention to markets and technologies, and its consequent neglect of the normative and ideological influences that shape managerial practices (Berggren 1992; Vallas and Beck 1996; Schoenberger 1997). Still others have questioned the body of evidence commonly used to support the post-Fordist paradigm, which they see as comprising little more than a "methodology of exemplars" (Penn and Sleightholme 1995:199). In light of these and other criticisms, many social scientists have come to view the tales of flexibility as an assemblage of organizational "just-so stories" (Sewell 1995), or as a refurbished variant on the theme of industrial society that surfaced more than a generation ago (Smith 1991).

Furtherance of the flexibility debate has if anything grown more difficult in recent years, apace with the proliferation of multiple variants of flexibility theory (the regulation school, the Atkinson model of the "flexible firm," arguments about post-Fordist society writ large, and so on), all of which lay claim to the mantle of "flexibility" (Wood 1989; Kalleberg forthcoming). Once combined with ambiguities that existed in the original formulations, the result has been a multiplication of accusations and disavowals in lieu of substantive theoretical and empirical exchange (see Hyman 1988, Pollert 1991; Hirst and Zeitlin 1991). The danger here is that despite its potential contributions, the debate over flexibility may collapse beneath its own weight, ultimately coming to be dismissed as yet another intellectual fashion whose product cycle has simply run its course (Pollert 1991).

In this article I seek to place the continuing debate over workplace flexibility on a stronger foundation by engaging in a two-pronged critique of the theory's claims. First, I scrutinize the post-Fordist argument in theoretical terms, identifying a set of silences, ambiguities, and tensions internal to its conceptual apparatus. Then, in a more empirical vein, I sift through recent research that flexibility theory has inspired, to see how far or in what respects the theory's premises find empirical support. I conclude that although post-Fordist theory has made an enduring contribution to the sociology of work, it has misspecified the nature of the workplace changes currently underway. Although support does exist for at least some of the theory's claims—most notably, its emphasis on the rise of interfirm collaborative networks—the bulk of the evidence confronts the theory with a set of empirical anomalies regarding the redesign of jobs and the restructuring of work organizations. Making sense of these anomalies forces us to search for a fuller and more comprehensive explanation of the new work relations than post-Fordist theory can provide. Evaluating alternative approaches, I briefly consider the application of neoinstitutionalist theory to the study of workplace innovation. Even more fruitful is a second approach: the theory of "flexible accumulation" recently advanced by Harvey (1989) and Rubin (1995, 1996; cf. Harrison 1994). The article briefly sketches an agenda for research based on this alternative that might fruitfully inform research on the disparate and often inherently contradictory organizational patterns currently unfolding in the United States.

A POST-FORDIST TURN?

The popularity of flexibility theory has by no means been limited to the United States; rather, in various incarnations the theory has found widespread appeal throughout the developed capitalist world. In France the "regulation school," chiefly associated with the work of Michel Aglietta and Robert Boyer, has developed a nuanced political-economic theory that speaks of the historical emergence of different "regimes of accumulation" and "modes of regulation"; using these concepts, regulation theorists have sought to identify the political and economic conditions that led to the rise of Fordism and now promote the development of either neo-Fordist or post-Fordist regimes in each of the developed nations (for discussion see Hirst and Zeitlin 1991; Amin 1994). In Britain, an intense debate has unfolded concerning Atkinson's theory of the "flexible firm," which makes considerable use of economic segmentation theory and thus foresees an increasing pattern of dualism within firms, industries, and economies (see Atkinson 1985; Wood 1989; Pollert 1991; and Kalleberg forthcoming). The German debate has centered on Kern and Schumann's arguments regarding the emergence of "new production concepts," which holds that employers have embraced an entirely new, less hierarchical conception of labor that begins to reverse the division of labor that has characterized Fordist industry since its inception (Kern and Schumann 1984, 1987, 1989, 1992). Although these contrasting uses of the flexibility concept do share certain common themes—most notably, the contention that the mass production model of routinized work has largely exhausted its economic potential—there are important distinctions among these formulations.

Even within the American context, at least two distinct strains of flexibility theory must be identified. The first can be termed a "posthierarchical" model of work.[1] Implicitly building on the older job redesign literature (McGregor 1960; Hertzberg 1973) and dissident strands of organizational theory (Burns and Stalker 1962), this approach stresses the accumulation of intraorganizational pressures, largely stemming from technological change, that compel firms to adopt new, decidedly more egalitarian structures than Fordism had allowed. According to this variant, the economic survival of technologically advanced firms hinges on their ability to forge new organizational structures that are capable of fully engaging the skills of their employees (Adler 1992a, 1992b; Hirschhorn 1984; Zuboff 1988). The second strain—that of "flexible specialization" theory—is largely congruent with the posthierarchical approach, but typically stresses extraorganizational influences (e.g., the shifting structure of product markets and consumer tastes) as the major sources of workplace change. Flexible specialization theory also places much greater emphasis on the nature of interfirm relations and the governance structures that develop within particular industrial locales. Despite these differences, both variants of the theory share certain common themes. Both portray the contemporary era in terms of historical discontinuity, likening the current restructuring to the massive upheavals that the first industrial revolution produced. Both see the present as holding great potential for the transcendence of the most onerous features of the industrial capitalist labor process. And, finally, both offer prescriptive suggestions regarding the new logic of production that is best adapted to the new economic conditions. Indeed, in these theorists' view, failure to embrace the new logic is likely to expose particular firms, regions, or even national economies to grave economic risk (Heckscher 1994; Zuboff 1988; Hirst and Zeitlin 1991).

[1]The term is found in Zuboff (1988). Somewhat different vocabularies are used, though to much the same effect, by Adler (1988, 1991, 1992), Hirschhorn (1984), Heckscher (1994), and others discussed further below.

The "Posthierarchical" Workplace

The posthierarchical view typically rests on two distinctive themes. The first is its emphasis on the causal efficacy (and, at times, even the emancipatory thrust) of new information technologies. The second is its emphasis on market competition as a mechanism that compels firms to adopt the newer and more efficient, post-Fordist conception of work.

The first of these themes is especially apparent in the work of Hirschhorn (1984). Tracing the history of machine design in manufacturing industries, Hirschhorn finds that the conceptions of work that arise in given historical periods are often implicitly shaped by the production technologies on which they rely. As he puts it (1984:15), industrial technology "always gives its stamp to the character of work," regardless of the social system in which production takes place. In this view, the nature of bureaucratic organizations is at least partly derived from the rigid, fixed-motion constraints that characterized machine design during the rise of industrial capitalism. With the eventual development of programmable machines, however, the bureaucratic structure of work gradually gives way to more flexible, innovative forms. Traditional organizational divisions, such as that between mental and manual work, begin to grow obsolete as new forms of knowledge arise that fuse the tacit skills of manual workers with the theoretical expertise formerly held only by professional employees. The postindustrial workplace therefore begins to place an increased emphasis upon learning and personal development as firms open up greater space for the exercise of judgment and discretion than ever before (Adler and Borys 1996).

Similar predictions were made by an earlier generation of upgrading theorists (e.g., Blauner 1964). What distinguishes the new generation of posthierarchical theorists from their predecessors, however, is an increased awareness of the social and political dynamics that unfold when new technologies are deployed.[2] Such dynamics emerge most clearly in the work of Zuboff (1988), whose research suggests that workplace change often provokes fear on the part of middle managers, who tend to resist what they perceive as threats to their traditional power and authority. In Zuboff's account, such managerial resistance is destined to be short-lived, for it prevents firms from utilizing the full economic potential of the new technologies they deploy. Here she invokes the second analytic theme characteristic of this genre: the belief that the restructuring of work is ultimately shaped by considerations of efficiency, rationality, and market competition, which compel firms to redistribute knowledge and authority more freely than before (1988:387–414). In a globally competitive marketplace, firms that embrace posthierarchical models of work will be able to use new technologies more productively than their competitors. Firms that resist the new forms of work organization will be left behind.[3]

Thus the first strain of post-Fordist theory expects that new process technologies and heightened levels of competition will compel bureaucratically organized firms to overturn the traditional division between mental and manual labor and to embrace new ways of designing workers' tasks. At a broader level, firms are expected to adopt organizational forms that base authority less on formal rank or credentials than on dialogue and "consensual legitimacy" (see Heckscher 1994). The emerging work regime, then, "is not merely *different* from bureaucracy" and Fordism; rather, it represents an "evolutionary develop-

[2]In this respect, the posthierarchical view bears a strong resemblance to the theory of the "new working class" of Mallet and Touraine, which posited a growing tension between new productive forces and old organizational forms.

[3]This emphasis upon competitive needs is especially prominent in the literature on lean production, which cannot be discussed here in any depth. Sympathetic accounts are in Womack, Jones, and Roos (1990), Kenney and Florida (1993), and Sabel (1994); for more critical treatments, see Dohse, Jurgens, and Malsch (1985), Best (1990), Berggren (1992), and Graham (1995).

ment *beyond* it, generating a greater capacity for human accomplishment" (ibid.:14 emphasis in original). Implied here is a view that foresees an increasing alignment of interests among firms, managers, and the workers they employ.

Flexible Specialization

If the posthierarchical variant of flexibility theory represents a reinvigoration of the American tradition of job redesign, the flexible specialization school has a different, more European origin. Partly inspired by Alfred Marshall's conception of industrial districts, and by the new economic regionalism in the "Third Italy" (Brusco 1982), flexible specialization theory aims to provide an historical theory of economic institutions that can avoid the pitfalls of both neoclassical and Marxist economics. Rather than viewing economic structures as the unmediated expression of either efficiency imperatives or the underlying mode of production, the flexible specialization school views them as inextricably bound up with social, political, and ideological influences. This "constructionist" impulse is most clearly apparent in Sabel's *Work and Politics* (1982). Examining the historical transformation of the factory system in varied West European contexts (including the Italian "hot autumn" of 1968-69), Sabel concludes that "within the broad limits imposed by competition in world markets, economic structure is fixed by political choices" (1982:231)—hence the title of his book. This constructionist theme is brought forward, although less consistently, in the broadly influential *The Second Industrial Divide* (Piore and Sabel 1984. See also Sabel 1991; Hirst and Zeitlin 1991; Sabel 1997).

Challenging the *post factum* logic of neoclassical economic theory, which has taught generations of thinkers that craft production is inherently incompatible with the demands of modern economic institutions, Piore and Sabel subject this deterministic conception to systematic critique (1984: ch. 2; Sabel and Zeitlin 1985). In so doing, they call attention to those social and political influences that underlay the triumph of the Fordist paradigm during this first "industrial divide." In fact, they observe, throughout much of the nineteenth century at least two technological paradigms—"craft" and "mass production" conceptions of work—were locked in ongoing competition (Scranton 1983, 1997). The eventual defeat of the craft paradigm was not due to its lesser efficiency; in fact, the craft paradigm provided the basis for vibrant industrial districts that flourished in many cities throughout England, France, Germany, and Italy. Rather, the triumph of Fordism was due to a wholly arbitrary set of influences that were most pervasively found in the United States—most notably, the absence of a guild tradition, the malleability of American tastes, and the scarcity of skilled labor (Sabel, Herrigel, Deeg, and Kazis 1989:384). Once victorious on the American terrain, the Fordist paradigm was quickly buoyed by the nation's rising economic tide, established itself as the reigning symbol of industrial modernity, and eventually provided the ideological lens through which its older craft rival would be viewed. Thus one of the principal objectives of *The Second Industrial Divide* is to denaturalize the Fordist paradigm, to contest its Darwinian claims of supremacy through superiority, and to reveal its nature as an arbitrary sociohistorical construct.

A second objective of their theory is to specify the conditions necessary for the reproduction of Fordism. Their general premise is that all technological paradigms require a regulatory or governance apparatus outside the firm that can secure the conditions needed to balance production and consumption (Hirst and Zeitlin 1991:2–8). The nineteenth-century industrial districts often rested on municipal political structures that provided the basis for interfirm cooperation, helped produce and retain skilled labor, and provided credit and social insurance. Fordism too has required an external system of institutional regulation, typically in the form of welfare states—Keynesian policies and industrial relations systems—that maintained demand for mass-produced consumer goods. As welfare states have increasingly betrayed their limits, however, Piore and Sabel have concluded that the conditions necessary for the reproduction of Fordism have grown increasingly problematic. Wracked by seismic shifts in its economic foundations, then, the developed capitalist world increasingly confronts a "second industrial divide."

At least three factors underlie this development. First, with the internationalization of trade, Keynesian policies can no longer effectively govern Fordist institutions, for policies that maintain aggregate demand within a given nation often only benefit foreign producers. The result saddles domestic firms with massive, vertically integrated structures that are plagued by excess capacity and falling market shares. This problem grows more severe as markets for mass consumer goods approach the point of saturation. Second, the ubiquity of mass production has itself stimulated increasing demand for quality goods through which consumers can express their distance from the "vulgar" world of mass taste (Bourdieu 1984; Brubaker 1985). These cultural dynamics, which are not easily squared with the logic of mass production, begin to alter the ground on which production stands. A third development stems from the widespread diffusion of sophisticated information technologies, which reduces the barriers to entry into production of quality "niche" commodities, rendering small-batch production increasingly profitable. These latter developments provide opportunities for small producers using neocraft models to produce diversified quality goods (Sorge and Streeck 1988).

Given these conditions, both large and small producers have experienced growing pressure to adopt new forms of production, giving rise to a process that Sabel terms "double convergence" (1989; Sabel et al. 1989). On the one hand, small firms have begun to form federated structures that provide them with decisive resources (research and development, training, credit) that were previously enjoyed only by large corporations. On the other hand, their giant competitors have sought to disaggregate their vertically integrated operations, seeking to "recreate among their subsidiaries and subcontractors the collaborative relationships" that smaller firms have historically enjoyed (Hirst and Zeitlin 1991:4). In this way, large and small employers begin to converge on a new technological paradigm—flexible specialization—which provides a powerful yet flexible engine of growth that is optimally suited to the new economic conditions. An important corollary is the belief that the old pattern of dualism between craft and mass production likewise begins to wane as the new paradigm takes root.

The outcome of workplace change is therefore understood at multiple levels of analysis. Since employers face increasingly volatile product markets, they begin to refrain from investing in fixed-purpose capital equipment, and instead rely more heavily on the skills and initiative of skilled hourly employees, whose knowledge becomes critical to the success of diversified production strategies. As Kern and Schumann (1989) put it, "new production concepts" begin to emerge that refashion the job structures of hourly employees, according skilled manual workers much greater centrality within the production process than the old Fordist paradigm allowed. Organizational structures, too, begin to shift, placing ever greater emphasis upon decentralized production units that can more rapidly respond to the flux and uncertainty of market trends. Finally, the boundaries among competing firms begin to blur, as collaborative networks take root in the new industrial districts. Enthusiastic over the possibilities opened up by such trends, Hirst and Zeitlin anticipate the "displacement of mass production by flexible specialization as the dominant technological paradigm of the late twentieth century" (1989:88).

Flexibility under Fire

These two strains of post-Fordist theory have made important contributions to the analysis of work and economic structures. Partly owing to their influence, a greater appreciation has emerged for the ways in which production processes depend on the nature of product markets, shifts in the economics of capital equipment, and the broader structure of governance institutions that surround the firm. Flexibility theory has also identified some of the dilemmas and contradictions that can emerge within organizations faced with rapid technological change. Finally, because these approaches lend themselves to the articulation of alternative strategies for the redesign of work and organizations, they have helped to shape the discourse of workplace innovation, thereby creating novel linkages between academic theory and managerial practices. These are no small contributions. Yet at the same time, important tensions, silences, and ambiguities hamper flexibility theory, impeding its capacity to map the organizational terrain that lies before it. In the following discussion, I identify three theoretical difficulties in the flexibility corpus: the unresolved tension it betrays between "power" and "efficiency" as the sources underlying workplace change; its failure to acknowledge the existence of multiple responses to the crisis of Fordism, several of which seem at odds with the post-Fordist paradigm; and its tendency to neglect the resurgence of labor market segmentation—the "*new* dualism"—which reserves the benefits of workplace flexibility for a small proportion of the workers in a given firm or industry.

The tension between power and efficiency. One of the features that unites the two variants of post-Fordist theory is their shared claim to have found the nascent form of work organization that is best adapted to the current era of economic development. Flexibility theorists do of course acknowledge the existence of managerial strategies that diverge from their prescribed path (e.g., Sabel 1989; Hirst and Zeitlin 1991; Womack, Jones, and Roos 1990). Yet, typically, they view such efforts as a temporary refuge or detour that holds out great risks for those firms, industries, or nations that pursue them. Failing to adopt the post-Fordist model, in short, dooms them to lag behind their more flexibly evolved rivals. The assumption that underlies such a view, then, is that the post-Fordist model (variously construed) provides a fuller and more viable framework for the organization of work than does any rival approach.

It is precisely this assumption that lends the post-Fordist argument its deterministic sweep, its historical power, and its confidence in offering guidelines for regional economic policy. Yet by embracing this assumption, flexibility theorists have run into an important theoretical dilemma concerning the primacy of "power" versus "efficiency" as factors governing the evolution of work organizations (Form 1980; Roy 1997; Vallas 1990).

This problem takes two different forms, in keeping with the two distinct strands of flexibility theory. With respect to the flexible specialization approach, it introduces a contradiction into the theory's logic. When discussing the *first* industrial divide—the triumph of mass over craft production—Piore, Sabel, and their colleagues are at pains to challenge the myth of economic inevitability with which Fordism has cloaked itself. Here the authors rely on a solidly constructionist logic that views the rise of mass production as the product of social and political conditions quite apart from efficiency or economic rationality (Block 1990). However, when Piore and Sabel turn their attention to the *second* industrial divide, they adopt a subtly different logic that views flexible specialization as the single most rational or efficient means of adapting to contemporary economic conditions. The problem here is that two opposing views of economic organization have been incorporated into the same perspective, introducing a fundamental tension into their analysis. Put differently, while Piore and Sabel *begin* by demystifying theological views of mass production, they nonetheless *end* by prophesizing the coming of flexible specialization.

If the flexible specialization variant of post-Fordist theory is torn between the two poles of power and efficiency, the posthierarchical approach (which stresses only the latter pole) manages to achieve much greater internal consistency. This seeming virtue, however, carries a steep price. Because the posthierarchical approach sees efficiency and market competition as the decisive factors shaping the outcome of work redesign, it systematically overlooks the bearing that social and political structures have had on either the origins or transformation of Fordist organizations. Moreover, by stressing market competition as the ultimate determinant of workplace change, the posthierarchical approach embraces a set of assumptions that fly in the face of sociological theory and research on economic institutions. Such scholarship has produced two general conclusions that have important bearing on the explanatory potential of the "efficiency" approach. First, studies conducted within work organizations have often found that economic action is rarely the work of rational, utility-maximizing actors. Typically, such research finds that it is the intraorganizational struggle for power, legitimacy, and the interests of particular occupational groups, and not the pursuit of efficiency, that shapes the character of new process innovations (Wilkinson 1983; Thomas 1994; Schoenberger 1997). Second, studies in the neoinstitutionalist genre of organizational research have consistently reported that the restructuring of work organizations depends more fully on the institutional demands of the wider environment rather than on the exigencies of market competition (see Baron, Dobbin, and Jennings 1986; Fligstein 1987, 1990; Edelman 1990; Sutton et al. 1994; Roy 1997; and below). To suggest that efficiency imperatives require firms to adopt one or another form of workplace organization, then, is to embrace a view that seems both sociologically naïve and empirically indefensible.

The multiplicity of market adaptations. Sabel and his colleagues have typically taken Fordism to mean a technological paradigm that rests on the deployment of unskilled labor, used in conjunction with specialized, single-purpose machinery, for the mass production of commodities. In other words, Fordism is a "low-trust system that separates conception of tasks from their execution: Once the routines are in place, subordinates are meant only to apply them" (Sabel 1982:210). The central claim of the theory is that such highly routinized work structures cannot survive in an era marked by increasingly volatile product markets and sharpening competition, for these conditions place a premium on rapid shifts in production methods that cannot quickly be reduced to routines. Under these conditions, the argument runs, employers have little choice but to adopt new, "high trust" conceptions of labor: To attain the levels of quality and flexibility they need to survive, they must learn to rely on the skills and judgment of the production workers they employ. The problem here is that this reasoning overlooks the *manifold* ways through which flexibility can be achieved, several of which need involve only minor shifts in the organizational logic of Fordist institutions.

This problem has emerged from time to time in the flexible specialization literature in particular. In the earliest formulations of his theory, Sabel (1982) was clearly quite mindful of the existence of multiple forms of market adaptation. In *Work and Politics*, while expressing admiration for the emerging industrial districts he found in Emilia-Romagna, he was careful to note that large corporations might in fact be tempted "to use a combination of innovative technologies and organizational devices to increase the flexibility of production *while holding to a minimum and sharply circumscribing discretion exercised at the workplace*" (1982:211, emphasis added). Yet in later, more fully elaborated versions of the theory, allusions to such possibilities have almost entirely disappeared. The result is a failure to acknowledge that the flexible specialization paradigm is but one among many ways of achieving the flexibility that employers strive to attain. For one thing, firms may in fact adopt flexible *product* innovations in ways that need imply few *process* innovations. This was the very principle that underlay the rise of the modern conglomerate: corporations sought to minimize their risks by developing a varied mix of product lines, each of which relied on Fordist techniques. Such changes provided flexbility with respect to *product* markets, without altering the logic of *production* in the least. The same may well apply to more recent practices such as just-in-time inventory systems, CAD/CAM systems, and Statistical Process Control.[4] Equally important, employers often seek to achieve greater operational flexibility by outsourcing or externalizing functions previously performed in-house (Pfeffer and Baron 1988; Appelbaum 1987; Callaghan and Hartman 1991; Harrison and Kelley 1993)—a pattern sometimes called "external" or "numerical" flexibility, which leaves the firm's hierarchical structures intact (Wood 1989; Sorge and Streeck 1988; Colclough and Tolbert 1992). Once we acknowledge the existence of *multiple* ways of attaining organizational flexibility, many of which seem to perpetuate deeply entrenched patterns, the rise of a post-Fordist conception of work seems much less probable or compelling than the theory allows.

Neglecting the new dualism. Post-Fordist theory typically fastens on stably employed manufacturing workers who enjoy the benefits of standard work arrangements. This relatively narrow focus introduces a third problem: even where the predicted changes in work structures *do* occur, they may in fact be bound up with other, far less sanguine developments that the theory neglects. For example, as firms "reengineer" their functions, they often introduce boundaries between workers whose functions are deemed essential to the firm and those whose functions are deemed expendable (whose jobs are therefore externalized). Since post-Fordist theory focuses one-sidedly on the former group—the "survivors" of the process—it overlooks the very different fate that may befall workers who become its "casualties." Inasmuch as the development of high-trust work relations in some workplaces or departments seems actually to depend on use of externalized or contingent labor elsewhere in the firm or its environment (Magnum, Mayall, and Nelson 1985; Barnett and Miner 1992; Davis-Blake and Uzzi 1993), focusing on a single facet of workplace change risks distorting an inherently complex phenomenon. Just as labor process theory was widely criticized for its reliance on a single, overly simplified model of workplace change (Braverman 1974, Burawoy 1979; cf. Edwards 1979, Burawoy 1985), so too has post-Fordist theory failed to note that the process of workplace change may involve *disparate* or *dual* tendencies whose interconnections must be taken into account. Rather than speaking of a single, unitary logic of post-Fordist work organizations, then, it seems more fruitful to acknowledge that the search for workplace flexibility is often an inherently contradictory, Janus-faced phenomenon, in which "radically different employment policies can be pursued for different groups" of employees (Atkinson 1985:18).[5]

[4]For example, the GM minivan plant in Doraville, Georgia, is able to reconfigure its production lines to adapt to changing levels of demand for its products, converting much of its facility to the production of trucks within less than 24 hours. Yet such product flexibility has virtually no impact on the structure of workers' jobs, which remains wholly Fordist. In such cases, management achieves the needed elements of product flexibility with none of the costs entailed by the redesign of workers' jobs. This phenomenon might best be referred to as "flexible Fordism."

[5]Precisely this point underlies the development of Atkinson's much-debated conception of the "flexible firm" in Great Britain (Atkinson 1985 and NEDO 1986; for discussion see Wood 1989; Penn, Lilja, and Scattergood 1992; Pollert 1991). For discussion of the new dualism, see Harrison 1994. For discussion of downsizing and shifts in the employment relationship, see Capelli 1995 and Capelli, Bassi, Katz, Knoke, Osterman and Useem 1997.

THE EMPIRICAL FOUNDATIONS OF FLEXIBILITY THEORY

Post-Fordist theory posits the existence of a broad historical shift in the organization of work: owing to changed economic conditions, firms can no longer rely on Fordist views of jobs and organizations, and must instead invoke new conceptions of labor, new patterns of organizational structure, and new relations with suppliers and subcontractors. From this general reasoning it is possible to glean three rough propositions concerning workplace institutions. First, at the level of job redesign, the theory leads us to expect that as firms grow more exposed to the new economic conditions—product and market flux, global competition, rapid technological change—they must replace routinized forms of production with autonomous, "high trust" models, thereby blurring the traditional division between mental and manual labor. A second proposition expects that the new economic conditions prompt firms to decentralize the structure of their operations and to embrace consensual forms of decision making in lieu of the traditional rule by command. To these two propositions we can add a third concerning interfirm relations, drawn largely from flexible specialization theory. Here the prediction is that firms confronting intensified market flux and uncertainty will progressively abandon vertically integrated structures of market control and instead participate in collaborative interfirm networks, whether via industrial districts, joint ventures or strategic alliances, or other patterns of interfirm cooperation.

Assessing the validity of these propositions is made difficult by the persistence of important methodological limitations in the existing research. For one thing, there have been abiding inconsistencies in the units of analysis that researchers have employed, with studies variously focused on the occupation, workplace, firm, industry, or region as the unit of analysis. Surveys of workplaces or firms have proved helpful in estimating the prevalence of the new work regimes, but have often been conducted at a considerable distance from the workplaces they seek to describe. Ethnographic analysts have succeeded in providing fine-grained analysis of work relations, but have failed to articulate clear criteria that must be met in order to satisfy the theory's expectations. Additionally, the great bulk of the literature has focused on industrial work processes; research on the transformation of service industries and occupations remains extremely rare (see Smith 1990, 1994; Manley, n.d.; and below). A final issue has been the difficulty inherent in drawing adequate inferences regarding changes that by their very nature are sometimes still in a formative stage (Heckscher 1994). Despite these difficulties, a small but substantial body of scholarly research has developed that enables us tentatively to assess the empirical validity of the propositions outlined above.

The following review is based on social scientific literature published since 1985 that directly engages the validity of post-Fordist theory as applied to workplaces in the United States (cf. Hodson 1996). Studies published before 1985 have been excluded on the assumption that managerial responses to the changed economic context were too protean to hold direct relevance for our purposes here. Also excluded have been studies that seem mainly normative in their concerns (e.g., managerial discussions of "best practice" firms), as well as studies whose focus is primarily on the effects of workplace change on firm performance. Table 1 identifies the twenty-four studies that remain and that bear on one or more of the propositions sketched above. My analysis reviews studies indicating changes at the level of the job, the organization, and interfirm ties, in that order.

The Restructuring of Jobs and Organizations

A sizable body of survey evidence has now been published that seeks to estimate both the prevalence of work redesign and the conditions that give rise to it. In the light of this

Table 1. Summary of Studies Bearing on Change in Workplace Institutions in the United States since 1985.

Study	Research Methods	Substantive Findings	Implications
Job-Level Changes			
Lawler, Mohrman, and Ledford 1992, 1995	Surveys of Fortune 1000 firms, 1987–93.	Consistent growth in EI and TQM programs, especially the latter	Mixed; corporate systems of control are rapidly changing, but unclear precisely how.
Osterman 1994	Cross-sectional survey of work establishments.	"Transformed" work organizations have grown more widely prevalent, esp. under pressure of competition.	Supportive; corporate experimentation with workplace transformation seems widespread.
Adler 1992a	In-depth interviews of NUMMI employees.	Toyota management has overcome adverse reactions to assembly line, even in a formerly hostile American plant.	Supportive; coercive aspects of Taylorism here have been overcome.
Zuboff 1988	Multifirm case studies in telephone, banking, insurance, paper, and pharmaceutical industries.	Information technologies are most effectively used in conjunction with posthierarchical work structures.	Mixed; rigid, centralized uses of new technology persist despite the greater performance of alternative models.
Thomas 1994	Multifirm case studies in aircraft and computer manufacturing, aluminum casting, and auto parts production.	The process of technological change is shaped by intrafirm conflicts, with varying effects on hourly workers' jobs.	Mixed; except under crisis conditions, intrafirm conflicts seem to reproduce the Fordist structure.
Shaiken, Herzenberg, and Kuhn 1986	Case studies of machine tool and metalwork plants.	Managers view neocraft work structures as threats to the quality and intensity of work.	Negative; only scattered evidence is found of the neocraft paradigm.
Taplin 1995	Interviews with plant managers in North Carolina apparel industry.	New technology and competition have led to labor intensification and adoption of neo-Fordist work structures.	Negative; workplace change has enhanced the position of firms but not their workers.
Smith 1994	Interviews with managers and workers of "Reproco," a national business service firm.	Flexible management practices incorporate contradictory themes, fusing participation with use of contingent labor.	Mixed; flexibility viewed as fundamentally Janus faced.

continued overleaf

Smith 1996	Interviews with workers of "Repro-co."	Low status workers embrace team systems as a means of acquiring valued skills.	Supportive; workers seem receptive to participative systems.
Taplin and Winterton 1995	Interviews, official statistics on North Carolina apparel industry.	Apparel firms are "squeezed" by textile suppliers, large buyers; hence seek to intensify labor.	Negative; little evidence of neo-craft paradigm or collaborative networks.
Vallas and Beck 1996	Interviews, ethnography at four pulp and paper mills in Southeast and Northwest, all using TQM and SPC.	Greater power of engineers enables them to shape technological change, defining craft knowledge as backward and prescientific.	Negative; skill requirements increase, but discretion of manual workers declines.
Hodson 1988	Interviews with workers and managers at 22 high tech firms in a southwestern city.	Skills increase, but sharp segmentation persists between mental and manual labor.	Negative; participation is limited to professional employees, despite rhetoric of involvement.
Grenier 1988	Participant observation at medical products plant owned by division of Johnson and Johnson.	Managers used quality circles as potent system of control, limiting workers' union capacities.	Negative; dismantling of bureaucracy deepens management control over labor.
Graham 1995	Participant-observation at a Japanese auto transplant.	Japanese work relations impose a new form of hegemony without altering Fordist work structures.	Negative; workplace change is largely a rhetorical mechanism of social control.
Barker 1993	Ethnography at small southwestern electronics firm.	Semiautonomous teams increase workers' obligations to the firm, establishing "concertive control" over labor.	Negative; worker discretion seems to increase, but within coordinates established by management.
Changes in Organizational Structure			
Prechel 1994	Interviews of managers of steel corporation.	Corporate management has moved to standardize managerial decision making at the plant level, using TQM and SPC.	Negative; trend is toward centralization of control.
Smith 1990	Interviews, ethnography at "American Security Bank" in California.	Top management moves to centralize control over various branches, using entrepreneurial rhetoric.	Negative; language of decentralization imposes regime of "coercive autonomy."

(Table continues)

Table 1. *Continued*

Study	Research Methods	Substantive Findings	Implications
Vallas 1993	Survey, fieldwork at Bell operating company.	Uneven increase in skill requirements, but trend is toward centralization of control.	Negative; professional and technical workers establish "algorithmic" control of work.
Kelley 1990	Cross-sectional survey of 1,015 plants in metalworking and machine tool industries.	Neocraft work structures are rare in large, unionized plants owned by multiplant firms.	Mixed; Fordist structures seem impervious to change in all but small establishments.
Manley, n.d.	Interviews and observations in health care organizations implementing TQM.	Mid-level administrators use TQM to standardize health care delivery, reducing the discretion of all but elite professionals.	Negative; polarization of health care employees unfolds.
From Firms to Networks			
Powell, Koput, and Smith-Doerr 1996	Secondary analysis of data on joint ventures and other ties among biotechnology firms.	Embeddedness in relational networks predicts subsequent performance of firm.	Positive; knowledge-intensive firms adopt the network form in lieu of vertical integration.
Uzzi, 1996	Ethnographic interviews with quality apparel firms in New York City.	"Embedded" ties are more vital to firm survival than "arms length" (market) ties.	Positive; firms located in volatile product markets depend upon collaborative networks.
Uzzi, 1997	Same, plus secondary analysis of economic ties among firms.	Embeddedness predicts subsequent performance of firms.	Positive; collaborative networks displace market ties.
Saxenian 1994	Interviews of managers in firms located in Silicon Valley and Route 128.	Collaborative network structure explains enduring performance advantage of Silicon Valley firms over their Massachusetts rivals.	Positive; collaborative networks encourage freer and more rapid flow of information and strategic personnel.

evidence, there seems little room for doubt concerning the increasing diffusion of corporate efforts to restructure work. The most revealing data on this point are the surveys of Fortune 1000 firms initiated by the Government Accounting Office in 1987 and then repeated in 1990 and 1993 by Lawler and his colleagues (Lawler, Mohrman, and Ledford 1992, 1995; see Appelbaum and Batt 1994; cf. Osterman 1994). These data indicate that such "employee involvement" (EI) practices as job redesign and enrichment, self-directed teams, quality circles, job rotation, and Total Quality Management (TQM) all found growing application within the largest American corporations during the late 1980s, and continued to gain favor even during the recessionary economic pressures of the early 1990s (Appelbaum and Batt 1994:63–64).

According to Lawler and his colleagues, in 1987 only 23 percent of all Fortune 1,000 firms included as many as 20 percent of their employees within any form of EI; six years later, that proportion had reached 43 percent. Osterman's study (1994), which used a broader sample of all U.S. workplaces with at least fifty employees, used somewhat different methods to assess the diffusion of "transformed" organizations. According to his data, by 1992, 37 percent of all private-sector establishments had involved at least 50 percent of their core employees in two or more forms of EI. Clearly, such surveys begin to suggest that *something* is happening to the traditionally Fordist structure of U.S. work organizations, for there does seem to be a widening pattern of interest in more participatory work structures among the largest corporations. Equally suggestive, both the Lawler and Osterman studies find that the pursuit of employee involvement is especially common among firms that have experienced sharp foreign competition and that produce goods for volatile product markets, precisely as both versions of flexibility theory predict (Lawler, Mohrman, and Ledford 1995: 87-91; Osterman 1994).[6]

Although helpful in estimating the prevalence of workplace change, these surveys provide few clues regarding the nature of the emerging job designs or the processes that gave rise to them. The remaining studies outlined in Table 1—especially those using ethnographic methods—are more helpful on this score. The thrust of these studies, it seems clear, provides relatively meager support for post-Fordist predictions regarding the redesign of workers' jobs.

Although much of this research reports an increasing level of conceptual skill requirements, only four of these case studies—those by Adler, Zuboff, Thomas, and to some extent Smith (1996)—find any appreciable increases in the level of discretion or control that workers exercise on their jobs, and even here the evidence is often weak or inconsistent. Adler's (1992a) study of the celebrated NUMMI plant, in which GM's Fremont, California, assembly plant was completely resurrected by the introduction of Japanese managerial practices, does support the first proposition: here, management virtually dismantled its industrial engineering staff and delegated its functions of task analysis and job design to production workers themselves, thereby shifting the locus of control to the shop floor. A similar pattern emerged in the studies presented by Zuboff (1988), however unevenly. Although several of her cases stand at odds with flexibility theory, in some—most clearly, in three highly automated pulp and paper mills—workers were assigned analytic responsibilities that had previously been reserved for middle management, much as post-Fordist theory expects. This case, which is especially relevant to the analysis, eventually proved

[6]The Osterman study provides the most generous estimation of the prevalence of workplace transformation available today. This may be due to limitations and ambiguities in the items this study used to measure particular forms of workplace change. The item measuring the existence of self-directed teams, for example, may well be construed as including traditional craft control under its purview, thereby inflating the proportion of workplaces deemed "transformed." For discussion see Appelbaum and Batt (1994).

abortive, as middle managers acted to reassert their bureaucratic authority, subverting the posthierarchical form before it could take root.

The studies by Thomas (1994) also offer a mixed verdict. Reporting on six instances of technological change, Thomas finds four that led to the reproduction of the Fordist pattern. In these cases, rival groups of engineers competed for status and autonomy, with each seeking to gain favor with corporate managers. The effect of this competition was the continued exclusion of manual workers from the decision-making process. In two of Thomas's cases, however, workplace change did face in the direction post-Fordist theory expects. In these cases, management fostered a "collaborative" pattern of job redesign that allowed manual workers full participation in the formulation of production strategy. (Interestingly, both of these cases involved dramatic crises in corporate control, much as was true in Adler's study of NUMMI.) Finally, in her study of a large business services firm she calls Reproco, Smith (1996) finds that employee involvement initiatives generally succeeded in winning support from even low-status workers, who viewed them as providing important career-enhancing skills (cf. Smith 1994).

Over against these four partially supportive studies, Table 1 reports nine accounts—those by Shaiken, Herzenberg, and Kuhn; Kelley; Hodson; Vallas and Beck; Taplin; Taplin and Winterton; Grenier; Graham; and Barker—that directly contradict the hypothesized trend toward greater job discretion. These studies brought varying research methods to bear on a wide range of manufacturing industries, including machine tools, apparel, pulp and paper, medical supplies, auto assembly, and electronics. The study by Shaiken, Herzenberg, and Kuhn (1986) explored the uses of Computerized Numerical Control and flexible manufacturing systems at thirteen different metalworking and machine tool plants. Few managers were receptive to the idea of expanding workers' discretion, viewing it as threatening both the quality of production and the intensity of the work effort. Kelley's (1990) survey of 1,015 manufacturing plants in the machine tool industry reports findings that are largely congruent with this pattern: large, multiplant firms provided especially unfavorable contexts for neocraft patterns of technology deployment. Hodson (1988) interviewed managers and workers at twenty-two high tech firms in a single southwestern city, and found evidence of increasing skill requirements, but little evidence of any post-Fordist pattern of job redesign. Team systems were in use, but mainly involved professional employees. A pattern of sharp segmentation existed between mental and manual workers, with managers viewing the latter group as a production factor to be controlled, not consulted.

The tenacity of hierarchical patterns also emerged in the study by Vallas and Beck (1996), who conducted ethnographic research at four large, highly automated pulp and paper mills. Although plant management had the opportunity to redistribute knowledge and control over the production process, few such changes in job design were actually made, largely because process engineers wielded organizational resources that helped them shape the restructuring of work in accordance with their own preferences. Succeeding in labeling craft workers' methods as symbols of a backward and unscientific disposition toward production, engineers were able to claim analytic functions and responsibilities as "naturally" belonging to them. They therefore came to enjoy even greater authority over production than before. This study begins to point toward the need for iterative models of workplace change, wherein the unequal distribution of material and symbolic resources under Fordism itself conditions the outcome of organizational change.

The studies of apparel manufacturing by Taplin (1995) and Taplin and Winterton (1995) are of particular interest in that they target an industry in which small firms must continually adapt to rapid changes in consumer tastes. Interviews revealed, however, that apparel firm managers experienced severe pressures at the hands of their suppliers (large textile firms) and their customers (department store chains), inducing them to use new technologies in ways that intensified the work process and diluted workers' skills. These trends emerged even within plants producing girls' and women's wear, the most volatile product markets in the industry.

The implementation of self-directed work teams has figured prominently in theories of workplace flexibility (Sabel 1994; Adler 1992a; cf. Womack, Jones, and Roos 1990; Kenney and Florida 1993), yet the separate studies by Grenier, Graham, and Barker, suggest that teams may implicitly enable firms to maintain hierarchical patterns of authority, if in subtle and distinctly nonbureaucratic ways. The study by Grenier (1988), conducted in an Albuquerque medical supply plant owned by Johnson and Johnson, found that quality circles in effect transmuted *vertical* conflicts between management and workers into *horizontal* conflicts among the workers themselves. In this respect, they proved particularly effective in management's struggle against the unionization of its employees. Graham's (1995) study of the Subaru-Isuzu transplant in Indiana found that team systems served mainly as a social control mechanism rather than as an instrument for the expansion of worker autonomy. Despite the circulation of much rhetoric concerning *kaizen* and the autonomous role of teams, few changes occurred in the traditional pattern of unilateral management control. Finally, Barker's (1993) study of a small electronics plant found that self-directed work teams tended to impose a regime of "concertive control" that actually extended managerial regulation of the labor process, chiefly by inducing workers to internalize their obligations to management and even to police one another's behavior (cf. Burawoy 1979). Although much more fully developed research is needed to understand the varied outcomes of team systems under different social and organizational contexts (cf. Appelbaum and Batt 1994), the available studies indicate that team systems often play a rather different role than flexibility advocates predict, especially in the absence of trade union organization (see Dohse, Jurgens, and Malsch 1985; Fantasia, Clawson, and Graham, 1988).

Changes in Organizational Structure

Studies directly bearing on the validity of the second post-Fordist proposition, concerning organization-level shifts in the distribution of control, are also outlined in Table 1. Although the evidence is less well developed here than with respect to the structure of workers' jobs, five studies exist that use varying methods and yet arrive at roughly the same conclusion. None provide clear indication of any decentralization of organizational structure or increased reliance on consensual decision-making. Indeed, although the evidence remains fragmentary, there is some suggestion of a trend toward *increased* centralization, as found in the studies by Prechel (1994) and by Smith (1990) in particular.

Exploring the transformation of the "managerial process" in a large steel corporation, Prechel found that the introduction of Total Quality Management programs served to tighten corporate control over operational methods and procedures that had previously been left to the discretion of plant managers. Smith's ethnographic account of new management practices within a California bank documents a shift from a decentralized system of branch management to much more centralized oversight of management practices (a contradictory system she dubs "coercive autonomy"). Interestingly, although corporate management's rhetoric was laced with antibureaucractic themes of entrepreneurialism and empowerment, in practice the new managerial system acted to constrain the choices that middle managers could make. In his study of a Bell operating company following the break-up of AT&T, Vallas (1993) found an emphasis on the use of "algorithmic" modes of control that likewise led toward a heightened centralization of control over each department's work (cf. Keefe and Batt 1997). Although some craft occupations enjoyed increases in the conceptual content of their jobs, the overall organizational structure took on a more tightly regulated cast. Kelley's (1990) study of machine tool plants, mentioned in the previous section, also has bearing here, insofar as it provides little indication of any "disaggregation" effect in large corporate organizations. Smaller and single plant firms *do* seem relatively open to flexible uses of programmable automation; but as Kelley makes clear, such firms lack the organizational, financial, and technological resources needed to play leading roles in the reshaping of managerial practices (Harrison 1994).[7]

The literature further suggests that some of the most influential management innovations—especially, Total Quality approaches in their various incarnations—seem to perpetuate the presence of centralized organizational structures (Klein 1994; Appelbaum and Batt 1994. Cf Hackman and Wageman 1995; Hill 1991). This point emerges in the studies by Prechel and by Vallas and Beck, where Statistical Process Control systems displaced the traditional knowledge of middle managers (Prechel) and craft employees (Vallas and Beck). Likewise, Manley's (n.d.) study of the organization of professional work in hospitals reports that after the introduction of Total Quality methods, middle-level administrators enjoyed greater control over the delivery of health care services, especially those provided by the subaltern professions.

Although much more social scientific research is needed on this point, the evidence suggests that Total Quality programs represent rather more than a passing fad. According to the Lawler et al. studies discussed above, the quality movement has consistently gained strength in large American corporations since the late 1980s; now, fully a quarter of Fortune 1000 firms report that *all* of their employees are involved in one or another facet of Total Quality. Moreover, Lawler et al. report that although employee involvement had previously been a central element within organizational redesign, managers have come to view it as merely part of the broader and more important thrust for control over product quality. Appelbaum and Batt (1994) suggest that an American version of "lean production" is developing that maintains a deeply hierarchical cast, not least because it implies a quantification of production standards and a standardization of work methods that place important limits on the discretionary powers of rank-and-file employees (cf. Moody 1997). Although Lawler et al. suggest that Total Quality can complement employee involvement, they acknowledge that "[t]he management role that is advocated is more directive" and "the utilization of problem solving tools [such as statistical process control] may introduce a certain rigidity that can be interpreted as *reducing* employee control and discretion" (1992:103, emphasis added). Pending further research, it seems plausible to suggest that continued growth of Total Quality approaches is likely to lead workplace change down avenues that lead further away from the post-Fordist path.

From Firms to Networks

Prompted by findings concerning the structure of the Japanese *keiretsu* and the European industrial districts (Gerlach 1992; Gomes-Casseres 1996; Brusco 1982; Sabel et al. 1989; Sabel and Zeitlin 1997; Harrison 1994), a small but highly suggestive literature has emerged that bears on the third post-Fordist proposition, involving the growth of collaborative networks among firms in given regions and industries. Although this line of analysis has only begun to find rigorous empirical application with respect to the United States, the studies outlined in Table 1 enable us to make the following observations.

[7]Kelley also found that unionized plants were especially likely to adopt centralized, hierarchical uses of programmable automation, even after other influences such as plant size and complexity were controlled. This point underscores the notion that corporate management in the U.S. remains strongly inclined to limit workers' discretion at the level of the establishment.

First, despite the small number of studies here, the evidence in support of the emergence of production networks as a central unit of economic activity seems both strong and consistent. All four of the studies identified in the table—by Powell, Koput, and Smith-Doerr; Uzzi; and Saxenian—conform to the collaborative network hypothesis. The study of biotechnology firms by Powell and his colleagues (1996) indicates that the network form of organization is especially pivotal in highly dynamic, knowledge-intensive branches of production. Under these conditions, firms must depend upon extremely complex and rapidly changing bodies of knowledge that no single firm can hope to control. In this study, firms that were more closely linked to other firms in collaborative networks enjoyed substantially greater performance is subsequent years. Although the span of time covered in the Powell et al. study is quite narrow, the findings do lend support to the broader argument that interfirm networks are becoming increasingly prevalent in the contemporary period (cf. Powell 1990; Powell and Smith-Doerr 1994).

The recent studies by Uzzi run directly parallel to Powell's. His research focuses on the "quality" sectors of the New York apparel industry, again a highly dynamic economic context. He distinguishes between "arms-length" (or market-based) transactions on the one hand, and "embedded" ties (based upon high levels of trust and cooperation) on the other. He finds that arms-length ties are the most frequent form of transaction, but that embedded ties hold far more strategic importance, for they provide managers with an array of practical resources and forms of mutual support that are especially vital in rapidly changing economic environments (Uzzi 1996). In a subsequent study of the same industry, Uzzi (1997) used labor union data on the frequency of interfirm transactions, and finds—much as did Powell and his colleagues—that participation in collaborative networks is highly predictive of subsequent economic performance. Thus both the Powell and Uzzi studies suggest that network ties comprise a critical resource on which small firms especially must rely.

A final study in this genre is Saxenian's (1994) comparative analysis of economic organization in Silicon Valley and Route 128 in Massachusetts. Basing her account on interviews with executives and technical specialists in these economic regions, Saxenian finds an enduring distinction between them: although firms in Route 128 are more heavily centralized and encumbered by bureaucratic traditions, their counterparts in Silicon Valley have more fully embraced the network form of organization. As a result, Silicon Valley firms are able to utilize knowledge, information, and technical personnel in a more flexible and innovative manner—advantages that are especially strategic in the high technology field. Studying each region's performance in such high technology markets as transistors and semiconductors, she finds repeated instances in which the industrial system on which Silicon Valley relied better equipped its firms to compete.

When these studies are placed alongside the broader literature assessing the centrality of the various forms of production networks (Romo and Schwartz 1995; Powell 1990), as well as studies documenting the decline of vertically integrated structures in industries such as motion pictures (Storper 1994), it begins to seem clear that production networks do play a decisive and perhaps an increasingly salient role within the U.S. economy, precisely as the third flexibility proposition expects. Yet many questions remain to be addressed. One concerns the temporal rhythms of network formation. Although the key proposition at issue here concerns an historic shift from vertically integrated structures toward collaborative, interfirm patterns of economic activity, evidence of a temporal shift has only indirectly been found. Moreover, since collaborative networks typically rest on strategic alliances and other joint ventures, their growth should be inversely related to the incidence of mergers and acquisitions, particularly in industries where production knowledge and product markets show the greatest flux. In the light of the explosive growth of mergers and acquisitions during the middle 1990s, it remains far from clear that this has been the case.[8]

Perhaps the most important issue yet to be addressed, however, concerns the morphology of network constellations. The communitarian imagery that flexible specialization theorists use leads them to emphasize the development of cooperative ties, Emilian-style, among the firms in a given industrial district or locale (Brusco 1982; Sabel et al. 1989). Such a collaborative or communal trend is certainly one conceivable outcome. Yet as Harrison (1994) contends, although the shift from vertically integrated firms to complex systems of production networks often lends the system of production a spatially decentralized cast, it can at the same time concentrate control over strategy, marketing, and finance in the handful of firms lodged at the center of each production network (a trend he calls "concentration without centralization"). The point is that although the rise of production networks seems to blur the boundaries among distinct firms, the structural patterns that such networks empirically assume remain as yet poorly understood.

Discussion

The available literature is insufficiently developed to support definitive conclusions regarding the three propositions at hand. For one thing, there is an obvious disparity between the relatively developed state of our ethnographic accounts and the far smaller pool of rigorous quantitative research. Moreover, there has been little effort to *conjoin* these disparate methods (cf. Uzzi 1997), drawing upon multiple sources of data that combine the strengths of each methodological approach. Perhaps because recent organizational theory has stressed the importance of firms' external environment, the literature offers only scattered suggestions regarding their changing internal structure (see Kalleberg et al. 1996:69–132). Little clear evidence exists concerning the consequences of downsizing, delayering, and other shifts for the structure of organizational control. Despite these absences, however, a number of observations can be made with respect to the empirical validity of the post-Fordist view.

First, as we have just seen, the literature does provide support for one important proposition that flexibility theorists have advanced: firms embedded in dynamic, uncertain economic environments do refrain from constructing vertically integrated structures, instead favoring the formation of strategic alliances with other firms. As production networks acquire increasing importance as economic actors, the boundaries among formerly distinct firms do seem to grow more blurred. The nature of this trend, however, is not yet well understood. Although it seems likely that a new system of interfirm governance is emerging, akin to previous epochs of organizational restructuring (Fligstein 1990), it is not yet clear that flexible specialization theory provides the most adequate model of these developments. Locally rooted industrial districts seem to represent one possible outcome of the "network revolution," but may well represent but one expression of a much broader development whose effects on the distribution of economic power are likely to be complex.

Second, although firms have been open to participating in new interfirm ventures and strategic alliances, they have seemed far less inclined to favor those forms of job and organizational redesign that flexibility theorists expect. With respect to the question of job design, for example, there is little evidence that firms have either placed greater stock in

[8]Discussing the scale of such mergers, which far transcends even the record-breaking pace of the early 1990s, Holson (1998) concludes that "there is a growing consensus among corporate executives that in a global economy, the only choice is between being one of the top two or three players in an industry or slowly bleeding to death."

the knowledge held by their manual employees or accorded workers increasing levels of discretion in their work. Instead, firms have preferred to place analytic functions in the hands of professional employees, setting sharp limits on the application of the neocraft paradigm. Employee involvement programs have spread rapidly, but only rarely seem to lead to structural change in jobs or organizations. Indeed, the use of EI or team systems seems either largely confined to rhetorical significance, or else subordinated to the thrust for greater production quality—a development that itself reflects a broader trend toward the quantification and standardization of work methods. And although post-Fordist theory predicts a trend toward decentralized and consultative patterns of work organization, the opposite seems to obtain: large corporations seem especially disposed toward *higher* levels of centralized organizational control, especially where the quality movement has taken hold. Taken together, these findings confront flexibility theory with significant empirical anomalies that it seems powerless to explain.

Such anomalies seem only to multiply when we consider developments about which post-Fordist theory remains silent, such as the growing use of nonstandard work arrangements (Pfeffer and Baron 1988; Tilly 1990; Kalleberg, Hudson, and Reskin 1997), as well as the effort to outsource or subcontract operations previously conducted in-house (Harrison and Kelley 1993). Insofar as these developments threaten to erode important elements of the postwar employment contract, they too seem difficult to square with flexibility theory. Indeed, the evidence suggests that these latter trends may well comprise a more salient feature of workplace change than the predicted post-Fordist turn. The rise of such trends, together with the empirical anomalies noted above, provides strong reason to suspect that the restructuring of work and organizations has followed a course that is not easily located on the dominant post-Fordist maps.

It might seem possible to argue that these empirical anomalies are largely an American phenomenon, and that the legal, political, and educational institutions necessary for the cultivation of genuinely post-Fordist work structures are simply less developed in the United States than in Western Europe or Japan. Indeed, a number of comparative analysts have made this argument in varying ways (Hartmann, Sorge, and Warner 1983; Lane 1988; Cornfield 1993; Lash and Urry 1987); and suggestions to this effect were to some extent incorporated into the argument developed in *The Second Industrial Divide*. The significance of this argument is that it provides an "escape hatch" for post-Fordism, in that it represents the American case (again) as an exception to the norms developing on other terrain. However, recent investigations of workplace change in European settings have begun to contradict this claim.

For one thing, several researchers have concluded that neocraft or post-Fordist work structures are much less prevalent in most Western European nations than the early, celebratory accounts allowed (Jurgens, Dohse, and Malsch 1986; Berggren 1989; Penn and Sleightholme 1995; Tagliabue 1996). Perhaps more important, the concept of "flexibility" itself seems to have undergone an intriguing metamorphosis in West European discussions, shifting from a vaguely worker-centered conception of the labor process to a different and far more managerial construct in which flexibility is defined as the willingness to conform to neoliberal economic strategies. This seems especially clear in the German debate, in which employers have appropriated the concept of "flexibilization," using it as a rallying cry with which to gain greater freedom from the postwar model of centralized collective bargaining and state regulation of employment (Lash and Urry 1987:154-65; Gottfried 1997; Flecker and Schulten 1997; Moody 1997). Although the outcome of economic restructuring in Western Europe is of course highly uncertain, it nonetheless seems clear that the remarkable growth of the American economy has reshaped economic thinking on the Continent, inclining European managers to favor neoliberal over neocraft conceptions of economic flexibility (Schumann 1998). Unless opposition to neoliberal policies can gain greater strength and endurance than is presently the case, it may be that "Europe's future will look more like that of North America than the other way around" (Moody 1997:126).[9]

FROM FORDISM TO FLEXIBLE ACCUMULATION

Thus far I have identified several theoretical and empirical reasons to conclude that post-Fordist theory provides a seriously flawed approach to the transformation of workplace institutions. The task then becomes one of articulating alternative theoretical models that afford greater purchase on the nature of the emerging regimes at work. Toward this end, this section briefly sketches the outlines of two alternative models. The first is provided by the neoinstitutionalist perspective toward work organizations (Meyer and Rowan 1977; DiMaggio and Powell 1983; Powell and DiMaggio 1991). The second, a less familiar but highly suggestive approach, is a nascent theory of "flexible accumulation" recently adumbrated in the works of Harvey (1989), Harrison (1994) and Rubin (1995, 1996).

Neoinstitutionalist theories have proliferated in recent years, rendering it difficult to construct any overall characterization (DiMaggio and Powell 1991; Hirsch and Lounsbury 1997). Yet one core feature of virtually all variants of this approach is their shared rejection of theories predicated on the functional independence of organizational forms. Equally important is their opposition to theories that stress economic considerations (technical efficiency or market competition) as the factors that shape organizational arrangements. In contrast with these approaches, neoinstitutionalists are wont to emphasize the manifold points at which organizations depend on the organizational fields and social structural contexts that shape organizational arrangements "from the outside." Further, many neoinstitutionalists stress the *symbolic* or *ceremonial* aspects of organizational arrangements, which are adopted not so much to enhance efficiency as to enhance the legitimacy organizations enjoy in the various arenas in which they operate. Indeed, when the adoption of widely accepted organizational patterns delivers economic rewards, neoinstitutionalists often suggest that the reasons lie not so much in the technical merits (if any) such innovations provide as in the legitimacy (and correlative resources) that firms gain by publicly embracing culturally established norms (Meyer and Rowan 1977; DiMaggio and Powell 1983; for more recent discussion, see Zucker 1987; Powell and DiMaggio 1991; Hirsch and Lounsbury 1997; Stinchcombe 1997).

Neoinstitutionalist research has drawn particular inspiration from the conceptual schema provided by DiMaggio and Powell (1983), who distinguished three ideal-typical ways in which organizational fields induce firms to adopt particular arrangements. "Mimetic" isomorphism is said to occur when managers or administrators are constrained to emulate the practices adopted by leading organizations in their fields. "Normative" isomorphism stems from the influences flowing from the shared outlooks and practices common to the professionals in given specialties (e.g., human resource managers and consultants). Finally, "coercive" isomorphism refers to the pressures imposed through legal mandates, court rulings, or contractual stipulations that explicitly define acceptable organizational prac-

[9]This is *not* to suggest that European managers' neoliberal inclinations are easily translated into practice. It *is*, however, to cast doubt on the claim that Western European societies provide an unproblematic sanctuary for the neocraft regime. Managerial conceptions of work are not formed in isolation from the ideas and strategies adopted in other countries, as post-Fordist theory so often assumes. The continuing power of the American economy, coupled with the erosion of Japanese firms, is sure to deepen the influence of neoliberal thinking internationally. For discussion of the "dominance" effect of successful economies, see Smith and Meiksins 1995.

tices. Operating singly or in combination (Meyer and Scott 1992:4), these pressures introduce powerful constraints on organizations, inducing them to adopt arrangements that are already embedded in their institutional environments.

Empirical applications of neoinstitutionalist theory have provided consistent support for its tenets. Studies have found that shifts in the legal or political environments—often mediated by the orientations and actions of professional employees—help explain the diffusion of key organizational innovations, such as the rise of the multidivisional structure of large corporations (Fligstein 1987, 1990), the adoption of due process arrangements and internal labor markets (Edelman 1990; Dobbin et al. 1993; Sutton et al. 1994), and modern personnel practices as well (Baron, Dobbin, and Jennings 1986). In all these cases, the technical and organizational characteristics of the firm emerge as less powerful predictors of organizational innovation than the demands of the firm's institutional environment.

Increasingly, neoinstitutional theorists have come to stress the ways in which the state impinges on the construction of organizational fields. This theoretical trend carries important advantages, not the least of which is the analytical clarity that results when we trace the origins of organizational innovations to particular aspects of the legal, judicial, or regulatory environment, rather than to free-floating customs or notions of "rationality." This political emphasis might seem questionable to some, however, on the grounds that the American political system has historically been too fragmented, weak, and limited to determine the outcome of workplace restructuring. Skeptics might contend that the major examples of state influence over organizational forms have occurred at the margins of organizations, via antitrust and EEO law (Fligstein 1987, 1990; Edelman 1990; Sutton et al. 1994). Outside of these realms, it might be claimed, the American state apparatus has played a more traditionally hands-off role. If so, then the institutions of civil and not political society might be more appropriate for emphasis in future research.

In response to such skepticism, theorists can point to the impact of labor law and federal oversight of collective bargaining, which have operated to shape the pattern of organizational governance that exists, even in firms that lack union representation (see Heckscher 1988; Brody 1980:215–55; Capelli 1995:571–573). Research by Wright (1985:223–24) suggests that the legal environment has significant effects on job structures: By exempting supervisory personnel and many professionals from coverage under the National Labor Relations Act, U.S. labor law gives firms an incentive to lodge analytical functions in the hands of supervisors, middle managers, and salaried employees (Kelley 1990), thereby perpetuating hierarchical approaches toward the restructuring of work. Moreover, the increasing deployment of temporary workers itself rests on shifts in legal and judicial definitions of the employment relationship, occurring on a state-by-state basis during the 1950s and 1960s, that enabled temporary help firms to function as surrogate employers of the workers they supply to client firms (Gonos 1997). Finally, although sociologists have largely neglected the subject, it seems reasonable to suggest that the American military's presence during the Cold War, as established via the Department of Defense's procurement processes, support for research and development, and other influences, acted to reinforce the hierarchical structure of industrial organizations in the United States.[10] These points begin to suggest that if the logic of Fordism has proved more tenacious than flexibility theorists

[10]Noble (1984) found that the military played a decisive role in the establishment of machine tool technologies that were based on complex, hierarchical programming languages that few small firms could afford. The effect was to weaken the hold of craft-based practices throughout the metal cutting and machine tool industries. I am endebted to my colleague, historian Philip Scranton, for pointing out the importance of the Cold War for the structure of American industry.

expect, the reasons may lie not so much in economic or organizational processes, but in aspects of the political and legal environment in which American firms are lodged.[11]

Although neoinstitutionalists have made occasional reference to the restructuring of work (DiMaggio and Powell 1983:151), the theory has yet to be applied to this empirical realm (for an exception, see Strang 1997). Yet clearly, institutionalism holds substantial promise in this field. Rather than taking organizational claims at face value, it develops a more fully sociological treatment of workplace change that is mindful of the social, cultural, and political influences so often brought to bear on organizational arrangements. It seems particularly well equipped to account for the often rhetorical nature of the new workplace initiatives: the adoption of Employee Involvement programs or Total Quality initiatives may be more outwardly than inwardly oriented, in effect providing a "badge of modernity" for the adopting firm. Further, institutionalist theory and research has frequently found evidence that professional specialists occupy increasingly influential positions in the diffusion of organizational innovations (see DiMaggio and Powell 1983; Edelman 1990). This emphasis also seems to square with the empirical literature reviewed above, and may further explain why workplace restructuring has so often failed to expand the autonomy of manual employees.

Despite such virtues, the neoinstitutionalist approach also carries inherent limitations. Precisely because the new institutionalists have concerned themselves with the outwardly oriented nature of so much organizational change—EEO provisions, due process arrangements, and personnel methods that are publicly visible—they have generally neglected the features of the firm's technical "core."[12] For this reason, they often fail to address issues that have been of enduring interest to students of the work process, such as the nature of informal work relations or the division between mental and manual labor. This failing is what presumably led Stinchcombe to observe that "the trouble with the new institutionalism is that it does not have the guts of institutions in it" (1997:17). Moreover, although the new institutionalism has shown how firms are deeply embedded in and constrained by the wider social structure, they have generally failed to account for the origins of the institutional influences they stress (Brint and Karabel 1991). More than this: they have largely overlooked the dialectical relations that unfold between large firms and their institutional environments (Perrow 1986:173-74). Given these difficulties, a second approach to workplace restructuring—that of the flexible accumulation theory—warrants especially close attention.

Under Fordism, corporations sought to control their economic environments by constructing massive, vertically integrated structures that essentially *internalized* strategic sources of raw materials, machinery, and labor power. In an era of market flux and uncertainty, however, such integrated structures can no longer be sustained, as "what were once viewed as accumulation-stabilizing structures increasingly become barriers to profitability" (Rubin 1995:298). Although initially firms responded by relocating Fordist plants to less developed countries—what Harvey calls the "spatial fix"—such steps only heighten the problem of overproduction and merely delay the structural shifts needed to adapt to the new economic conditions. Recasting the strategies and structures they employ, firms have

[11]A further point, which cannot be developed here, concerns the American failure to develop credentialing institutions in support of manual workers' skills. Given the atrophy of apprenticeship programs since WWI, employers seeking certified forms of knowledge have understandably turned to professional employees, whose credentialing systems are much more strongly institutionalized than is the case for manual employees. For a discussion of the credentialing systems established in Germany, see Littek and Heisig (1991).

[12]Indeed, theorists have been unsure whether manufacturing firms are as susceptible of institutional influences as are educational and service organizations. See Meyer and Rowan ([1978] 1992), which seems to introduce important qualifications into their initial (1977) theory.

been forced to *externalize* many of the resources they have historically amassed (see especially Pfeffer and Baron 1988), dismantling their bureaucratic hierarchies by downsizing, "delayering," and reducing their reliance on inventories, internal labor markets, and other fixed costs. In addition, firms have been forced to reduce the time needed to reap returns on their capital investments, accelerating the pace of product and process innovation (often through CAD/CAM and just-in-time inventory systems), leading some to speak of the emergence of "fast capitalism" (Gee, Hull, and Lankshear 1996). As firms have adopted new and more flexible ways of accumulating capital—or, put differently, removed the inherited barriers to profitability—they have at the same time refashioned the structure of work, labor markets, and the employment relation itself.

One such effect is the redrawing of the boundaries that distinguish different groups of employees from one another. Especially critical is the sharp boundary that firms begin to draw between "core" and "peripheral" employees. Those in the former group perform functions that are deemed essential to the firm, while the work of the latter group is deemed dispensable. In many cases, the functions of peripheral workers are outsourced or subcontracted to smaller, satellite firms (shifting work into the secondary labor market), or contracted out to temporary help firms (expanding the contingent economy). In cases where peripheral workers remain on the payroll, they nonetheless find themselves denied key benefits of the standard employment relation, such as access to internal labor markets, job security, and autonomy in their work. Increasingly, such rewards are reserved for core employees, whose control over strategic knowledge places them in a distinctly more favorable position within the firm. Hence one major feature of flexible work regimes is the resurgence of dualism and disparity in the distribution of job rewards.

Related to the above is a second characteristic of the new production regimes: their tendency to privilege the formal knowledge of professional employees over manual workers' skills (Aronowitz and DiFazio 1995). This trend has many sources. As employers struggle to introduce the latest product innovations, meshing consumer and marketing research with Total Quality methods, they impose a greater standardization and quantification of work methods that expand the centrality and authority of professional knowledge and expertise (Burris 1993; Derber and Schwartz 1991; Dudley 1994). Likewise, the firm's increased reliance on sophisticated process technologies provides professional and technical employees with abundant opportunities to expand their autonomy, often defining their own "formulaic" knowledge as more rational, modern, and scientific than the local knowledge of manual employees (Vallas and Beck 1996). Especially in firms employing advanced process technologies, credential barriers begin to emerge that reserve even lower level supervisory jobs for employees holding technical qualifications. The result—directly contrary to the post-Fordist view—is a *sharpened* division between mental and manual labor and a *strengthening* of the hierarchical relation between the labor of "head" and "hand."

A third feature of the flexible work regimes grows out of their increased reliance on outsourcing and subcontracting relations, which generate ever more complex forms of interfirm ties. This trend may occasionally yield strategic alliances or industrial districts, much as flexible specialization expects. Yet from the standpoint of flexible accumulation theory, the *collaborative* form of production networks is inherently ephemeral: it tends to give way to domination by a single large firm (as in the case of Benetton), to provide attractive targets for acquisition by large outside firms, or else to expose firms to catastrophic failure once they seek to shed the locally based orientation from which they drew their strength (cf. Harrison 1994). More common is the *hierarchical* form of production network—the equivalent of the *kereitsu* or *chaebol*, which provides a functional equivalent of the vertically integrated enterprise, or a collectivized form of capital (Carchedi 1975; Roy 1997) that provides de facto financial control over satellite firms, yet with few of the drawbacks of direct ownership.

The proliferation of hierarchical production networks has important consequences for the internal structure of the firm. Insofar as firms embedded within hierarchical networks often place professional employees in strategic positions as the conduits of highly valued knowledge, while at the same time increasing the need for administrative, legal, and financial coordination of complex network ties, the rise of such networks tends to *reinforce* the division between mental and manual labor, acting powerfully to countermand any putative shift toward post-Fordist job designs (Reich 1992). The prediction here, then, posits an *inverse* relation between network embeddedness and the autonomy enjoyed by manual employees. This prediction perfectly captures the empirical pattern identified above, in which firms increasingly participate in production networks, yet prefer to vest analytic functions and operational authority in workers wielding professional expertise.

A last consideration concerns the relation between flexible work regimes and the cultural environment in which they exist. Recall that flexible specialization theory suggests that the increasing volatility of consumer demand, coupled with the rising interest in diversified, quality goods, induces firms to embrace the neocraft paradigm. Flexible accumulation theory rejects this approach and the assumptions on which it is based. Rather than viewing consumer tastes as an exogenous influence to which firms must react, the theory suggests that consumer tastes have themselves become subject to the new regime's accumulation needs. Because of the saturation of demand for many mass consumer goods in the advanced capitalist nations, firms must increasingly appropriate cultural forms and use them to accelerate the pace with which product cycles unfold. Hence marketing strategies place an increasing emphasis on "quick-changing fashions and the mobilization of all the artifices of need inducement and cultural transformation that this implies" (Harvey 1989:156). Although firms have used such "artifices" at least since the days of Alfred Sloan, the deployment of marketing research and advertising campaigns has now grown so intensified as to weaken or even destroy the relatively stable boundaries of mass culture. In their place there arises a postmodern aesthetic that, insofar as it "celebrates difference, ephemerality, [and] spectacle" (ibid.), is uniquely equipped to multiply demand for luxury or near-luxury commodities whose superior qualities attest to the virtues of the consumer. Note that unlike neoinstitutionalist theory, which sees the firm's normative environment as an autonomous realm, the flexible accumulation approach emphasizes the ways in which firms can utilize marketing research and the media as instruments with which to harness and shape cultural trends in accordance with their accumulation needs.

This sketch of flexible accumulation theory begins to identify several important virtues that the theory brings to bear on the debate over workplace change. First, it seems capable of addressing the empirical anomalies that the post-Fordist approach seemed powerless to explain. Such developments as the resurgence of economic dualism and segmentation, the expanded centrality of professional knowledge, the drift toward standardization and quantification of work methods, and the tenacity of hierarchy are all quite explicable from the flexible accumulation point of view. Second, the failure of team systems and EI programs to effect structural changes in workers' jobs may reflect their predominantly ideological functions: they provide management with mechanisms with which to maintain social cohesion despite the increasing disparities found among occupational groups.

Third, rather than viewing the transformation of work organizations in terms of a single uniform tendency (such as the spread of the neocraft regime), this alternative theory instead approaches workplace change as a complex and often contradictory phenomenon involving two opposing trends: one that expands the discretion and authority of some employees (typically, core professional and salaried employees), and a second, opposing tendency that restricts the job rewards accorded other groups (hourly workers, but increasingly many salaried employees as well). By acknowledging this inherent duality of workplace change—patterns that Smith (1994) terms its "enabling" and "restrictive" aspects, respectively—flexible accumulation theory makes possible a fuller and more adequate representation of the new forms of work than has previously been achieved. Finally, in contrast with the institutionalist approach, flexible accumulation theory provides a concrete conception of the internal character of the emerging production regimes. In so doing it opens up a novel line of analysis regarding the consequences of flexible work regimes for the structure of work organizations, internal labor markets, and the character of the employment relation as well.

CONCLUSION

I have suggested here that the dominant theory of workplace change today suffers from serious theoretical and empirical flaws. Perhaps because of their eagerness to attract policy audiences, advocates of the post-Fordist view often embrace "efficiency" conceptions of economic institutions, construing their preferred form of workplace change as the *telos* toward which innovation inevitably leads. Although such views allow analysts to justify their prescriptive suggestions for the restructuring of work, they stand at odds with sociological analysis of economic institutions. They also conflict with the theory's effort to transcend the Taylorist and Fordist legacy it seeks to overcome. Further, because post-Fordists typically view the transformation of work in terms of a single, undifferentiated tendency, they overlook the multiple ways in which firms can respond to the new economic conditions, many of which seem either to reproduce the hierarchical cast of Fordist institutions or to lend workplace change a complex and inherently contradictory character.

Review of the empirical literature in the field casts further doubt on the theory's validity, bringing to light several anomalies that it seems powerless to explain. In contrast with post-Fordism's predictions, the evidence provides only weak indications that "new production concepts" have been incorporated into work processes, or that the traditional division between mental and manual labor has begun to give way. To the contrary: the evidence suggests that job designs accord professional employees increased centrality and authority, especially where the quality movement has held sway. Programs for more participative work structures have spread rapidly throughout American business, but seldom seem connected to enduring patterns of job redesign. Finally, organizational restructuring seems in fact to lead in the direction of *greater* centralization of control and standardization of work methods—again, contrary to the predicted trends. In the light of these considerations, it seems safe to conclude that post-Fordist theory provides a seriously distorted guide to the changes currently gripping work and organizations in the United States.

Searching for theoretical alternatives to the post-Fordist approach leads to consideration of two models in particular: the neoinstitutionalist and flexible accumulation approaches. The former is well equipped to account for the diffusion of workplace innovations, viewing them as managerial efforts to enhance or protect the firm's legitimacy by conforming to the demands of its institutional environment. Yet institutionalism encounters important limits when applied to the restructuring of work organizations. Because it concerns itself mainly with the outward appearance of the firm, institutionalism seems reluctant to address questions concerning the technical "core" of production processes. By viewing the characteristics of firms as simple expressions of environmental demands, moreover, neoinstitutionalism misses the complex, iterative relations that occur between organizations and their environments. For these reasons, the theory of flexible accumulation—although itself less firmly institutionalized—may ultimately hold greater promise for the study of workplace change.

In this second view, the thrust for flexibility is likely to produce an expansion of discretion and autonomy for many employees, much as post-Fordism has claimed. Yet these benefits of workplace change are typically reserved for professional employees who occupy core positions within the firm. Others—most hourly workers, but increasingly salaried employees as well—are likely to encounter the thrust for flexibility in a very different form, that is, the effort to externalize their tasks or otherwise consign them to peripheral economic positions. Hence the new flexible work regimes produce a resurgence of dualism, as the boundaries within and among firms are recast. They also increasingly vest authority in the formulaic knowledge of professional employees, and engender complex, hierarchical production networks that further contribute to the resurgence of economic dualism.

The image flexible accumulation theory provides of workplace change is one of sharpening disparities in occupational and employment conditions, whether at the level of the workplace, firm, region, or national economy. At times, such disparities may be obscured by spatial and/or organizational dispersion (Pfeffer and Baron 1988), or by social devices such as production teams (Smith 1996). At other times, however, the reconfiguration of jobs and organizations seems likely to produce tensions that cannot be contained, generating conflicts that assume a variety of forms (Hodson 1995). Thus important questions have yet to be addressed concerning not only the structure of the flexible production regimes but also the agency and actions of different occupational groups.

Flexible accumulation theory expects that the rise of the new dualism has transformed key aspects of the Fordist employment relation, in effect rescinding internal labor markets, benefits, and job stability for many workers while retaining them for others. It also suggests that the contradictory tendencies within the new flexible work regimes—its "enabling" and "restrictive" trends—are systemically interconnected, with instances of participative work relations often presupposing the growing effort to externalize or peripheralize the functions of other employees. Although evidence has emerged in support of this view—as when the use of temporary workers gives permanent employees access to more rewarding jobs (Magnum, Mayall, and Nelson 1985; Barnett and Miner 1992; Davis-Blake and Uzzi 1993:219–29)—the organizational mechanics of the new dualism remain as yet poorly understood. Much of the rhetoric in the business press leaves readers with the impression that a "new employment contract" is emerging that in effect rescinds the implied promise of secure employment and access to job ladders (see Powell forthcoming; DiTomaso 1996; Boyle 1997), yet there is little evidence concerning the impact of such trends on different occupational groups within the same firm. Moreover, it remains unclear whether the dual tendencies immanent within the flexible work regimes flow from an explicit human resource strategy (as Appelbaum [1987] and others have claimed), or are instead the unintended by-products of incremental, ad hoc measures taken to keep production costs down. Further questions concern the efficacy of trade union efforts to modify flexible work regimes (e.g., by contesting the outsourcing of work and/or the use of part-time labor) and the impact of tight labor markets on the uneven distribution of job rewards within the firm. Clearly, the most basic questions concerning the nature of flexible work regimes remain to be addressed.

Interestingly, both institutionalist and flexible accumulation theory contend that the course of workplace innovation increasingly hinges on the actions of professional employees. Although our knowledge of the role played by professional and managerial employees in the restructuring of work is still incomplete, studies generally depict such workers as enjoying substantial capacity to modify managerial initiatives. Thus the literature suggests that middle managers and professionals can subvert efforts to increase worker participation (Zuboff 1988), resist or reshape top management strategies (Smith 1990), engage in informal campaigns for greater occupational autonomy (Thomas 1994), and influence the criteria used to define legitimate knowledge at work (Vallas and Beck 1996). The picture we have of hourly workers, by contrast, is one that endows them with little or no capacity to challenge the authority of professional employees or otherwise shape the structure of their jobs (Burawoy 1979; Willis 1977; Halle 1984). Although these representations may of course reflect the greater power work organizations bestow upon professionals, or else the historical context of labor union decline, they may also stem from researchers' attachment to research methods (such as surveys of human resource managers) that systematically neglect the informal capacities of hourly employees. Conceivably, aspects of the new production regimes may prove to have unanticipated consequences, as when teams enable workers to form solidary work groups, or just-in-time inventory systems provide workers with increased powers of disruption. If, as flexible accumulation theory suggests, organizational disparities and patterns of segmentation are likely to grow increasingly pronounced, then the question of how varied groups of employees shape or respond to the new work structures will stand in urgent need of systematic research.

As firms increasingly distinguish between core and peripheral categories of employees, an important question arises regarding the criteria on which managers base this critical distinction. The materialist bent of flexible accumulation theory leads its advocates to stress economic considerations (e.g., the value added by distinct occupational groups, the cost of benefit packages, or the price and availability of labor power on the external market). Such considerations are of obvious importance. Yet social and cultural influences, too, may play a critical role, affecting the ways in which managers draw boundaries between core and peripheral functions. Conceivably, the commitments, identities, and normative influences found among particular ranks or divisions of management may incline corporate officials to favor certain organizational configurations over others (Schoenberger 1997; cf. Fligstein 1990). The unequal distribution of social capital (i.e., linkages to high-status personnel) may differentially enable actors to shape the process of organizational change (Thomas 1994; Child et al. 1984). Moreover, the relative value placed on particular occupational functions is likely to depend on a wide array of normative factors. Firms in which professional employees are well entrenched (e.g, high tech firms in Silicon Valley) may be especially likely to externalize the functions of manual employees. Assuming further that workers' racial and gender status will affect the value managers impute to their work, jobs with high proportions of women and minority workers may stand at special risk of externalization. The point here is that rather than taking the core/periphery boundary for granted as an economic fact, it will prove wise to problematize this distinction, inquiring into the processes through which this boundary is itself defined.

Inspired by the economic conditions of the 1980s—a moment of profound crisis for American manufacturing in particular—post-Fordist theory founded an ambitious intellectual movement whose signal contribution was its recognition that the mass production paradigm represented a historically specific phenomenon that was likely to undergo dramatic change. Yet in depicting the new work structures, post-Fordism viewed them as implying a radical "break" with the logic of mass production. The evidence that has steadily accumulated over the last decade has suggested otherwise, indicating that the legacy of the Fordist era has by no means been dispelled. Inflexibly adhering to the post-Fordist approach, however, researchers have allowed a widening gap to open up between sociological theories of workplace change and the empirical realities they seek to explain. Closing this gap will require that we adopt newer and more sophisticated conceptual approaches than post-Fordist theory has itself produced. Above all, it will require that we view the concept of "post-Fordism" in a rather different light than theorists have done: as a reminder that work organizations have been formed in the shadow of the Fordist tradition from which they trace their descent.

REFERENCES

Adler, Paul. 1988. "Automation, Skill and the Future of Capitalism." *Berkeley Journal of Sociology* 33:1–36.

———. 1992a. "Introduction." Pp. 3–14 in *Technology and the Future of Work.* edited by Paul Adler. New York: Oxford University Press.

———. 1992b. "The 'Learning Bureaucracy': New United Motor Manufacturing, Inc." *Research in Organizational Behavior* 15:111–94.

Adler, Paul, and Bryan Borys. 1996. "Two Types of Bureaucracy: Enabling and Coercive." *Administrative Science Quarterly* 41:61–89.

Amin, Ash. 1994. *Post-Fordism: A Reader.* Oxford: Blackwell.

Appelbaum, Eileen. 1987. "Restructuring Work: Temporary, Part-Time, and At Home Employment." Pp. 268–310 in *Computer Chips and Paper Clips: Technology and Women's Employment*, edited by Heidi Hartmann, Robert E. Kraut, and Louise A. Tilley. Washington, DC: National Academy Press.

Appelbaum, Eileen, and Peter Albin. 1989. "Computer Rationalization and the Transformation of Work: Lessons from the Insurance Industry." Pp. 246–65 in *The Transformation of Work? Skill, Flexibility and the Labour Process*, edited by Stephen Wood. London: Unwin Hyman.

Appelbaum, Eileen, and Rosemary Batt. 1994. *The New American Workplace: Transforming Work Systems in the United States.* Ithaca NY: ILR Press.

Aronowitz, Stanley, and William DiFazio. 1995. *The Jobless Future.* Minneapolis: University of Minnesota Press.

Atkinson, John. 1985. "The Changing Corporation." Pp. 13–34 in *New Patterns of Work*, edited by D. Clutterbuck. London: Gower.

Barker, James R. 1993. "Tightening the Iron Cage: Concertive Control in Self-Managing Teams." *Administrative Science Quarterly* 38:408–37.

Barnett, William P., and Anne S. Miner. 1992. "Standing on the Shoulders of Others: Career Interdepedence in Job Mobility." *Administrative Science Quarterly* 37:262–81.

Baron, James N., Frank R. Dobbin, and P.D. Jennings. 1986. "War and Peace: The Evolution of Modern Personnel Administration in U.S. Industry." *American Journal of Sociology* 92:2:350–83.

Berggren, Christian. 1989. "New Production Concepts in Final Assembly—The Swedish Experience." Pp. 171–203 in *The Transformation of Work? Skill, Flexibility and the Labour Process*, edited by Stephen Wood. London: Unwin Hyman.

———. 1992. *Alternatives to Lean Production: Lessons from the Swedish Automobile Industry.* Ithaca, NY: ILR Press.

———. 1995. "Japan as Number Two: Competitive Problems and the Future of Alliance Capitalism after the Burst of the Bubble Boom." *Work, Employment and Society* 9:1:53–95.

Best, Michael. *The New Competition.* Cambridge MA: Harvard University Press.

Blauner, Robert. 1964. *Alienation and Freedom: The Factory Worker and His Industry.* Chicago: University of Chicago Press.

Block, Fred. 1990. *Post-Industrial Possibilities: A Critique of Economic Discourse.* Berkeley: University of California Press.

Bourdieu, Pierre. 1984. *Distinction: A Social Critique of the Judgment of Taste.* Cambridge: Harvard University Press.

Boyle, Maryellen. 1997. "The New Employment Contract: A Study of the Symbolic Functions of Education on the Job." Paper presented at the Annual Meetings of the American Sociological Association, New York.

Braverman, Harry. 1974. *Labor and Monopoly Capital: The Degradation of Work in the Twentieth Century. New York: Monthly Review Press.*

Brint, Steven, and Jerome Karabel. 1991. "Institutional Origins and Transformations: The Case of American Community Colleges." Pp. 337–60 in *The New Institutionalism in Organizational Analysis.* Chicago: University of Chicago Press.

Brody, David. 1980. *Workers in Industrial America.* New York: Oxford University Press.

Brubaker, Rogers. 1985. "Rethinking Classical Theory: The Sociological Vision of Pierre Bourdieu." *Theory and Society* 14:723–44.

Brusco, Sebastiano. 1982. "The Emilian Model: Productive Decentralization and Social Integration." *Cambridge Journal of Economices* 6:167–84.

Burawoy, Michael. 1979. *Manufacturing Consent: The Labor Process under Monopoly Capitalism.* Chicago: University of Chicago Press.

———. 1985. *The Politics of Production: Factory Regimes under Capitalism and Socialism*. London: Verso.

Burns, Tom, and G. M. Stalker. 1962. *The Management of Innovation*. Chicago: Quadrangle Books.

Burris, Beverly. 1993. *Technocracy at Work*. Albany, NY: State University of New York Press.

Callaghan, Polly, and Heidi Hartmann. 1991. *Contingent Work: A Chart Book on Part-Time and Temporary Employment*. Washington DC: Economic Policy Institute.

Capelli, Peter. 1995. "Rethinking Employment." *British Journal of Industrial Relations* 33:4:563–602.

———, Laurie Bassi, Harry Katz, David Knoke, Paul Osterman and Michael Useem. *Change at Work*. NY: Oxford.

Carchedi, G. 1975. *The Economic Identification of Social Classes*. London: Routledge.

Child, John, R. Loveridge, J. Harvey and A. Spencer. 1984. "Microelectronics and the Quality of Employment in Services." Pp. 163–90 in *New Technology and the Future of Work and Skills*, edited by P. Marstrand. London: Pinter.

Colclough, Glenna, and Charles Tolbert, II. 1992. *Work in the Fast Lane: Flexibility, Divisions of Labor, and Inequality in High-Tech Industries*. Albany, NY: State University of New York Press.

Cornfield, Daniel. 1993. "Labor and the Participative Potential of New Technology." Paper presented at the Conference on Industrial Relations and Human Resource Management in an Era of Global Markets. Seoul, Korea: Korea Labor Institute.

Davis-Blake, Alison and Brian Uzzi. 1993. Determinants of Employment Externalization: A Study of Temporary Workers and Independent Contractors. *Administrative Science Quarterly* 38:195–223.

Derber, Charles, and William Schartz. 1991. "New Mandarins or New Proletariat? Professional Power at Work." *Research in the Sociology of Organizations* 8:71–96.

DiMaggio, Paul, and Walter W. Powell. 1983. "The Iron Cage Revisited: Institutional Isomorphism and Collective Rationality in Organizational Fields." *American Sociological Review* 48, 2:147–60.

DiTomaso, Nancy. 1996. "The Subcontracting of Everyone? The Loose Coupling of Jobs." Paper presented at the Annual Meetings of the American Sociological Association, New York.

Dobbin, Frank, John R. Sutton, John Meyer, and W. Richard Scott. 1993. "Equal Opportunity Law and the Construction of Internal Labor Markets." *American Journal of Sociology* 99:2:396–427.

Dohse, K., U. Jürgens, and T. Malsch. 1985. "From 'Fordism' to 'Toyotism'? The Social Organization of the Labor Process in the Japanese Automobile Industry." *Politics and Society* 14:2:115–46.

Dudley, Kathryn Marie. 1994. *The End of the Line: Lost Jobs, New Lives in Post-Industrial America*. Chicago: University of Chicago Press.

Edelman, Lauren. 1990. "Legal Environments and Organizational Governance: The Expansion of Due Process in the American Workplace." *American Journal of Sociology* 95:6:1401–40.

Edwards, Richard. 1979. *Contested Terrain*. New York: Basic

Fantasia, Rick, Dan Clawson, and Gregory Graham. 1988. "A Critical View of Worker Participation in American Industry." *Work and Occupations* 15:4:468–88.

Flecker, Jorg, and Thorsten Schulten. 1997. "The End of Institutional Stability: What Future for the German Model?" Paper presented at the 15th Annual International Labour Process Conference, University of Edinburgh, Scotland.

Fligstein, Neil. 1987. "The Intraorganizational Power Struggle: Rise of Finance Personnel to Top Leadership in Large Corporations, 1919–1979." *American Sociological Review* 52:1:44–59.

———. 1990. *The Transformation of Corporate Control*. Cambridge: Harvard University Press.

Form, William. 1980. "Resolving Ideological Issues on the Division of Labor." Pp. 140–55 in *Sociological Theory and Research: A Critical Appraisal*, edited by H.M. Blalock. New York: Free Press.

Gee, James, Glynda Hull, and Colin Lankshear. 1996. *The New Work Order: Behind the Language of the New Capitalism*. Boulder CO: Westview.

Gerlach, Michael. 1992. *Alliance Capitalism: The Social Organization of Japanese Business*. Berkeley: University of California Press.

Gomes-Casseres, Benjamin. *The Alliance Revolution: The New Shape of Business Rivalry*. Cambridge: Harvard University Press.

Gonos, George. 1997. "The Contest over 'Employer' Status in the Post-war U.S.: The Case of Temporary Help Firms." *Law and Society Review* 31:1:81–110.

Gottfried, Heidi. 1997. "Mapping Neo-Fordism Across Six Countries." Unpublished manuscript, Department of Sociology, Purdue University.

Graham, Laurie. 1995. *On the Line at Subaru-Isuzu*. Ithaca, NY: ILR Press.

Grenier, Guillemo. 1988. *Inhuman Relations: Quality Circles and Anti-Unionism in American Industry*. Philadelphia: Temple University Press.

Hackman, J. Richard, and R. Wageman. 1995. "Total Quality Management: Empirical, Conceptual and Practical Issues." *Administrative Science Quarterly* 40:309–42.

Halle, David. 1984. *America's Working Man*. Chicago: University of Chicago Press.

Harrison, Bennett. 1994. *Lean and Mean: The Changing Landscape of Corporate Power in the Age of Flexibility*. New York: Basic.

Harrison, Bennett and Maryellen Kelley. 1993. "Outsourcing and the Search for 'Flexibility'." *Work, Employment and Society* 7:2:213–35.

Hartmann, G., I. Nicholas, A. Sorge, and M. Warner. 1983. "Computerized Machine Tools, Manpower Consequences and Skill Utilization: A Study of British and West German Manufacturing Firms." *British Journal of Industrial Relations* 21 (July): 221–31.

Harvey, David. 1989. *The Condition of Postmodernity*. London: Blackwell.

Heckscher, Charles. 1988. *The New Unionism: Employee Involvement in the Changing Corporation*. New York: Basic.

———. 1994. "Defining the Post-Bureaucratic Type." Pp. 14–62 in *The Post-Bureaucratic Organization*, edited by Charles Heckscher and Anne Donnellon. Thousand Oaks, CA: Sage.

Hertzberg, Frederick. 1973. *Work and the Nature of Man*. New York: New American Library.

Hill, Stephen J. 1991. "Why Quality Circles Failed, but Total Quality Management Might Succeed." *British Journal of Industrial Relations* 29:4 (December): 541–68.

Hirsch, Paul M., and Michael Lounsbury. 1997. "Ending the Family Quarrel: Toward a Reconciliation of 'Old' and 'New' Institutionalism." *American Behavioral Scientist* 40:4:406–18.

Hirschhorn, Larry. 1984. *Beyond Mechanization*. Cambridge MA: MIT Press.

Hirst, Paul, and J. Zeitlin. 1989. "Flexible Specialization and the Competitive Failure of UK Manufacturing," *Political Quarterly* 60:3:164–78.

———. 1991. "Flexible Specialization Versus Post-Fordism: Theory, Evidence and Policy Implications." *Economy and Society* 20:1:1–56.

Hodson, Randy. 1988. "Good Jobs and Bad Management: How New Problems Evoke Old Solutions in High Tech Settings." Pp. 247–79 in *Industries, Firms, and Jobs: Sociological and Economic Approaches*. New York: Plenum.

———. 1995. "Worker Resistance: An Underdeveloped Concept in the Sociology of Work." *Economic and Industrial Democracy* 16:79–110.

———. 1996. "Dignity in the Workplace Under Participative Management: Alienation and Freedom Revisited." *American Sociological Review* 61:5:719–38.

Holson, Laura M. 1998. "Magnetic Mania: In This Merged, Merged World, Anything Goes." *New York Times*, 26 June, C1.

Hyman, Richard. 1988. "Flexible Specialization: Miracle or Myth?" Pp. 48–60 in *New Technology and Industrial Relations*, edited by R. Hyman and W. Streeck. Oxford: Basil Blackwell.

Jürgens, U., K. Dohse, and T. Malsch. 1986. "New Production Concepts in West German Car Plants." Pp. 258–81 in *Between Fordism and Flexibility: The Automobile Industry and Its Workers*, edited by S. Tolliday and J. Zeitlin. Cambridge: Polity.

Kalleberg, Arne. Forthcoming. "The Advent of the Flexible Workplace: Implications for Theory and Research." In *Working in Restructured Workplaces: Challenges and New Directions for the Sociology of Work*, edited by Daniel Cornfield, Karen B. Campbell, and Holly J. McCammon. Thousand Oaks, CA: Sage.

Kalleberg, Arne, Ken Hudson, and Barbara Reskin. 1997. "Bad Jobs in America: Non-Standard, Contingent and Secondary Employment Relations in the United States." Paper presented at the Annual Meetings of the American Sociological Association, Toronto.

Kalleberg, Arne, David Knoke, Peter Marsden, and Joe L. Spaeth. 1996. *Organizations in America: Analyzing Their Structures and Human Resource Practices*. Thousand Oaks, CA: Sage.

Keefe, Jeffrey H., and Rosemary Batt. 1997. "Technology and Market-Driven Restructuring: The United States." Pp. 31–88 in *Telecommunications: Restructuring Work and Employment Relations Worldwide*, edited by Harry C. Katz. Ithaca, NY: ILR/Cornell University Press.

Kelley, Maryellen. 1990. "New Process Technology, Job Design and Work Organization: A Contingency Model." *American Sociological Review* 55, 2:191–208.

Kenney, Martin, and Richard Florida. 1993. *Beyond Mass Production: The Japanese System and Its Transfer to the U.S.* New York: Oxford University Press.

Kern, Horst and Schumann, Michael. 1984. *Das ende der arbeitsteilung? Rationalisierung in der industriellen produktion*. Munich: Verlag C.H. Beck.

———. 1987. "Limits of the Division Of Labour: New Production and Employment Concepts in West German Industry." *Economic and Industrial Democracy* 8:151–70.

———. 1989. "New Concepts of Production in West German Plants." Pp. 87–110 in *Industry and Politics in West Germany: Toward the Third Republic*, edited by Peter Katzenstein. Ithaca: Cornell.

———. 1992. "New Concepts of Production and the Emergence of the Systems Controller." Pp. 111–48 in *Technology and the Future of Work*, edited by P. Adler. New York: Oxford University Press.

Klein, Janice. 1994. "The Paradox of Quality Management: Commitment, Ownership, and Control." Pp. 178–94 in *The Post-Bureaucratic Organization: New Perspectives on Organizational Change*, edited by C. Heckscher and A. Donnellon. Thousand Oaks, CA: Sage.

Lane, Christel. 1988. "Industrial Change in Europe: The Pursuit of Flexible Specialization in Britain and West Germany." *Work, Employment and Society* 2:2:141–68.

Lash, Scott, and John Urry. 1987. *The End of Organized Capitalism*. Madison: University of Wisconsin Press.

Lawler, Edward, Susan Mohrman, and G. Ledford. 1992. *Employee Involvement and Total Quality Management: Practices and Results in Fortune 1000 Companies*. San Francisco, CA: Jossey Bass.

———. 1995. *Creating High Performance Organizations: Practices and Results of Employee Involvement and Total Quality Management in Fortune 1000 Companies*. San Francisco, CA: Jossey Bass

Littek, Wolfgang, and Ulrich Heisig. 1991. "Competence, Control, and Work Redesign: Die Angestellten in the Federal Republic of Germany." *Work and Occupations* 18:1:4–28.

Magnum, Garth, Donald Mayall, and Kristen Nelson. 1985. "The Temporary Help Industry: A Response to the Dual Internal Labor Market." *Industrial and Labor Relations Review* 38:4:599–611.

Manley, Joan. n.d. "The Quality Premise: The Contradiction of TQM in Professional Organizations." Unpublished manuscript, Department of Sociology, Louisiana State University.

McGregor, Douglas. 1960. *The Human Side of Enterprise*. New York: McGraw Hill.

Meyer, John, and Brian Rowan. 1977. "Institutionalized Organizations: Formal Structure as Myth and Ceremony." *American Journal of Sociology* 83:2:340–63.

———. [1978] 1992. "The Structure of Educational Organizations." Pp. 71–98 in *Organizational Environments: Ritual and Rationality*, edited by John Meyer and W. Richard Scott. Thousand Oaks, CA: Sage.

Meyer, John, and W. Richard Scott (eds.). 1992. *Organizational Environments: Ritual and Rationality*. Thousand Oaks, CA: Sage.

Moody, Kim. 1997. *Workers in a Lean World*. London: Verso.

National Economic Development Office. 1986. *Changing Working Patterns: How Companies Achieve Flexibility to Meet New Needs*. London: NEDO.

Noble, David. 1984. *Forces of Production: A Social History of Industrial Automation*. New York: Knopf.

Osterman, Paul. 1994. "How Common Is Workplace Transformation and Who Adopts It?" *Industrial and Labor Relations Review* 47:2:173–88.

Penn, Roger, and David Sleightholme. 1995. "Skilled Work in Contemporary Europe: A Journey into the Dark." Pp. 187–202 in *Industrial Transformation in Europe: Process and Contexts*, edited by E. J. Dittrich, G. Schmitdt, and R. Whitley. London: Sage.

Penn, Roger, Kari Lilja, and Hilda Scattergood. 1992. "Flexibility and Employment Patterns in the Contemporary Paper Industry: A Comparative Analysis of Mills in Britain and Finland." *Industrial Relations Journal* 23:3:214–23.

Perrow, Charles. 1986. *Complex Organizations: A Critical Essay*. New York: Random.

Pfeffer, Jeffrey and James Baron. 1988. "Taking the Workers Back Out: Recent Trends in the Structuring of Employment." *Research in Organizational Behavior* 10:257–303.

Piore, Michael, and Charles F. Sabel. 1984. *The Second Industrial Divide: Possibilities for Prosperity*. New York: Basic.

Pollert, Anna. 1991. "The Orthodoxy of Flexibility." Pp. 3–31 in *Farewell to Flexibility?*, edited by Anna Pollert. Oxford: Blackwell.

Powell, Walter W. 1990. "Neither Market nor Hierarchy: Network Forms of Organization." *Research in Organizational Behavior* 12:295–336.

———. Forthcoming. "The Capitalist Firm in the 21st Century: Emerging Patterns." In *Firm Futures*, edited by Paul J. DiMaggio. Princeton: Princeton University Press.

Powell, Walter, and Paul DiMaggio, eds. 1991. *The New Institutionalism and Organizational Analysis*. Chicago: University of Chicago Press.

Powell, Walter, and Laurel Smith-Doerr. 1994. "Networks and Economic Life." Pp. 368–402 in *The Handbook of Economic Sociology*, edited by N. Smelser and R. Swedberg. Princeton: Princeton University Press.

Powell, Walter, Koput, K., and L. Smith-Doerr. 1996. "Interorganizational Collaboration and the Locus of Innovation: Networks of Learning in Biotechnology." *Administrative Science Quarterly* 41:116–45.

Prechel, Harland. 1994. "Economic Crisis and the Centralization of Control over the Managerial Process: Corporate Restructuring and Neo-Fordist Decision-Making." *American Sociological Review* 59:5:723–45.

Reich, Robert. 1992. *The Work of Nations*. New York: Vintage.

Romo, Frank, and Michael Schwartz. 1995. "Structural Embeddedness of Business Decisions: The Migration Behavior of Plants in New York State between 1960 and 1985." *American Sociological Review* 60, 6:874–907.

Roy, William G. 1997. *Socializing Capital: The Rise of the Large Industrial Corporation in America*. Princeton: Princeton University Press.

Rubin, Beth A. 1995. "Flexible Accumulation: The Decline of Contract and Social Transformation." *Research in Social Stratification and Mobility* 14:297–323.

———. 1996. *Shifts in the Social Contract: Understanding Change in American Society*. Thousand Oaks, CA: Pine Forge.

Sabel, Charles. 1982. *Work and Politics: The Division of Labor in Industry*. New York: Cambridge University Press.

———. 1989. "Flexible Specialization and the Reemergence of Regional Economies." Pp. 17–70 in *Reversing Industrial Decline*, edited by P. Hirst and J. Zeitlin. New York: St Martin's.

———. 1991. "Moebius-Strip Organizations and Open Labor Markets: Consequences of the Reintegration of Conception and Execution in a Volatile Economy." Pp. 23–53 in *Social Theory for a Changing Society*, edited by P. Bourdieu and J. Coleman. New York: Russell Sage.

———. 1994. "Learning by Monitoring: The Institutions of Economic Development." Pp. 137–165 in *The Handbook of Economic Sociology*, edited by N. J. Smelser and R. Swedberg. Princeton/New York: Princeton University Press/Russell Sage Foundation.

Sabel, Charles and Jonathan Zeitlin. 1985. "Historical Alternatives to Mass Production: Politics, Markets and Technology in Nineteenth-Century Industrialization." *Past and Present* 108:133–76.

———. 1997. "Stories, Strategies, Structures: Rethinking Historical Alternatives to Mass Production." Pp. 1–33 in *World of Possibilities: Flexibility and Mass Production in Western Industrialization*, edited by Charles Sabel and Jonathan Zeitlin. New York: Cambridge University Press.

Sabel, Charles, Gary Herrigel, Richard Deeg and Richard Kazis, 1989. "Regional Prosperities Compared: Massachusetts and Baden-Wurttemberg in the 1980s." *Economy and Society* 18:4:374–404.

Saxenian, AnnaLee. 1994. *Regional Advantage: Culture and Competition in Silicon Valley and Route 128*. Cambridge MA: Harvard University Press.

Schoenberger, Erica. 1997. *The Cultural Crisis of the Firm*. London: Blackwell.

Scranton, Philip. 1983. *Proprietary Capitalism: The Textile Manufacture at Philadelphia, 1800–1885*. New York: Cambridge University Press.

———. 1997. *Endless Novelty: Specialty Production and American Industrialization*. Princeton: Princeton University Press.

Sewell, Graham. 1995. "How the Giraffe Got Its Neck: An Organizational 'Just So' Story." Paper presented at the Sixth APROS International Colloquium, Universidad Autónoma Metropolitana-Iztapalapa, Cuernivaca, Mexico.

Shaiken, Harley, Stephen Herzenberg, and Sara Kuhn. 1986. "The Work Process under More Flexible Production." *Industrial Relations* 25:2:167–83.

Smith, Chris. 1991. "From 1960's Automation to Flexible Specialization: A *Deja Vu* of Technological Panaceas." Pp. 138–157 in *Farewell to Flexibility?*, edited by Anna Pollert. Oxford: Blackwell.

Smith, Chris, and Peter Meiksins. 1995. "System, Society and Dominance Effects in Cross-National Organizational Analysis." *Work, Employment and Society* 9:2:241–67.

Smith, Vicki. 1990. *Managing in the Corporate Interest: Control and Resistance at an American Bank*. Berkeley: University of California Press.

———. 1994. "Institutionalizing Flexibility in a Service Firm: Multiple Contingencies and Hidden Hierarchies." *Work and Occupations* 21:3:284–307.

———. 1996. "Employee Involvement, Involved Employees: Participative Work Arrangements in a White-collar Service Occupation." *Social Problems* 43:2:166–79.

———. 1997. "New Forms of Work Organization." *Annual Review of Sociology* 23:315–39.

Sorge, Arndt, and Wolfgang Streeck. 1988. "Industrial Relations and Technical Change: The Case for an Extended Perspective." Pp. 19–47 in *New Technology and Industrial Relations*, edited by R. Hyman and W. Streeck. New York: Oxford University Press.

Stinchcombe, Arthur. 1997. "On the Virtues of the Old Institutionalism." *Annual Review of Sociology* 23:1–18.

Storper, Michael. 1994. "The Transition to Flexible Specialization in the US Film Industry: External Economies, the Division of Labour and the Crossing of Industrial Divides." Pp 195–226 in *Post-Fordism: A Reader*, edited by Ash Amin. Oxford: Blackwell.

Strang, David. 1997. "Cheap Talk: Managerial Discourse on Quality Circles as an Organizational Innovation." Paper presented at the Annual Meetings of the American Sociological Association, Toronto.

Sutton, John R., Frank Dobbin, John Meyer, and W. Richard Scott. 1994. "The Legalization of the Workplace." *American Journal of Sociology* 99:4:944–71.

Tagliabue, John. 1996. "In Europe, A Wave Of Layoffs Stuns White Collar Workers." *New York Times*, 20 June, p. A1.

Taplin, I. 1995. "Flexible Production, Rigid Jobs: Lessons from the Clothing Industry." *Work and Occupations* 22:4:412–38.

Taplin, Ian, and J. Winterton. 1995. "New Clothes from Old Techniques: Restructuring and Flexibility in the US and UK Clothing Industries." *Industrial and Corporate Change* 4:3:615–38.

Thomas, Robert J. 1994. *What Machines Can't Do: Politics and Technology in the Industrial Enterprise*. Berkeley: University of California Press.

Tilly, Chris. 1990. *Short Hours, Short Shrift: Causes and Consequences of Part Time Work*. Washington, DC: Economic Policy Institute.

Uzzi, Brian. 1996. "The Sources and Consequences of Embeddedness for the Economic Performance of Organizations: The Network Effect." *American Sociological Review* 61:4:674–98.

———. 1997. "Social Structure and Competition in Interfirm Networks: The Paradox of Embeddedness." *Administrative Science Quarterly* 42:1:35–67.

Vallas, Steven. 1990. "The Concept of Skill: A Critical Review." *Work and Occupations* 17:4:379–98.

———. 1993. *Power in the Workplace: The Politics of Production at AT&T.* Albany, NY: State University of New York Press.

Vallas, Steven, and John Beck. 1996. "The Transformation of Work Revisited: The Limits of Flexibility in American Manufacturing." *Social Problems* 43:3:501–22.

Wilkinson, Barry. 1983. *The Shop Floor Politics of New Technology.* London: Heinemann.

Willis, Paul. 1977. *Learning to Labour: How Working Class Kids Get Working Class Jobs.* New York: Columbia University Press.

Womack, J., P. Jones, and D. Roos. 1990. *The Machine That Changed the World.* New York: Rawson.

Wood, Stephen. 1989. "The Transformation of Work?" Pp. 1–43 in *The Transformation of Work? Skill, Flexibility and the Labour Process,* edited by Stephen Wood. London: Unwin Hyman.

Wright, Erik Olin. 1985. *Classes.* London: Verso.

Zuboff, Shoshana. 1988. *In the Age of the Smart Machine.* New York: Basic.

Zucker, Lynne G. 1987. "Institutional Theories of Organization." *Annual Review of Sociology* 13:443–64.

[10]

Anna Pollert

Dismantling flexibility

● 'Greater flexibility in the labour market is no panacea, but it is nevertheless a means for quite a substantial amelioration in current problems' (OECD, 1986: 6).

● 'It was noticeable during 1985 that many managements were seeking greater flexibility in the use of labour' (ACAS, 1985: 9).

● 'The practices of Japanese companies . . . are understood in the managerial community in terms of a notion of flexibility very different from that which provokes the employment debate. Managers attribute the success of these firms to an unusual degree of responsiveness to market conditions' (Piore, 1986a: 158).

Two distinct ideas – that the labour force is becoming more flexible and that flexible specialisation offers a new means for capital to exert control – are critically evaluated. The evidence for both fashionable theses is found to be wanting; their increasing acceptability is linked to a retreat from class politics and from a confrontation of inequalities of race and sex. Indeed they herald a new form of imperialism.

'Flexibility' has engulfed conceptions of employment structure and restructuring since the early 1980s. To call it 'the flexibility debate' is over-complimentary. For most of the flexibility literature tends to be either prescriptive, or assumes there is a 'new' trend, which it then proceeds to describe and generalise. There has been no discussion on the origins of the term, to unravel its many connotations, to question what is indeed 'new', and to set this against the ideological processes which have unleashed it. Such a situation has political dangers: it defines the agenda of debate, assumes a radical break with the past, conflates and obscures complex and contradictory processes within the organisation of work, and by asserting a sea-change of management strategy and employment structure, fuses description, prediction and prescription towards a self-fulfilling prophecy. A deconstruction of 'flexibility' as an extremely powerful term which legitimises an array of policy practices is long overdue.

Within management literature, 'flexibility' has been a handy legitimatory tool precisely because of its all-purpose resistance to precise definition. Production flexibility, technical flexibility, organisational flexibility, labour process flexibility, time, wage, financial, marketing flexibility – all these issues are presumed to have a connection, and this connection is implicitly understood to reside in the behaviour of labour. Such a slippage has allowed the discussion of restructuring to veer away from the global issues of capital structure, investment, exchange rates, trade relations, to the homely, and apparently more manageable 'problem' of labour. The argument runs that the changeability of markets is the supreme challenge facing management, and its solution lies in the flexibility of adaptability of *labour*, both in the workplace, and in the labour market. Other dimensions of production, such as organisation and design, in which labour costs are not a prime issue, and other managerial driving forces, such as marketing, are swept off the table (Rubery *et al*, 1987). Such an ideological assumption is far from new; it accompanied the concern with the British worker's 'productivity' some twenty-odd years ago:

> . . . it had always been there, a sort of background music full of familiar melodies . . . it had always been accompanied by a more or less constant refrain – something about British workers and productivity. (Nichols, 1986: xi)

On a wider level, the whole 'flexibility' ethos is notable for its futurological discourse. With the growth of insecure, or irregular forms of work, and unemployment, social scientists have 'discovered' what casual wage labourers and those outside paid employment – very largely women and youth – already knew, that not all 'work' is stable and secure. Libertarian futurologists have heralded work in the small business, self-employed, informal and domestic economy as a 'new' form of work, not only more competitive (because more 'flexible' in adjusting to markets), but offering autonomy, 'flexible' hours, and in general 'flexible working lives'. A body of futurological writing based on a post-industrial conception of a 'radical break' in 'industrial society' and a largely technically determinist prophecy of the 'collapse of work' and 'discovery of leisure' (Jenkins & Sherman, 1979) has meshed with a dualist construction of the economy. In essence, a demise of regular wage labour, indeed, of wage labour itself, is given a libertarian legitimation, because it is both inevitable and progressive.

The left-reformist writing on 'flexibility', most notably articulated by Piore and Sabel (1984), explicitly distances itself from the concern with labour market flexibility (Piore, 1986a). It is also on quite a different intellectual level of theorisation from the management and 'manpower studies' literature (e.g. Handy, 1984; Atkinson, 1984). Nevertheless, the panacea of 'flexible specialisation' (FS) indicates an overwhelming consensus between left and right on the sovereignty of markets as the defining principle of restructuring. For the neo-classical revivalists and their management practitioners, this is obviously the strategy for capital. For the FS school, the restructuring of capital by the restructuring of production and *labour*, is once again the route forward, only this time, as a reform for capital *and* labour. The convergence between the right and FS dissemination of 'flexibility' is a central theme of deconstruction in the following analysis.

There are further conceptual affinities between the reformist FS perspective and the managerial literature, stemming from the prime place given to markets. This is the elevation of the fragmentation and decentralisation of production, and of the small firms sector, as the economic saviours of an ailing system. Behind this resurrection of economic dualism in both approaches, is the 'post-industrial' vision of a radical historical cleavage and branching point in social development, the projection of trends and a 'grand theory' futurology of technical and productive paths, from 'Fordism' to 'neo-Fordism'.

Before a detailed examination of these two 'flexibility' literatures and their relationship, two areas of mystification implicit in both need clarifying. First are the dimensions of labour flexibility under discussion, and second, the assumption that these are new concerns.

Flexibility in employment and in work – what's new?

The 'flexibility' debate has concentrated on two aspects of labour flexibility: flexibility in *employment* and flexibility in *work*. The neo-classical revival has provided the cue for the international concern with 'Eurosclerosis', the solutions for recession and unemployment sought in wages and labour market flexibility. Labour market deregulation in the form of the legislative offensive on labour organisation and employment rights and protection (Lewis, 1987; Hepple, 1987), the ideological and practical sponsorship of private enterprise and the small business sector, in the form of privatisation and competitive tendering, and the maintenance of a vast reserve army of the unemployed set the material agenda for using the competitive labour market as the solution to economic regeneration. Of state employment deregulation policies, the clearest statement of the virtues of labour market flexibility is the 1985 White Paper, *Employment, the challenge for the Nation*, a document which was deliberately modelled as the antithesis of the 1944 White Paper on employment, which embodied the post-war consensual commitment to full employment (Standing, 1986: 45).

Linked to the concern with labour market flexibility, the implicit focus on the flexibility or rigidity of labour within *work*, perpetuates the concern with flexibility manifest in British productivity dealing in the 1960s and 1970s, albeit with a crucial shift in bargaining power away from labour. A typical productivity deal in the 1960s concerned 'Increased flexibility in the use of manpower . . . [and] changes in working practices, e.g. tighter scheduling of transport operations, shift work changes and flexibility over numbers and types of machines supervised' (NBPI, 123, 1969: 4).

Concern with both forms of flexibility is not new. It is the flexibility of human labour which creates the elastic commodity of labour power and allows its extension and intensification in the extraction of surplus value. Capital has always required flexibility of labour; the struggle over its control has structured management development, the capitalist labour process, and forms of labour organisation.

In the early days of industrial capitalism, the emphasis was on defining rules and demarcations, both in time and task. The tension between the discipline of wage labour and the rhythms of life outside this social relation has been a source of conflict throughout the development of capitalism. E P Thompson charts the harnessing to time discipline through the 18th and 19th centuries, when the employers, far from wanting flexibility, imposed rigidity on the 'irregularity of the working day and week . . . framed until the first decades of the 19th century, within the larger irregularity of the working year, punctuated by traditional holidays and fairs. Eventually, the rigid wall between work and non-work was built':

> The first generation of factory workers were taught by their masters the importance of time; the second . . . formed their short time committees . . . the third struck for overtime or time-and-a-half. They had accepted the categories of their employers and learned to fight back within them. They had learned their lesson, that time is money, only too well. (Thompson, 1963: 76, 86).

The debate from this period about the benefits and costs of regular employment, organised or 'flexible' labour markets, was a perennial issue. Thompson (1963: 78) notes 'the debate between advocates of regularly-employed wage-labour and advocates of 'taken-work' (contract labour, AP). Casual work has always contributed to the process of capital accumulation, alongside the creation of a more organised proletariat. Berg (1985) emphasises the continuation of domestic and other forms of non-wage labour, particularly among women workers in the mass-consumption industries, until well into the 19th century. Littler's (1982) discussion of the decline of internal sub-contracting also points out its uneven continuation, which was seen until comparatively recently in the docks, and characterises (perhaps increasingly) the British construction industry. And the use of cheap out-workers, both in traditional industries and expanding new firms is not new (Hakim, 1985), nor is the deliberate use of married women as a low-paid, stable, yet unprotected labour force, in the small firms sector, or the 'toleration' of high turnover in sectors with unstable product markets (Craig *et al*, 1984). Casual labour may or may not have increased, but more important, it has become ideologically visible.

Only when 'rigidity' of time and task become the basis of skill, or of occupational ownership and bargaining strength did management unequivocally stress 'flexibility'. In engineering, for example, this became the centre of the 'craft question'. During the 1890s, the recently-founded engineering employers' association came in direct conflict with the Amalgamated Society of Engineers over the demarcation of work and the right to control the entry of skilled labour, and the use of new machines (Clegg, 1983: 20). The lockout of 1897 was precisely about wresting flexibility from craft workers, and gave 'freedom to employers in the management of their works, and set forth such conditions as to employment . . . the rating of workmen, the training of operatives . . . as the Trade Unions would not have accepted had they not been beaten' (*Cassier's Magazine*, 1902, quoted in McGuffie, 1986: 39). The struggle over the 'right to a trade' is a basic organising principle of trade unionism on the one hand, and of managerial prerogative on the other. To isolate it as a contemporary, or 'new' phenomenon, obscures this historical continuity and serves as a mystifying veil cutting off collective memory of past experience.

Having said this, however, there *is* something new about the current articulation of an imperative towards labour flexibility. This is its construction within a *dualist* employment and industrial perspective. It is to this question that we now turn.

Flexibility and dualism: the 'flexible firm' and 'flexible specialisation'

The distinctive way in which the new emphasis on labour flexibility has been mediated within a dualist perspective lies in the relationship between two apparently unconnected literatures: an elegant management policy model and a scholarly historical 'grand theory' developed as a strategy for economic restructuring. Two highly influential uses of 'flexibility' as a basis for labour and production reform, the British 'flexible firm' model (Atkinson, 1984, 1985a), and left reformist perspective of 'flexible specialisation' (Piore & Sabel, 1984; Sabel & Zeitlin, 1985), have both to be addressed, to gain leverage over the meaning of the term.

Both these restructuring expositions are grounded in dualist and segmentation perspectives; the first treats the simple dual labour market as a deliberate, conscious policy goal, and the second is a reversed application of dual labour market analysis by at least one originator of the dualist segmentation model. However, what is common to both is the legitimation of dualism as a dynamic and progressive force for the current period.

The 'flexible firm' model: an outline

The 'flexible firm' model asserts there is an 'emerging manpower system within the large primary sector firm; a 'new' polarisation between a 'core' and a 'periphery' workforce, the one with functional and the other with numerical flexibility, to provide adaptability to changing product markets (Atkinson, 1984, 1985a, 1985b). The multi-skilled 'core' group offers 'functional' flexibility in the labour process, by crossing occupational boundaries. It also offers flexibility by time, in terms of adjusting more closely to production demands. The 'periphery' provides 'numerical flexibility'; workers may be insecurely or irregularly employed, or not have a direct relationship with the firm at all, being, for example, sub-contracted or self-employed. Here, it is the precariousness of the employment relationship and competitiveness of the labour market which provides the employer with a numerically variable workforce.

The economic and social basis of the privileged sector is allegedly its importance to the 'organisation's key, firm-specific activities' (Atkinson, 1984a); thus, the model rests on the assumption of a direct relationship between primary employment conditions and the business concept of managerial 'strategic choice' of an organisation's 'core activity'. The 'peripheral' workers, by contrast, are less central (i.e. they are important, but not part of the 'core business'), making them more easily recruitable from the open labour market, and therefore less protected from its competitive pressure. It is their disposability which matters: 'numerical flexibility is sought so that headcount can be quickly and easily increased or decreased in line with even short-term changes in the level of demand for labour' (Atkinson, 1984).

The appeal, but also the weakness, of this model is both its comprehensiveness, in covering a wide number of employment situations in one model, and its simplicity. In a later reference to the model, Atkinson and Meager (NEDO, 1986) describe it as a tool with which to analyse a wide variety of changes in employment practice, rather than as a description of any real firm. There are others who share the view that the model is a rough and ready guide to understanding key changes in employment (e.g. Hakim, 1987: 93), and this is reflected in the wide currency of common usage of the terms 'core' and 'periphery', and of the two kinds of flexibility. Although Atkinson registers the insecurity of the secondary sector labour force, its deliberate expansion is nevertheless seen as either inevitable, or desirable from capital's point of view, and it is up to labour to make the best of this.

The 'flexible firm' was widely reported in the press (e.g. *Guardian*, 18 April 1984; *Industrial Relations Review and Report*, 7 August 1984; *Personnel Management*, August 1984; *New Society*, 30 August 1985), and 'core' and 'periphery' are common parlance in management conferences (e.g. IRS, 1985, 1986) and academic language. It has also entered trade union debate; its incursion into bargaining language has made 'functional' and 'numerical' flexibility common currency, and has forced a critical response from the labour movement (TUC, 1985; LRD, 1986). But besides evoking some critical commentary on the left, it has also frequently been incorporated as evidence of a 'new' management strategy in radical texts (CAITS, 1986; Mitter, 1986; Nichols, 1986; Potter, 1987). To the extent that it is fact or has become a material force as practice, this inclusion is justified. Yet this is far from proven, and its reportage as a 'fait-accompli' hardly clarifies the matter. It is therefore long overdue that the model, its factual evidence, its conceptual base, its policy implications are put to rigorous scrutiny.

Flexible specialisation: an outline

The strategy of 'flexible specialisation' is accomplished by a new form of artisan production made easily adaptable with the aid of programmable technology. It supplies customised goods to satisfy a plethora of individual tastes and needs and revitalises markets in a way similar to Atkinson's 'flexible firm', the labour process resembling that of the 'functionally flexible' 'core' workers. However, Piore and Sabel's scenario is not simply a resurrection of a legitimised dual labour market, but of a benign one.

The case for flexible specialisation as a strategy of production restructuring is based on the concept of technological paradigms, and an explanation of 'industrial' development based on competing systems of industrial technology: craft production and mass production. The dominant productive paradigm had been craft production, but became mass production as a result of historical contingencies – a 'chain of accidents compounded by mistakes' (Piore & Sabel, 1984: 193). The dominant 'system of industrial technology' (ibid: 5) is now the special purpose machine organised on an assembly line, operated by a semi-skilled labour force, mass producing standardised products for a mass market.

But the current economic crisis is a crisis of the dominant technological paradigm, mass production. It has reached the limits of expansion; mass markets for standardised goods are saturated, and have given way to a more fragmented and discerning pattern of demand, in which quality competition is more important than the price advantage of mass production. In such a crisis of production and markets, an alternative technological paradigm can break the deadlock, and offer the chance for a 'second industrial divide'. The signs are provided by the resiliance of industrial regions based on specialist artisan production which have successfully exploited microtechnology for new opportunities of more flexible production and the opening of new, specialised and variable market niches. Simultaneously, a managerial shift in larger firms to smaller units of more flexible production, including technical changes to flexible specialisation and organisational changes to just-in-time inventories, suggests a shift away from mass production and the sclerosis of saturated mass markets.

Basically, it is both *technology* and the *mass market* which previously determined the dynamism of the 'primary' sector, with its capital-intensive advantage, and held back the 'secondary' sector, with its low capital investment and inability to use economies of scale. And it is *new technology* and the *fragmentation of market demand* which reverses this relationship, making available to the small firm sophisticated, small-scale, technical flexibility. With the alleged change in the consumer market, this gives small-scale production the competitive edge over the vertically integrated 'primary' sector firm, where labour and technical stability are now rigidities, and where economies of scale are no longer cost effective. This *reversal* of dualist analysis can be seen at two levels: as a logical continuation of a *technological and market demand* about sectoral relations, and as a *political* shift from a major concern with the experience of underprivilege, to a commitment to the health of markets.

Flexibility as description, prescription and prediction

One of the distinctive traits of both the model of the 'flexible firm' and the FS thesis, is that the writing on flexibility is couched in a complex form of 'post-industrial' discourse. This asserts that the current period is a radical break from the past; the 'flexibility' managerial imperative is 'new', as is its accompanying segmentation. 'Flexible specialisation' brings the possibility of a 'new industrial divide', because the crisis of mass production brings a crucial branching point in history (Piore & Sabel, 1984). Yet the evidence for both these sweeping assertions is thin. To the extent that the empirical basis can be challenged, there is justification in arguing that both approaches are refractions through *an ideological lens*, which embraces a 'radical break', a new role for dualism, and a particular context for the emphasis on labour flexibility as progressive. To analyse this question, we have to examine both the concrete evidence for change, the ideological implications of this re-cast dualism, and conceptual underpinning which gives such force to the complex fusion of description, prescription and prediction characteristic of the revivalism of 'flexibility'.

Why is this so important politically? The empirical evaluation of the 'core-periphery' model is essential in the British context, since it has gained extremely wide currency and is accepted by many as an accurate description of new division in the working class. The empirical validity of flexible specialisation is no less important, both in relation to its basic premise that mass markets are indeed breaking up (an assumption also underlying the need for 'flexibility' of the 'flexible firm'), and in relation to the actual impact of programmable technology on the organisation of work. Such analysis cannot be divorced from the political implications of this preoccupation with labour flexibility and dualism. The novelty of the polarisation between a privileged core and a marginalised periphery is one of the elements of the view that the working class as an active force is dead. It is irretrievably divided, and the 'core' can only consolidate itself by working *with* capital. This is the case for the new realism, the case for the broad alliance of the soft left with any 'progressive' group, and the relegation of a large part of the working class – the periphery – to an unorganisable underclass. It is a conception built on a new accommodation with 'flexible patterns of work' legitimising pliability, insecurity, unemployment, and 'getting by' with self-employment. Here 'flexible specialisation' appears to square the circle for labour, since the artisan worker can now recreate the strengths of the craft tradition in the secondary sector with the invigorating breath of new technology and new markets. Both positions depend ultimately on a market perspective, and so, not surprisingly, converge with the neo-classical revival which sets the political agenda of the day.

We begin, then, with an empirical evaluation both of the 'flexible firm' model and the FS analysis, as the precursor to unravelling the common threads running through both literatures.

The 'flexible firm'

If the 'flexible firm' is indeed the 'new' trend in employment structure, it should be possible to show evidence both for an expanding 'periphery', and for a consolidating 'core' (Pollert, 1987). On existing evidence, there is little support for either.

The evidence for a growing 'periphery'

The 'core'-'periphery' dichotomy has been picked up and broadcast by some observers as a way to emphasise the importance of casual employment in the economy. Hakim has argued that, 'By the mid-1980s the labour force divided neatly into two-thirds "permanent" and one-third "flexible" . . . the importance of the "flexible" sector has clearly been underestimated; it is hardly a narrow and insignificant fringe on the edge of the labour market' (Hakim, 1987: 93). Unfortunately, this conceptualisation has always been a muddy conflation of legally disadvantaged, but long-term work, such as part-time work, and temporary work, so that, from the start, 'numerical flexibility' has added to a confused understanding of employment. This has been further complicated by the ahistorical 'discovery' of insecurity, as though it were radically new, and the calling of this a 'flexibility' trend.

The influential NEDO report, 'Changing Working Patterns', for example, claims that 'nine out of every 10 respondents had introduced changes to manning practices since 1980 designed to increase numerical flexibility' (NEDO, 1986: 6). However, the conclusion of 'radical change' is undermined by the conflation of standard practices to vary production with changes in employment. Besides part-time work, the inclusion of overtime working in 'numerical flexibility' confuses a traditional practice for permanent employees with varying the numbers of workers.

But if 'peripheral' work is taken to mean 'non-standard' contractual status, deviating from a full-time 'norm', there is little evidence that there has been a recent increase.

Temporary work. Hakim (1987) originally claimed, on the basis of the 1985 Labour Force Survey, that 'the most dramatic increase' in employment was in temporary work; this enthusiasm was dampened when the figure of a 700,000 growth between 1981 and 1985 had to be revised downward to 70,000 (*Employment Gazette*, April 1987: 218). Seen as a cyclical phenomenon, the 'increase' recedes even further:

> the proportion of temporary workers in the workforce peaked in 1979, before declining rapidly in the recession and resuming an upward trend in 1982-1983, such that by 1985 it has regained the level of the mid-1970s. (Meager, 1985: 4)

Surveying the private sector in general, the impression is that the greatest use of temporary workers occurs where this is already a well-established practice. The Warwick IRRU Company Level Survey (Marginson *et al*, 1988) found that the majority of enterprises, and just over half establishments, reported no change in policy towards the use of temporary contracts.

The most recent study of temporary work (Casey, 1987) provides further evidence of the absence of significant change in the use of temporary work. Excluding the 12.5 per cent on special employment measures (1984 Labour Force Survey), there was 'no real growth in the period 1983 to 1985. In 1983 temporary workers not in special schemes made up 5.5 per cent of the labour force; in 1985 they made up 5.5 per cent'.

Ironically, the private sector focus of the 'flexible firm' model excludes the one area where the use of temporary work may be most significant – the *public sector*. In 1984, fixed-term contracts were more common in the public sector services than elsewhere and the education sector was responsible for half of these (Millward & Stevens, 1986: 210). The Civil Service, particularly the DHSS, and the Water Authorities, are two other services found to be introducing 'new' temporary workers (Potter, 1986, 1987).

There is thus little evidence that there has been an overall increase in temporary work, and its appearance may well be the cyclical response witnessed in previous periods. Sectoral variation gives little support to a view of overall restructuring, and also highlights what the generalisation obscures: *the importance of changes in the public service sector.*

Part-time work. Between 1981 and 1986 the proportion of part-timers in the workforce grew from 21 to 23 per cent. From 1983, the only employment growth was an increase in female employment in the service sector, which was almost all part-time work. Between 1979 and 1986, 62 per cent of the rise in women's service sector jobs was in part-time work.

At the same time, the aggregate increase in part-time employment obscures the fact that it *declined* from 7.6 to 6.7 per cent of the manufacturing workforce. For women, this was a 40 per cent loss in manufacturing part-time jobs between 1979 and 1986 (Lewis, 1987). The 1984 Census of Employment and 1985 Labour Force Survey cast further doubt on a simplistic view of increasing part-time work. At an aggregate level, the proportion of the total workforce who worked part-time (defined here as working fewer than 30 hours per week) increased only from 14 per cent to 16 per cent between 1980 and 1984. As previously, it remains concentrated in the non-manufacturing sector, where it has increased from 18 per cent to 22 per cent of the workforce between 1980 and 1984 (Millward & Stevens, 1986: 205). For women, part-time work rose only from 40 per cent to 42 per cent of all women workers in the service sector between 1979 and 1984, and somewhat surprisingly, in retail it increased only from 34 per cent to 36 per cent of female employees. This could mask huge differences, by firm size, for example. It could also miss increases in the number of women earning below the National Insurance threshold. But it may well be that there are contradictory processes of growth and decline, as well as, what may be more qualitatively significant, changes in the type of part-time work (hours, length and regularity of shifts), rather than a dramatic increase.

Finally, turning to the public sector, we find once more that it is highly relevant: 'Two fifths of the manual workforce in the public services sector consisted of part-time workers in 1984, compared with just over a quarter in the private services sector' (Millward & Stevens, 1986: 207). And its main growth between 1980 and 1984 appears to be 'amongst the largest workplaces in the public services sector' (ibid: 209). These findings demonstrate that the 'flexible firm' conflates *changes in the public sector, and the services in general, with management strategy* in the private sector.

Outworking, freelancing and homeworking. This is a notoriously difficult area to research. Homeworkers, particularly ethnic minority women, are vulnerable to state or employer harassment and more likely to remain invisible in surveys, while the small firm sector, perhaps the most important for these forms of employment, is one of the most difficult areas to study. But in large firms (the subject of the 'flexible firm' model), Millward and Stevens (1986) found a decline in all these forms of employment between 1980 and 1984. This is explained mainly by the contraction in manufacturing. There is thus little evidence of the growth of a 'periphery' composed of these workers.

Subcontracting. The alleged increase in subcontracting is perhaps one of the most topical areas of the 'flexible firm', since it relates most directly to the interest in Japanisation, in the fragmentation of production and in the growth of small firms. But there is little evidence at an aggregate level of a dramatic increase in subcontracting in the private sector. The Warwick IRRU Company Level Survey found that 61 per cent of establishment respondents reported no change in the level of subcontracting over the previous five years (Marginson *et al*, 1988). Rather, subcontracting of services was *already* extremely widespread; 83 per cent of establishment managers reported that they subcontracted out at least one service, and 39 per cent of this group contracted out three or more services.

The NEDO report similarly found the long-standing nature of subcontracting of ancillary or service activities in the private sector (NEDO, 1986: 56), with many of the companies surveyed reporting that 'they started down this road in the early 1970s'. There appears to be a move towards more outside purchasing of parts in engineering; on the other hand, in food and drink, both the NEDO and Marginson *et al*, found a decline in subcontracting. Rubery *et al*'s study of consumer goods industries (1987: 140) points to differing rationales in the use of subcontracting. While some firms bought in and subcontracted components to reduce costs and increase flexibility of production, 'there was no evidence of a widespread increase in subcontracting'. Some subcontracted as an industrial relations tactic to avoid workplace conflict. Others had 'bought-in' production previously subcontracted, in response to 'major problems of co-ordination, design and quality control, particularly because of the growth of subcontracting on an international scale' (ibid).

The blinkered concern with *increases* in subcontracting masks crucial changes at the qualitative level. The ESRC-funded Steel Project, which investigated the experience of redundant British Steel Corporation workers in Port Talbot between 1980 and 1985, found that while BSC expanded the range of work put out to subcontract, what was new was newcomer subcontractors' undercutting of established subcontract firms. This was achieved through breaking union membership, lowering wage rates, ignoring health and safety precautions, and organising work 'off the cards' (Fevre, 1986: 23). An explanation of these varied and often contradictory trends would provide a more penetrating analysis than the imposition of a managerial imperative to 'peripheralise'.

Self-employment. Estimates of the increase in self-employment since 1979 also vary enormously, from between 30 per cent and 45 per cent since 1979. Overlooking this problem, it appears that 62 per cent of this increase was in the service sector, 20 per cent in construction, 10 per cent in agriculture and 8 per cent in manufacturing (TUC, 1986c). Most of this appears to be due to an increase in single-person businesses: between 1981 and

1984 the number of people who were self-employed in their main job grew by 442,000 to 2.6 million, or from 9.2 to 11.2 per cent of total employment (Creigh *et al*, 1986). While this may be associated with redundancy, it does not seem to be associated with flexibility strategies in large companies. Atkinson and Meager (NEDO, 1986: 58) found the growth in self-employment 'negligible', except in financial services and the food and drink sector, and here the rationale did not appear to be to increase 'flexibility'.

The evidence for a 'core'

The 'core' defined by recruitment criteria. To start with the problem of defining a 'core', the labelling of all permanent employees as the 'core' workforce is either tautologous or highly debatable if it overlooks segmentation in status, terms and conditions within the permanently employed. Marginson *et al* (1988) therefore suggest special recruitment procedures as one criterion of 'core' work. They start 'from the straightforward postulate that if any group of employees constitutes a "core" workforce then that must include managers.' Recruitment of such a core would normally involve higher-level management.

The general conclusions of their survey are that the majority of firms do not involve higher-level management in the recruitment of non-manual and manual employees. In the main, recruitment of non-managerial employees remains a local, establishment matter. There are also marked sectoral patterns, with enterprises in finance most likely to recruit employees at a central level, and those in mechanical engineering, textiles, clothing and footwear and distribution (for manuals) recruiting at local level only. Using the recruitment criterion, it would appear that the deliberate fostering of a 'core' workforce is at best patchy, and sector-based rather than generalised.

Functional flexibility as the basis for a 'core'. Other research has looked in detail at the developments in 'functional flexibility' as chief ingredient of the 'core'. But the extent of multiskilling appears limited. Cross's (1987) research indicated that most occupational flexibility was confined to the production/maintenance boundary, with little change to multiskilling in engineering and craft areas. IDS Study 360 of capital-intensive manufacturing industry, questions whether, in fact, multiskilling is as widely desired as a 'flexibility' trend suggests:

> Companies have widely different aims. Competitive pressures vary between sectors. Skill requirements are dependent on widely different technologies. More skill, different skill, less skill, can all be legitimate objectives.

> Generalisations about moving towards a small, highly skilled and highly flexible workforce may fit one company and be contradicted by the pressure towards a quite different workforce in another. (IDS, 1986: 1)

Similarly, a trade union based study of 'flexibility' concludes that, 'in spite of the impression created by a small number of widely publicised case studies . . . a full range of flexible working practices are not frequently implemented' (LRD, 1986a: 5). Many 'flexibility' agreements were only 'enabling' not faits-accomplis. Nor was there, apart from rare exceptions, evidence of a 'core' in terms of improved job security.

Even the NEDO (1986) study, which used the 'flexible firm' as an analytical model, found extremely limited evidence for 'functional flexibility'. In manufacturing there was little 'radical job enlargement' (NEDO, 1986: 45) which crossed 'core trade' boundaries; while three-quarters of respondents had achieved a limited overlap between maintenance craftsmen, only a third had achieved dual skiling even within electrical and mechanical trade groups and only 15 per cent had achieved it across the electrical/mechanical divide (ibid: 46). Significantly, since the introduction of technical change 'appeared to have the effect in most manufacturing companies of reducing the number, and skill level, of tasks required by process operatives, the tendency was to reduce the numbers of operatives and/or spread them more thinly across the plant, adding a number of deskilled tasks to their job' (ibid: 50). Further, the greatest obstacle to increasing 'functional flexibility' was the cost of re-training, and shortage of training resources (ibid: 50). And since a short-cut to skill shortage was often the buying in of specialist labour, the development of flexibility in the 'core' workforce was actually hindered.

In Britain 'core' status thus has a hollow ring, especially when 'employment security boiled down in practice to a reduction in the threat of job loss rather than anything more positive' (ibid: 79). This is hardly a convincing case for a consolidating 'core' workforce.

The flexible firm: conclusions on the evidence

From this review of the changes in the workforce and in employment practices, the 'flexible firm' model is left standing with few clothes. Where there has been most major restructuring, this has been led by the state as employer. But in the private sector, sectoral continuity is far more in evidence than change, with little evidence of polarisation beween an (ill-defined) 'periphery' and a privileged 'core'. Despite this, the 'core'-'periphery' model has become a conventional wisdom of today's alleged 'flexibility' trend.

Before trying to explain this at a political and ideological level, the radical restructuring programme of flexible specialisation should be similarly exposed to empirical analysis.

Flexible specialisation

Flexible specialisation is part of a much wider debate on early industrialisation which cannot be addressed here. As a method of analysis and strategy, it has been criticised on a number of different levels (e.g. Hyman, 1986; Smith, 1987; Williams *et al*, 1987). For the present purposes, the assessment of the FS thesis will be judged on the validity of its key premises:

1) the concept of the 'technological paradigm' based on the equation of a 'system of industrial technology' and a dominant system of production.

2) the alleged decline of mass production.

3) the alleged decline of mass markets.

4) the alleged new opportunities of new technology for small firms.

5) the actual spread of skill-upgrading FS as the prevalent use of programmable technology in large firms.

Much of the evaluation of FS, particularly the assertion of the fragmentation of markets and the 'new' importance of the multiskilled worker and new technology also applies to the 'core-periphery' employment restructuring model.

1) Mass production as a technological paradigm

The concept of Fordist mass production as a dominant system of 'industrial technology' is a crude and inaccurate delineation of the main developments of production systems in 20th-century capitalism. Types of technology, their organisation into different systems of production, and types of labour processes cannot be conflated into a single paradigm; the empirical evidence is far more complex. Within the large, technologically advanced firm associated with primary sector production and which Piore and Sabel identify with mass production, a variety of technologies and production systems exist, depending on the nature of the product market. Thirty years ago, Joan Woodward's research on firms employing over 100 employees found differences in productive systems 'could not be related to the size of firm', or to technology, but:

> on their detailed objectives which depended on the nature of the product and the type of customer. Thus some firms were in more competitive industries than others, some were making perishable goods that could not be stored, some

produced for stock and others to orders; in fact, marketing conditions were different in every firm. (Woodward, 1958)

These different production systems have long been recognised to fall into three standard categories: 1) small batch and unit production, 2) large batch and mass production, and 3) process production. Woodward also identified a lack of neat fit between type of technology and a single productive system: 'Automatic control can be applied most readily to mass production and continuous flow-process production, but even in unit and small-batch production devices for the control of the individual machine can be used' (ibid).

Some firms also remained outside the threefold classification, simply because they were 'too mixed or changing'.

To set up a false dichotomy between 'mass production' on the one hand, and 'craft production' on the other, and discover product diversity as proof of an alternative 'paradigm', is thus a distortion of a much more complex reality.

2) *The decline of mass production*

To demonstrate a basic opposition between two types of production – mass production relying on mass markets, dedicated equipment and semi-skilled workers making standardised goods, versus flexible equipment and skilled workers producing specialist goods – poses a false problem. No such clear opposition exists.

But even if we confine the argument to developments in assembly-line production, there is substantial evidence that even within flow-line technology, product diversification takes place without prolonged delays or increased skills. Multi-model and mixed-model lines can be used either to change the products or vary a basic model (Bennett, 1986: 89). The resilience of mass production to greater product diversity and responsiveness to markets has been discussed elsewhere. Murray (1983) points to improvements in stock control. Gough (1986: 64) observes that even the much quoted example of Benetton, 'produces most of its clothes bleached, the colour variation being added as the final stage'.

The crude distinction between 'dominant' production trajectories also obscures the complexity and contrary directions of production organisation: there are trends *away* from customised *towards* mass production in the production of machinery (Piore & Sabel, 1984: Ch. 4); and decentralisation of production through subcontracting cannot be conflated with a return to artisan production. It can simply be another mode of division of labour and standardised production.

But a major, and much neglected, problem with the theory of 'Fordism' and the decline of mass production is the omission of 'Fordism before Ford' – the female gendering of most semi-skilled assembly workers. The focus has either been on the car industry – which perhaps stands out because the Fordist assembly line is the exception, rather than the rule, for male production jobs – and engineering occupations, which are rarely part of mass assembly production anyway.

As Glucksman (1986) has demonstrated, the new mass production industries of the inter-war years producing the consumer goods for the new mass markets were constructed as *women's work*. The assembly line, the moving conveyor belt and flow processes created a new division of labour in which women performed the semi-skilled work of repetitive standardised production. Taking electrical engineering as typical of the new industries, women were concentrated in the assembly of electric lamps, batteries, telephone and wireless apparatus, valves, electric light accessories, heating, cooking and other apparatus. Jobs became sex-typed, for women were 'unquestionably' more patient, dexterous and reliable but 'wherever new methods of production were introduced, men acquired, or retained, control over the technical process' (Glucksman, 1986: 29).

The sexual division of labour in manufacturing is no different today. Where semi-skilled women workers have not been rationalised away and remain a cheaper, or more reliable option than machinery, they remain in the labour-intensive production, assembly and packing jobs of the mass production industries (Herzog, 1980; Pollert, 1981; Cavendish, 1981; Armstrong, 1982; Westwood, 1984). Yet consumer durable industries, and the assembly lines in them, are absent from the FS discussion.

Existing work on the social construction of skills and occupations, and the gendering of technology, confirms women's isolation from machinery, and the likely displacing and deskilling impact of new technology (Cockburn, 1983, 1985). Since social differentiation, by sex, race or age, is a key process in the formation of occupational and skill identity, where restructuring does take the form of upgrading to flexible specialisation, the gains in control are likely to be gendered. Without conscious feminist struggle, this would reinforce existing segregation. The alleged trajectory of a decline in mass production and rise of FS is thus a complacently male sex-blind view as well as empirically inaccurate.

The decline of mass markets?

If there is no clear opposition between mass production and

batch production, and if product markets have always been a mixture of standardised and differentiated goods, then what is the evidence for the break up of mass markets? With the concept of 'mass markets', we enter the same problems of definition as with mass production. As Williams *et al* (1987) ask: at what point do we cross the rubicon of *diversified* mass markets and enter the fragmented market for the specialised goods of flexible specialisation?

The much discussed fragmentation of the market in the clothing industry is a further case in point. For example, the proliferating variety of 'Next' retail outlets, including the takeover of Grattan mail-order (*Observer Magazine*, 10 January 1988), is a planned strategy of carving the consumer market into predictable chunks. The capturing of markets by the deliberate cultivation of finely-tuned consumer tastes naturally becomes more urgent as competition intensifies, but this is not evidence of a decline, but of a more sophisticated *manipulation* of the mass market. Indeed, one can look at a whole range of industries which are based on mass and large batch production and continue to sell well to large markets: food, drinks, flat-pack furniture, DIY goods, toiletries, records, toys – the list covers most consumer goods.

The Japanese consumer durable industry has been quoted in support of the rationale for FS:

> Managers attribute the success of these firms to an unusual degree of responsiveness to market conditions, responsiveness which eludes the traditional, tightly integrated, hierarchical corporations. (Piore, 1986a: 156)

But the Japanese success can hardly be described as a shift away from mass markets; televisions sell not because new models keep appearing, but because they are well designed, reliable and cheap. New products also create markets, while the need for replacements likewise means that the market for domestic equipment has no more reached saturation today than before (Williams *et al*, 1987: 425).

As Gough has argued (1986), the focus on a fragmentation of demand as a root of crisis, mistakes cause for effect. Fragmentation is part of marketing strategy, which is one way (along with price competition, which cannot be said to be obsolete for all the claims of the FS school) to deal with the problem of overaccumulation. This leads to the other major misconception in the analysis of a decline in mass markets: the substitution of a totally arbitrary conception of saturation in *use* values, for a crisis in the realisation of *exchange* values. The crisis of overaccumulation is reduced to a crisis in mass markets. Markets, and the tastes of the 'sovereign consumer' (Hyman, 1986) are as crucial to the FS analysis as they are to the neo-classical world view.

Production flexibility: new opportunities for small firms?

One of the lynchpins of FS as both the explanation of the persistence of artisan production and facilitator of decentralised restructuring is the suitability of the new, flexible technology for small firms (Zeitlin, 1985: 13, 21).

The evidence does not support this case. In the British clothing industry, as supporters of decentralisation have conceded, the major items of equipment for more flexible production, such as computer-aided design (CAD) and computer-controlled cutting systems, are very expensive, and available only to the large clothing manufacturers (Zeitlin, 1985: 14). Any future reduction in cost is speculative (Zeitlin, 1985: 13). In the meantime, it is the *large* producers who are best able to implement flexible production of relatively short runs, threatening of London's small, specialist garment manufacturers (GLC, 1985: 13). Thus, the reality is that large firms can invest, where smaller ones cannot – hardly a revelation. The 'solutions' for the small firms, such as developing a federated structure for sharing equipment and management advice, would face the formidable competitive advantage of the larger (75-250 worker) firms.

Paradoxically for the advocates of decentralisation, quality competition (the driving force for more flexible production) leads retailers to rely more heavily on larger manufacturers; far from decentralising, there has been growing pressure on suppliers to move production 'in-house' (Zeitlin, 1985: 19). This has added further advantage to the larger firms (GLC, 1985: 3).

The agument thus returns to the basic point of analysis, that flexible production does not necessarily mean small production units (GLC, 1985: 4). One cannot conflate technology, productive system and organisational form into a technologically determinist form of macro-restructuring. Similarly, one cannot assume a technologically driven restructuring of semi-skilled work to craft revival – the final point of our critique of FS.

Production flexibility and the rise of flexible specialisation?

The continuation of debate over the skill upgrading or deskilling utilisation of microprocessor technology, challenges any assertion of a dominant trajectory towards the revitalisation of craft skills in FS. The 'post-Taylorist' view of Piore and Sabel can be contrasted with the evidence of Shaiken *et al* (1986). This demonstrates the use of new technology to wrest control from production workers and further centralise management control.

Elger's (1987) review of recent British studies of restructuring using new technology for more flexible production, reinforces a growing body of evidence that both management and labour can wrest gains and suffer costs in the negotiation of change. Uneven development also emerges as the most marked result of recent research on consumer goods industries (Rubery *et al*, 1987). This shows that while more flexible production and ease of product differentiation is increasingly important with heightened competition, this is not necessarily achieved by new technology, and where it is, does not necessarily lead to the creation of a polyvalent craft workforce (Rubery *et al*, 1987: 145). Indeed, the outcomes for labour of shifts to more flexible production vary between continuing dependence on traditional skills, deskilling, skill increases and skill polarisation – a conclusion similar to Kelley's review of the debate (1987). What does seem to emerge repeatedly is that there is no clear development towards an upgraded craft revival.

The critiques of both the managerial and the 'radical' applications of both labour and labour market flexibility highlights one common characteristic: both cases are based on 'ideal type' models of reality, which do not bear up well to empirical scrutiny. This leaves us with a fundamental question: is there a deeper conceptual and ideological affinity between these apparently distinct constructions, which serves to explain why, in spite of their shaky factual foundations, they are pursued so intently? We must turn then to a conceptual and ideological analysis.

Dualism, the 'informal sector' and 'post-industrialism'

In discussing the 'flexible firm' and 'flexible specialisation', we are juxtaposing very different literatures, but they have a distinctive conceptual alliance. One is a rooting in, and re-application of dual labour market analysis. The other is in a post-industrial conception of a radical social break, forecasting new trajectories into the future. Both employ a method of argument, fitting complex facts into 'ideal type' models, and basing global sweeps on selective evidence. With the 'flexible firm', this is a future trend in management strategy. In 'The Second Industrial Divide', there are a number of possible futures.

In both literatures, economic dualism is in different ways the basis of a new set of policies and new social structure. The dual labour market resurrection of the 'flexible firm' as a 'new' segmentation is a crudely ahistorical transformation of an academic analysis into a practical policy. The very basis of dual labour markets were not homogeneous, but segmented between an internal labour market and an external one (Doeringer & Piore, 1971). The 'flexible firm' model ignores this long-standing segmentation, and starts from a false premise of initial homogeneity. It also reproduces the lack of clarity of dual labour market analysis in providing no clear criteria for its dichotomisation (Pollert, 1987).

It is an irony that Piore, who in the early 1970s pioneered dual labour market analysis as an explanation of the underprivilege of black workers (Doeringer & Piore, 1971), should recast his perspective in a reversed form, the 'secondary' industrial sector becoming the beacon of progress, the ills of 'secondary' labour market conditions finding a cure.

Looking back to this early period, Piore recalls:

> For us at the time the different fates of black and white workers in the economy were the critical factors. 'Dualism' captured the perceptions of black workers and of many white workers and employers as well as the structure of differences between black and white employment opportunities. (Berger & Piore, 1980: 16)

One can begin to detect a shift of interest in the 1980s, with the theorisation for what was largely a descriptive account of dualism (Berger & Piore, 1980). This looked towards the technological and political basis of economic uncertainty as the source of dualism. Returning to Adam Smith's analysis of the relationship between the division of labour and the size of the market, Piore lays the foundations for the later elaboration of FS, in a model of industrial structure driven by the relationship between the structure of market demand and technological development and organisation (Berger & Piore, 1980: 24).

The growing interest in flexibility and uncertainty was already apparent in Berger's account of the resilience of the traditional sector in France and Italy (Berger & Piore, 1980) echoing the enthusiasm of Italian segmentation theorists for the small firms sector (e.g. Brusco, 1982; Solinas, 1982).

Ultimately, this shift in priorities can be seen in political terms. The segmentation tradition was not alone in turning its attention to the secondary sector in this period. Recession, growing unemployment and deepening crisis at the end of the 1970s meant that the small firm sector became both ideologically and politically more significant, as a sector which might potentially shoulder the risks and costs of 'economic recovery', and generate some national employment growth (Gerry, 1985: 300). Although the small business sector has 'proliferated' by high turnover, rather than growth – 'the birth and death rates of small enterprises appear to have more or less matched one another in recent years' (Gerry, 1985: 308) – the ideological impact of the small-business/self-employment focus is vital. It has fostered competitive individualism, and obscured the extent of exploitation in this sector with the illusion that the self-employed are all independent, profit-making entrepreneurs.

There is a heritage to the dualist focus on the small business sector in the discussion of the 'informal' sector in the Third World. The 'informal' or 'marginal' sector constructs were inspired by 'the need to explain the impoverishment of increasing numbers of people, simultaneous with economic growth' (Connolly, 1985: 60). Dualism served to mystify the connections between unemployment, poverty and the *integration* of small-scale production into capital accumulation, by presenting two 'economies' as separate. It became the corner-stone for separate 'solutions' for both 'economies', either as a celebration of the informal sector as a vanguard of enterprise, or as a social problem for philanthropic concern. Either way, the linkage of poverty to the underlying processes which make it a necessary part of the global system of production, appears broken.

Connolly argues that in the case of Mexico in the early 1980s, the informal sector served to legitimise policies which reinforced the extreme inequality of wealth. Massive state subsidies to the most profitable capital-intensive industries were justified on the grounds that 'modern development' was the panacea to unproductive backwardness. Complementing this, welfare expenditure and 'selective aids' to labour-intensive activities using 'adequate' technologies could be justified, if only theoretically, to encourage an autonomous informal sector, without threatening the underlying concentration of capital (Connolly, 1985: 67).

Similar policies underpinned the use of the 'formal', 'informal' nomenclature elsewhere; in the ILO missions to Colombia in 1970, to Sri Lanka in 1971 and Kenya in 1972 (Gerry, 1985: 296). As crisis deepened, and it became clear that the alleged 'trickle down' effect from the formal to the informal sector was not to be, the more the informal sector became ideologically transformed from a social blight, to the vanguard of enterprise (Pearson, 1987).

This ideological celebration of the 'informal' economy can also be seen in the recessionary phase of 'post-industrial' theory, a perspective which has found strong expression in the current obsession with 'flexibility' as a return to small business. The post-war boom inspired a number of 'post-industrial' perspectives of a radical break from 'industrial society', rooted in assumptions of guaranteed growth, general affluence and satisfaction of material needs. Bell (1974), the most influential exponent of this optimistic projection, held that the break consisted in a fundamental employment shift from manufacturing, to the provision of services, and a new 'service society', with a 'knowledge elite' as its basis of power.

As the end of the post-war boom forced close questioning of these full-employment projections, Bell's benignly voluntarist 'service society' came under scrutiny. Gershuny (1978) pointed out that service occupations were not the same as services for people, since most were integral parts of manufacturing organisation. This successfully undermined the basis of a 'post-industrial' break. Yet Gershuny continued the 'qualitative break' tradition, but now in the context of recession and unemployment. His analysis pursued the problem of how services were actually provided, and focussed on the household economy and, increasingly, on work in the 'informal' sector. Kumar (1978) sought an ideological break with the 'growth' ethic of 'industrial' society, towards a libertarian, 'small is beautiful', world. Veering away from concern with large-scale employment and wage labour, a new literature of 'recessionary' post-industrialism emerged, predicting the demise of wage labour in the formal economy. Emphasising the economic viability of production in the informal sector, it has been influential in the broadcasting of a polarisation between formal and informal economies, core and peripheral employment.

The 'post-industrial' emphasis on the informal sector as a radical break has a parallel in Piore and Sabel's recasting of the secondary sector as not only economically dynamic, but beneficial for labour. The argument behind Gershuny's vision of the self-service economy comes close to Piore and Sabel's analysis of the demise of mass production, and vision of alternative trends. Both advocate a 'third way' out of crisis, which is neither a Keynesian bolstering of demand, nor a monetarist solution, but a furtherance of a dual economy. Both also invest prime importance in consumer demand, and the transformative powers of new technology to supply this in new ways. For Gershuny,

> The third strategy, instead of working against the trends, works with them . . . [it] is to start to consider the redesign of the household, or rather to consider the social basis of final consumption . . . The 'dual economy' strategy would not seek to discourage the continuing drive for efficiency, with accompanying unemployment, in the formal sector. Instead, it would seek to improve both work and leisure in the informal sector; indeed, since in this sector production and consumption activities are based on the same social unit, the distinction between work and leisure might itself become less clear-cut. As a result of this strategy, the informal sector might become a viable alternative to

employment in the formal sector. (1978: 160-1)

Piore and Sabel's arguments for the resurgence of craft production is based on a historical analysis of competing productive paradigms, not on complementary systems (except at the international level). But the resonance between their own celebration of the craft tradition, and the 'post-industrial' economy of self-employment is unmistakable.

The political implications of the 'post-industrial' embrace of the small business sector reveal themselves in their convergence with enterprise capitalism's enthusiasm for deregulation. Piore remarks 'there is considerable truth in at least Reagan's vision of American prosperity . . . we are more dependent on entrepreneurial activity than we were in the past' (1986b: 211). It is no coincidence that the 'post-industrial' vision embellishes the management literature on 'flexible' work. 'New' types of working offered by new technology and non-standard forms of work offer possibilities of flexible working time, and the much sought blurring between work and leisure. The 'radical break' brings a new quality of inevitability and liberal appeal to the 'post-industrial' economy of the demise of collective wage labour.

Clutterbuck and Hill (1981: 4) assert that, '. . . many forecasters believe we are on the edge of a shift in attitudes and behavioural patterns comparable to that involved in the transition from the Middle Ages to the Renaissance'. This, together with the decline in manufacturing, new technology, unemployment, and the growth of the informal economy, is all moulded to the need for 'flexibility': 'As will emerge time and again throughout the following pages, the key to meeting the challenge of tomorrow's workplace is flexibility of response. . . . In theory at least, those companies that develop the flexibility of response . . . will be the ones to prosper' (ibid: 22).

Handy (1984: 8) predicts: '. . . Jobs will be shorter . . . jobs will be difficult, more dispersed, and, in many cases, more precarious.' As a consequence, he forecasts the growth of the 'dispersed' or 'contractual' organisation (ibid: 74, 79): 'Whether we like it or not (and there are many who don't), the contractual organisation is with us, is growing and is likely to grow faster' (ibid: 81).

The 'flexible firm', with its emphasis on decentralisation and 'new' management strategy, is part of this 'radical break' management literature with a similar fusing of assertion, prescription and prediction (Pollert, 1987). The strain between the model's assertions and unsupportive empirical facts expresses itself in a duality between sweeping generalisation on the one hand, and immediate qualification or retraction, on the other. In

IMS Report No. 89, a radical change is first announced, then negated:

> In the UK we are beginning to witness important changes to orthodox ideas about work organisation and the deployment of labour. As yet such changes cannot be said to add up to new and coherent employment strategies on the part of employers. (Atkinson, 1985a: 1)

The report contains this tension throughout (Pollert, 1987). Eleswhere we are told, 'flexibility has become an important theme in emerging corporate thinking', yet, 'research suggests in practice, relatively few UK firms have explicitly and comprehensively reorganised their labour force on this basis' (Atkinson, 1985b).

Trends are both asserted and denied, contradictions obscured by the interpretation of numerous practices as 'radical breaks', even if the 'small print' shows they are not. This prophetic style has unleashed a method of policy dissemination, where the prescriptive message is carried in the assertion of a trend. A distinctive language has been created, based on the compelling force of asserting trends which will leave you behind if you do not join in. But the arguments of FS are similar, except here the projection of trends is built on an elaborate historical edifice. In both, the edifice is less secure than the confident style admits.

The common ground between the radical restructuring programme of reviving artisan production, and of management strategies promoting greater labour market flexibility leads to further questions about the relationship between both 'flexibility' literatures at the policy level. Some argue that the Atkinson model is quite different from the FS scenario (Williams *et al*, 1987: 434). While there are obviously differences, there are more similarities than meet the eye.

The 'flexible firm' and flexible specialisation: Political convergences

Labour flexibility and 'core-periphery' as a new state of capitalist integration

The policy implications of the 'flexible firm' model and of FS are neither identical in scope nor content. The first is a managerial strategy for controlling labour, the second, a strategy for restructuring both the labour process and capital as a whole. In the first, the concept of 'functional flexibility' at the 'core' allows for the multiskilled craft worker, but also the more pliable employee and the removal of demarcations. But in both models, centrality in the labour process strengthens labour not against capital as the most advanced, organised section of the working class, but to work *with* capital in a new partnership. Japanese lifetime employment is implicitly – (or with FS explicitly) – the reference point.

But there is also a much wider ideological message of social integration. The 'core' and 'periphery' model is one of organisational balance, labour process flexibility in the one supplemented by labour market flexibility in the other. The FS analysis is likewise based on dual labour market analysis as a model of dynamic equilibrium. It rejects an analysis of capitalism as a system based on contradictory class interests, and wholeheartedly supports market regeneration. It poses a new equation of sectoral and productive balance, which is healthier for markets. In both projections, dualism suggests a functionalist balance, whereby economic and social equilibrium is maintained by labour market division.

The trade unionism of FS (and 'core') workers co-operates with capital, apparently as the *realistic* 'Challenge for Union Survival' (Piore, 1986b). In new-realism-speak:

> The hope for unions lies in articulating an alternative vision, one which is both more realistic and effective in providing an operational understanding of the socio-economic system and which, at the same time, provides an organic and central place for worker organisation. (Piore, 1986b)

For the car industry, Katz and Sabel argue:

> Unions will have to exchange the rights to impose uniform conditions for rights in decision making . . . the unions will have to find ways of tying the interests of particular companies to the interests of their industry and even of the economy as a whole. (1985: 314)

Such a scenario resonates with the New Realism of certain union leaderships in Britain (the AEU and the EETPU for example), eager to secure their 'core' status in the 'flexible firm'.

Core-periphery and flexible specialisation as social division

In FS, the benign version of economic dualism, the question of underprivilege fades into silence. The 'flexible firm's' explicit divisiveness is paradoxically a more candid admission of the concrete applications of fragmentation strategies, than all the sophisticated caveats of welfare protection in the FS case. But the stubborn persistence of inequality in the FS utopia will not go away (GLC, 1985; Gough, 1986; Murray, 1983, 1987; Jenson, 1988). The craft ideal is worrying in itself:

> Hints that women may not do very well under flexible specialisation begin to niggle at feminists while reading

> Piore and Sabel's fond references to Proudhon . . . embedded at the heart of his understanding of the independent, craft based working class – which he celebrated – was the belief that women should only work in the home and that any self-respecting working man ought to be able to support his non-waged wife and children. (Jenson, 1988: 4)

Murray (1983, 1987) has shown the seamier side of decentralised production in the Third Italy: racial, gender, skill and age divisions are essential to its success. Competition between firms results in a survival of the fittest which appears in no way ameliorated by local state intervention. Geographical fragmentation between firms and phases of production creates 'maximum wage differentials between different groups of workers', undermining solidarity, and making union organisation 'an uphill task'. Meanwhile 'much work can in no way be described as prized and non-alienating craft labour' (Murray, 1987).

Sharp differences between workforces and firms extends to geographical fragmentation, so that privileged and successful industrial regions, in the Third Italy, for example, can export the insecurity, poor pay and working conditions needed to provide 'flexibility' with fluctuations in demand, to surrounding areas (Solinas, 1982: 334, 350; Brusco, 1982: 177).

Piore and Sabel rightly feel the need to defend FS from being compared to 'the old Bourbon kingdom of Naples, where an island of craftsmen, producing luxury goods for the court, was surrounded by a subproletarian sea of misery' (1984: 279). Their answer rests on a voluntarist vision of co-ordination of market relations by local community structures and by national social-welfare regulation. But the international perspectives are not quite so rosy, or if so, only on the basis of harmonious inequality. While it *may* be possible 'to modernise the burgeoning craft sector of third world countries along the lines of flexible specialisation – rather than urging these countries to imitate the mass-production history of the advanced countries – (who does the urging? – AP) – Piore and Sabel continue,

> It is conceivable that flexible specialisation and mass production could be combined in a unified *international economy*. In this sytem, the old mass-production industries might migrate to the underdeveloped world, leaving behind in the industrialised world the high-tech industries and the traditional dispersed conglomerates in machine tools, garments, footwear, textiles, and the like – all revitalised through the fusion of traditional skills and high technology

> . . . To the underdeveloped world, this hybrid system would provide industrialisation. To the developed world, it would provide a chance to moderate the decline of mass production and its de-facto emigration from its homelands. (1984: 279)

This blithely imperialist perspective is a further dimension of the unconsciously divisive and elitist complacency implicit in the FS school, for all its utopian collectivist aspirations. It further confirms the compatability between this programme of restructuring for *labour*, with the advocacy of labour flexibility as a panacea for *capital alone*, in the 'core-periphery' model as a managerial strategy.

Conclusions

From this analysis of 'flexibility' presented as a 'radical break', spearheading something 'new', it is worth reiterating that flexibility is far from new. It was argued earlier that capital's harnessing of the inherent flexibility of human labour is the defining characteristic of labour power. It is, and always has been, essential to capital accumulation. How this flexibility was organised has been, and is, part of class conflict. In the early days of the factory system, division and rigidification of time and task were the central management priorities, while flexibility on the work-people's terms was attenuated. Today, flexibility on *labour's* terms remains an area of struggle as it was before; in spite of unemployment and the appropriation of 'flexibility' as a managerial and market concept, the organisation of work and employment continues to be socially negotiated (whether through formal industrial relations or informal means), and labour can, as in previous periods, recoup and transform managerial initiatives to its own ends. However, such outcomes depend, as 'participation' exercises, productivity deals and other managerial initiatives to gain worker co-operation show, on the groundwork of worker organisation and self-confidence. Whether the libertarian legitimations of casualisation and 'flexible' working lives can thus become re-appropriated by labour as positive improvements in terms of working hours and arrangements, paid for by capital, not labour, is an open question. But it is not a new terrain of struggle, just as the 'flexible workforce' is not the radically new phenomenon which it has been perceived to be.

The deconstruction of 'flexibility' as a celebration of labour market dualism in fact reveals a strong convergence in practice between FS, management policies for labour fragmentation, and neo-classical policies of labour market deregulation. All these approaches place the major responsibility for economic recovery on changes in *labour*, with 'new technology' as companion to a select few. The focus on labour is itself ideological, foreclosing perspectives on international movements of capital, capital concentration, money markets, or the social distribution of wealth. 'New technology' too is conceived ideologically, as progressive, neutral and inevitable.

The use of *labour* and *new technology* as part of the 'flexibility' restructuring programme has allowed a benign recasting of dual labour market theory. The 'flexible' workforce, and economy, is to be composed of a technically advanced, but pliable, 'core', and a disposable, elastic periphery. With the 'flexible firm', labour market segmentation is resurrected as though it were new, with no allusions to existing divisions by race, sex and age. With Piore and Sabel, the secondary, small firms sector is re-interpreted as the dynamic part of the economy, becoming, in this sense, the 'core' and ousting the ailing 'primary' sector firms based on declining mass production. New, flexible technology thus becomes the saviour and economic transformer of what had been the subordinate sector. In each approach, all labour flexibility is celebrated as work enhancing, while decentralising and fragmentation are embraced. As such, the informal and secondary economies are legitimised, and the significance of the disjuncture between organised, collective and directly employed labour, and isolated, atomised production is masked.

The convergence between these perspectives, and the neoclassical revival of the enterprise economy, individualised competition and policies of employment deregulation and attacks on trade unionism lead one to question why such a broad ideological consensus should have developed. Capitalist crisis, and the lack of control by nation states over the system may predispose the relinquishing of policies of control, towards an emphasis on the primacy of 'markets', raising economic flexibility to be the panacea. At the same time, the concern of governments with small businesses may itself be the product of the 'contradiction between increasing involvement in the international economy and political dependence upon a nationally/territorially defined social and political structure' (Gerry, 1985: 303). That this perspective should find its way into management writing is not surprising. Nor is it new for academic discourses to reinforce, reproduce and legitimise a political consensus.

For 'radical' restructuring strategies to turn to what are fundamentally market-dominated solutions of working *with*, not *against* capital, suggests an abandonment of class analysis and class action as the basis for change. This is one reflection of the more general disorientation and demoralisation of the left. Fortunately there is now a growing swell of opposition to such fatalistic defeatism, accompanied by the theoretical critique of its policies.

The effect of 'flexibility' as both material strategy and ideology is complex. There can be no doubt that workplace and trade union organisation are both under attack. And yet, as the survey of the empirical evidence shows, the extent of change in the direction of casualisation is more uneven and complex than the 'radical break' perspective implies. In this sense, the 'discovery' of the 'flexible workforce' is part of an ideological offensive which celebrates pliability and casualisation, *and makes them seem inevitable*.

This paper has therefore attempted to analyse the interweaving of 'flexibility' both in its concrete applications, and as ideology. Seen in this way, the language of 'flexibility' reveals itself as the language of social integration of the 1980s: how to live with insecurity and unemployment, and learn to love it.

Acknowledgements

Thanks to the following people for their helpful comments and suggestions on earlier drafts of this paper: Peter Armstrong, Irene Bruegel, Fergus Murray, Peter Nolan, Ruth Pearson, Chris Smith and John Ure.

References

ACAS (1985) *Annual Report*. Advisory, Conciliation and Arbitration Service, London.

Armstrong, P. (1982) 'If it's only women it doesn't matter so much', in West, J. (ed.), *Work, Women and the Labour Market*. Routledge and Kegan Paul, London.

Atkinson, J. (1984) 'Manpower Strategies for Flexible Organisations', *Personnel Management*, August.

Atkinson, J. (1985a) *IMS Report* No. 89. IMS, Falmer, Sussex.

Atkinson, J. (1985b) 'Flexibility: Planning for an Uncertain Future', *Manpower Policy and Practice*. Vol. 1, Summer. IMS.

Bell, D. (1974) *The Coming of Post Industrial Society: A Venture in Social Forecasting*. Heinemann, London.

Bennet, D. (1986) *Production Design Systems*. Butterworths, London.

Berger, S. & Piore, M.J. (1980) *Dualism and Discontinuity in Industrial Societies*. Cambridge University Press, Cambridge.

Brusco, S. (1982) 'The Emilian model: productive decentralisation and social integration', *Cambridge Journal of Economics*, No. 2.

Building Businesses not Barriers (1986) Government White Paper, May. Cmnd 9794. HMSO, London.

CAITS (1986) *Flexibility: Who needs it?* CAITS, Polytechnic of North London.
Casey (1987) 'The extent and nature of temporary work in Great Britain', *Policy Studies*, Vol. 8, July.
Cavendish, R. (1981) *Women on the Line*. Routledge and Kegan Paul, London.
Clegg, H.A. (1983) *The Changing System of Industrial Relations in Great Britain*. Blackwell, Oxford.
Clutterbuck, D. & Hill, R. (1981) *The Re-making of Work, Changing Patterns of Work and How to Capitalise on them*. Grant McIntyre, International Management, London.
Cockburn, C. (1983) *Brothers: Male dominance and technical change*. Pluto, London.
Cockburn, C. (1985) *The Machinery of Dominance*. Pluto, London.
Connolly, P. (1985) 'The politics of the informal sector: a critique', in Redclift and Mingione (eds).
Craig, C., Garnsey, E. & Rubery, J. (1984) *Payment structures in smaller firms: women's employment in segmented labour markets*. Research Paper No. 48, Department of Employment.
Creigh, S., Roberts, C., Gorman, A. & Sawyer, P. (1986) 'Self employment in Britain. Results from the Labour Force Surveys', *Employment Gazette*, June.
Doeringer, P. & Piore, M. (1971) *Internal Labour Markets and Manpower Analysis*. Heath, Mass.
Elger, T. (1987) 'Flexible Futures? New Technology and the Contemporary Transformation of Work', *Work, Employment and Society*, Vol. 1, No. 4, December.
Employment, the Challenge for the Nation (1985) Government White Paper, March. Cmnd 9474. HMSO, London.
Febre, R. (1986) 'Contract Work in the Recession', in Purcell, K., Wood, S., Waton, A. & Allen, S. (eds), *The Changing Experience of Employment*. Macmillan, Basingstoke.
Gerry, C. (1985) 'The working class and small enterprises in the UK recession', in Redclift, N. & Mingione, E. (eds).
Gershuny, J. (1978) *After Industrial Society? The Emerging Self-Service Economy*. Macmillan, London and Basingstoke.
GLC (1985) *Strategy for the London Clothing Industry: A Debate*. Economy Policy Group, Strategy Document No. 39, May.
Glucksman, M. (1986) 'In a Class of their Own? Women Workers in the New Industries in Inter-war Britain', *Feminist Review* No. 24, October.
Gough, J. (1986) 'Industrial policy and socialist strategy: Restructuring and the unity of the working class', *Capital & Class* 29, Summer.
Hakim, C. (1985) *Employers' use of outwork*. Research Paper No. 44, Department of Employment.
Hakim, C. (1987) 'Homeworking in Britain', *Employment Gazette*, February. Department of Employment, London.
Handy, C. (1984) *The Future of Work*. Blackwell, London.
Hepple, B. (1987) 'The Crisis in EEC Labour Law', *Industrial Law Journal* 16, No. 2, June.
Herzog, M. (1980) *From Hand to Mouth*. Penguin, Harmondsworth.
Hyman, R. (1986) *Flexible specialisation – miracle or myth?* Paper to EGOS/AWG on Trade Union Research, Warwick Colloquium on

Trade Union Research, June. University of Warwick.
IDS (1986) 'Flexibility at Work'. *IDS Study* 360, April. Income Data Services, London.
IRS (1985, 1986) *Flexibility – The Key Concept of the 1980s.* IRS Conferences, March, October 1985 and October 1986. Industrial Relations Services, London.
Jenkins, C. & Sherman, B. (1979) *The Collapse of Work.* Eyre Methuen, London.
Jenson, J. (1987) 'The Talents of Women, the Skills of Men: Flexible Specialisation and Women', first draft for inclusion in Wood, S. (ed.) (1989) *The Transformation of Work?* Hutchinson.
Kelley, M. (1987) 'Programmable Automation, Management Strategy, Flexibility, and Skills', first draft for inclusion in Wood, S. (ed.) (forthcoming, 1989).
Kumar, K. (1978) *Prophecy or Progress? The Sociology of Industrial and Post-Industrial Society.* Penguin, Harmondsworth.
Lewis, J. (1987) *The Changing Structure of the Workforce.* Unpublished paper to 'Change at Work and Flexibility: The Challenge for Trade Unions', Northern College/TGWU Region 10 Conference, 27-28 March. Northern College, Yorkshire.
Lewis, R. (1987) 'Reforming Labour Law: Choices and Constraints', *Employment Relations* 9.4.
Lifting the Burden (1985) Government White Paper, July. Cmnd 9571. HMSO, London.
Littler, C. (1982) *The Development of the Labour Process in Capitalist Societies.* Heinemann, London.
LRD (1986) 'Flexbility Examined', *Bargaining Report.* Labour Research Department, London.
Marginson, P., Edwards, P., Martin, R., Pucell, J. & Sisson, K. (1988) *Beyond the Workplace.* Blackwell, Oxford.
McGuffie, C. (1986) *Working in Metal: Management and Labour in the Metal Industries of Europe and the USA, 1890-1914.* Merlin Press, London.
Meager, N. (1985) *Temporary Work in Britain.* IMS Manpower Commentary, No. 31. Institute of Manpower Studies, Falmer, Sussex.
Millward, N. & Stevens, M. (1986) *British Workplace Industrial Relations, 1980-1984.* The DE/ESRC/PSI/ACAS Surveys. Gower, Aldershot.
Mitter, S. (1986) *Common Fate, Common Bond. Women in the Global Economy.* Pluto, London.
Murray, F. (1983) 'The Decentralisation of Production and the Decline of the Mass-Collective Worker?' *Capital & Class* 19.
Murray, F. (1987) 'Flexible specialisation in the "Third Italy" ', *Capital & Class* 33, Winter.
NBPI (1969) 'Productivity Agreements', *National Board for Prices and Incomes Report* no. 123. HMSO, London.
NEDO (1986) *Changing Working Patterns.* Report prepared by the Institute of Manpower Studies for the National Economic Development Office in association with the Department of Employment. NEDO, London.
Nichols, T. (1986) *The British Worker Question.* Routledge and Kegan Paul, London.
OECD (1986) *Labour Market Flexibility: Report by a High Level Group of Experts to the Secretary General.* OECD, Paris.

Pearson, R. (1987) 'What can we learn from the debates about the Informal Sector in the Third World?' Unpublished paper given to Warwick CSE Group, June.
Piore, M.J. & Sabel, C.F. (1984) *The Second Industrial Divide: Possibilities for Prosperity*. Basic Books, New York.
Piore, M.J. (1986a) 'Perspectives on labour market flexibility', *Industrial Relations*, Vol. 25, No. 2, Spring.
Piore, M.J. (1986b) 'The decline of mass production and the challenge to union survival', *Industrial Relations Journal*, Vol. 19, No. 3, autumn.
Pollert, A. (1981) *Girls, Wives, Factory Lives*. Macmillan, Basingstoke.
Pollert, A. (1987) *The 'Flexible Firm': A Model in Search of Reality (or a policy in search of a practice?)*. Warwick Paper in Industrial Relations, No. 19. IRRU, University of Warwick.
Potter, T. (1986) *Temporary Benefits, a Study of Temporary Workers in the DHSS*. West Midlands Low Pay Unit, Birmingham.
Potter, T. (1987) *A Temporary Phenomenon: Flexible Labour, Temporary Workers and the Trade Union Response*. West Midlands Low Pay Unit, Birmingham.
Redclift, N. & Mingione, E. (eds) (1985) *Beyond Employment, Household, Gender and Subsistence*. Blackwell, Oxford.
Rubery, J., Tarling, R. & Wilkinson, F. (1987) 'Flexibility, Marketing and the Organisation of Production', *Labour and Society*, January. International Institute for Labour Studies, Geneva.
Sabel, C. & Zeitlin, J. (1985) 'Historical Alternatives to Mass Production: Politics, markets and technology in nineteenth century industrialization', *Past and Present*, No. 108, August.
Shaiken, H., Berzenberg, S. & Kuhn, S. (1986) 'The Work Process Under More Flexible Production', *Industrial Relations*, Vol. 25, No. 2, Spring.
Smith, C. (1987) 'Flexible Specialisation and Earlier Critiques of Mass Production', 5th Annual UMIST/Aston Organisation and Control of the Labour Process Conference, UMIST, March.
Solinas, G. (1982) 'Labour market segmentation and the workers' careers: the case of the Italian knitwear industry', *Cambridge Journal of Economics*, No. 6.
Standing, G. (1986) *Unemployment and labour market flexibility: the United Kingdom*. International Labour Office, Geneva.
Syrett, M. (1985) *Temporary Work Today*. Federation of Recruitment and Employment Services, London.
Thompson, E.P. (1963) 'Time, Work Discipline, and Industrial Capitalism', *Past and Present*, No. 38.
TUC (1985) *Flexibility: A Trade Union Response*. TUC, London.
TUC (1986) 'Labour Force Trends'. Unpublished paper. TUC, London.
Westwood, S. (1984) *All Day Every Day*. Pluto, London.
Williams, K., Cutler, T., Williams, J. & Haslam, C. (1987) 'The End of Mass Production?' Review article on Piore, M. & Sabel, C. (1984) in *Economy and Society*, Vol. 16, No. 3, August. Routledge and Kegan Paul, London.
Woodward, J. (1958) 'Management and Technology', in Burns, T. (ed.) (1969) *Industrial Man*. Penguin, Harmondsworth.
Zeitlin, J. (1985) 'Markets, Technology and Collective Services: A Strategy for Local Government Intervention in the London Clothing Industry', in GLC (1985).

[11]

LABOUR & INDUSTRY, Vol. 1, No. 2, pp. 187-209, June 1988. 187

The Flexibility Debate: Industrial Relations and New Management Production Practices

Thomas Bramble

Abstract

There is increasing interest in Australia and overseas about the implications of the changing product market and technological environment for the skills and responsibilities of the shopfloor workforce in manufacturing industry. Broadly, there are two approaches. The first, based on Piore and Sabel's work (1984), suggests that the advent of niche-based product market strategies and small-batch production will herald a new era of enriched work roles and employee involvement practices. The second approach, based on a more pessimistic interpretation of changing management practices, stresses the possible intensified managerial control and employment losses associated with the new environment. Making use of a series of seven case studies of metal industry plants in Victoria and NSW, this study investigates these different approaches and concludes that the latter framework offers a more realistic interpretation.

INTRODUCTION[1]

There is an increasing amount of evidence to suggest that management practices in the large firm sector of the Australian metal and engineering industry have been undergoing significant if not widespread changes in the last decade. Perhaps the most important features have been the slow diffusion of practices designed to stimulate motivation and company loyalty amongst shopfloor employees and the enhancement of employee involvement in low level decision making on the shopfloor, (Bramble, 1986, 97-99). The data reveals the growing incidence of such schemes. For example, the MTIA Survey of Management-Employee Consultation/ Participation in the Metal and Engineering Industries (reported in Frenkel, 1986, 12-13) finds that approximately 30 per cent of the respondent plants (representing just over one-half of total employment) had some form of employee participation scheme. Growth has been particulartly rapid in recent years in direct schemes. These comprise both 'top-down' briefing groups and 'bottom-round' forums, such as quality circles and autonomous work groups. In nearly twenty per cent of cases there were meetings between management and employees; in 13 per cent productivity improvement groups, and in nearly ten per cent some form of quality circles.

Related to employee involvement has been the slow development of a 'two tier' approach to the employment relationship. To the extent that wage determination has been outside the control of individual employers in recent years, due to the 'no further claims' clause of the wage fixing principles, and before that, the Metal Trades Agreement of 1981, managers have sought to improve labour utilization by non-wage mechanisms (Rimmer, Plowman and Taylor, 1986, 77). So, although the trend is not as obvious as in the clothing industry, there is some evidence to suggest a moderate shift away from conventional patterns of employment in the metal and vehicle industry. This has involved a rise in the incidence of part time and casual employment. For example, the ratio of part time workers in the manufacturing labour force as a whole rose from 4.6 to 7.0 per cent between 1970-84 (Kirby, 1985, 38). This figure undoubtedly underestimates the true extent of such work patterns (see Lever-Tracy in this issue).

The contention of this paper is that the stimulus for reform in Australia has come not from any national-specific factors, and certainly not from any kind of behavioural revolution on the part of management, but because of factors that lie outside the immediate domain of industrial relations. The argument is that it is the combination of *changing product markets* and the *new production practices*, coming on stream to cope with the new economic environment, that are the critical factors that explain management's changing approach to the manual workforce in the Australian metal industry.

THE THEORETICAL DEBATE: FLEXIBLE SPECIALIZATION

There is a growing literature in Britain and the United States that stresses the link between the adoption of new production methods and forms of employee participation. One of the most important works in this field has undoubtedly been Piore and Sabel (1984), whose basic propositions have done much to set the terms of debate in recent years, just as Braverman (1974) had done a decade earlier. Piore and Sabel, following in the footsteps of Sabel (1982), argue that the traditional approach to employee participation in Western manufacturing has its roots in the adoption of mass production and the creation of standardized mass markets in the first thirty years of this century. With the routinization of production and the introduction of inflexible capital equipment requiring a high capital outlay, acute labour specialization and deskilling became an integral part of management, on the lines advocated by F.W. Taylor. This specialization involved a division of labour between deskilled material manipulating and machine feeding assembly workers on the one hand, and relatively highly skilled maintenance, technical, planning and supervisory workers, with a certain degree of discretion, on the other (Curtain et al, 1986, 6). In essence, mass production gave rise to, and was predicated upon, the 'low trust' strategy, involving a

> ....zero sum approach to this bargaining process; a tendency to perceive and interpret the behaviour, communications, policies, or values of the other side as being antipathetic to one's own interests and concerns; and such forms of behaviour as indifferent performance, clock-watching, and high absence, sickness, wastage or turnover rates (Fox, 1974, 102).

What this meant in terms of management's approach to the workforce in large manufacturing operations such as GM and Ford in the USA by the outbreak of World War Two is described by Harris (1982, 102):

> Workers then were seen as unsatisfactory, unreliable and refractory agents for the achievement of management's purposes. They had to be driven. Their behaviour had to be governed by strict rules, minutely enforced. They had to be controlled by loyal, obedient foremen and supervisors. Their autonomy had to be limited by the design of production processes.

Piore and Sabel go on to argue that the conscious exclusion of the workforce from any kind of involvement in decision making at the workplace could continue so long as the material basis for large-batch or mass production prevailed. However, by the late 1970s, several factors had come to bear which made the continuation of traditional practices impractical. In particular, the onset of the recession, and competition from the low wage Newly Industrialising Countries and from Japan, forced management to reassess their mass-produced low-margin approach to production, based on economies of scale. There was growing awareness of the costs involved in such an approach, particularly those associated with inventories and fault detection and rectification (Sayer, 1986, 48-50).

Central to the restructuring that has resulted has been the attempt to combine the advantages of Japanese production systems with Western technological expertise. The emphasis is now on operational flexibility and market responsiveness. As Piore (1986, 162) writes:

> The tightly integrated, hierarchical corporation was able to control its market; and in which, as a consequence, long run planning was feasible and heavy financial commitments to particular products were encouraged. The new environment is much more unstable and uncertain. Markets can no longer be effectively managed and controlled. The organisation must therefore be able to respond quickly and 'flexibly' as market conditions change.

This has not been confined to the USA. Atkinson (1987, 88) writes:

> If the past decade can be characterised as a period of transition for all

European economies, then it was surely a transition to economic and social configurations that will permanently embody many of the features of that transition; for example, greater competitiveness in product markets, an increased rate of technological change embodied in both product and process innovation, and greater immediacy between the recognition of market opportunities (or pressures) and the need to respond to them.

Recognition of such changes has spawned increasing interest in the question of socio-economic flexibility. Boyer (1987, 108) establishes five main principles of flexibility: the adaptability of the productive equipment to variable demand, both in terms of output and composition; the adaptability of workers to a variety of tasks; the possibility of varying employment and hours of work; the sensitivity of wages to the fortunes of both the company and the economy at large; and the reduction of social security contributions by companies.

At the individual company level, employer interest has been keenest in the first three areas. In some of the more advanced cases, the drive for production flexibility has taken the form of the adoption of the 'Just In Time' (JIT) inventory control philosophy. The JIT approach has been taken up with some vigour in the West in the 1980s, especially in industries that lend themselves to small-batch production. The essence of it is that instead of producing 'just in case' the product is needed, (i.e. producing large batches for stock), the JIT production schedule is based on producing directly for market demand. The key element involved is the reduction of set-up times, which offsets the economies of long production runs and which therefore help to reduce buffers and wasteful inventory. Since it is associated with producing only so much as there is demand for at any one time, it is also critical that defects should be minimized, otherwise a breakdown occurs between orders and delivery. Without standby stock in warehouses, production managers have to keep production flowing through smoothly. As a result, JIT is often associated with Total Quality Control (TQC), the idea of 'building quality in at source', rather than taking defects out at the end of the line.[2]

The resulting configuration of market and technology has had profound implications for management's approach to employee relations and the employment relationship. Piore and Sabel's 'flexible specialisaton' hypothesis suggests that the Western economies are witnessing the re-emergence of a 'craft oriented' approach to production. They argue that the advent of flexible, market-responsive manufacturing, based on generalized, rather than product-specific machinery and manpower, has led to increased workforce flexibility, including the upgrading of skills and responsibilities of the shopfloor workforce (Sabel, 1982, 194). Atkinson (1987) outlines a useful typology of such form of 'flexibility'. In his description of the 'flexible firm', Atkinson suggests that there is increased interest at company and government level in flexibility in both the internal and the external labour markets. The former comprises *numerical flexibility*, the ability of employers to vary the number of workers, or the level of worked hours, in response to changes in the workloads; *functional flexibility*, the ability of firms to reorganize jobs, so that the worker can deploy his or her skills across a broad range of tasks; and finally *distancing*, involving subcontracting. In terms of their frequency, Atkinson (1987, 91) cites a British Government report which states that out of the 72 firms considered, 90 per cent have raised their numerical flexibility, 70 per cent have increased their distancing operations and over half have been involved in raising functional flexibility in recent years. In this article, we are particularly interested in the growth of functional flexibility. Shaiken et al (1986, 167) outline three ways in which the new production processes call forth new demands on the manual workforce: because workers have to play a critical role in debugging programs or intervening when things go wrong in small-batch production; because skilled workers' knowledge of the production process is essential to product and process innovation and, thirdly, because workers have to master new responsibilities through broader skills when firms repeatedly change product lines.

Piore and Sabel argue that it is changes such as these, involving the upgrading of process workers concomitant with the new regime, that have led to the abandonment of Taylorist methods of work design and the low-trust, zero-sum bargaining relationship associated with these. Piore (1986, 160), after having outlined changes relating to production equipment and to organizational forms, explains that:

> Both co-operation and flexibility of work assignment are seen as critical in an environment where production is continually being reorganised, to adapt to the market or to incorporate technological change. Thus, they are inevitably accompanied by other distinct practices which bind the worker more closely to the firm: quality circles, profit sharing and in union firms, often worker representatives are included on company boards.

This product market-driven approach to changing labour relations stands in contrast to the 'control-driven' labour process theories. In place of the previous emphasis on a sharp demarcation between mental and manual labour, management have been forced to introduce employee relations policies that are based on co-operation and information sharing. Consensus management replaces management by diktat, or as Piore and Sabel (1984, 278) argue: 'Flexible specialisation is predicated on collaboration'.

The above argument is buttressed by the optimistic interpretation of the implications of computerized technologies. Writers such as Sorge et al (1983, 96-9) point to the experience of management in Japan and West Germany where automated production systems have been introduced in such a way as to involve low-level operators in reasonably sophisticated tasks such as computer programming. With more valuable employees, it is argued, management have been encouraged to develop forms of employee involvement, such as quality circles and semi-autonomous work groups. In summary, in so far as Taylorist methods of job design adopted by management were in response to the technical dictates of mass production, so employee involvement and job redesign have been their response to the technical dictates of flexible specialization (Wantuck, 1985, 10).

The growth of direct form of employee involvement has also been associated in practice in both Britain and America with changes to the employment relationship. In order to give senior managers the flexibility necessary to respond to changing market conditions, and yet maintain shopfloor commitment to the management perspective, there is some evidence that they have attempted to develop employment security for a core of the most valued workers, coupled with the employment of a periphery of casual or part time workers (Atkinson, 1987, 94; Bramble, 1986, 88-90). With the development of company-specific skills, it is in management's interest to enhance functional flexibility at the expense of numerical flexibility for the core group, since the training costs that are associated with high labour turnover would otherwise become prohibitive. The central features of employment for the core workers are security, retraining and the cultivation of company loyalty. The core forms the basis on which team-based work production systems are founded, each team being made responsible for quality control, set-ups, machine and tool maintenance, inventory control and task rotation (Katz and Sabel, 1985, 302).

There is a strong relationship between employment security and retraining on the one hand, and the development of employee involvement on the other, in those experiments that have been tried in the USA and Britain. The main thrust behind the introduction of employment security 'guarantees' is therefore motivational, secure workers being more likely to accept changes in work practices accompanying the introduction of new production methods. Where the task-mobility of the primary workforce is insufficient to give management the flexibility needed, the evidence from Britain and America is that they have also established a periphery of casual workers, who are usually unskilled and who enjoy none of the benefits of those in the primary sector (Atkinson, 1987, 91-94).

Table 1: The Key Features of the First and Second Industrial Divides

	FORDISM	FLEXIBLE SPECIALIZATION
Market	stable	uncertain
Product Variety	standard	differentiated
Production Regime	mass production, 'just in case'	small batch, 'just in time'
Organizational Form	large corporations	small firms, decentralized
Skill Trends	deskilling, Taylorism	upskilling conception and execution reunited
Employee Relations	confrontationist	co-operative

Source: based on Piore and Sabel (1984).

The key elements of Fordism and Flexible Specialization, as described by Piore and Sabel (1984), are summarized in the ideal-type table above (see Table 1), while Figure 1 illustrates the inter-relationships between the market-technology configuration, forms of work organization and the nature of employee relations, both under the Fordist regime and Flexible Specialization.

CRITIQUE OF THE FLEXIBLE SPECIALIZATION HYPOTHESIS

Although the flexible specialization hypothesis is still relatively new and undeveloped, it has already attracted attention from critics. In particular, the 'neo-Fordists' argue that the new production methods herald not the withering away, but the accentuation of the basic principles of scientific management. They contend that the rationale behind management's new found interest in employee participation lies elsewhere.

Several criticisms of flexible specialization can be made. Firstly, there is some doubt as to the historical schema of mass production as laid out by Piore and Sabel. There is some argument as to the pervasiveness of the classic model of mass production. General Motors, for example, in the heartland of the American auto industry, the ideal-type Fordist industry, never followed Henry Ford's maxims regarding product standardization. Instead, they preferred to focus on product variety and customer choice. As a result, GM were never able to bring about such an intense division of labour as at Ford (Sayer, 1986, 50). Similarly, Tolliday and Zeitlin (1986) argue that, far from being based on a standardized mass market, the major European car producers of the inter-war years were restricted by the poverty and the small size of the market to a highly specialized luxury demand.

Secondly, once established, it is highly debateable that Western manufacturing is being, or can be, transformed *in toto* from mass production to small-batch, artisanal production methods. Indeed, more significant than increased product variety and the reduction of batch sizes for the vehicle industries of Britain, the USA and Australia, for example, may be the strategy of internationalizing the production of such items as engines, transmission and conponents, as part of the 'world car' strategy (Williams, Williams and Haslam, 1986; Wood, 1986, 9). This involves an even greater emphasis on large-batch manufacturing based on the extensive use of economies of scale. Furthermore, in so far as small-batch production methods have been introduced into these three countries, there is little evidence that management have used the potential of innovations such as JIT and computerized technology to broaden skills and reintroduce artisanal production. Instead, the emphasis has been on the intensification of managerial control and cost reduction. For example, American and British managers have used automated production systems primarily to reap cost economies on restricted product lines and have emphasized rather than undercut, the hierarchical nature of work organization (Jones, 1986, 12; Shaiken et al, 1986, 169-73).

Similarly with JIT, the evidence is that, contrary to the claims of its more enthusiastic proponents, the philosophy represents not the refutation but the intensification of Taylorist methods of work organization (Schonberger, 1983, 193). The elimination of buffer stocks (through the kanban system) and the need to stop the line in case of defects (jidoka) enables management to pinpoint areas of slack production much more easily than when large stocks of work-in-progress hid worker inefficiency. Shaiken et al (1986, 176) argue that JIT offers management the inestimable advantage of being able to introduce assembly line pacing into batch production areas, which were hitherto subject to a degree of operator control over the pace of work.

The neo-Fordists also argue that we should treat the claims made regarding upskilling with a certain scepticism. First of all, the skills that are involved do not possess one of the key features of established trade skills; transferability between companies. The broadening of internal flexibility, by potentially reducing the marketability of the skilled workforce's skills, may actually serve to undermine external flexibility (Hyman, 1986, 9). While more extensive company-specific skills may make the workers concerned more valuable to the company at a particular point in time, these also tie them to the company if it goes bankrupt, taking the value of their skills down with it. Hyman is also fairly sceptical about the autonomy that is supposed to accrue to production workers as a result of team-based systems. As he remarks: 'Delegated management does not equal self management nor does an expanded portfolio of competences necessarily equal enhanced skill' (Hyman, 1986, 10).

Indeed, the key to management's interest in employee participation in recent years may lie not so much in any spurious upskilling of the workforce accompanying the new production methods as the fact that the new methods are extremely dependent on workforce co-operation. The experience of mass stand-downs in the Australian car industry in August and October 1986, resulting from strikes by handfuls of storemen and packers, illustrates the vulnerability of JIT. Since the plants downstream of the facilities directly affected held no spare supplies of components, a small strike very quickly led to the disruption of the whole industry. Because JIT places individual work teams in positions of immense strategic advantage, due to the tightness of the line and the lack of buffer stocks, the winning of workforce commitment has become one of management's central concerns.

We might surmise from this that management's recent adoption of direct forms of employee participation in the Australian manufacturing industry may be due to their desire to co-opt their workforces. By impressing on the workforce the centrality of competition and the need to accept an intensification of work effort, by means of top-down

Figure 1: The Flexible Specialization Hypothesis

THE FIRST INDUSTRIAL DIVIDE
Mass Production Era

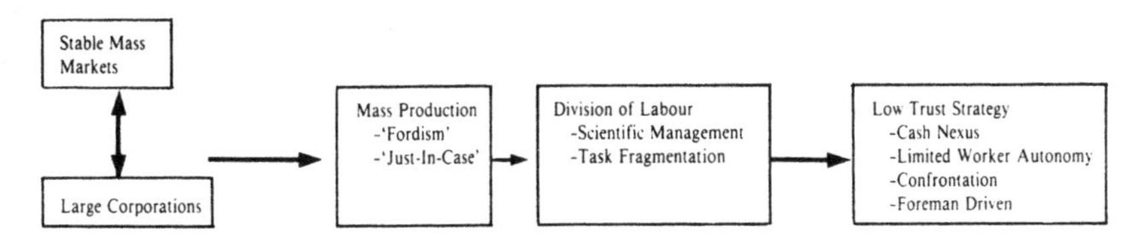

THE SECOND INDUSTRIAL DIVIDE
Flexible Specialization

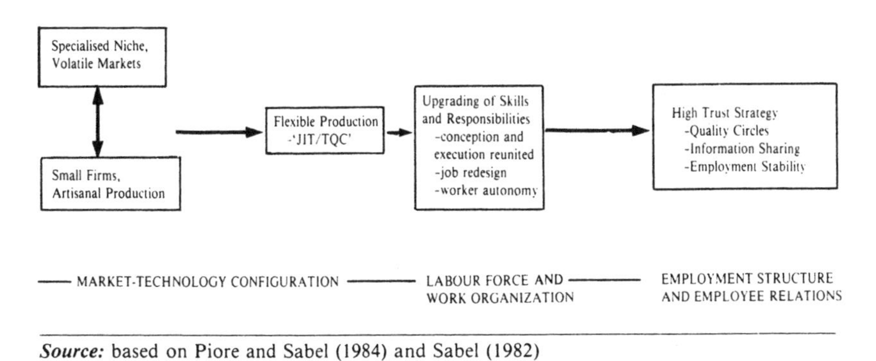

Source: based on Piore and Sabel (1984) and Sabel (1982)

Figure 2: A Neo-Fordist Interpretation of the Flexible Specialization Hypothesis.

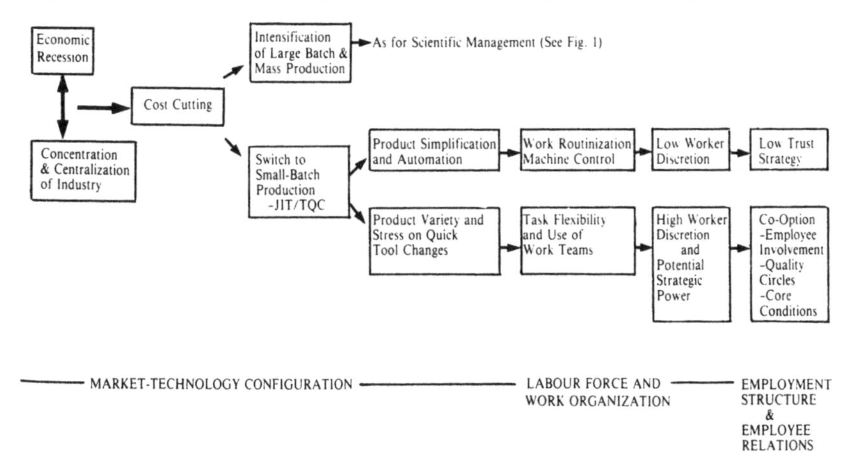

briefing groups and quality circles, management attempt to centralize control and reduce the threat of industrial action by gaining a more complete ideological hegemony (Garrahan, 1986; Shaiken et al, 180; Giesecke, 1985, 23). This is not to suggest that management seek to enhance labour control as as end in itself, as might be implied by some of the more crude interpretations of labour process theories, but that such control is a necessary prerequisite for the main task of the capitalist enterprise, the generation of surplus value.

Furthermore, while the motivational factor is probably primary in explaining the adoption of direct forms of employee involvement it is not the only one. Quality circles may also be useful in so far as they allow management to make use of the detailed day-to-day knowledge of the workforce in the diagnosis and solution of production problems or quality defects. For example, Wood (1986, 9) comments that Japanese quality circles represent:

> not an alternative to Taylorism but rather a solution to the classic problem of the resistance of workers to placing their knowledge of production in the service of rationalisation. It is the attempt to combine and harness the specialist and developing knowledge of the professional engineer with the day to day concrete knowledge of the worker.

To summarize, the contention is that the adoption of employee involvement schemes such as quality circles and briefing groups, and the restructuring of employment relationships in the large firm sector of the Australian metal industry in recent years, can be attributed to management concern about the control implications associated with new production methods. Where such methods have led to the strengthening of the strategic position of the workforce and/or individual work teams, we hypothesize that employee involvement practices will be widespread. The corollary of this is that where automated production systems have been used progressively to eliminate manpower, labour relations will be characterized by the intensification of the sharp hierarchical differentiation noted by Bill Ford (1976, 49). Figure 2 shows the relationship between the market-technology configuration, work organization and employee relations as envisaged in the 1980s by the neo-Fordist approach.

Of course, this is not to suggest that the changes in managerial practices above are the *only* way in which management can or have responded to the deterioration of the economic climate in the 1980s. Faced with a decline in profit margins, firms may be better served by taking over their competition, by switching out of production and into financial activities, by relocating to greenfield sites, or by employing the entire labour force on a subcontract basis. Many American firms in the 1970s for example, ran down their productive activities inside the US and opened up factories in South East Asia or Latin America, and, if anything, intensified the traditional Fordist methods of production. As a result, we have to be aware that management can change its work and corporate organization without changing its treatment of labour and vice versa. This area of research is beyond our ambit here, however, and we will concentrate on areas where such reorganization has taken place.

A priori, it might be thought that several factors might limit the applicability of the flexibility debate to the Australian metal industry. The first relates to the limits placed on the autonomy of plant management by centralized arbitration (Bramble, 1986, 92-93), although with the restructuring of the Metal Industry Award and the associated breaking down of established skill and union demarcation barriers as envisaged in 'Australia Reconstructed', this may become less important. The second concerns the relatively small scale of Australian manufacturing operations compared to factories overseas. In this situation, we might expect the degree of labour specialization to be relatively undeveloped. However, although it is true that companies in the Australian metal industry have been unable to achieve the same kind of production economies as in the USA, production management have nontheless sought to emulate Fordist principles of mass production. For example, in a 1984 survey of 315 medium to large (over 50 employees) metal industry plants, only 16 per cent of respondents reported engaging in custom-built or jobbing production, compared with 27 per cent involved with continuous or large volume production (Frenkel, 1987, 43). The third limiting factor for case study research concerns the relative newness of the new production methods that have been implemented. While it is true that these factors might well have limited both the extent of deskilling evident in the era of scientific management and the rate of change in more recent years, it was felt that an investigation of the empirical trends in the Australian context would be worthwhile.

METHODOLOGY

In order to assess the validity of the competing approaches, a study of seven metal industry plants in Victoria and NSW was made in 1985. The details of these plants are summarized in Table 2. Perhaps the most important thing to note is that the plants surveyed were amongst the leaders in the application of JIT and Total Quality Control schemes in the metal industry. Several were involved in pilot programmes being run jointly by the NSW and Victorian State Governments and the Technology Transfer Council over the course of 1985. There is no attempt to suggest, therefore, that the experience of management in these plants is in any way typical of the metal industry at large. Rather, they were investigated because it was thought that they might reveal the nature of the change between traditional and the new practices most starkly. The plants represent the main manufacturing operations of subsidiaries (Plants C, D, and E) or divisions (Plants A, B and F) of larger companies, except for Plant G which is a small satellite operation of a company based in Melbourne. This was incorporated into the study because, although a small plant by itself, all the major production and

198 *LABOUR & INDUSTRY*, Vol. 1, No. 2, June 1988.

labour relations decisions are made at the Melbourne HQ which, until the early 1980s, employed over 350. As a result, the degree of professionalism and specialization of the labour relations practices that were evident in this plant were more typical of large plants.

Table 2: Characteristics of the Plants Under Consideration

Plant	Labour Force	Of Which Casual or Part Time	Industry	Ownership
A	482	1%	Industrial Appliances	Division of Aust. PLC
B	1200	25%	Electrical Durables	Division of Aust. PLC
C	945	15%	Electrical Durables	WOS of US PLC
D	493	0	Metal Fasteners	Joint WOS of 2 Aust. PLC
E	440	5%	Metal Fasteners	WOS of Aust. PLC
F	380	12%	Electric Motors	Division of Aust. private co.
G	40	5%	Metal Fasteners	WOS of US PLC

Key: WOS = wholly owned subsidiary

The product markets served by the plants have historically been regional if not national or international. Where possible, product variety was minimized and production fed to the market via inventories, two central features of the mass production model outlined by Piore and Sabel. This strategy was based on the steadily growing post-war market for industrial and consumer goods. Relatively inefficient production and working practices were subsidized by high levels of tariff protection and the absorption or liquidation of rivals.

Labour relations in Plants B-F were characterized by relatively low level of management specialization, by confrontation with the unions and frequent recourse to employers' associations and to the industrial tribunals. Relations between management and the workforce were extremely distant, with employment insecurity and seasonal instablility highlighting a high level of mistrust and suspicion. For example, an internal report by management at Plant B complained in 1982 that:

> The trade union membership is predominately AMWU, whose strategy using well trained young organisers calls for precipitating as much agitation as the workforce will tolerate, more especially tempering whatever action is taken by management to improve morale with effective communication... To say the least, the year-by-year history of industrial relations at the plant, from the inception of operations in 1946 through to the mid 1970s was interspersed with industrial disharmony in all forms, short of rioting.

The Industrial Relations Manager at Plant C described a typical confrontation with the workforce in the early 1970s over a log of claims: 'Senior management said no to everything... the men went on the grass... we were hit by a two day strike, and we then gave in on every point'. He lamented the total lack of communication that existed and the detrimental results that resulted: 'The Organiser would get up on his box at stop-work meetings and say "The boss is a bastard", and we wouldn't be trying to put our side of the case at all'. This was also accompanied in the early to mid 1970s by a situation in which members' loyalty to their union was outweighed by loyalty to workgroup. As the ETU Organiser for Plant F complained: 'When anything went wrong, they were all out, regardless of whether or not the issue affected their department'. At times, the union machinery lost control over when and over which issues to strike. The AMWU Organiser for Plant D remarked that: 'They'd go out over anything... sometimes it would be over really silly little things; things that we could easily sort out without having a strike over it. Sometimes, I think they just did it to get back at management.' The situation was similar in terms of the basic management approach to the workforce at Plants A and G, but these two differed from the other five in that Plant G was non-union as a result of a deliberate management policy to exclude union activists from the company. Similarly, in Plant A the union was very weakly organized on the shopfloor despite having a relatively high membership.

THE NEW PRODUCT MARKET ENVIRONMENT AND MANAGEMENT'S RESPONSE

By the mid-1970s the established practices of employee relations at these plants came under pressure from the economic deterioration that took place in the metal industry. The second downturn of 1982-3 hit the metal industry with devastating results. Employment in the industry fell by 17 per cent in this period (Cashman, 1985). This was accompanied by a sharp increase in the competitive threats perceived by management. Import penetration rose from 22 to 37 per cent of the domestic fasteners market between 1977-8 to 1983-4, and from 27 to 43 per cent of the electrical durables market between 1981 and 1985 (Metal Fabrication Industry Council, 1985). In this situation, management in the case study plants have been forced to respond under threat of extinction.

The most obvious reponse has been the rationalization of operations, involving the closure of sister plants and the reduction of employment. In every case, the workforce at each plant is considerably smaller than in the mid-1970s, sometimes by as much as one half. Despite the reduction in employment, production has been more or less maintained at previous levels. The result has been a major increase in productivity. This has been achieved most recently by the adoption of small-batch methods of production, usually within the auspices of the JIT philosophy. This has, in turn, had major implications for the workforce.

The main feature of the new production methods has been the installation of production cells (group technology) into key areas. The emphasis has been on reducing set-up times, quick tool changes and mixed modelling, in place of the old system based on large-batch production and 'single streaming'. As a result, the work load for operatives has been made more steady and evened out. 'Kanban' signals between different work stations have been established and work-in-progress between work stations reduced dramatically. At Plant D, the growing emphasis on non-standard lines has encouraged management to introduce link-line production methods into the screw department. Half of the entire plant has been turned over to eight production cells, each comprising a full range of primary and secondary processes, linked together by conveyors, and each staffed by separate work teams. In some cases, management have got over the possible clash between the economies of long production runs and the need for high variety, by adopting 'late conversion', in which standardized components are produced and only at the last moment customized at the end of the line.

Quality control (QC) responsibilities have been shifted from full time QC Inspectors to the operatives themselves. At the new line in Plant C, management are introducing a 'stop and flow' assembly line, whereby if a process worker identifies a particularly serious fault, he or she can stop the line and bring production to a halt while the appropriate action is taken to prevent its recurrence. At least partly as a result of such changes, the number of quality control inspectors employed at this plant has fallen from 66 to 25 since the early 1980s. The operatives have also been made responsible in several cases for the limited organization of the division of responsibilities within each cell. At Plant A for example, management give the team leader, (each team comprising four or five operatives), the responsibility for turning out a certain number of products by the end of the day. How the basic machining and assembly tasks are then assigned is up to the team leader, under the broad supervision of production management.

THE IMPLICATIONS FOR THE LABOUR FORCE

What have been the implications of these changes for the workforce? The flexible specialization approach would suggest that the introduction of 'bottom-round' forms of employee participation has resulted from the job enrichment and greater autonomy of process workers under the new production regime. In fact, the evidence from these plants suggests that, by simplifying the production process, by physically opening up the shopfloor and by removing large amounts of work-in-progress, the implementation of the JIT philosophy has actually allowed management to enhance their surveillance and intensify control of the workforce. The Production Management at Plant D, for example commented that:

> There are no kingdoms anymore. The system shows up the problem areas, both technical and human. All operators are now open to easy surveillance... They can't hide behind their machinery. It's highlighted those workers who are worth our while.

The main implication of JIT for production workers has actually been the steady intensification of work, by the widening of routine operations performed by each worker, and the associated reduction of manning levels. The manning rate on the different production lines dropped, often from one worker per machine to three workers running a cell of eight to ten machines, this being accompanied by a sharp rise in productivity and a steady fall in the size of the workforce. There seems to have been no clear tendency on management's part to attempt to develop broadly skilled job roles. When asked about the prospects for upskilling at Plant A, the production manager said that: 'There's definitely no move towards upgrading the process worker. We don't want them tampering with anything else', while at Plant F, management expressed their fears that: 'A problem might occur if they become too independent... if you give them an inch, they take a mile'. Perhaps the best summary of the tensions of the new approach was given by the production manager at Plant C who was well versed in JIT and TQC techniques:

> Fordism tended to assume that employees were mindless and fools. This led to intense deskilling. Now, we identify the person as a human, able to make a contribution, so we try to focus all human resources, not just their hands... What we're doing is removing the chaos, by the trend to simplification and the reduction of discretion on the shopfloor.

When fully implemented, the JIT approach requires that process workers take responsibility for routine set-ups and the maintenance of machinery and tools. In practice, however, the blurring of demarcation between trades and process work has been very slight. Although the potential is increasingly being opened up, it is not being taken advantage of. For example, the process workers at Plant C are unable to use the tools that have been distributed to the shopfloor after the disbanding of the maintenance shop. The reasons for the lack of change in this regard are several; the relatively recent introduction of the programmes, management hesitancy, the demarcation problems associated with occupational unionism and health and safety considerations.

Despite this gradual tendency towards work intensification, in circumstances where innovations such as production cells and 'stop and flow' production lines have been introduced, the line workforce has clearly been endowed with greater strategic power than previously. The direction of change for tradesmen is presently uncertain. It is clear that some have been upgraded and re-trained to cope with the technical complexity of computerized control, as envisaged by Bill Ford (1986, 31-33). There has been a push within the metal industry in recent years to break down demarcation between the mechanical and electronic trades, although there had been little progress in this direction at the time of the fieldwork (mid-1985). Despite the upgrading that has occurred for some, however, the majority have probably experienced a reduction in the use of their technical skills, as a result of the simplification of tooling and die changes. For example, the Production Manager at Plant F commented that: 'We want to turn the tool change into a meccano set... preventive maintenance can now be learned in one month and not six'.

The adoption of JIT has also resulted in the need for greater technical skills and human relations awareness amongst foremen. Their technical skills have been upgraded to include not just the maintenance of section efficiency, but also an understanding of the sequence of production as a whole, and the relationship between production and stock control, so that they are able to diagnose the more common problems. Because line rebalancing has become more frequent as a result of the change to small-batch production, poor quality supervision becomes more obvious because it is not disguised by reserve stocks of work in progress in the event of production hitches.

In short, the new production methods adopted at these plants require heightened workforce commitment to managerial goals, functional flexibility and a sense of responsibility to maintain production, qualities that were largely irrelevant with the traditional approach to production based on large batches and minimum interaction between management and the workforce. We might expect, therefore, that employee relations practices at these plants would have been substantially transformed. This was the case in six of the seven plants surveyed. However, in the seventh, the reorganization of production took a different form. At Plant G, production management have introduced an almost entirely automated production line, based on the attempt to remove operator discretion and physical intervention. This has been premised on different approach to marketing: instead of product diversification, management at Plant G have reduced product range from 23 very different products to five fairly similar products. Automated magnetic conveyors shift the product from one work station to the next. Micro-electronic monitors have been installed, thereby removing the need for operator quality control. As a result, the task of the single operative on the line is simply one of patrolling and observation. The Production Manager remarked that:

> More or less anyone can come in off the street now and do the setting and heading. They just have to slot in the requisite tooling and press the button. The system is much less dependent on specialist or operator expertise than formerly. Most of the subjective aspects of the job have been lost and isolated to the office instead... The operators aren't required to exercise judgement anymore.

The fact that production has been reorganized in such a way in this plant has profound implications for employee relations. It is to these that we now turn.

IMPLICATIONS FOR EMPLOYEE RELATIONS

Even before the introduction of the new production methods, there was in several of the plants some move away from the old style of management. The general concern is well expressed by an internal report commissioned by corporate management at the parent company of Plants A and B in 1982 which noted:

> Our employees have the ability and desire to contribute more on issues directly related to their work. Old fashioned autocratic management styles do not encourage this contribution. Workers become dissatisfied, frustrated, restless and militant. They then seek excessive and unreasonable wage rises, conditions or benefits. They, like most of us, do not really

understand the basis of their frustrations, and consequently, head-on confrontation results.

Such concerns led to a certain amount of experimentation. However, progress has been most obvious in those areas where the stability and reliability of the workforce is critical and where their technical knowledge is important. There was a clear link in each of the plants surveyed between the decision to introduce small-batch methods and production cells and the development of widespread forms of direct, shopfloor employee involvement. This started with the very introduction of the new methods and continues today in their operation. According to the plant newsletter at Plant A: 'At every stage, employees are being encouraged to participate in the project by offering suggestions on layout and methods'.

One comment by the Production Manager at Plant C summed up the significance of the changes presently taking place. According to him, three basic principles underpinned the new approach:

> .... the reduction of waste, improvements in quality and the encouragement of employee involvement ... This involves a total rethinking of the shopfloor in a way that runs contrary to 25 years training. The part of the programme that will have the biggest impact and the greatest push for improvement is employee participation.

The chief means by which management have sought to develop a sense of company identification amongst those on the JIT production cells has been by the introduction of formal channels of direct communication and consultation with the manual workforce. For example, on the production cell at Plant A, a weekly problem solving group has been established, comprising the foreman, the production managers and the manual workforce. According to management, the purpose of these meetings is to encourage the workforce to contribute to productive efficiency and to make them more aware of the needs of the company. At Plant B, two levels of teams have been established. Firstly, there is a co-ordinating team of middle and staff management who meet once a week and who supervise the introduction and maintenance of the JIT programme. Alongside this, there are weekly shopfloor problem-solving groups, involving operatives, tradesmen, technicians and foremen. Having been trained in statistical quality control techniques, these teams have now been set to work to identify, diagnose and solve some of the more common problems on the line. Participation in these groups is voluntary in order to elicit, as one manager put it, the 'unbiased and enthusiastic co-operation' of the workforce. The industrial relations manager at Plant B remarked:

> While the longer term objectives of reduced stocks, improved quality and shorter lead times were clear, the short term objective was to achieve a high level of worker involvement through the use of group meetings.

While at Plant C, the Production Manager commented that: 'JIT requires absolute commitment and dedication by management and the involvement of every company employee'. The need for a high level of worker commitment in the context of JIT programmes was commented on by managers in each plant surveyed. At Plant F, the Production Manager argued that: 'The new production processes require much more lateral communication between workers and vertical communication with management'. The message was hammered home at Plant C by the presence of colourful posters carrying graphs tracking the reduction of defects, exhortational messages on boards hanging from the ceiling and small boards describing the basic tenets of the JIT process: 'Simplification, Cleanliness, Discipline, Organisation and Participation'.

Management have gone some way to developing commitment by offering tangible rewards in the form of increasingly sophisticated manpower policies. In almost every plant, management have at least attempted to offer retraining and assistance to a minority of those affected by the restructuring. This has been accompanied by the increasing segmentation of the labour force into a core, comprising workers with job security and the focus of employee participation programmes, and a band of peripheral employees, exposed to the external labour market. The advantage of employment stability had already been realized by management at Plants A and B. The Project Team Report noted that:

> Job security is that basic ingredient for sound industrial relations, and whilst it is not possible to give an unequivocal guarantee, our objective should be to offer some degree of job security to a base multi-skilled workforce... Some of the benefits arising from having a stable workforce with reasonable job security are that employees develop a sense of belonging and therefore a more co-operative attitude, improved morale, reduction in training costs, higher efficiency, better quality products and few retrenchment payouts.

However, it was not until the introduction of the JIT production line that much was done in this regard. It was felt by management that the new production arrangements require certain attributes that might not be forthcoming if workers found that their co-operation resulted in being laid off. Some managers have, therefore, taken steps to retain labour during downturns even at the cost of temporary over-manning. They have opted for long term productivity and workforce loyalty at the expense of short term production flexibility. The second aspect of this policy has been the development of a periphery of casual or part time employees, employed on 'Time Limited Contracts'. In several of the plants these accounted for up to 25 per cent of the total workforce, (see Table 2). The advantage of such employees is that they can be taken on and disposed of according to product market variations.

Although management have been largely successful in implementing JIT programmes, except at Plant E, (where union concern over the lowering of manning levels precluded radical experimentation), they have not been especially successful in converting acquiesence on the part of either lower line management or the workforce more generally into explicit commitment to the process of restructuring. Passive obedience is more evident than enthusiasm. This problem was commented on by many managers, who contrasted their predicament with how they perceived the situation in Japan. The predominant underlying characteristic of relations between management and the workforce continues to be one of mistrust. Management have, so far, been incapable of overcoming the heritage of generations of low-trust relations and in this situation 'Mutual trust or mistrust, among members of a group, are likely to be reinforced, unless there is marked or prolonged disconfirming behaviour' (Zand cited in Fox, 1974, 67). Management have found it difficult to cross the threshold into a high-trust dynamic.

The change from large to small-batch production and the associated development of formal channels of employee communication has altered the qualities looked for by management in their production workforce. Such qualities include flexibility and the willingness to perform a range of different tasks, and to assume a certain responsibility over the performance of the job at hand. Once employed, management have to be careful not to lay off workers possessed of the communication skills and flexibility required. The overall stimulus for these changes, to repeat, was because changes to the technical and production process made traditional approaches to employee relations unsuitable.

The adoption of the unmanned production cell system at Plant G has also had important implications for managing labour. Just as the greater task variety and responsibilities of workers on the manned production cells in Plant A-F encouraged management to consider methods of breaking from the old low-trust relationship with the workforce, so the removal of process worker discretion and physical input at Plant G gave management the opportunity to consolidate rather than reverse their traditional approach. Although manual employees are kept informed of changes to production methods, in the absence of trades unions they are given no opportunity to negotiate the nature or pace of change. Management's efforts to maintain employee loyalty has meant that a strong emphasis has been placed on top-down communication direct to the workforce. The fact that production has been reorganized in such a way as to weaken both the skills and the strategic power of the workforce at this plant has meant that such communication has not been accompanied by the development of bottom-round forms of employee involvement. Indeed, management efforts were geared increasingly to the steady elimination of labour from the production line. In this context, there was clearly no incentive to develop employee participation.

CONCLUSIONS

The central proposition of this study was that the labour relations practices adopted by management in the Australian metal industry in recent years will have been strongly affected by the nature of reorganized production processes. Such reorganization involved the intensification of work effort, the promotion of workforce flexibility, large scale redundancies, natural wastage, the attempted development of dual labour markets and/or sharp reductions in manning levels. The sheer depth of the restructuring that has been involved has led management in every plant to devote greater effort to 'enlighten' their workforces as to the need for sacrifice in the new environment. Hence the use of either formal or informal top-down briefing sessions and small-group meetings. Overall, management in the case study plants are fairly satisfied with the rate at which they have been able to extract work practice concessions from the workforce.

That enhanced employee involvement and greater employment stability has been so noticeable for those workers on the new production lines is perhaps the best indication of the close connection between the new production practices and the restructuring of employee relations. Contrary to the flexible specialization argument, such forms of employee participation were not introduced because of any upgrading of the intrinsic skills of blue collar production workers. In fact, all the available evidence would suggest that the introduction of JIT and manned production cells has been done in such a way as to further reduce the autonomy and transferable skills of production workers. The main incentive behind their introduction is the fact that they allow management to extend their control over the pace of work. The greater task flexibility of the workforce represented not so much a benefit for the workforce in terms of alleviating boredom, but a burden in terms of greater workload and extra responsibilities. As Hyman (1986, 18) comments, the flexible specialization approach may potentially represent a humane development, but only 'if liberated from the dominance of capital: only if it serves to loosen, not reinforce the hold of commodity production and the wage-labour relationship'.

Ideally, further study in this area would involve an investigation of a wider range of plants in a wider spread of industries, both in the public and private sectors. Fairbrother (1987) points to the far-reaching moves towards employee involvement in the British public sector. While these are not motivated by product market changes, they are responses to tighter and more decentralized financial controls as a reaction to the same economic difficulties that are affecting the manufacturing sector. Management in the public sector still face the problem of how to raise section efficiency and enhance worker co-operation in a period of extensive change in the political and economic environment and it is for this reason that attention is being paid to industrial democracy initiatives.

This study examined the relationship between labour relations and changes to work organization. In both Plants A to F and Plant G, the nature of work organization changed quite sharply. As indicated above, however, there are other options open to management than changes to internal production practices. Further study into this relationship would examine changes to labour relations practices in the absence of production changes. This would require examination of strategies such as relocation, production for export rather than the home market, or an emphasis on leasing and less on sales. In addition, the force of the findings from Plant G is tempered by the fact that at least some of the differences in management's approach must have been due to its lack of unionization. Further study would obviously have to eliminate other possible explanatory variables such as this, which only succeed in clouding the main issues. Further, despite the complicating factors raised above (the limitations placed on management autonomy by the arbitration system, the small size of Australian workplaces compared to those in the UK and America, and the relative novelty of the experiments currently being attempted), this study has demonstrated that an investigation of the argument between the flexible specialization hypothesis and that of the 'Neo-Fordists' can yield interesting observations. In particular, the argument should focus our attention away from ad hoc research into the effects of new technology, by stressing the broader dynamics of the present era of restructuring and the inter-relations between production methods and new employee relations practices.

One final consideration should be the degree to which the outcomes described above in the case studies merely reflect the absence of any strategic union intervention in the process of plant level restructuring. It has been argued that the union movement may be able to use JIT as a window of opportunity for improving its bargaining position and, with it, job security and industrial democracy. The dominant position amongst, for example, the AMWU is that the present conjuncture of political dominance of the ALP and the technical needs of JIT and TQC means that the union movement is faced with an unprecedented opportunity of pushing the frontier of control forward at workplace level. This view is expressed by, amongst others, the Industrial Democracy Officer for the AMWU, who, although recognizing that: 'By becoming involved we run the risk of being incorporated and doing management's job for them' (Ogden, 1986, 39), still argues that there are great opportunities for the union movement to exploit. The problem, as Ogden (ibid, 39) explains, is that: 'If we simply go for direct confrontation on wages and conditions, then in the context of the global reorganization of manufacturing, it's likely to mean a further rapid fall in the number of jobs in manufacturing'. Instead, the union movement should seek to extend the Accord process down to the shopfloor, where, in return for union co-operation with the implementation of JIT and new technology, the breaking down of old skill demarcations and suitable restraint on wage demands, management should agree to fully consult with the union representatives about local content agreements, set employment targets and pay for time-off for stewards to attend union-run courses on JIT. Frenkel (1986, 28) likewise suggests that the participation that results may be 'a vehicle for enhanced employee satisfaction and greater trade union influence' and which is 'essentially a political concept for it foreshadows changes in power relations within the enterprise and by implication changes in social relationships outside the factory'.

The problem with this strategy however, is that the union movement may indeed become incorporated, not only in terms of its base level organizations in joint productivity improvement groups, and quality control committees, but also, (and more critically in terms of the long term future for independent trade unionism in the metal industry), incorporated into accepting managerial logic about the efficiency-employment trade-off. By accepting that a trade-off exists between employment and efficiency, by accepting the logic of the Accord approach to economic management at shopfloor level, the union movement is ideologically disarmed. With this disarmament, managerial logic prevails.

The union movement in the metal industry is therefore faced with considerable problems. It either accepts management's contention that there is a logical connection between higher productivity and work rule concessions (such as are being negotiated at the second tier at present) and job security, and therefore accompanies management on the road to further work intensification, or it takes a stand against management logic and stands accused of trying to 'sabotage industry'. Ultimately, the union movement is faced with a crisis of politics, between attempting to eke out reforms within the present system or posing a fundamental challenge.

ACKNOWLEDGEMENTS

I gratefully acknowledge the assistance and comments from Vic Taylor, Steve Frenkel and the anonymous Labour & Industry referees. However, all responsibility for errors of ommission and commission remain mine.

NOTES

1. This article is based on a paper presented to the Colloquium on Industrial Relations Management at the Australian Graduate School of Management, August 1987.
2. For a full description of JIT and TQC, refer to Schonberger (1982).

REFERENCES

Atkinson, J. (1987) 'Flexibility or Fragmentation? The United Kingdom labour market in the eighties', *Labour and Society*, 12, 1, 87-105.

Boyer, R. (1987) 'Labour flexibilities: Many forms, uncertain effects' *Labour and Society*, 12, 1, 107-129.

Bramble, T. (1986) *Management Strategy and Industrial Relations in Selected Manufacturing Establishments*, MCom (Hons.) Thesis, University of New South Wales.

Braverman, H. (1974) *Labor and Monopoly Capitalism*, Monthly Review Press.

Cashman, K. (1985) *Metal and Engineering Industry: Overall Cost Competitiveness – Major Influencing Factors*, paper presented to the MTIA National Forum, Canberra.

Curtain, R. Krbavac, L. and Stretton, A. (1986) *Skill Formation in Australia: In Search of an Agenda*, Draft Conference Paper, BLMR, Canberra.

Deery, S. (1987) *Trade Union Involvement in the Process of Technological Change*, PhD Thesis, University of Melbourne.

Fairbrother, P. (1987) *Restructuring Management, Flexible Work Patterns and Trade Unionism: A Study of the British Public Sector*, paper presented to the Industrial Relations Colloquium, AGSM, August.

Ford, G.W. (1976) 'A Study of Human Resources and Industrial Relations at the Plant Level in Seven Selected Industries' in *Policies for the Development of Manufacturing Industry*, Volume 4 (The Jackson Report), AGPS, Canberra.

Ford, G.W. (1986) 'Transforming manufacturing: A trilogy of internationalisation, innovation and integration', *Work and People*, 12, 3, 31-33.

Fox A. (1974) *Beyond Contract: Work, Power and Trust Relations*, London, Faber.

Frenkel, S. (1986) *Employee Participation in Decision Making in the Metal and Engineering Industry*, paper prepared for the Green Paper on Industrial Democracy, AGPS, Canberra.

Frenkel, S. (1987) 'Managing Through the Recession: An Analysis of Employee Relations in the Australian Engineering Industry', *Labour and Industry*, 1, 1, 39-60.

Garrahan, P. (1986) 'Nissan in the North-East of England', *Capital and Class*, 27.

Giesecke, T. (1985) 'Industrial Relations and Management in Japan: Other Views', *Work and People*, 11, 3, 21-26.

Harris, H. (1982) *The Right to Manage*, Wisconsin, University of Wisconsin Press.

Holloway, J. (1987) 'The Red Rose of Nissan', *Capital and Class*, 32, 142-164.

Hyman, R. (1986) *Flexible Specialisation: Miracle or Myth?*, mimeo.

Jones, B. (1986) *Flexible Automation and Factory Politics; Britain in Comparative Perspective*, paper presented to the Conference on Industrial Structure and Industry Policy, London.

Katz, H. and Sabel, C. (1985) 'Industrial Relations and Industrial Structure in the Car Industry', *Industrial Relations*, 24, 3, 295-315.

Metal Fabrication Industry Council (1985) *Fasteners Working Group, Final Report*, mimeo.

Ogden, M. (1986) 'Industrial Democracy and New Management Techniques in the Australian Metal Industry', *Journal of Australian Political Economy*, 20, 37-43.

Piore, M. (1986) 'Perspectives on Labor Market Flexibility', *Industrial Relations*, 25, 2, 146-166.

Piore, M. and Sabel, C. (1984) *The Second Industrial Divide*, New York, Basic Books.

Report of the Committee of Inquiry into Labour Market Programs, (The Kirby Report), 1985, AGPS, Canberra.

Rimmer, M., Plowman, D. and Taylor, V. (1986) *Industrial Relations and Structural Change*, University of New South Wales, mimeo.

Sabel, C. (1982) *Work and Politics*, CUP.

Sayer, A. (1986) 'New developments in manufacturing: the just-in-time system, *Capital and Class* 30, 43-72.

Schonberger, R. (1982) *Japanese Manufacturing Techniques: Nine Hidden Lessons in Simplicity*, New York, The Free Press.

Shaiken, H., Herzenberg, S. and Kuhn, S. (1986) 'The Work Process Under More Flexible Production', *Industrial Relations*, 25, 2, 167-183.

Sorge, A., Hartmann, G., Warner, M., Nicholas, I. (1982) *Microelectronics and Manpower in Manufacturing*, Aldershot, Gower.

Tolliday, S. and Zeitlin, J. (eds.) (1986) *The Auto Industry and Its Workers: Between Fordism and Flexibility*, Polity Press.

Wantuck, K. (1985) *JIT for America*, Technology Transfer Council, Sydney.

Williams, K., Williams, J. and Haslam, C. (1986) *Flexible Specialisation and the Future of British Manufacturing*, paper presented to the Conference on Industrial Structure and Industry Policy, London.

Wood, S. (1986) *Technological Change and the Cooperative Labour Strategy in the U.S. Auto Industry*, paper presented to the Colloquium on Trade Union Research, Warwick.

THOMAS BRAMBLE — Lecturer, Department of Industrial Relations, University of New South Wales.

[12]

Flexible specialization versus post-Fordism: theory, evidence and policy implications

Paul Hirst and Jonathan Zeitlin

Abstract

There is widespread agreement that something dramatic has been happening to the international economy over the past two decades: rapid and radical changes in production technology and industrial organization, a major restructuring of world markets, and consequent large-scale changes in the policies of economic management at the international, national and regional levels. At the same time there is a great deal of confusion about how to characterize these changes, the mechanisms at work, and the policy implications for different groups of economic and political actors. One way of accomplishing these tasks is to postulate a change of basic manufacturing organization from a 'Fordist' pattern that prevailed in the years of the long post-1945 boom to a 'post-Fordist' successor in the later 1970s and 1980s. Many people habitually conflate three approaches to industrial change under this heading: flexible specialization, regulation theory, and a more diverse body of explicitly 'post-Fordist' analyses. The resulting problem is that significant differences of approach are concealed by a superficial similarity between the proponents of flexible specialization and a set of apparently similar but underlyingly divergent ideas. The purpose of this paper is to examine systematically the differences between flexible specialization, regulation theory, and other variants of 'post-Fordism' with respect to their fundamental assumptions and theoretical architecture, their methodological approach and use of evidence, and their policy implications.

There is widespread agreement that something dramatic has been happening to the international economy over the past two decades: rapid and radical changes in production technology and industrial organization, a major restructuring of world markets, and consequent large-scale changes in the policies of economic management at the international, national and regional levels. At the same time there is a great deal of confusion about how to characterize these changes, the mechanisms at work, and the policy implications for different groups of economic and political actors. One way of accomplishing these tasks is to postulate a change of basic manufacturing organization from a 'Fordist' pattern that prevailed in the years of the long

Economy and Society Volume 20 Number 1 February 1991

post-1945 boom to a 'post-Fordist' successor in the later 1970s and 1980s. Many people habitually conflate three approaches to industrial change under this heading: flexible specialization, regulation theory, and a more diverse body of explicitly 'post-Fordist' analyses. The resulting problem is that significant differences of approach are concealed by a superficial similarity between the proponents of flexible specialization and a set of apparently similar but underlyingly divergent ideas. The purpose of this paper is to examine systematically the differences between flexible specialization, regulation theory, and other variants of 'post-Fordism'.

The paper is organized as follows. In sections 1–3 we outline the basic arguments of the three approaches, focusing on their fundamental assumptions and theoretical architecture. In section 4 we deal with the problem of evidence, firstly by examining the general methodological approach of each theory, and secondly by examining the substantive issues raised by their different attempts at evidentialization. In section 5, finally, we consider the policy implications of the three theoretical approaches for the manufacturing sector.

1. Flexible specialization: technological paradigms and possible worlds

Despite their apparent similarities, flexible specialization and post-Fordism represent sharply different theoretical approaches to the analysis of industrial change. Where post-Fordism sees productive systems as integrated and coherent totalities, flexible specialization identifies complex and variable connections between technology, institutions and politics; where post-Fordism sees industrial change as a mechanical outcome of impersonal processes, flexible specialization emphasizes contingency and the scope for strategic choice. The distinctiveness of flexible specialization as a style of analysis can best be appreciated by examining the way in which its theoretical architecture builds upwards from simple ideal types to a complex and multi-leveled system of concepts applicable to a diverse range of empirical cases.[1]

The central building block of this approach is its distinction between mass production and craft production or flexible specialization as technological paradigms, ideal-typical models or visions of industrial efficiency. Mass production for these purposes can be defined as the manufacture of standardized products in high volumes using special-purpose machinery and predominately unskilled labour. Craft production or flexible specialization, conversely, can be defined as the manufacture of a wide and changing array of customized products using flexible, general-purpose machinery and skilled, adaptable workers.[2]

Neither mass production nor flexible specialization on this view is inherently superior to the other. Each model is theoretically capable of generating a virtuous circle of productivity improvement and economic growth. Under mass production, sub-divided labour and dedicated equipment can reduce unit costs through economies of scale, extending the market for standardized goods and facilitating new investments in special-purpose technologies, which further reduce costs, extend the market and so on. Under flexible specialization, conversely, versatile labour and universal equipment can reduce the cost of customization through economies of scope, extending the market for differentiated goods and facilitating new investments in flexible technologies, which narrow the price premium for customized products, extend the market and so on. But the practical realization of either possibility depends on a contingent and variable framework of institutional regulation at the micro-level of the firm or region and the macro-level of the national and international economy. Hence the technological dynamism of each model and its potential for sustained development cannot be evaluated outside of definite institutional and environmental contexts. Thus just as there may be technologically innovative forms of both mass and craft production, so too are there stagnant variants of each model in which firms compete through squeezing wages, working conditions and product quality, practices as common in large, declining enterprises as in small sweatshops.

The structural properties of each technological paradigm define a set of micro and macro-regulatory problems whose resolution is crucial for their long-term economic success.[3] In each case, however, similar problems may be solved in different ways, and a plurality of institutional frameworks can therefore be observed within both mass production and flexible specialization alike. For mass production, the crucial micro-regulatory problem is that of balancing supply with demand in individual markets: co-ordinating the flow of specialized inputs through the interdependent phases of production and distribution; and matching the output of productive resources that cannot easily be turned to other uses with the normal level of demand for each good. But as Piore and Sabel argue in *The Second Industrial Divide* (1984), these common goals may be pursued through a range of individual strategies, such as market segmentation, inventory variation and superficial product differentiation, while the institutional framework provided by the large, hierarchical corporation likewise varies considerably both within and across national economies. Thus the organization of mass-production firms in the United States, West Germany and Japan, to choose some notable examples, differ significantly along key dimensions such as levels of administrative centralization and vertical integration, relationships with financial institutions, and systems of shop-floor control.

For flexible specialization, by contrast, the crucial micro-regulatory problem is that of sustaining the innovative recombination of resources by balancing co-operation and competition among productive units. Two major types of institutional framework may be identified for the performance of these functions: 'industrial districts' of small and medium-sized firms; and large, decentralized companies or groups. In the industrial districts, geographically-localized networks of firms subcontract to one another and share a range of common services which are beyond the capacity of individual enterprises to provide for themselves, such as training, research, market forecasting, credit and quality control. Successful districts are also typically characterized by collective systems of conflict resolution which encourage firms to compete through innovation in products and processes rather than through sweated wages and conditions. Within any particular district, however, there may be substantial differences in the roles played by specific institutions, from trade or employers' associations and co-operative banks or credit unions to trade unions, churches and local government; and the political complexion may also vary sharply from 'red' regions such as Tuscany or Emilia to 'white' ones such as the Veneto or Baden-Wurttemberg.

In large, decentralized companies, on the other hand, the relatively autonomous productive units often resemble small, specialized firms or craft workshops, while obtaining services such as research, marketing and finance from other divisions of the parent enterprise. As in the case of mass production corporations, however, large, flexible firms may also differ significantly from one another – for example, in their relationship to banks or trade unions – depending on their individual histories and the national institutional context. There are signs, too, that the extended period of volatility in international markets since the mid-1970s is giving rise to what Sabel calls a 'double convergence' of large and small-firm structures, as small firms in the industrial districts build wider forms of common services often inspired by large-firm models, while the large firms themselves increasingly seek to recreate among their subsidiaries and subcontractors the collaborative relationships characteristic of the industrial districts.[4]

As the Great Depression of the 1930s demonstrated, the market stabilization strategies of large corporations by themselves could not solve the central regulatory problem of mass production: how to ensure a continuing balance between consumption and production across the national economy in order to amortize lumpy investments in product-specific equipment. While the Keynesian welfare state emerged as the dominant form of macro-regulation during the postwar period, here too the differences among national economies remained striking: differences, for example, in the methods of managing budgetary aggregates, in the commitment to counter-cyclical deficit finance and public welfare provision, and in the role of collective bargaining agreements and other 'private' means for relating purchasing power to productivity growth. Like mass production itself, Keynesian macro-regulation was as much a project as an accomplished fact; and nowhere was this more true than at the international level. Despite Keynes's own postwar proposals, no effective institutional mechanisms were created to ensure a steady expansion of global demand in line with productive capacity or recycle purchasing power from surplus to debtor countries in the world economy.

If the macro-regulatory requirements of mass production are relatively well-defined, those of flexible specialization remain the least developed aspect of the model. Thus Piore and Sabel argued that the superior capacity of flexibly-specialized firms to accommodate changes in the level and composition of demand makes macro-regulation less vital than in mass production, giving the price mechanism a greater role in equilibrating supply and demand as in the nineteenth-century competitive economy. But Piore and Sabel also emphasized the micro-regulatory need for such an economy to take wages out of competition and maintain welfare services in order to avoid debilitating breakdowns of solidarity among economic actors, distinguishing possible low and high consumption variants of a flexible specialization regime through the contrasting images of Bourbon Naples and a Proudhonian artisan republic.

More recently, Sabel has developed these ideas by treating macro-economic regulation as a problem of reinsurance: whereas for mass production, the key problem is that of reinsuring firms against unpredictable fluctuations in the level of demand through macro-economic management, the problem for flexible specialization is that of reinsuring regional economies against large-scale shifts in its composition by establishing inter-regional mechanisms to facilitate structural adjustment. On this basis, in turn, he distinguishes two possible futures for the welfare state under a regime of flexible specialization: an exclusive or dualist variant in which regional economies increasingly opt out of the national welfare state while remaining vulnerable to unpredictable external shocks as well as to disruption from those left outside the system; and an inclusive variant which would integrate firms and industrial districts into national systems of training and reinsurance, extend flexible specialization to less successful regions and social groups, and build on existing trends towards the decentralization of the welfare state itself.[5]

A final macro-regulatory issue concerns the implications of flexible specialization for the international division of labour. In *The Second Industrial Divide*, Piore and Sabel suggested that one possible scenario might be the emergence of new forms of interdependence in the world economy, as mass production migrated to underdeveloped countries, while advanced economies increasingly shifted over to flexible specialization. Under these conditions, the First and Third Worlds might also come to share a common interest in a new institutional framework of multinational Keynesianism to regulate world demand and ensure macro-economic stability (Piore and Sabel 1984: 279–80). But flexible specialization might also be conceived as an alternative development strategy for parts of the Third World itself. Such a strategy might build on existing forms of small-scale enterprise concentrated in the substantial 'informal' sectors of many developing economies; and it might also build on the unavoidable flexibility of pre-existing forms of mass production imposed by the constraints of narrow markets and shortages of appropriate skills and materials. Either way, flexible specialization might offer an attractive route to economic development for such countries in which 'appropriate technologies' were not necessarily inferior and modern forms of industrial organization could more easily be adapted to local conditions.[6] Like any development strategy for the Third World, flexible specialization would clearly be advanced by the creation of effective mechanisms of international macroeconomic co-ordination, but unlike mass production it could also be successfully pursued under the more likely conditions of continued volatility in the world economy. Which of these possible worlds may in fact be realized, and to what extent, like the institutional frameworks of micro and macro-regulation, cannot be derived from flexible specialization as an abstract model, but depends instead on the outcome of strategic choices and political struggles.

From this account it should be clear that flexible specialization is at once a general theoretical approach to the analysis of industrial change, and a specific model of productive organization whose micro and macro-regulatory requirements may be satisfied through a variety of institutional forms. But in no sense can this general approach be understood as an evolutionary teleology in which the triumph of flexible specialization as a specific model is a necessary consequence of some immanent logic of economic or technological development. Much of the debate over flexible specialization has in fact missed the mark by construing the latter as a similar type of theory to post-Fordism in its many variants.

Contrary to what many critics have supposed, for example, mass production and flexible specialization are ideal-typical models rather than empirical generalizations or descriptive hypotheses about individual firms, sectors, or national economies.[7] As the original formulations made clear, neither model could ever be wholly predominant in time or space. Thus mass production requires a continuing role for skilled workers and craft production both inside and outside the large firm, to design, set up and maintain special-purpose machinery on the one hand and to manufacture goods for which demand is too small or unstable to justify investments in dedicated equipment on the other. Conversely, some standardization of intermediate goods and components is a necessary condition for the flexible manufacture of diversified final products.[8] Hence the persistence of firms, sectors and even whole national economies organized on alternative principles does not in itself undercut the notion of a dominant technological paradigm in any given period.

At a deeper level, moreover, the analytical distinction between mass production and flexible specialization is also compatible with the empirical finding that hybrid forms of productive organization are the rule rather the exception. As historical research conducted within this framework shows, firms in most countries and periods deliberately mix elements of mass production and craft or flexible production because they are acutely aware of the dangers involved in choosing an unalloyed form of either model. Thus economic actors' understanding of the pure models paradoxically leads them to hedge against risks in ways that blur the lines between them. The resulting interpenetration of elements of flexible and mass production also means that firms often find it easier to shift strategies from one pole to another than an abstract consideration of the two models might lead one to expect.[9]

Contrary to another widespread misconception, therefore, flexible specialization is neither a technological nor a market determinism.[10] Just as trajectories of technological development in this approach are shaped by competing visions of production, so too are patterns of demand shaped by competing visions of consumption. Thus, for example, the realization of either virtuous circle between investment, productivity and the extension of the market depends not only on the creation of an appropriate institutional framework but also on the relative success of flexible and mass producers in persuading consumers to accept or reject a price premium for differentiated goods over their standardized counterparts. This dynamic interaction between production and consumption means that the flexible specialization approach regards market structures not as fixed parameters which impose a uniquely appropriate form of conduct on economic actors, but rather as contingent historical constructs which reflect the competitive strategies of the actors themselves. Hence, for example, current trends towards the diffusion of flexible specialization as a productive strategy result not only from the pervasive volatility of demand but also from the conscious efforts of firms organized along these lines – most notably in Japan – to fragment the mass market still further through the constant introduction of new speciality products.[11]

While flexible specialization strategies may be pursued within a plurality of productive and institutional forms, the range of variation is neither infinite nor arbitrary. Thus, for example, the regulatory requirements of flexible specialization are incompatible with a neo-liberal regime of unregulated markets and cut-throat competition. In each of its institutional forms, flexible specialization depends for its long-term success on an irreducible minimum of trust and co-operation among economic actors, both between managers and workers within the firm and between firms and their external subcontractors. And as we have already noted, such co-operation depends in turn on the establishment of rules limiting certain forms of competition such as sweated wages and conditions, as well as on collective institutions for the supply of non-market inputs such as technological information or trained labour. Hence flexible specialization should not be conflated with opposed conceptions of 'flexibility' as labour-market deregulation which have become common currency not only among businessmen, policymakers and trade unionists, but also among post-Fordists and their critics.[12]

If flexible specialization depends on trust and co-operation, finally, this does not imply the absence of any conflict. On the contrary, flexible specialization, like any system of production, is prone to potentially debilitating conflicts among economic actors, not only between employers and workers, but also between firms and their subcontractors or subsidiaries. The reproduction of social consensus within these systems, though it may build on formative experiences in the past, can only be sustained in the longer term through the creation of institutional mechanisms for the resolution of disputes whose operation is broadly satisfactory to all the parties concerned. While the maintenance of consensus is always provisional and contingent, so too is the crystallization of particular conflicts into durable antagonisms between social groups: neither outcome is predetermined by flexible specialization as a technological paradigm without reference to a definite social and institutional context. For flexible specialization, unlike post-Fordism and its Marxist antecedents, social and political identities cannot be derived from the structure of production through the ascription of objective interests to abstract categories or classes of actor; and therein lies another fundamental difference between these two contrasting approaches.

2. 'Post-Fordism': totalities and social types

The term 'post-Fordism' has such a wide currency and is used so indiscriminately that some demarcations are necessary if different approaches grouped under the term are not to be confused. It has often been used, along with 'Fordism', to characterize aspects of the work of the flexible specialization and regulation schools. Here it will be used to specify a body of work quite distinct from both of them, although in many cases drawing eclectically on their concepts. We must emphasise that 'flexible specialization' is not coterminous with attempts to characterize the international economy since the early 1970s as a 'post-Fordist' era. We will consider three approaches here:

A. the work of the key contributors to the British political magazine *Marxism Today*[13] which has used the concept as part of the thesis that we are living in 'New Times' which mark a decisive break with the economic and social patterns of this century and call for a new left politics;

B. the work of Scott Lash and John Urry (1987) claiming that 'organized capitalism' has been replaced by a new phase and new forms of institutionalization, 'disorganized capitalism' [we will consider this separately only to the degree that it differs from *Marxism Today*, since Urry's work has featured in the magazine and is included in the collection *New Times* (Jacques and Hall 1989);

C. the very different work of Christopher Freeman and Carlota Perez, included here for convenience since it is both more technologistic and economistic than the above and stresses the autonomous role of cycles of technical innovation.[14]

A. *Marxism Today* relies heavily on a stereotyped version of the technical, economic and social relations that preceded 'New Times' – typified by the concept of 'Fordism'. As its leading members are mainly neo-Gramscian Marxists, this concept owes much to Gramsci's essay 'Americanism and Fordism' in the *Prison Notebooks* (1971) and also to works like Harry Braverman's *Labor and Monopoly Capital* (1974) which emphasize Taylorism as a form of production and labour organization characteristic of the Fordist era. Ford's name is used to sum up a series of innovations in manufacturing introduced in the first two decades of this century in the USA and supposedly generalized as a model of industrial production world-wide thereafter. Fordism is mass production on the assembly line model, using special-purpose machinery and mainly unskilled labour in a division of labour based on the increasing fragmentation of tasks. The Fordist era is characterized by the dominance of mass markets and long runs of standardized goods. Fordism is based on the technological efficiency of planned production arising from the separation of conception and execution, and on the economic efficiency of large-scale plants. It comes to dominate by economic logic, the logic of competitive advantage and market performance. However, it was only when Fordist production was coupled with appropriate macro-economic policies by national governments that this production system could find the stability to exploit its full potential; Fordist growth was possible when the mass purchasing power to sustain the mass markets necessary for mass production was assured. Fordism and Keynesianism were responsible for the great postwar boom from 1945 to the first oil price crisis of 1973.

The Fordist system implied a definite type of society, industrial society based on a homogeneous, male, full-time working class, concentrated in large plants in large industrial cities. The assembly line, the concentration of labour in large plants and a full-employment economy promoted the central role of the unions and workers' parties in politics. Traditional social democracy was underwritten by Fordism and with it the primary role of the welfare state.

What is wrong with the Fordist stereotype? Firstly, that it ascribes the dominance of Fordism to economies of scale, to a narrowly economic explanation without reference to actual markets, plant sizes or specific forms of production organization. Taylorism is taken for granted, without reference to the actual complexities of work organization or the role of labour. The division of labour has never been a purely management prerogative: strongly institutionalized patterns of industrial relations mean that it is impossible to read off work organization from stereotyped visions of management objectives. Moreover, these objectives have never been so stereotyped or free of other complex constraints of economic calculation as the notion of Taylorism supposes. Distinct national patterns and phasings of adoption of standardized mass production are ignored.[15] Secondly, it takes Keynesianism for granted – ignoring the complex questions of what macro-economic policies were pursued in different national economies. Thus, for example, neither West Germany nor Japan followed 'Keynesian' policies in their major periods of postwar expansion, adopting instead orthodox fiscal and monetary policies.[16] Close attention to the different national experiences in the great postwar boom reveals quite different manufacturing strategies and institutional underpinnings to growth. *The Second Industrial Divide*, far from conforming to the Fordist stereotype, has the great merit of carefully considering these different national routes to economic organization.[17]

'Fordism' involves a simplified view of manufacturing and macro-economic management, homogenizing the postwar world to over-stress the social and political differences after 1973. 'Fordism' is the coherent concept in *Marxism Today*'s analysis; it provides the rigid point of departure against which to project the changes from the early 1970s onwards. 'Post-Fordism' is a much less coherent concept. It is a way of bundling together a series of economic and social changes. Just as Fordism is supposed to have prevailed throughout the industrial world – other countries following the American model – so 'post-Fordism' is a world-wide phenomenon, with, it appears, Japan leading the way. *Marxism Today* argues that Britain exemplifies the new 'post-Fordist' pattern of economic relations.

Not only are national and regional complexities obliterated here, but there is no real explanatory core to the post-Fordist case other than the decomposition of Fordist structures. The post-Fordist case involves, by and large, borrowing and radically simplifying the flexible specialization approach to manufacturing. However, this borrowing is accompanied by very little attention to the wide range of forms and hybrids of flexibly-specialized production and their social and institutional conditions. The primary role in explaining change is assigned to the collapse of mass markets. Market differentiation and volatility are taken as purely external factors enforcing new patterns of manufacturing. The switch from Fordism to post-Fordism is seen in conventional economic terms as market-led. Classic capitalist entrepreneurs of the conventional type and individual firms respond with new intra-firm strategies to the new market conditions. Strategies are reactive and firm-initiated, decisions draw production techniques, product ranges and labour market changes after them. This violates the basic assumptions of the flexible specialization approach, that effective strategies are anticipatory more than reactive, and that not merely the firm but inter-firm and collective regional and national patterns are crucial in the balancing of competition and co-operation necessary for their more progressive institutionalizations.

Marxism Today characterizes Britain as a 'post-Fordist' society without enquiring further into the extent to which British firms have, indeed, adopted new forms of production technique and manufacturing organization. The evidence for such adoption is scant. British firms appear to be reproducing many of their worst faults of the postwar decades: inattention to training, poor delivery times, misuse of technology, misconceptions of competitors' strategies, etc.[18] How then can Britain's economy be typified as 'post-Fordist'? Only by our realising that 'post-Fordism' for *Marxism Today* is not a rigorous concept but a loose sociological metaphor. Britain is 'post-Fordist' because of the presence of the rhetoric of 'flexibility', because the labour force has changed in its composition and because traditional 'Fordist' national economic management and political patterns have collapsed (even though the latter are largely constructs of *Marxism Today*'s own vision of the past).

Most of *Marxism Today*'s authors are Marxists or post-Marxists. The use of a concept drawn from production organization to characterize a whole era would thus seem to be following on the traditional Marxist methods of assigning a determining role to the forces of production. But technological determinism is one of the main things *Marxism Today* and its key authors like Stuart Hall have been striving to avoid for over ten years. The use of features of the production system to characterize the wider society is no longer a rigorous, if rigid, causal determinism; rather it has slipped to the level of a casual metaphor. The 'post-Fordist' analysis is primarily interested in society, politics and culture, not in rigorously exploring the actual changes in the economy. Indifference to such analysis is coupled with staggering generalizations on a flimsy empirical basis. Stuart Hall says, for example, 'post-Fordism. . . is not committed to any prior determining position for the economy' and, indeed, 'it could just as easily be taken in the opposite way - as signalling the constitutive role which social and cultural relations play in relation to any economic system' (Jacques and Hall 1989: 119). We are not sure what a 'constitutive role' is, since clearly social and cultural relations have done little to change the faults of British manufacturers. What is striking about this reference is not its ritual anti-economism, but its indifference to the investigation of the actual ways in which economic organization and strategies are embedded in social relations. Flexible specialization has drawn repeated attention to the social and institutional context of manufacturing, although the results of its analyses do little to confirm the 'post-Fordist' conception of social organization. *Marxism Today* is happy to bowdlerize the concept of flexible specialization as an ideal type of manufacturing whilst studiously ignoring what its exponents have to say about the routes to the construction of an appropriate social context for this type of production.

Post-Fordism's analysis of 'New Times' is little more than pop sociology combined with a tendency derived from classical Marxism to think of societies as coherent types. The 'post-Fordist' concept is linked to that of post-modernism to produce a view of modern society as fluid and changing, dominated by a shift from collectivism to individualism, from production towards consumption and the service sector, from substance toward style, and to new political issues and new social movements. Pointing out that a traditional collectivist socialism based on the assumption of the electoral primacy of a mass, homogenous manual working class is dated can hardly count as either politically innovative or sociologically novel. Eduard Bernstein said as much in 1899; C. A. R. Crosland repeated the point in 1956.[19] Only the most dogged classical Marxists rising like Rip Van Winkle from their illusions of a parliamentary road to socialism could be impressed by the novelty of this claim. The characterization of the 'Fordist' era in these terms is an absurdity, no one attentive to the politics of European labour movements could ever have believed in this unitary working class, this hegemony of collectivism or this assumption of previous political and electoral success.

When we turn from this fantastic history of the era before 1973 and the illusory post-modernist contrast with the world afterwards we can see how unlike 'New Times' are the most successful regional and national economies that have followed flexible specialization strategies.[20] Neither the Third Italy nor Japan are models of post-modernist social and cultural fluidity. In West

Germany and Italy flexible specialization has been based on legacies of labour skill and institutions of regional and social co-operation that persisted into the supposedly 'Fordist' era. In each of these countries conservatism in social relations and attitudes has provided the basis for social co-operation and economic co-ordination necessary to develop innovative strategies in manufacturing. Social continuity has provided the stability and support to absorb and promote radical change in technology and economic organization. One can overstress the role of conservatism and traditional 'bourgeois' attitudes in these cases, since all three societies were radically disrupted by the experience of fascism. Nevertheless, all three rebuilt their societies and economies after the war, combining innovation and conservatism, and drawing strength for co-operation from the experience of the antagonisms and divisions of the fascist era. Other regions have found different routes to the patterns of co-operation and institutional collaboration necessary to promote flexible specialization, but none of the major industrial districts looks remotely like 'New Times'. The best thing one can say for *Marxism Today* is that it has sought to discard obsolete socialist shibboleths, but the means it has chosen to do so are almost as damaging as the illusions from which it seeks to escape. In particular it has portrayed Britain as a rapidly changing and innovating country, with Mrs Thatcher's Conservative Party in the lead because it has recognized these changes and sought to exploit the new individualism consequential upon them. Not only has British manufacturing not changed in the way *Marxism Today* supposes, not only is the British economy re-running on a new basis the macro-economic failures of the 1960s, but British social attitudes stubbornly refuse to conform to the stereotype of 'New Times'. *Marxism Today*'s analysis comes close to celebrating Thatcherism: seeing it as a breakthrough when Britain is teetering on the edge of economic failure.

B. Scott Lash and John Urry's *The End of Organized Capitalism* (1987) has been widely praised as an innovative work of historical and comparative sociology. It can broadly be classed in the 'post-Fordist' camp for three reasons. First, it draws a dramatic contrast between the era of organized capitalism that prevailed until the 1960s or 1970s, depending on the country, and the era of disorganized capitalism that succeeds it and represents the new phase of capitalist post-modernity. As for *Marxism Today*, the two phases of capitalism are distinct social types. Second, it uses, like *Marxism Today*, the concept of 'post-modernity', emphasizing wider social and cultural changes as well as economic and political ones. Third, it borrows the thesis of flexible specialization to characterize the new technical and organizational forms of modern capitalism.

Lash and Urry certainly attempt to be more rigorous in their sociological concepts and specification of the two distinct types than *Marxism Today* and they attempt to marshall evidence for their national cases – considering the UK, USA, West Germany, France and Sweden. It would be churlish to deny their attempt to demonstrate their argument by comparative analysis, but ultimately their case stands or falls on the consistency of the two types they have outlined. It is here that the problems start. Organized capitalism is seen as a phase of institutionalization that responds to and resolves some of the contradictions of the classic concept of the capitalist mode of production advanced by Marx and Engels. They cite as their major sources for this type the Marxist theorist of monopoly and finance capitalism Rudolf Hilferding and the social historian Jürgen Kocka.[21] The classic Marxist concept of the capitalist mode of production as a totality with certain necessary tendencies (subscribed to by Hilferding) is converted into a list of economic and institutional features making up a type. However, the relationship between the Marxist concept of totality and the list of features or empirical generalizations making up the modified organized capitalist type that develops from the latter phase of the nineteenth-century capitalist economy is left unexplored. Their fourteen points of specification of disorganized capitalism is likewise a list of features consisting in empirical generalizations, of different phenomena grouped to form a whole. Even if most of the features thus identified do indeed occur we cannot treat organized or disorganized capitalism as social types of the same order of those of classical social theory, since they lack a rigorous specification of the necessary causal connections which make the types an operative social whole. The tendency to think in terms of successive social types persists, even though their theoretic status and causal force differ. This does not stop Lash and Urry doing what classical social theorists have always done, which is to draw necessary social and political consequences from the types.

Like *Marxism Today* and the use of the notion of 'Fordism', Lash and Urry see organized capitalism as dominated by class, by collectivism and by class-based parties. Yet, as we have argued in the case of *Marxism Today*, this is to ascribe to the industrial working class and to the labour movement a saliency that can be questioned for all periods of capitalism.

We cannot enter here into a detailed criticism of the way Lash and Urry use their different national cases or with the wider problems of comparative analysis. Instead, we will confine ourselves to three major critical points which relate to their overall conceptual scheme and which undermine the contrast they seek to draw between organized and disorganized capitalism. The first major critical point concerns the use of the concept of flexible specialization. Lash and Urry see flexible specialization as developing from three causes: changes in technology, changes in taste which break up mass market demand, and competition from Third World producers (Lash and Urry 1987: 199–200). The problem with this is, again like *Marxism Today*, it conceives flexible specialization strategies as reactive and the product of exogenous causes. Technology here is being treated as if it were an independent causative agent and as if its role could be specified independently of how it is integrated in production organization and business strategy. We now know, however, that technologies are not given, and that the very same techniques and even machine systems can be used in quite different ways. Buying the technology used by flexible specialists does not automatically give one the advantages of flexible specialization, as many British firms have found to their cost.[22] Likewise, the same point can be made about markets; markets are not created by taste *per se* but also by strategy. At the risk of repetition, central to flexible specialization are strategies of product development that seek to shape markets as well as rapidly adapting production to expressed patterns of demand. Lash and Urry have taken the concept of flexible specialization as if it specified a given technical/economic pattern, and have not understood the primary role of strategy.

Secondly, Lash and Urry's thesis of disorganized capitalism tends to prioritize national, macro-economic and purely intra-firm forms of organization, minimizing the importance of regional and inter-firm patterns of regulation and co-operation. The thesis of disorganized capitalism tends to ignore the emerging regional forms of regulation of production centring on public/private collaboration in providing key inputs that unregulated markets cannot deliver, on local co-operation of industry and labour in such areas as training and manpower policy, and on regional provision of welfare and the development of a confederal welfare state.

Thirdly, Lash and Urry have chosen studiously to ignore two economies where their thesis of disorganized capitalism simply does not fit, viz. Italy and Japan. Italy has had weak national governments and ill co-ordinated macro-economic policies throughout the postwar period. It is a classic example of the emergence of specific patterns of regional collaboration and regulation, notably in the 'Third Italy'. Japan involves a more complex process of organization, centered on a coalition of big industry and a strongly developmental state, with informal and quasi-formal relations between key decision-makers and collaboration between labour and management at the level of the enterprise. One must also stress a neglected feature of the Japanese case: the importance of industrial districts and local economic management.[23] In the USA and West Germany too, emerging patterns of industrial districts and regional governance of the economy are downplayed by Lash and Urry. No serious observer can ignore the development of patterns of organization of advanced industry in regions such as Baden-Württemberg and Massachusetts.[24] Lash and Urry, refer to such developments as 'decentralization' without dwelling on the forms of organization in the new regimes that thus emerge. Their socio-geographic emphasis on spatial patterns leads them to treat these changes as significant in themselves, without looking much further into the specific components of such new emerging regions. Thus apparently similar forms of 'decentralization' in West Germany and the UK have led to very different consequences for their manufacturing firms.

All in all, the main thrust of the flexible specialization argument has passed Lash and Urry by, in that they see it predominantly as a generalization about forms of industrial technology rather than as a distinctive approach to the analysis of modern economies and their forms of governance. Lash and Urry are trapped into using the model of Fordist-Keynesian national economic regulation as the model of organized capitalism and seeing all divergence from this as evidence of 'disorganization'.

C. The writings of Christopher Freeman and Carlota Perez offer a third, distinct and far more academically sophisticated variant of 'post-Fordism'. Their work in contrast to that of *Marxism Today* or Lash and Urry is marked by two key differences, a sophisticated attention to economic theory (particularly Joseph Schumpeter's work on investment, technological innovation and growth cycles, and Kondratiev's work on long waves of technological innovation linked to periods of upswings and downswings in growth) and a more detailed and specific attention to technology.

Freeman and Perez identify five successive waves of innovation linked to Kondratiev cycles. In each of these waves the upswing is dominated by the generalization of a new 'technological paradigm', which emerged from the crisis in the preceding downswing and which in turn leads to a new downswing as the impact of technical innovations and the investment cycle falters. Each successive wave involves new institutional forms which support the phase of economic and technical innovation; these institutions then fall into crisis during the period of downswing, and then involve a period of institutional transition or crisis as the conditions for the next wave are prepared.

The five waves are set out schematically in Freeman and Perez (1988): 1) the period from the 1770s–80s to 1830s–40s, divided into an upswing called the 'Industrial Revolution' and a downswing called 'Hard Times' and characterized by early mechanization; 2) the period from the 1830s–40s to the 1880s–90s, divided into an upswing of 'Victorian Prosperity' and a downswing of the 'Great Depression' based upon steam power and railway technologies; 3) the period from the 1880s–90s to the 1930s and 1940s divided into an upswing called the 'Belle Epoque' and a downswing of the second 'Great Depression' in the 1930s and based upon electrical and heavy engineering technologies; 4) the period from the 1930s–40s to the 1980s and 1990s and divided into an upswing called the 'Golden Age of Growth and Keynesian Full Employment' and a downswing called the 'Crisis of Structural Adjustment' and based upon Fordist mass-production technologies; and 5) an emerging fifth period beginning in the 1980s and 1990s based upon new information and communications technologies, basically a 'post-Fordist' phase.

The work of Freeman and Perez involves major and crippling theoretical problems despite its seriousness and its attempts to avoid some of the major pitfalls of explaining growth through cycles of innovation. They are careful to try to specify the technologies underpinning the major waves, and to look for evidence for their periodization and diffusion. Unlike other 'post-Fordists' they are careful to remain at the level of techno-economic evolution and not to generalize from this into social types, which are ensembles of economic, social and cultural relations. They are also careful to avoid drawing simplistic and overblown political conclusions from their successive waves of technological innovation. However, they still subscribe to the 'Fordist'/'post-Fordist' dichotomy and to roughly the same periodization as the other two bodies of work considered here.

The first major theoretical problem is that Freeman and Perez's analysis is closely tied to the cycles of long waves of technological innovations proposed by the Russian economic statistician Nikolai Kondratiev (Kondratiev 1979). Kondratiev's cycles do not merely illustrate their thesis, they are central to its periodization and to its explanation of successive phases of growth, crisis and innovation. If Kondratiev's claims are false then so is the framework of Freeman and Perez. Kondratiev's work is statistically based and depends on inferences from economic data on growth – particularly on relations between prices and output. As such it offers no convincing causality for a 50–60 year cycle, or why the pattern of technical innovation should 'bunch' in this way. It is essentially a work of statistical economic history and its periodizations are dependent on certain relationships being shown to be statistically significant and to correspond to the chosen cyclic periods. It has stimulated a great deal of debate and it has been comprehensively tested and found wanting by Solomos Solomou (1987). Solomou found that there was no significant correlation between production trends and Kondratiev's waves in Britain, France, Germany or the USA, and no correlation of innovation clusters in the way Kondratiev supposed. In the absence of a 50–60 year cycle divided into roughly equal periods of upswing and downswing, shorter and more contingent cycles of growth or technological innovation will not sustain the Freeman-Perez framework. In the absence of a rigorous statistical proof of the Kondratiev cycle Freeman and Perez must fall back on a weaker version of Schumpeter's account of the investment cycle.

The second major theoretical problem is that although Freeman and Perez are careful to refuse a simple teleology of autonomous technical change, recognizing the role of economic and conjunctural factors in the reception and diffusion of technologies, they still tend to see change in productive systems as driven by technology rather than by business organization and strategy. As we have seen, the flexible specialization approach – far from foregrounding technology – emphasizes the adaptation of techniques to distinct forms of productive organization. Thus the adoption of microelectronics, the core technology of Freeman and Perez's fifth ('post-Fordist') wave, does not of itself lead to flexible specialization strategies and production methods. Perez's discussion of 'flexible production', despite its relative sophistication, remains predominantly at the level of technical processes and potentialities and does not give due weight to production organization, market transformations and institutional changes at the supra-firm level (Perez 1985).

Thirdly, the concept of a 'techno-economic paradigm' is very different from the 'technological paradigm' of flexible specialization, despite the apparent similarity of the terms. A 'techno-economic paradigm' is a diffusible pattern of technology linked to changes in cost-structures. It remains the case that this paradigm is the core of an evolutionary phase of a technical-economic system that comes to dominate every branch of the economy, and it prevails because of the intrinsic gains of the technologies involved and the availability of generalizable low-cost inputs, corresponding to classical economic ideas of competitive efficiency. It is not an ideal-typical model of a new form of productive efficiency, effective because adopted by key economic actors, rather it is a new form of technological efficiency *per se*.

Fourth, technical change is evolutionary in Freeman and Perez's analysis and prevails because of its inherent adaptational advantages, after giving due allowance for delays and crises. It is, despite qualifications, much closer to a form of technological-economic determinism than flexible specialization theory. Evidence of this is the tendency to view successive waves of innovation as necessary and to view the corresponding institutional adaptations of the economy to the new technologies in a functionalist manner. Technical change is conceived much in the manner of Marx's inherent dynamic of the forces of production. The 'socio-institutional framework' needed to accommodate such technologies is seen much in the manner of Marx's relations of production, as a social linkage that is functionally necessary to a certain phase of technology and which then moves into a crisis of functional mal-adaptation when the potential for innovation in the prevailing technology is exhausted. This can be illustrated by a passage from Carlota Perez which could be a translation from *Capital*:

> The prevailing pattern of social behaviour and the existing institutional structure were shaped around the requirements and possibilities created by the previous paradigm. This is why, as the potential of the old paradigm is exhausted, previously successful regulating or stimulating policies do not work. In turn, the relative inertia of the socio-economic framework becomes an insurmountable obstacle for the full deployment of the new paradigm. (1985: 455)

Freeman and Perez are more careful in building their social types than *Marxism Today* or Lash and Urry, the types are techno-economic systems not ensembles of all possible socio-cultural relations, and they offer a more coherent and general causality than do the other 'post-Fordists', but the result is an evolutionary theory with a strongly technical necessitarian cast. As such, although their theory is not classically Marxist, it has many of the same faults as the technological determinism explicit in the dynamic role of Marx's forces of production. The flexible specialization approach has been constructed to avoid just these faults, and so, despite some similarities of terminology and despite some very interesting points about technical change and the specific technologies involved, Freeman and Perez's views are fundamentally opposed to its main theoretical thrust.

3. Regulation theory: middle way or blind alley?

These theoretical contrasts emerge most starkly from a comparison of flexible specialization with the more simplistic versions of post-Fordism put forward by *Marxism Today* or Freeman and Perez. But between these two poles stands a large but elusive 'middle ground' which seeks to combine the openness and contingency of the flexible specialization approach with a continuing insistence on the systematic nature of capitalism as a mode of production and the centrality of class struggle in its development. Undoubtedly the most important and developed representative of this approach is the French 'regulation school' in its 'Parisian' variant.[25] Despite the many subtle differences among themselves, the Parisian regulationists share a conceptual framework and methodological approach which appears to mark them off from flexible specialization and post-Fordism alike. This group of heterodox economists, many of whom are associated with the French state planning apparatus, draw their inspiration from a distinctive combination of Althusserian structuralist Marxism, post-Keynesian macro-economics, and *Annales*-school historiography of the *longue durée*. On this basis they out to execute a 'slalom between the orthodoxies' of neo-classical general equilibrium theory and classical Marxism to produce a rigorous but non-deterministic account of phases of capitalist development which leaves considerable scope for historical variation and national diversity.[26] As in the case of post-Fordism and flexible specialization, this section will examine the theoretical architecture of the regulation approach before going on to consider critically the relationship between this conceptual edifice and its putative methodological foundations.

Like post-Fordism, but unlike flexible specialization, the regulation school takes as its point of departure the concept of capitalism as a mode of production. Capitalism, in this view, is a contradictory and crisis-ridden economic system which requires some form of institutional regulation for its continued reproduction; but in contrast to orthodox Marxism, the operation of these crisis tendencies and their resolution is 'underdetermined' by the abstract properties of the mode of production itself. Social and political struggles therefore play a crucial role in the creation of the regulatory institutions which sustain each new phase of capital accumulation; and like post-Fordism, but unlike flexible specialization, the central actors involved in these struggles are conceived essentially in class terms.

At a lower level of abstraction, the regulation school analyses successive phases of capitalist development in terms of a series of modes of development based on a combination of regimes of accumulation and modes of regulation.[27] A regime of accumulation is a relatively stable and reproducible relationship between production and consumption defined at the level of the international economy as a whole. Each regime of accumulation encompasses a number of interrelated elements, from the pattern of productive organization and the time horizon for investment decisions through the pattern of income distribution and effective demand to the relationship between capitalist and non-capitalist modes of production. Within any regime of accumulation, however, each national economy may have its own distinctive mode of growth depending on its insertion into the international division of labour. Thus the regulation school distinguishes four major regimes of accumulation in the history of capitalism since the eighteenth century: extensive accumulation; intensive accumulation without mass consumption (Taylorist); intensive accumulation with mass consumption (Fordist); and an emergent post-Fordist accumulation regime whose contours have yet to be fully determined.

A mode of regulation, on the other hand, is a complex of institutions and norms which secure, at least for certain periods, the adjustment of individual agents and social groups to the overarching principles of the accumulation regime. Like regimes of accumulation, modes of regulation are complex ensembles composed of a number of interrelated elements: the form of monetary and credit relationships; the wage-labour nexus; the type of competition; the mode of adhesion to the international regime; and the form of state intervention. On this basis, the regulation school likewise distinguishes four major modes of regulation over the past two centuries: old regime regulation (*régulation l'ancienne*); competitive regulation; monopolistic regulation; and an emergent semi-flexible mode of regulation whose contours again remain to be determined. Within any national economy, however, each mode of regulation may be realized in very different ways depending on the pre-existing institutional context and the outcome of domestic social and political struggles.

When a regime of accumulation comes together with an appropriate mode of regulation, the resulting mode of development makes possible a sustained period of technological progress and economic growth. Under these circumstances, crises are essentially cyclical and perform an equilibrating function for the economic system as a whole. But each mode of development also contains its own internal limits – reflecting the deeper contradictions of capitalism as a mode of production – and eventually comes to decay over time. The consequence is the onset of a structural crisis in which regulatory institutions no longer perform a positive function for the regime of accumulation, and the economic system loses its self-equilibrating character. Only the creation of a new relationship between accumulation and regulation can break the impasse and revive the growth process, but such a breakthrough in turn depends on the strategic choices and political struggles of the major social actors.

Despite differences of emphasis between individual writers, for example, the regulation school attributes the crisis of Fordism during the 1970s and 80s primarily to two structural tendencies within this regime of accumulation. First, the progressive exhaustion of productivity gains from Fordist forms of work organization undercut the virtuous circle between consumption and investment, and precipitated a clash between corporate profitability and the institutional mechanisms which sustained aggregate demand within each national economy. Second, the gradual erosion of American hegemony in the world system undercut the role of the dollar as an international currency, destabilizing the implicit mechanisms which had ensured a steady expansion of global demand during the postwar period despite the absence of more formal institutions of international macro-economic co-ordination. The effects of these tendencies were then amplified by conjunctural shocks such as the oil price rises of 1973 and 1979 and by the strategic responses of the major actors: internationalization of production and marketing on the part of large firms; defence of real wages on the part of trade unions; and adoption of deflationary policies on the part of the state. If the roots of the crisis of Fordism for the regulation school are at once structural and conjunctural, there is nothing automatic about the emergence of a new mode of development, and a number of future configurations are conceivable, from neo-Taylorism plus neo-liberalism through flexible specialization in the narrow sense to flexible mass production plus international Keynesianism. Which of these possible outcomes will in fact be realized is determined, on this view, both by the compatibility between the various components of the emergent regime of accumulation and mode of regulation, and by the strategic choices of large firms, trade unions and the state.[28] At first glance, therefore, the regulation school appears to avoid the theoretical pitfalls of the more simplistic versions of post-Fordism while maintaining a close link between empirical analyses of industrial change and the structural dynamics of capitalism as a mode of production. Like flexible specialization, the regulation school avowedly rejects both determinism and functionalism, while enthusiastically embracing historical contingency and national diversity. Neither the onset of structural crises nor their resolution are automatic processes dictated by the laws of motion of capitalism itself; regimes of accumulation do not secure their own regulatory requirements; and neither the capitalist class nor the state plays the role of a 'system engineer' consciously ensuring the dynamic stabilization of the economy. The emergence of an appropriate mode of regulation is thus always 'miraculous', provisional and potentially unstable: at best, one can speak of an *a posteriori* functionalism in which a particular set of regulatory institutions once constituted prove compatible with the demands of a viable mode of development.[29] Finally, as we have seen, different national modes of growth may co-exist within the same international regime of accumulation, while each mode of regulation may be realized through a variety of institutions, giving rise to a wide range of distinctive national experiences in any historical epoch.

But how consistent are these methodological principles with the theoretical architecture of the regulation approach? On closer examination of the key texts, a series of problems can be identified which highlight the tensions between such methodological declarations of intent and the practical application of regulation theory to the explanation of empirical phenomena.

The first set of problems concerns the nature of regulation itself. What precisely is to be regulated within this approach? Is it the general contradictions of capitalism as a mode of production – redefined in neo-Keynesian terms as the need to maintain a dynamic balance between production and consumption – or the specific dilemmas of an individual regime of accumulation? What is the relationship between the various elements of the mode of regulation and which level is most crucial to the achievement of sustained growth? Despite formal declarations that modes of regulation are complex structures each of whose elements must be compatible with the others, most regulationist analyses in practice tend to privilege a single component of the system – usually the wage-labour nexus or the form of competition – and use it to characterize an entire phase of development as Fordist or post-Fordist, competitive or monopolistic. How far do the objects of regulation such as the wage relationship or the form of state intervention precede the emergence of a particular mode of regulation, and how far, as Jessop suggests, do the objects and mode of regulation form a structural coupling whose contours can only be determined under a definite set of historical circumstances? Despite their methodological reservations about functional explanations, most regulation school analyses reason in practice as if the persistence of the objects of regulation could be accounted for by the development of a smoothly functioning mode of regulation.[30]

A second set of problems concerns the relationship between higher and lower levels of analysis or between theoretical abstractions and empirical cases. How far can the historical succession of regimes of accumulation and modes of regulation be separated from the structural properties of capitalism as a mode of production? To what extent is the existence of a plurality of national modes of growth genuinely compatible with the postulation of a dominant international regime of accumulation during any given period? How far can divergent national configurations of institutions such as collective bargaining or the welfare state be legitimately considered as limited variations of a single mode of regulation?

When the regulation school seeks to characterize broad historical periods in terms of dominant regimes of accumulation and modes of regulation, its exponents tend to elide the diversity of national experiences acknowledged in more detailed case studies.[31] Despite the many empirical caveats scattered through their work, therefore, regulation theorists, as other critics have charged, systematically overstate the dominance of Fordist modes of regulation during the postwar period, whether in terms of the pervasiveness of Taylorist work organization, institutionalized collective bargaining, Keynesian demand management or the welfare state.[32] Conversely, the application of a regulation approach to national case studies, typically involves not only a severe 'stylization of the facts' to fit its theoretical categories, but also *ad hoc* modifications of the categories themselves to accommodate observed variations. The result is the multiplication of hybrid formulations poised uneasily between theory and empirical description, such as 'flex-Fordism' (West Germany), 'blocked Fordism' (UK), 'state Fordism' (France), 'delayed Fordism' (Spain, Italy), 'peripheral Fordism' (Mexico, South Korea, Brazil) or 'primitive Taylorization' (Malaysia, Bangladesh, the Philippines).[33] When regulation theorists discuss the possible shape of an emergent post-Fordist mode of development, finally, they often fall back illegitimately on general tendencies of capitalism as a mode of production – such as the concentration and centralization of capital or the development of economies of scale – to privilege the likelihood of certain outcomes over others. Thus, for example, in arguing against the viability of flexible specialization as the basis for sustained growth, Boyer revealingly asks, 'Is it plausible that the trends that have operated since the Industrial Revolution could be reversed? Does it not imply a complete shift in the balance of power and social and economic structures? Is it conceivable that all the indivisibilities that characterize contemporary society could thus be reduced. . .?'[34]

A third set of problems concerns the role of classes as social actors in the emergence and decay of successive modes of development. At a methodological level, regulation theorists reject the idea of classes as collective subjects whose interests can be derived from the abstract structure of capitalist relations of production, recognize the existence of a plurality of social actors, and assign considerable autonomy to the state as something more than a bulwark of capitalism. When it comes to more specific regulationist analyses, however, social forces are typically analysed in class terms with little attempt at empirical justification. Thus, for example, the declining rate of productivity growth under Fordism is largely attributed to class struggle at the point of production, assuming without argument that particular disputes between managers and work groups can be unproblematically aggregated into more general conflicts between capital and labour. Similarly, national modes of regulation are characteristically treated as class compromises, with little specific attention to the identity or organization of the actors concerned: to what extent, for example, can trade unions, business firms or employers' associations in countries as different as the US, France and Japan all be treated as direct representatives of wider social classes? Finally, despite occasional references to the importance of 'new social movements', regulation school analyses of the political forces and social constituencies which may shape a possible post-Fordist future are overwhelmingly couched in class terms.[35]

Each of these problems highlights the inherent tensions between the totalizing ambitions of regulation theory – taken over from classical Marxism as well as from neo-classical economics – and its putative commitment to the recognition of national and historical variation. The regulation school set out to discover a middle way between general theory and empirical analysis, but their approach has run into a blind alley in which its conceptual holism must alternatively override or be undermined by the diversity of particular cases. Despite its apparent methodological sophistication, therefore, the theoretical architecture of the regulation approach is ultimately little different from the more simplistic versions of post-Fordism, while contrasting quite sharply with that of flexible specialization.

4. Problems of evidence

A. Methodological questions

So far we have considered the theoretical structure and major substantive claims of the three main bodies of work which address the problem of changes in methods of manufacturing organization, and the approaches to technology and business strategy that accompany them. We have shown that each of these approaches stems from a distinct theoretical background and that each makes quite distinct claims about the type and level of changes that are taking place in contemporary manufacturing industry. These very real differences tend to be masked by widespread assumptions in political and academic debate that the theories are all saying something similar about the same basic, if changing, reality. This is a commonsense mistake that it is essential to rebut if we are to be able to assess the respective claims of the theories to validity and to consider the adequacy of their means of demonstrating the very distinct knowledge-claims that they advance.

It is an illusion to assume that each of them can be tested by reference to some common set of changes occurring 'out there' in the real world. Each of these bodies of work sets up a very different world, each pre-constructs it in theory, and each specifies in a particular way what sort of evidence and research strategy will count in showing what they claim to be happening is taking place. Theory, in the sense of a set of concepts, the construction of a set of objects on the basis of such theory, and the postulation of a domain of evidence relevant to such concepts and propositions are all particular to the style of work in question. We cannot construct a theory-neutral domain of evidence that will suffice to adjudicate between the claims of the three styles.

Rather we must look first at each of the styles of work and see what theory-evidence relation it seeks to establish. Then we must question the adequacy of that relation. Is the process of evidentialization consistent with the claims of the theory? Does the evidence proposed and gathered actually justify these claims? Is it practically possible to build up a body of evidence? If not, are there substitute forms of evidentialization that will provide weaker but indicative or persuasive support for the theoretical claims advanced?

All these questions relate to the relationships possible between theoretical concepts, the objects they construct and the type of evidence considered appropriate. These relationships are complex. We assume that there is no one 'correct' set of epistemological doctrines or methodological protocols that will cover them all. We are far from wishing to legislate proscriptively about epistemological doctrines or styles of research. Clearly, we have little time for the naive empiricism or positivist phenomenalism that would refer each of these theories to some common body of given facts or constructed operationalizations of observable phenomena. Equally we are by no means satisfied that a realist epistemological premise actually enables one to capture 'reality' more effectively than conventionalist premises about the theory-dependent nature of the objects of investigation. In practice, all demonstration and evidentialization is more complex for any definite body of work than general epistemological protocols or methodological textbooks suppose. We are sceptical about the merits of generalized doctrines in epistemology and the philosophy of science as guides to social-scientific research. But we are not indifferent to questions of validity and evidence. Our concern with these questions is governed by two suppositions.

The first is that justice must be done to distinct claims of a theory about the evidence relevant to it. It is illegitimate to seek to 'refute' a theory solely on the basis of evidence external to the theory and in disregard of the arguments it may make about why that evidence is inappropriate. However, serious contradictions between theory and evidentialization – exemplified by inconsistency, uncertainty of claim, and constructing *ad hoc* arguments to make evidence fit ambiguous propositions – all tend to diminish the possibility of taking the tests proposed by a theory seriously or to accept its evidence at its own value. We shall claim that there are such contradictions between theory and evidence implicit in regulation theory, and that they stem from its basic objectives and intellectual starting points. We shall also claim that the 'post-Fordist' view as advanced by *Marxism Today* in particular, is too loose theoretically to be evidentialized in any rigorous way.

The second supposition is that ultimately the only way to determine the validity of these theories and their very distinct constructed objects is not to refer each of them to a common external 'reality', but to relate each of the theories one to another and judge them, in terms of the plausibility of the arguments that can be advanced within one of the theories – flexible specialization – against the others. This claim is not theory-neutral and its outcome rests not on some general methodological protocol, but on the explanatory power of the specific argument advanced. Ultimately the test of a theory is its intellectual productivity, not its elegance, nor its conformity to established methodological canons, nor its relation to pre-given political expectations or desirable policy outcomes. That productivity rests solely on its capacity for arguing and showing what it claims is the case with a reasonable degree of probability.[36]

We shall now consider each of the main styles of work in turn in order to examine the general theory/evidence relation it constructs before going on to consider substantive questions about that evidence.

Flexible specialization

The flexible specialization approach takes its start from a criticism of social theories which assume that society is a 'totality', a set of relationships governed by a single general principle and consistent in their character with such a principle. It also entails the criticism that such theories frequently presume a process of necessary social development or evolution based on certain fundamental 'tendencies' operative in such a totality.[37] These criticisms apply to both Marxist and non-Marxist general sociological theories alike. Flexible specialization emphasizes the contingency and complexity of the connections between social relations, it insists on the distinctiveness of the national and regional routes to the establishment of such connections between social relations, it recognizes the crucial role of strategy and bodies of ideas in constructing such routes, and it is aware that things could have been otherwise. It is, therefore, alert to the specific conditions producing certain outcomes and to the possible coexistence of several distinct sorts of outcomes. The variety of possible outcomes that can be constructed from the basic ideal-typical concepts of flexible specialization is, therefore, considerable and each establishes a different relation between concepts, constructed social objects, and type of evidence that will demonstrate whether or not the social outcomes connected to such objects are operable.

To illustrate this complexity we shall outline three kinds of relation between theory and evidence in flexible specialization.

First, flexible specialization can be used in a mode we call the normative-empirical. Flexible specialization emphasizes that each social 'world' contains a number of possibilities. A prevalent technological paradigm and the typical modes of social organization connected to it arises for a complex variety of factors, and has predominance over other possible outcomes for reasons that fall far short of social-structural or historical logics or necessities. This means that we must be alert to competing strategies and assess outcomes in terms that do not predetermine which of them will prevail. Hence the attention given to historical alternatives to mass production and the search for other reasons for the saliency of mass production strategies than an assumption of their inherent efficiency due to economies of scale.[38] Part of the role of evidence here consists in showing that other alternatives were possible, that they coexisted with the dominant paradigm, and that they offer distinct routes to innovation and change should the specific complex of conditions favouring the dominant paradigm cease to apply.

Thus flexible specialization is concerned to rewrite history in order to show that the complexity of the past helps us to recognize that there are a variety of options in the present. The relation of theory and evidence in such historical work is complex, in particular such claims cannot be refuted by pointing to the importance of mass production; rather they depend crucially on the conditions under which it came to prevail and national and regional variations in the forms of mass production strategies themselves. The same complexity of evidence about the coexistence and possibility of a number of worlds relates to present debates as well. Part of the role of theory is to identify certain instances or cases of progressive flexible specialization strategies, to show that such things are socially possible and to investigate whether they can be generalized given appropriate policy commitments and satisfactory conditions. If flexible specialization strategies are possible, if their conditions are not too difficult to satisfy, and if certain of their policy consequences and social outcomes are attractive from a certain normative standpoint then the role of evidence here is to serve as a support for advocacy and a means of generalizing the process of learning from certain national, regional or enterprise experiences.[39] Simply showing that flexible specialization strategies have not been generalized, that they exist only in certain cases, and that they do not exist in a pure ideal-typical but in a hybrid form thus does not constitute a refutation of flexible specialization as advocacy. All the advocate of flexible specialization as a normative approach has to do is to show that such strategies are possible and that they can be expanded beyond given cases, even if in a hybrid form. Thus much of the 'empirical' criticism of flexible specialization analysis is beside the point, since the use of such concepts is not confined to the hypothesis that flexible specialization is the prevailing or generalized mode of manufacturing organization.

Second, flexible specialization serves as a positive heuristic. Thus flexible specialization theory includes a battery of concepts drawing attention to a number of distinct ideal types of production systems, progressive and stagnant variants of the same, possible forms of hybridization, and also ways in which these various forms can be combined in large and small companies, national and regional economies. The result is a very large range of possible situations and complex cases, a wide variety of types and hypotheses. Thus it will not do to select one of these types or hypotheses and seek to 'refute' it without reference to the others. This is a common failing of the critical literature which tends to operate as if flexible specialization were a theory which gives necessary prominence to small firms over large ones, or which supposes that the industrial district based on small firms is the sole or major form of flexible production. The ideal-type is not to be taken as an empirical generalization, and, therefore, it should not be treated as if it consisted in a proposition that the majority of firms in a given national economy would conform to its features. Moreover, the simple ideal-type is just a part in a more complex and multi-layered process of theorization. This process emphasizes the importance of social context, the complexity of coexisting strategies and structures of manufacturing, the contingent nature of their conditions of existence, and the variety of possible outcomes. Flexible specialization cannot be reduced to a few simple hypotheses. At the same time, this theoretical complexity is not the result of *ad hoc* argumentation and incoherence.

Third, flexible specialization serves as a negative heuristic. Flexible specialization is a theory about the nature of manufacturing as a form of social organization. Even if flexible specialization were not widespread, the concept would still be valuable. For the analysis of mass production and its conditions of existence could not remain the same, the specific social routes to its generalization as a paradigm and the variant forms of its institutionalization would gain in saliency as against traditional claims that mass production prevailed because of technological necessities and economies of scale. Flexible specialization is thus not merely an hypothesis about one type of production but is part of a much wider theory about production systems in general and their socio-political conditions of existence.

'Post-Fordism'

The advocates of 'post-Fordism' – *Marxism Today* and Lash and Urry in particular – are trying to respond to and to register certain perceived social changes. Their point of departure is the product of a general theory of a mode of production having certain necessary tendencies. Late capitalism and 'post-Fordist'/'post-modernist' contemporary reality are entities of a different socio-theoretical type. This is clearest in Lash and Urry's *The End of Organized Capitalism* (1987) where organized and disorganized capitalism are characterized by two lists of phenomena: most of the features in the list relating to organized capitalism are derivations from a theory of the capitalist mode of production in its monopoly phase, elements of a totality, whereas the other list is a collection of merely co-present facts. The status of these co-present facts, their coherence, and how they could evolve out of a phase of late capitalism are all left unaddressed. As we have seen, the tendency is to argue back from the bundle of features postulated in 'post-Fordism' to demonstrate the presence of the phenomena, to assume productive flexibility by reference to, for example, the differentiation of the labour force and labour markets. Metaphor and *ad hoc* generalization have replaced rigorous concepts.

'Post-Fordist' arguments like that of *Marxism Today* simultaneously under and over-totalize. They over-totalize by aggregating together phenomena into a social type: deriving social relations from a metaphor based on production organization. 'Post-Fordism' is a new type of society and despite claims about fluidity of social relations, the co-present elements are not permitted to vary except to the most limited degree. Thus the inability to see that those areas in which the most 'advanced' production exists (the most 'post-Fordist' manufacturing organization) are frequently socially and institutionally conservative (the least 'post-modernist' social relationships).

At the same time these arguments under-totalize because they are incapable of postulating consistent causal relationships between the phenomena in this new social type, its elements are merely co-present. For the 'post-Fordists' economic determinism is *passé*, so is the essentialism of the mode of production and its necessary tendencies. But no new explanatory schema has been developed consistent with the level of generality of the social typology. Like regulation theory, 'post-Fordism' is a retreat from classical Marxism and classical sociological theory that has not fully settled its accounts with the latter.

The result is a form of argument that is *ad hoc* in its basic substance, rather than just forced into *ad hoc* arguments to reconcile specific contradictions and ambiguities as is the case with regulation theory. It is difficult to see how the 'post-Fordists' could construct a domain of evidence since there is no coherent theory to be evidentialized. Merely pointing to certain phenomena, showing things have changed, or showing that the bits of a list do indeed exist will not do. 'Post-Fordism' is simply not coherent enough as a theory to merit more specific concern about the evidence that might justify it. We shall not, therefore, go on to discuss it when we consider the specific issues about evidence below.

For Freeman and Perez, on the other hand, definite questions of evidence do arise. The most important of these is the question of statistical proof of the existence of Kondratiev waves. As we have covered this issue in our theoretical discussion of the 'techno-economic paradigm' approach we will not take it further in the following section on substantive problems of evidence, and as this issue is the one on which their analysis stands or falls we will not consider lesser issues of evidence further here.

Regulation theory

The regulation theorists share the same basic concepts, though their specific methodological pronouncements differ. At the same time two methodological positions are common to most regulationists: first, a desire to move away from the more rigidly economic determinist, historicist and conceptual realist features of classical Marxism; second, a willingness to use forms of evidence, such as econometric tests, quite alien to classical Marxism.

Alain Lipietz expresses this methodological opposition to conceptual realism particularly forcefully in *Mirages and Miracles* (1987). He argues that concepts are only valuable insofar as they enable us to grasp phenomena, and that one should not make a fetish of them. Conceptual realism, the hypostatization of concepts into entities, is a major fault of much Marxist social analysis and an epistemological obstacle to concrete analysis. Lipietz uses Umberto Eco's nominalist monk William of Baskerville to make his point.[40] The rejection of conceptual realism has been a commonplace, at least since Whitehead's day.[41] But conceptual realism has been a persistent feature of Marxism precisely because of its strong Hegelian legacy. Marxism views society as a totality driven toward certain states of affairs by tendencies inherent within it. This totality can be apprehended in general concepts because it exists as a determinate generality; concepts thus partake of the nature of this social reality. Conceptual realism in Marxism is not misplaced concreteness: it is rather the apprehension of the concrete generality in thought.[42]

Conceptual realism makes methodological sense only within a strongly structuralist or historicist conception of Marxism. Yet this is exactly the sort of determinism Lipietz, Boyer and the other regulationists wish to avoid. They want to insist on the concrete variability of capitalism, on its crisis-ridden nature, and on the absence of historical inevitability. Yet they wish to retain the concept of capitalism as a mode of production and the existence of certain fundamental tendencies within it. How can this be done? The answer it appears is to postulate levels of analysis, and to use the higher levels of generality to explain the specific conditions which are the only real states of existence of the capitalist system; capitalism exists not as generality but as specificity.

In fact it is much more difficult than the regulationists think to be a nominalist and yet to retain some major elements of classical Marxism. Three main problems present themselves. Firstly, conceptual nominalism prevails in regulation theory only at the formal level, and conceptual realism is overcome in practice by injecting elements of classical Marxist concepts into the phenomena. Nominalism is compatible with Marxism only by the 'Marxianization' of the phenomena to be explained. The very problem of regulation is of this nature - it is a consequence of 'Marxianizing' the economy. Regulation theory involves taking a version of the 'problem' of reproduction outlined in Vol. II of *Capital* as a necessary economic problem, and seeking to show how production and consumption can be interconnected by specific institutional forms. Concrete economic relations are treated as if they must actually resolve at least temporarily the contradictions in the capitalist mode of production. This is only a 'problem' if you read *Capital* in a certain way, and believe that concrete economies actually have to answer in practice the problems posed by this reading.

Secondly, there is a theoretical failure to explain how the more general concepts and abstract levels of analysis derived from classical Marxism can actually be carried over into and made compatible with the more specific concepts and levels of analysis sought after by regulation theory. This problem arises from rejecting as unacceptable the classical Marxist conception of the capitalist mode of production as a totality (a concrete generality) which subsumes particular capitalisms as mere instances of the workings of its laws. Levels of abstraction are needed because this direct presence of the totality in the concrete simply doesn't work: it leads to the postulation of certain necessary states of affairs, and yet these states of affairs stubbornly fail to materialize. Capitalism, far from collapsing, survives and proves capable of institutional innovation and national variation within a prevailing international regime of accumulation. Hence the withdrawal of the totality to a higher level of abstraction, and the postulation of the concrete as the specific domain in which the general 'problems' of the capitalist mode of production are worked out in specific spatio-temporal forms. The problem is not merely one of levels of abstraction, how capitalism is present in and as its specific spatio-temporal conjunctures, but also involves that of concealed conceptual realism. The more general concepts are supposedly merely a means to get at a reality that is always and inevitably concrete, and yet somehow these concepts get 'behind' reality and particular national economies become exemplars of variant forms of phases in the evolution of capitalism.

Thirdly, there is the problem of the status of social actors, and social classes in particular, created by the complexity of the regulation school's relation to the classical concept of the capitalist mode of production. In classical Marxism classes are both categories of economic agents, defined by their relationship to the means of production, and social forces active in politics. The evolution of the capitalist mode of production, its contradictory tendencies, and the emerging crises of capitalism lead the social groups active in politics to crystallize around the opposed social poles of bourgeoisie and proletariat. Marxism has been in retreat from this simplistic social teleology of economic and political class relations ever since the *Communist Manifesto* was written. But regulation theory has a special problem since, for it, classes as categories of economic agents can only relate to capitalism at the highest level of abstraction. At more concrete levels, economic and political actors are seen to be specific and contingently constituted agencies. However, regulation theory has devised no means to relate classes at the different levels of abstraction any more than it can relate any other social phenomena. Thus it tends to continue to treat classes as stereotyped entities and to refer to them as active social agents, alongside or underlying more specific social groupings and organizations. The result is schematic analysis and the tendency to over generalize the political consequences of and responses to forms of economic relations both nationally and internationally – treating capital and labour as if they were a coherent social actors.[43]

Thus the main methodological problem of regulation theory is not some specific difficulty with evidence, but the much more basic issue of whether the basic entities and processes it seeks to evidentialize exist at all. As these entities are of a more general nature than specific bodies of evidence, such as statistical series, the latter cannot serve to demonstrate their existence but presuppose it if they are to count as evidence. Indeed, the use of evidence in regulation theory presupposes the general truth of its conceptions of capitalism.

Unlike flexible specialization, regulation theory is tied to certain general concepts that must be present in reality; it cannot operate in an ideal-typical mode. Flexible specialization does not suffer from this problem of different levels of generality; from a need for the device of concretizing abstractions through the construction of specific states of their realization in definite conditions. Flexible specialization neither needs to posit a totality nor does it need to limit the range of variation of regional and national economies to fit in with some prevailing phase of its actualization. We should be clear that the postulation of a social totality (in however complex and qualified a form) is not inherently more virtuous in explanatory terms than theories like flexible specialization which stress the variation, contingency and complexity of phenomena. We should also be clear that anti-totalizing theories are not more 'empiricist' than theories which seek to make totalizing general concepts operative in reality. There are other routes to conceptual rigour than totalization or quasi-totalization: flexible specialization uses concepts as ideal-types to organize analysis and explain how in the concrete case contingent phenomena are causally connected one to another.

The problem with totalization is that it stakes too much on the validity of concrete generalities. It is forced to limit possible phenomena to fit in with the general principles of social organization that make social relations consistent wholes. The problem with regulation theory is not its commendable drive to break away from the worst features of totalizing Marxism; rather it consists in the belief that they can be purged by using the device of levels of abstraction whilst retaining elements of the concept of mode of production. Like most half-way houses, or third ways between alternatives that are perceived as unacceptable, it suffers in practice from the faults of that from which it seeks to retreat. The problems of evidentialization in regulation theory are thus effects of its basic theoretical strategy, and it is difficult to see how they can be overcome or that strategy reformed. Only a complete break with classical Marxism would resolve these problems of evidence, but that would explode the general problem of 'regulation' in terms of which these questions of evidence arise.

B. From theories to evidence

Regulation theory

Of all the competing approaches, regulation theory has undoubtedly devoted the most explicit attention to the problem of empirical verification. Among its central methodological objectives is the reconciliation of Marxist analytical concepts with modern methods of formal economic modelling and econometric testing. Hence the regulation school characteristically seeks to verify its hypotheses by translating key concepts into appropriate micro and macroeconomic models, which can then be assessed both for their internal consistency and for their ability to explain comparative and historical variations in statistical data sets. What were the mechanisms of wage formation in particular countries during different periods? What were the central causes of the crises of the 1930s and the 1970s for individual countries and the international economy as a whole? What are the prospective relationships between demand, investment, productivity and growth under different possible configurations of regimes of accumulation and modes of regulation? These are the types of question which the regulationists' methods of empirical testing are designed to answer.

There can be little dispute about the ingenuity deployed by the regulation school in constructing such models, and the assumptions involved are often more realistic than those of more conventional economic analysis. But a number of fundamental problems can none the less be discerned. The first concerns the nature of the modelling exercise itself. However sophisticated the model, the theoretical conclusions which follow are no better than its initial assumptions. Thus, for example, Boyer's conclusion that flexible automation (combined with international Keynesianism) offers better prospects of sustained growth than flexible specialization follows directly from his assumption of an inferior trade-off between flexibility and productivity under the latter, an assumption which would not be accepted by theorists of flexible specialization themselves.[44]

The second problem concerns the quality of the data to which the models are applied. As with much quantitative economic history, the statistical evidence is rarely robust enough convincingly to support the econometric tests to which it is subjected by regulation school analyses. Thus as Boyer acknowledges in his long-term study of French wage-formation, the unreliability of nineteenth-century statistics on unemployment and inter-sectoral wage variations make his conclusions 'suggestive rather than definitive', while it is unclear how far the apparent reinforcement of competitive processes of wage formation between the wars is a real phenomenon rather than a result of changes in the method of constructing the industrial production series.[45] Even more seriously, his discussion of real wage movements, as in most such studies, appears to be based on indices of money wage *rates*, although detailed sectoral research typically reveals significant differences between rates and *earnings*, reflecting the influence of fluctuations in hours worked across the business cycle as well as of variations in bargaining power and the incidence of state regulation.[46]

The third problem concerns the conclusions which can be drawn from the confrontation of these models with empirical data. Like all econometric tests, those deployed by regulation theorists can only demonstrate statistical correlations rather than causal relationships between the variables investigated. Thus, for example, it is a central claim of the regulation school that the major cause of the depression of the 1930s in countries such as France and the United States was an imbalance between intensive accumulation and competitive regulation, as demonstrated by the disparity between the rate of growth of productivity and of effective demand during the preceding decade.[47] But the regulation theorists simply infer the transition to intensive accumulation from rapid growth of French productivity during the 1920s, despite considerable evidence that the diffusion of Taylorism and scientific management in French industry remained quite limited throughout this period.[48] And when it comes to the crisis of Fordism, Boyer himself admits that the explanations put forward by the regulation school, at least for the moment, 'are both too numerous and insufficiently articulated amongst themselves' (Boyer 1986a: 103–4).

Such problems are common to all forms of economic modelling and econometric analysis, and are in no sense peculiar to regulation theory. But they present special difficulties for the regulation school because of the distinctive nature of its theoretical project. For orthodox economists, following the methodological precepts of Friedman and others, the inability of econometric tests to verify causal hypotheses is unimportant because the value of economic models lies in their predictive power rather than the descriptive realism of their assumptions.[49] For the regulation school, on the other hand, econometric tests are intended to demonstrate the truth of its models as descriptive hypotheses about the real world (Boyer 1989: 1403–5), and their inadequacy for this purpose opens up an unbridgeable disjunction between the theory and its preferred mode of evidentialization. This disjunction, in turn, is simply a further sign of the regulation theorists' continued but unacknowledged attachment to conceptual realism, together with the impossibility of escaping from the deterministic consequences of classical Marxism while retaining key elements of its underlying conceptual framework.

Flexible specialization

For flexible specialization, by contrast, the relationship between theory and evidence is significantly more complex. Unlike regulation theory, the central problem for flexible specialization is not to demonstrate the truth or falsity of its basic concepts, since these are explicitly conceived as ideal-types instead of real forces operating behind the observable phenomena themselves. The appropriate criterion for the assessment of such ideal-types is not their truth value but rather their heuristic productivity: how far does the conceptual framework of flexible specialization illuminate observable processes of industrial change? As we have seen in previous sections, flexible specialization as a general theoretical approach is compatible with a broad spectrum of possible forms of productive organization – including the continued predominance of mass production. But a hypothesis whose validity does depend on empirical evidence is whether – as much of the literature argues – current manufacturing practice is moving in the direction of flexible specialization as a specific model of productive organization, taking account both of the plurality of institutional forms within which it may be pursued and of the possibilities of hybridization.

What sort of evidence might permit us to test this hypothesis? Both macro and micro levels of analysis are in principle relevant. At the macro level, one would ideally like to establish statistical indicators of cross-sectional variations and changes over time in the distribution of different forms of production across industrial sectors, national economies and the international economy as a whole. Thus one would need large-scale data about such issues as product diversification (number of distinct products manufactured, rate of introduction of new models, average batch size); productive flexibility (costs of product changeover, nature of equipment used, minimum efficient scale of operation); workforce versatility (skill composition, job content, training); inter-firm relationships (extent and nature of subcontracting, reliance on collective services); and geographical agglomeration of economic activity.

But there are good reasons, both practical and theoretical, why reliable macro-data of this type are likely to be difficult if not impossible to obtain. The first arises from the nature of the available industrial statistics. Like all official statistics, the classification systems used by industrial censuses in different countries reflect particular sets of theoretical assumptions and administrative practices. Thus as Storper and Harrison rightly point out, national statistical accounts cannot be used to analyse the operation of real input-output systems or industrial sectors 'since they tend to classify whole firms or establishments according to their "principal" activities' (Storper and Harrison 1990: 6). And even within these limitations, as Luria has shown, there are fundamental difficulties in measuring inter-temporal variations in product diversity because of widespread inconsistency in classification, frequent code changes and suppression of data to protect proprietary information (Luria 1989: 16). The more rapid the rate of product diversification, moreover, the more serious these difficulties become, since earlier classifications become obsolete more quickly and incommensurability of data from different periods increases.

For many of the other empirical questions thrown up by the flexible specialization hypothesis such as product batch sizes, workforce versatility or patterns of subcontracting, little large-scale data is available because official statistics have not been compiled with these issues in mind.[50] But even were industrial census classifications rewritten with an eye to flexible specialization, significant conceptual problems would still remain in the interpretation of such evidence. The key issue is the context and strategy-dependence of each element of a productive system within the flexible specialization approach. Each industrial branch or sector, for example, has its own specific market and technological characteristics against which any particular indicator must be assessed. Thus a given product batch size will have a different significance in, say, clothing, steel or automobiles, and there is no obvious way to aggregate such data across the economy as a whole. At a deeper level, moreover, the same component or practice may have a different significance depending on its place in the broader strategy of the individual firm: thus as a substantial body of research has demonstrated the same equipment such as numerically-controlled machine tools or flexible manufacturing systems can be used in contrasting ways in different national and industrial settings.[51] Similarly, even the most determined mass producer may be obliged to manufacture some speciality lines in small batches, while the best-selling lines of a successful flexible specialist may likewise be turned out in significant volumes.

For all these reasons, such macro-level indicators can only provide a suggestive guide to broad trends in industrial reorganization rather than a definitive test of the flexible specialization hypothesis. The preferred form of evidentialization for flexible specialization is instead the analytical case-study conducted at the micro-level of particular firms, regional economies or industrial sectors. Only detailed case-studies permit the close attention to context and strategy which is the hallmark of a flexible specialization approach; and only this method makes possible the comparative analysis of relationships between forms of production and institutional frameworks which is central to its theoretical architecture.[52]

Two major problems arise from this strategy of evidentialization through case-studies: interpretation and representativeness. As we have already noted, most firms or regional economies characteristically combine elements from both flexible and mass production rather than embodying pure examples of either model. How then can one assess the precise balance between the two models in any given case from the standpoint of the flexible specialization hypothesis? In principle, the solution might appear to lie in a search for objective indicators of flexible specialization such as those discussed at the macro-level which could more successfully be applied to less heterogeneous micro-data. Considerable mileage can undoubtedly be obtained through this route in documenting the spread of flexible specialization, as a number of suggestive studies have indicated.[53] In practice, however, the conceptual problems raised by such indicators remain the same: their precise meaning in any case cannot be determined without reference to the strategies of the actors concerned. Hence the case-study method necessarily entails an ineradicable element of subjective interpretation, in which there is considerable scope for legitimate disagreement among different observers. But if there is an unavoidable degree of indeterminacy about the interpretation of case-study evidence, this does not mean that there is no valid basis for discriminating among competing views, contrary to what current fashions in literary theory might appear to suggest.[54] Competing interpretations, like competing theories more generally, can properly be ranked in terms of their plausibility in accounting for agreed features of a common body of evidence according to internally consistent criteria.

These considerations can be illustrated more fully through a brief examination of current debates about the nature of Japanese manufacturing practices. How far can the success of Japanese firms be properly interpreted as evidence of the diffusion of flexible specialization, and how far instead as evidence of the development of increased flexibility within mass production itself? At one level, as Sayer points out, this question may be considered largely semantic, since there is considerable agreement among apparently conflicting interpretations about key features of the Japanese system: the rapid pace of model renewal and new product development; the productive flexibility obtained through organizational innovations such as just-in-time component supply, quick die-changes or mixed-model assembly lines; the prevalence of job rotation, teamworking and other forms of functional flexibility among large sections of the labour force; and the importance of 'relational subcontracting' between large and small firms.[55] At another level, however, this question is crucially important, given the centrality of Japanese manufacturing to the characterization of current trends in competitive strategy and productive organization in the international economy.

Some differences between the two interpretations are based on a mis-specification of the opposed view: thus it is no objection to the flexible specialization hypothesis properly understood to emphasize the continued role of large firms in Japan, nor the high overall volume of different products manufactured within the same firm or plant. Others arise from the results of new empirical research, such as David Friedman's (1988) demonstration of the limited role of MITI and the importance of small-firm industrial districts in the development of key Japanese export industries such as machine tools. But many differences of interpretation arise from ambiguities in Japanese industry itself. There can be little doubt that most Japanese innovations in sectors such as automobiles or consumer electronics originated in domestic firms' adaptation of mass production methods to local conditions during their postwar drive to catch up with the West.[56] And important features of more recent Japanese practice can still be legitimately interpreted in this light, from the limited range of variation on certain models and the long production runs of key components through the continued importance of hard automation and the relatively narrow skills required for many jobs to the dominance of large firms over their subcontractors. But there are also signs that many Japanese manufacturers are pushing these innovations in a more radical dimension in order to trade on their competitive advantage in catering for fragmented markets and volatile demand. Thus leading firms in these sectors appear to be dramatically increasing the pace of product innovation, expanding the range of distinct models which can be manufactured with a given combination of workers and machinery, and devolving responsibility not only for component supply but also for final assembly and product development to suppliers whom they often encourage to work for other manufacturers as well. Such strategies may not lead Japanese manufacturers to converge on a pure model of flexible specialization, but like many Western experiments at industrial reorganization, they have already progressed too far to fit comfortably into an alternative conception of neo-Fordism or flexible mass production.[57]

But even if it were agreed that particular cases could legitimately be interpreted as examples of flexible specialization, a significant problem of representativeness would still remain. As Sabel himself remarks, 'for example is not a proof' (Sabel 1989a: 23), and no quantity of case-studies however convincing could demonstrate the validity of the broader flexible specialization hypothesis. In the absence of a comprehensive macro-level map of the relative importance of competing models of productive organization, this difficulty may appear insuperable. But the Japanese case discussed above suggests an alternative strategy – common to most approaches to the analysis of industrial change – of focusing on those national economies, regions and firms which have proved most successful in the current phase of international competition. Beyond Japan itself, for example, flexible specialization analyses have concentrated on regions such as Emilia-Romagna and Baden-Württemberg whose technological dynamism, export competitiveness and importance to the national economy are relatively well documented.[58] Other analyses, conversely, have concentrated on countries such as Britain in which flexible specialization has been weakly developed, highlighting its role in explaining their poor performance in manufacturing competition during the 1980s.[59] In either case, finally, the claim is not that international competition imposes a single form of productive organization on economic actors, given the plurality of institutional frameworks and the possibilities of hybridization, but rather that tendencies can be observed towards the displacement of mass production by flexible specialization as the dominant technological paradigm of the late twentieth century.

5. Policy implications

Our discussion of policy must begin with a number of caveats that make clear our approach to this issue. In discussing the policy implications of the three broad perspectives outlined above it is necessary to consider which policies actually follow from these ways of conceptualizing forms of manufacturing organization and their wider socio-political conditions. It would be all too easy to slip into discussing the general views of the authors, confusing the policies that can strictly be justified by their approach to manufacturing with their political opinions and wider social concerns. Hence we need to consider the specific types and levels of policy intervention that have some direct connection with the explanatory and analytic frameworks in question.

Thus we must avoid over-generalizing policy into politics. But we must also avoid considering policy implications too specifically, as if the sole relevant domain of policy is 'industrial policy' in the narrow sense.[60] It by no means follows that an approach to manufacturing will, by reason of its own conceptual apparatus, foreground industrial policy in the sense of targeted interventions in particular industries. Broader policies for education, training, industrial finance and macro-economic management may be considered the most effective ways of securing the conditions for advanced forms of manufacturing organization.

A. *'Post-Fordism'*

Of the various 'post-Fordist' approaches little need be said. There are obvious political implications of, for example, *Marxism Today*'s arguments but few policy implications. Given the inattention of *Marxism Today* to specifying and analyzing the economic relationships involved in 'post-Fordism', this is hardly surprising.

Freeman and Perez's work, by contrast, does have definite policy implications. Both authors make comments on many policy areas, including education and training, but here we are concerned with the strict implications of their basic thesis about the role of technical change. Given the primary role they assign to the diffusion of the new core technologies that form the basis of a new 'techno-economic paradigm', then the key areas of intervention are technology policy and R&D policy. The aim of policy should be to identify the emerging technologies and adopt programmes that favour their diffusion through the whole economy. Freeman (1987, 1988) comments favourably on the role of Japan's Ministry of International Trade and Industry (MITI) in this context. One could characterize his view as a variant of the strategy of 'picking winners', which we discuss further and criticize in section C below. In his case, the primary object of such a policy of selective state intervention is not support for potentially successful firms (which is what the phrase often means in US debates) but support for technologies whose generalization is capable of producing growth across the whole economy.

B. *Regulation theory*

The regulation school's approach to policy is developed in relation to its general theory of capitalism as riven by periodic and specific systematic crises that are far more than the normal cyclical crises recognized by conventional economics. Regulationists claim that the present period is one of crisis and transition in the structures sustaining capitalism on a world-wide scale. The regulationist approach to policy can best be seen as a combination of a neo-Keynesian expansionary programme at the international level on the demand side and a neo-Marxist conception of the restructuring of the supply side, involving a new settlement between capital and labour in order to stabilize current economic volatility and class conflict. The expansionary policies are relatively clear, if both relatively conventional and very difficult to implement on either a global or a national scale. The policies that further the emergence of a new mode of regulation are less clear, and to the extent that they are developed in the political programmes of regulationist authors they have little to do with the original neo-Marxist terms in which the problem was set up.

Regulation theory must thus seek policies that both secure a new international regime of accumulation and a new mode of regulation, with the various particular institutional settlements being thrashed out as the result of social and political struggles in the national capitalist economies. Regulationist policy must operate across a number of different levels and be relatively coherent between these levels. These levels are: the international economy, the major economic blocs like the European Community (EC), and the relation of the major social forces within the national economies. The regulationists' policy domain is thus far more extended than those of flexible specialization and its policy requirements are inherently far more ambitious.

The regulationists argue that the current period of crisis within capitalism creates both major conflicts between the main components of the world economy, between the First and Third Worlds, between the major industrial powers and between the major social interests within the different national capitals. The previous international regime of accumulation based on American hegemony has broken down. The hitherto successful combination of Keynesianism sustaining demand and Fordism generating the productivity gains on the supply side to make an expansionary policy possible is no longer stable or sustainable. Reaching for classic Keynesian or Fordist solutions will not do. We are in a transitional period characterized by instability and crisis in the regime of accumulation and the mode of regulation. There are different routes out of this crisis.

The regulationists' value standpoint is to resolve the crisis on the terms most favourable to the left – to the peripheral economies of the Third World, and to labour in rich and poor countries alike. This resolution, however, will be within the limits set by the continuation of global capitalism. The regulationists reject the old classical Marxist breakdown theory of capitalism as a global system. In policy, as in theory, the regulationist approach claims to represent a 'middle way'; in this case between the illusions of internationalist revolutionary socialism and the insularity of reformist social democracy in the national economies of the advanced countries. Regulation theory remains concerned with capitalism as a global system and it seeks policies for the international economy that will promote stability and growth.

Lipietz (1987, 1989) is clearest on this international programme. He argues passionately that conventional policies will only exacerbate the plight of the Third World, driven into accelerated crises by the burden of debt. Ultimately an international expansionary policy that permits the First and Third Worlds to grow would be of mutual benefit, and in the interest of the advanced economies. While he is aware of the improbability of 'world Keynesianism' (1987: 192), Lipietz argues (1989) for a new international order founded on multilateralism, involving a 'new Bretton Woods' agreement. The latter would involve a new international credit money, the cancellation of Third World debt and the financing of aid to the Third World though the institutions of the new agreement, and a system of open trade not defensive protectionism. One might agree with most of this: it would be hard to defend First World banks or to argue that development aid should be less than the trickle it is. Our main point of criticism is not just that a new Bretton Woods is unlikely, rather it is that it is a classically Keynesian radical-liberal position. The regulationists' problems are drawn from Marxist theory, and yet their solutions, when they are clear, are fundamentally neo-Keynesian. This is not intended to disparage neo-Keynesianism. On the contrary, it is to insist that such policies can be adopted without reference to the theoretical problems from which the regulationists started.

For the developed countries, and the EC in particular, the regulationists' arguments for re-vitalizing the economy are less clear and less emphatic. Both Lipietz (1989) and Boyer (1988) argue for the further development of the EC. Lipietz (1989) argues for a new wages pact between labour and capital. This would be based on a trade-off of realistic wages for a measure of control over new technology, security of employment and greater free time. It would also involve a new form of the welfare state involving in particular a guaranteed minimum income and a 'third sector' of socially useful subsidized work. Boyer (1988) argues that a variety of scenarios are possible for the restructuring of wage/labour relations and labour markets. But ultimately he appears to consider a form of neo-Fordism incorporating elements of flexibility the most likely. This combined with neo-Keynesian demand management, a suitable regime of international trade, and restructuring of production may be sufficient to resolve the crisis. Again, as with policies to stimulate world demand, the difficulty is that the neo-Marxist emphasis on the class struggle leads in practice to policy proposals that, while radical, are in no sense tied to the regulationists' own theoretical problems. Many of Lipietz's ideas are widely scouted by post-marxists like André Gorz or advocates of 'associative democracy' based on flexible specialization like John Mathews (1989).

The regulationists seem unclear whether or not the Fordist era in either production organization or wage/labour relations is really at an end. They appear to see the most likely future as one in which elements of flexible specialization and new information technologies and robotics are incorporated into Fordism to produce a neo-Fordist synthesis. Which of the several possible outcomes or scenarios prevail will depend ultimately on the struggles of the key players: state decision makers, capitalist managers and trade union leaders. Policy in this sense is largely advocacy and shades into a general political programme. It is striking how little industrial policy in the narrow sense can be found in regulationist texts and that, in respect of the wider policies for re-regulation, how little there is that is specific to them or dependent on the theoretical problems of regulation. This may well be, to be fair, because figures like Lipietz are effective and pragmatic political advocates. It may also be because the regulationists' starting point in the problems of classical Marxism is quite irrelevant to the actual problems of economic policy.[61]

Our aim here is not to counter the regulationists' actual political ideas, many of which are perfectly reasonable. The point we want to make is that the political positions taken by the regulationists are not strictly policy implications of their underlying theory, or rather, that even if they are compatible with that theory they are not necessary to it and can be substained on other theoretic bases. Insofar as regulation theory has a distinct policy dimension, it is to give prominence to the need to stabilize the international economy and achieve a more equitable balance between its different components. That, however, is the most difficult level at which to achieve co-ordinated policies between the various agencies and actors.

C. *Flexible specialization*

We may begin our consideration of the policy implications of flexible specialization by noting that it involves a critique of traditional economic and industrial policy instruments. Flexible specialization challenges four widely canvassed ways of promoting the manufacturing sector through specific forms of intervention:

1) classic Keynesian strategies for promoting effective demand in particular national economies, the claim being that state action to promote effective demand will support the manufacturing sector by sustaining the overall level of economic activity and therefore maintaining or accelerating demand for manufactured goods;
2) strategies based on state-directed planning, promoting manufacturing output by co-ordinating investment policies of firms and concentrating development on advanced technologies and sectors – the three variants of such a policy are socialist command economies directing investment and orchestrating industry through administered prices, indicative planning on the French model, and the state direction of industry through analogues of Japan's MITI;[62]
3) state intervention through deregulation and measures to improve the

free working of market mechanisms, active competition policies, and the promotion of the prerogatives of enterprise management through industrial relations and labour market reform;
4) a contrary policy of state intervention to promote concentration of ownership and through it to exploit both organizational economies of scale at the level of the firm and production economies of scale at the level of the plant (both also objectives of state planning strategies).

Flexible specialization is not anti-Keynesian in principle or opposed to demand management for general economic-theoretic reasons, such as those implied in monetarist and free-market doctrines. Rather it argues that there are severe practical limits to Keynesianism as a national policy to support manufacturing under current conditions. The internationalization of trade in manufactures between advanced industrial countries and the liberalization of international trade through GATT and economic blocs like the EC means that domestic manufacturing sectors can only survive if they meet the conditions of international competition. This can only be achieved through appropriate supply-side policies and business strategies. Stimulating domestic demand may merely promote the ingress of foreign manufactured goods and even accelerate de-industrialization if the manufacturing sector is not able to meet international standards of competition. The experiences of France under the first Mitterand government and the consumption and credit expansion based policies of the Thatcher government since 1983 (although the latter is not formally 'Keynesian')[63] only serve to confirm this. Moreover, the increasing volatility of international markets and rapidly shifting patterns of demand in the period since the early 1970s favour more specific strategies of response like that of flexible specialization. Keynesianism should not be seen to be obsolete because it is linked in some necessary way with 'Fordism'. Demand management will not work to secure the home market for long runs of standardized goods by domestic producers under the current conditions of international trade, since foreign 'Fordists' might be able to be more competitive. Equally, were Keynesianism possible and flexible specialization widely developed, demand management would not automatically favour 'Fordist' strategies, since flexible specialists might be able to exploit production, marketing and design advantages in a climate of rising demand and growth. Keynesianism is problematic for macro-economic reasons, not because 'Fordism' is obsolete. Flexible specialization is thus not tied to a view of the functional obsolescence of Keynesianism, as some 'post-Fordist' ideas clearly are, but to a view of its conjunctural and structural inappropriateness in manufacturing economies under current conditions of international trade and competition.

Planning of a centralized and state-directed kind is seen to be inappropriate by flexible specialization theory for a variety of reasons. Firstly, command economies tend to be ineffective, whatever the prevailing form of manufacturing organization. The Soviet model has not proved effective at organizing production on 'Fordist' or mass-production lines. It supposes that the major economic parameters can be held stable and that shifts in output and accelerated growth of sectors can be pre-established by planning norms. This is difficult in a semi-closed economy where demand is subordinated to programmed supply, it is virtually impossible in an open economy that is forced to match foreign economics, and that is necessarily required to respond to consumer demand. Add to this the volatility of international currency and equity markets, the changing patterns of demand for manufactured goods and the rapidity of product innovation and it becomes difficult to hold any of the major national macro-economic parameters stable enough for them to be planned. Central planning has failed in command economies, and it is clearly impossible for Western-style economies.

Similar objections also hold for more limited forms of interventionist planning. State direction of investment and the anticipation and promotion of leading technologies presupposes that state officials, with or without the collaboration of business, have the information necessary to 'pick winners' and to concentrate national resources on key future technologies. There is little evidence that government bureaucracies are flexible enough and well-informed enough to do this. On the contrary, they will tend to equate economic success with promoting key 'over the horizon' technologies; these technologies are frequently not the most commercially successful. Officials will tend to pick major projects capable of state management, and not families of technologies capable of rapid state-of-the-art diffusion. There is a great deal of evidence both that MITI is a 'peculiar institution', heavily dependent on the close and informal links between senior civil servants and corporate executives that would be regarded as unacceptable in a Western democracy, and that it has a more mixed record than its most enthusiastic Western advocates allow. Moreover, there is much evidence for both orthodox company-level innovation in Japan and for the success of the local regulation, industrial districts, clusters of small and medium-scale firms, and major company organizational structures and manufacturing strategies which are characteristic of flexible specialization.[64]

Deregulation and the promotion of 'free' markets are becoming decidedly less fashionable among policy makers now. Flexible specialization is strongly opposed to the free-market model precisely because it emphasizes the importance of social relationships that secure crucial inputs and vital collective services for firms. Such inputs that cannot be guaranteed by the model of sovereign enterprises purchasing the factors of production in open markets include: trained labour, low-cost finance, market and export information, the diffusion of technical information, and ongoing relationships based on trust with sub-contractors and partner firms. The high costs to firms of low-trust relationships, the commodification of information, and the absence of ongoing collaboration with labour involve real competitive disadvantages. We have called these factors 'dis-economies of competition' and they are most evident in those countries and regimes that have failed to balance co-operation and competition, conflict and co-ordination (Hirst and Zeitlin 1989b). These 'dis-economies of competition' can offset the economies of scale supposedly available to large firms free to choose how best to secure the factors of production in open markets. Some of the most important 'factors' are intangible and are also not readily tradeable – trust relationships with other firms and co-operation with labour are two good examples.

One aspect of free-market advocacy is the celebration of small firms. Flexible specialization is not a theory predicated on preferring small firms to large ones. Small firms have no necessary economic attributes, other than some arbitrary cut-off point that defines their size (number of employees, value of capital, etc.).[65] 'Large' firms, indeed, multi-product, multi-national companies, can adopt flexible specialization strategies and so can small workshops.

But flexible specialization is clearly opposed to the policy objective of promoting concentration of ownership to secure economies of scale. This policy was widely adopted in the 1960s and 1970s, linked to views of the inherent efficiency of large-scale mass production, in order to promote firms of the size necessary for international competition – 'national champions'. In the UK, for example, the Labour governments of 1964–70 supported such concentration and created an agency – the Industrial Reorganization Corporation – to promote it. The results varied from ghastly failures (British Leyland) to stagnant survivors (GEC).[66] Such ideas are once again current in the EC, in response to the present merger wave. Policy makers take for granted the need to concentrate capacity in a few Euro-companies in each major industrial sector.[67] However, just as small firms are not inherently more responsive or innovatory in the way that free-market advocates often suppose, so large firms are not inherently more stable or successful, nor are they necessarily able to exploit economies of scale.[68] Size divorced from strategy is no inherent advantage, as the examples of General Motors or US Steel demonstrate. In practice active concentration policy or the state indulgence of merger waves has tended to prevail over free-market inspired anti-trust and pro-competition policies, even in the recent period of widespread belief in the virtues of the market.

There are good reasons to suppose that large firms which have rapidly concentrated by a process of merger and/or acquisitions are not particularly efficient and that they do not possess the means adequately to direct their subsidiaries. Those major multinationals that have sought functionally to decentralize into constellations of sub-units, able to co-operate or compete as needful, seem to have judged the matter better and to be one of the complex routes of industrial efficiency outlined by flexible specialization theory.

So far we have considered the negative criticisms made by flexible specialization theorists of other major approaches to policy; we now turn to positive proposals. Flexible specialization in its policy implications is a radical supply-side policy. It starts from the relationships necessary to effective manufacturing strategies and performance, and is concerned with wider macro-economic and social policies on this basis. It is not a comprehensive economic and political programme as such, although flexible specialization concepts can be integrated into such broader advocacy of strategies for reform.

Flexible specialization emphasizes the wide variety of possible institutional forms. There is no flexible specialization equivalent of the 'workers' fatherland', a chosen national or regional path that offers a given model to policy makers in other regions or countries. West German regions, or Japan or the 'Third Italy' offer examples of institutions that sustain flexible specialization, but they cannot be slavishly copied. Strategies have to be adopted that take account of national and regional conditions. Important in this respect are the institutional and political conditions that favour or hinder the patterns of co-operation and co-ordination which support flexible specialization strategies. Those countries most dominated by liberal competitive politics, by the stress on the sovereignty of the firm, and by antagonistic competition between social interests are least likely to be able to introduce such patterns.[69] The UK offers the most obvious example, and it is further hampered by weakly differentiated and institutionalized regional economies. The broader components of flexible specialization strategies cannot be directly transferred in the way technologies can, precisely because they are not centred on a model of a technology or a form of corporate organization, but on a more complex ensemble of socio-political conditions.

Despite the wide variety of institutional routes to flexible specialization they have one major factor in common, the existence of political, normative and organizational means of creating relationships which foster co-operation and co-ordination. Flexibly specialized economies at regional or industrial district level are embedded in patterns of social relations that go beyond the market and the formal structures of democratic government. Such embeddedness may arise from pre-existing institutional and political legacies that promote solidarity and interest co-operation. These patterns may involve more or less developed forms of 'corporatist' intermediation of organized interests. Crucial to the development of such co-operative relationships, either from existing legacies, or by explicit strategies of reform, is an agency capable of exercising social leadership, of providing the conditions for a pact which cements a consensus between firms and between the major interests. This agency is not fixed: it may be a political party, a public body like a regional economic development agency, or a strongly-led trade association.

Trust needs to be institutionalized and co-operation presupposes forums in which it can be developed. This is the most important lesson for policy makers seeking to revitalize declining regions or to sustain previously successful industrial districts. This 'political' dimension of economic policy is more important than any specific economic doctrine or any particular conception of the leading technology. Politics in this sense is broader than the policies of public agencies; it involves the creation of a regional or sectoral 'public sphere' in which firms, labour interests, officials and politicians can interact and co-operate. Such a 'public sphere' is essential if co-operation and competition are to be constructively balanced in the interaction of firms, and if co-ordination and conflict are to be constructively balanced in the relationship of management and labour. These balances are crucial if economic fluctuations and technical changes are to be handled effectively, if continuity and innovation are to be given their respective roles. Those districts that can change and adapt are most likely to be successful and to weather the effects of volatility in international markets.[70]

It is the socio-political conditions in which manufacturing is embedded that form the core of the flexible specialization approach to policy. There are two main routes for developing and institutionalizing flexible specialization: one may be called the strategy of building up, linking firms with collaborative institutions to form and cement industrial districts, and seeking to generalize and link such districts so as to form the dynamic core of a national economy; the other is the strategy of building down through the reorganization of major multinational firms into constellations of semi-autonomous sub-units that may co-operate one with another or with other firms in an industrial district. Ultimately, these two strategies can converge. Constellated firms mirror internally the patterns of co-operation and trust that are found in the most developed and cohesive industrial districts. Such firms are also less likely to be invasive or destructive of the industrial districts in which they are located, they find it easier to link horizontally with other firms and co-operative institutions in a district and they are more likely to collaborate than centrally directed and hierarchically organized firms.

Flexible specialization is not just a new management philosophy. It emphasizes those aspects of economic life that cannot be bought on the market and it emphasizes relationships with other firms, public bodies and labour that cannot easily be accommodated to the management prerogatives of the 'sovereign' firm. At the same time flexible specialization is not anti-management and anti-business; the purpose of co-operation and co-ordination is to make firms more productively efficient and more commercially competitive. It involves managements learning lessons and adopting a style of operation not widely taught in Anglo-Saxon business schools.

The core of a flexible specialization policy is thus to create and sustain those institutional patterns that lead firms to co-operate one with another as well as compete. These policies may be informal and firm-centred, or they may be more formal and involve firms collaborating with trade and industry associations, regional governments, and organized labour. In both cases the purpose of collaboration is to secure the inputs firms cannot easily purchase on the market – such as suitably trained labour, low-cost finance and commercial information. Such collective services enhance the efficiency and competitiveness of firms. Flexible specialization policies thus place considerable emphasis on training, since broadly skilled workers are a core component of such a strategy. Such training cannot be undertaken by firms alone if each firm is competing for labour and most firms adopt the attitude of a free-rider by poaching personnel from those firms that bear the costs.[71] Equally importantly, broad-based training involves co-operation with organized labour to set mutually acceptable standards and employment policies. Flexible specialization is often highly profitable for the firm but it is by no means the case that such firms can bear individually the full costs of R&D, marketing information and new investment at prevailing commercial costs. Co-operative institutions lower and spread these costs; allowing smaller firms to gain economies of scale outside of production by pooling the costs of supporting developed networks of collective services for marketing, consultancy, technological information and so on.

If flexibly specialized firms are profitable they will often be targets for asset-rich major companies that seek to acquire them. Alongside collaborative institutions there is the need for public policy to protect firms from unwelcome acquisitions and to preserve the autonomy of management in successful industrial companies of less than major multinational size. Flexible specialization, therefore, requires something akin to a competition policy, that is, a policy that enables firms to retain autonomy. Competition policy has too often been linked to the dual assumptions that free markets are the most efficient allocative mechanisms and that the more intensely firms compete in open markets the better. Such free-market policies in practice tend towards further concentration. This is because their anti-monopoly aspects cut against the primary objective of increasing the freedom of firms as market actors. Deregulators are in practice more concerned with freeing management from restrictions than with ensuring effective competition. Recognizing that competition must be balanced by co-operation involves giving greater emphasis to protecting firms from predators through company law and policies regulating equity markets.[72]

Flexible specialization strategies thus do not emphasize an active industrial policy in the narrow and traditional sense – that is, targeting key sectors or firms for state aid and intervention. Flexible specialization theory assumes that a wide variety of sectors can be successful components of an advanced economy; varying from high-tech sectors like advanced machine tools or computers to traditional sectors like clothing or furniture.[73] Traditional sectors may utilize advanced technology to gain competitive advantage, and may need public support to introduce and pay for such investment. Public policy should be confined primarily to providing broad-based support for sectoral and regional initiatives, rather than seeking to favour a few major firms in a few key industries. Support for training, investment, the building of collaborative networks, the protection of company autonomy, and so on, does not involve public agencies directly choosing which private firms to favour. *All* firms in a district or sector may benefit from such broad-based policies, if they have the initiative and energy to do so. This policy aspect of flexible specialization is thus more compatible with the requirement for liberal democratic states to be neutral and not to bestow public favours on private agencies by administrative discretion. This is not a free-market or anti-interventionist point, it concerns the forms and objects of a policy of active public support for industry. *Dirigiste* industrial policies favour centralism and large bureaucracies. Far from being a policy of *laissez-faire*, flexible specialization is an interventionist policy that favours democratic accountability. The close link between centralized state and major business corporations is a real threat to democracy: it is a source of corruption, and, all too often, a means of protecting big but inefficient companies. Westerners have been so seduced by Japanese economic success that many industrial policy advocates see MITI as a model to be copied. Japanese radicals do not, and the unhealthy links between big business, the Liberal Democratic Party and the state are now a key target of democratic criticism and one of the sources of support for opposition politics.

Flexible specialization emphasizes the effectiveness of regional institutions of economic co-operation. It points to the need to build regional autonomy and to foster the collaboration of industry, labour and public bodies at the regional level. This is important since regional economic policy offers the best way of compensating for the lessened effectiveness of national macro-economic policies. Flexible specialization offers the prospect of radical supply-side policies, vital in a conjuncture where rightist free-market supply-side policies have manifestly failed at the national level and Keynesian demand management at the national level has severe limits. Moreover, regional intervention is not confined to policies for manufacturing alone, but also involves the health, education and welfare policies necessary to reinforce them. In many countries the regionalization of intervention and the development of a federal welfare state are proceeding hand in hand (Sabel 1989b). Some countries, like the UK, have continued to maintain a national centralized welfare system in social security, health and education. In the case of the UK this goes along with highly centralized government, weak institutions of regional economic management and a declining manufacturing sector. Britain is the major industrial country where flexible specialization strategies have had the least impact, and which faces the greatest problems of de-industrialization and poor international competitiveness. British top executives are strongly committed to retaining exclusive company-level management control, seeing little need to build co-operative relationships with other firms or an ongoing partnership with labour. Britain is thus the acid test of whether flexible specialization can be consciously encouraged by public policy initiatives, since the UK both desperately needs such strategies to become competitive and yet has few of the institutional resources needed to develop them.

Birkbeck College
University of London

Notes

1 This exposition is based primarily on Piore and Sabel 1984, Sabel and Zeitlin 1985, Sabel 1989a and Sabel 1990.
2 It will be evident that this definition of flexible specialization as a form of craft production entails a revaluation of the conventional stereotype of the latter as the manufacture of luxury goods in tiny volumes using hand tools and obsolete methods. The historical basis for a more positive interpretation of the technological dynamism of craft production is presented in Sabel and Zeitlin 1985. For a related approach which seeks to distinguish between craft production and 'diversified quality production' on the basis of the volumes involved, see Streeck 1987 and Sorge and Streeck 1988.
3 As Piore and Sabel acknowledge (1984: 4–5), their notion of regulatory requirements of the technological paradigm is borrowed from the French regulation school, but used in very different ways.
4 For fuller discussions of the reorganization of large corporations and the process of 'double convergence', see Sabel 1989a, Sabel *et al.* 1989, and Sabel 1990. This process implies, as Sabel 1989b observes, not only that corporate operating units are coming to resemble the constituent elements of the industrial districts, but also that there are increasing numbers of exchanges and alliances between large-firm subsidiaries and their small-firm counterparts in the districts themselves.
5 For an extended account of these ideas, see Sabel 1989a: 53–9 and Sabel 1989b; and for the potential role of a reorganized labour movement in the transformation of the welfare state, see Kern and Sabel 1990 and Sabel 1990.
6 For a fuller account of flexible specialization as a development strategy, see Sabel 1986; and for a thoughtful critical discussion, see Schmitz 1989. For an extended attempt to apply this approach to the problems of a small semi-developed economy on the European periphery, see Murray 1987.
7 c.f. Williams *et al.* 1987; Pollert 1988; Wood 1989. For an insightful discussion of the critical debates surrounding flexible specialization, see Badham and Mathews 1989.
8 Piore and Sabel 1984: 26–8, 219, 279–80; Sabel and Zeitlin 1985: 137–8; Sabel 1989: 40.
9 These arguments draw on the work of the international working group on 'Historical Alternatives to Mass Production' sponsored by the Maison des Sciences de l'Homme in Paris. See Sabel and Zeitlin, forthcoming; and for a discussion of contemporary problems of technological hybridization, Sabel 1990.
10 c.f. for example Elam 1990. Thus flexible specialization is not, among other things, an optimistic general theory of the labour process which can be counterposed to Braverman's deskilling thesis: c.f. Thompson 1989, pp. 218–29; Wood 1989.
11 Piore and Sabel 1984: 261–3; Sabel 1989a: 37–40; Sabel *et al.* 1990; Sabel 1990.
12 c.f. especially Pollert 1988.
13 See the collection of articles from *Marxism Today* edited by Jacques and Hall 1989, especially section 1.
14 See in particular Freeman and Perez 1988; Perez 1983, 1985; and Freeman *et al.* 1982.
15 Even in the automobile industry, for example, national responses to the Fordist model varied considerably: see Tolliday and Zeitlin 1986.
16 For different national responses to 'Keynesianism', see Weir and Skocpol 1985 and Hall 1989.
17 See Piore and Sabel 1984: ch. 6.
18 For a review of the evidence, see Hirst and Zeitlin 1989b.
19 See Bernstein 1961 and Crosland 1964.
20 A variant of the 'post-Fordist' argument which gives great emphasis to the concept of 'post-modernity' is David Harvey's *The Condition of Postmodernity* (1989).

We have not considered this text in detail since it adds little to the arguments found in *Marxism Today* and Lash and Urry.
21 See Hilferding 1981 and Kocka 1974.
22 For the misuse of flexible manufacturing systems by British firms, see Jones 1989 and for a wide-ranging survey of managers which highlights the failure of British firms to obtain significant gains from new technologies, see New and Myers 1986.
23 For the case of Italy, see Piore and Sabel 1983, 1984: esp. chs 6 and 9; Goodman *et al.* 1989; Brusco 1982; Trigilia 1986; and Bagnasco 1988. For the case of Japan, see Dore 1986, 1987; and Friedman 1988.
24 c.f. Sabel *et al.* 1989a.
25 For a valuable overview of the many variants of the regulation approach, see Jessop 1990. In our account of the Parisian regulationists' ideas, we have concentrated on the recent work of Robert Boyer, Alain Lipietz and their collaborators. Key texts we have used include: Boyer: 1979, 1986a, 1986b, 1988a, 1988b, 1988c, 1988d, 1989; Lipietz: 1987, 1989; Leborgne and Lipietz 1988.
26 See especially Boyer 1986a, 1986b: ch. 1, and 1989.
27 See particularly Boyer 1986 and 1988a; Lipietz and Leborgne 1988; and Jessop 1990: 42–3. Lipietz also identifies the technological paradigm as a third constituent of the mode of development, defined as 'the general principles which govern the evolution of the division of labour' across the economy as a whole. But Lipietz does not specify how this relates to the overall framework of the regulationist analysis, and it remains unclear whether the concept is used in the manner of flexible specialization or of Freeman and Perez: see Leborgne and Lipietz 1988: 264–5.
28 For detailed discussion of the nature of possible post-Fordist modes of development, see Boyer 1988b, 1988c: chs 11–12, and 1988d; and Leborgne and Lipietz 1988.
29 On the regulationists' attempts to avoid teleology and functionalism, see Lipietz 1987: 16; Boyer 1986a: 59, 95; and Jessop 1990: 67–75.
30 See Jessop 1990: 49–50, 67–9.
31 See particularly the general formulations of their approach in Boyer 1986a, Lipietz 1987: ch. 2, and Lipietz 1989: ch. 2.
32 Compare, for example, the nuanced discussion of the development of Taylorism in France in Boyer 1984 with the balder formulation in Boyer 1988a: 82.
33 Boyer 1986b, 1988c; Lipietz 1987.
34 Boyer 1988c: 270–1; c.f. also ibid.: 230–2; Boyer 1988d: 404–6; Leborgne and Lipietz 1988: 267–8.
35 c.f. for example Boyer 1988d; Lipietz 1989.
36 For a fuller discussion of these issues, see Hirst 1985 and 1990.
37 See for example the theoretical arguments in Sabel 1982.
38 See Sabel and Zeitlin 1985 and forthcoming.
39 Good examples of such normative-empirical advocacy using flexible specialization concepts are Mathews 1989a and 1989b.
40 Lipietz 1987: 9–10. William of Baskerville is a character in Eco's novel *The Name of the Rose*.
41 See Whitehead 1926.
42 For an argument that tries to show that conceptual realism is central to Marx's epistemology in *Capital* and that this position is theoretically unsustainable, see Cutler *et al.* 1977.
43 For a penetrating account of the theoretical objections to treating classes as social actors, see Hindess 1987.
44 See Boyer 1988b: 623–7; 1988c: 229–32; 1988d: 404–15.
45 Boyer 1978: 107–8, 111.
46 For British studies highlighting the differences between wage rates and

earnings for the period before 1939, see McClelland and Reid 1985: 157; and Gillespie and Whiteside 1989.
47 Aglietta 1976; Boyer 1979, 1986a: 65, 1986b: 227; Boyer and Mistral 1978. On the basis of similar calculations, however, Mazier *et al.* 1984 argue that the transition to intensive accumulation in France did not occur until after 1945.
48 See, for example, Fridenson 1987, and Montmollin and Pastré 1984, pt I. Even for the United States, the impact of Taylorism on work organization and labour management can easily be overstated: see Nelson 1975 and 1980. The same methodological problems apply to the otherwise extremely stimulating study of historical changes in the nature of French unemployment by Salais *et al.* 1986, which uses principal component analysis rather than more conventional econometric tests. See Salais *et al.* 1986: 85–94 and Salais 1989: 274–6; and compare the much more cautious and historically sensitive discussion by Benedicte Reynaud in Salais *et al.* 1986: 132–7.
49 The classic statement is Friedman 1953.
50 For an ingenious but ultimately unsatisfying attempt to use existing data to test the flexible specialization hypothesis for the US manufacturing, see Luria 1989. Thus, for example, Luria uses share of value-added in manufacturing output (VA/M) as an indicator of product batch-sizes, on the assumption that VA/M rises proportionately as batch sizes fall, and explores its relationship with labour productivity (value-added per employee) for SIC industries over the past two decades. But there may be many reasons for an industry to be characterized by a high VA/M ratio besides small-batch production, while the use of this indicator also depends on the absence of any significant shift in the relative productivity of small and large-batch production, the very question to be examined.
51 See, inter alia, Sorge *et al.* 1983; Jones 1982 and 1989; Maurice *et al.* 1986; and Adler 1989.
52 For a selection of case studies written from a flexible specialization perspective, see: Best 1989; Brusco and Sabel 1981; Friedman 1988; Herrigel 1989; Hirst and Zeitlin 1989b; Katz and Sabel 1985; Lorenz 1989; Lyberaki 1988; Michelsons 1987, 1989; Piore and Sabel 1983; Regini and Sabel 1989; Sabel 1984; Sabel *et al.* 1989a, 1989b; Sabel and Zeitlin forthcoming; Storper 1989; Tolliday and Zeitlin 1987; Zeitlin and Totterdill 1989. The major statements of the flexible specialization approach such as Piore and Sabel 1984, Sabel and Zeitlin 1985, and Sabel 1989, are all built up from a comparative analysis of such case studies.
53 See Michelsons 1987; Storper and Christopherson 1987; Christopherson and Storper 1989; Salais and Storper 1990.
54 For a discussion of current debates about the interpretation of texts in the history of political ideas, see Tully 1988.
55 Compare, for example, Piore and Sabel 1984, Tolliday and Zeitlin 1986, Friedman 1988 and Sabel 1989 with Williams *et al.* 1987, Wood 1988, Sayer 1986 and 1989, Kenny and Florida 1988 and 1989.
56 For a major case study, see Cusumano 1985.
57 See Tolliday and Zeitlin 1986; Sabel 1989: 37–9; Regini and Sabel 1989: 33–44; Sabel *et al.* 1989b; and Sabel 1990. But c.f. also Sayer 1989: 685–9; Kenney and Florida 1988 and 1989.
58 On Emilia-Romagna and the 'Third Italy' more broadly, see, in English, Brusco 1982; Sabel 1982: 220–6; Piore and Sabel 1983; Zeitlin 1989b; Goodman *et al.* 1989; Pyke *et al.* 1990. Contrasting interpretations can be found in Murray 1987 and Amin 1989. On Baden-Württemberg, see Sabel *et al.* 1989, and Herrigel 1989. For a more general debate on the interpretation of regional case studies in relation to the flexible specialization hypothesis, see Amin and Robins 1990 and the responses by Michael Piore, Charles Sabel and Michael Storper in Pyke *et al.* 1990: ch. 12.
59 See Hirst and Zeitlin 1989a and b; Lane 1988. Attempts to use predominately

British evidence to criticize the flexible specialization hypothesis therefore badly miss the mark: see, for example, Pollert 1988.
60 For a survey of debates on industrial policy primarily in this narrow sense, see Thompson 1987.
61 For Boyer's response to criticisms of regulation theory for its lack of a coherent policy programme, see Boyer 1986a: 105–9.
62 For criticism of Keynesianism and state planning strategies, including the use of MITI as a model for 'picking winners', see Hirst and Zeitlin 1989a: Introduction. For an example of advocacy that Britain adopt a MITI-style strategy, see Smith 1984.
63 For a characterization of the British Conservative's approach to macro-economic policy between 1982 and 1988 as 'electoral Keynesianism', see Hirst 1989: ch. 4.
64 On MITI, beyond the classic account of Johnson 1982, see Friedman 1988 and Okimoto 1989.
65 For a useful discussion of the definitional problems of the category 'small firms' in different national contexts, see Sengenberger and Loveman 1988.
66 For an account of the failure of British industrial policy based on concentration into 'national champions' in the 1960s and 1970s, see Williams *et al.* 1983.
67 For critical discussions of the likely impact of 1992 on economies of scale, see Giroski 1989 and Thompson 1991.
68 Prais 1976 argued persuasively that the concentration of firms in the UK in the period 1909–70 had exhausted any possible production economies of scale at *plant* level and, moreover, that UK firms often concentrated by bringing under common control constellations of less than efficient plants.
69 This argument for the failure of free-market economies like the UK to create 'developmental states' is powerfully articulated by Marquand 1988.
70 This argument is developed more fully in Hirst and Zeitlin 1989b and Hirst 1990a.
71 This is of course, the current position in the UK – see Campbell *et al.* 1989.
72 This argument is developed in Hirst 1989: ch. 6.
73 See Zeitlin and Totterdill 1989 and Best 1989.

References

Adler, Paul and **Borys, Brian** (1989) 'Automation and Skill: Three Generations of Research on the NC Case', *Politics and Society*, 17(3): 353–76.

Aglietta, Michel (1976) *Régulation et crises du capitalisme: l'expérience des Étas-Unis*, Paris: Calmann-Levy.

Amin, Ash (1989) 'Flexible Specialisation and Small Firms in Italy: Myths and Realities', *Antipode*, 21(1): 13–34.

Amin, Ash and **Robins, Kevin** (1990) 'Industrial Districts and Regional Development: Limits and Possibilities', in Pyke *et al.* 1990: ch. 11.

Badham, Richard and **Mathews, John** (1989) 'The New Production Systems Debate', *Labour and Industry*, 2(2): 194–246.

Bagnasco, Arnaldo (1988) *La costruzione sociale del mercato*, Bologna: Il Mulino.

Bernstein, Eduard (1961) *Evolutionary Socialism*, trans. Edith C. Harvey, New York: Schocken.

Best, Michael (1989) 'Sector Strategies and Industrial Policy: The Furniture Industry and the Greater London Enterprise Board', in Hirst and Zeitlin 1989a: 191–222.

Boyer, Robert (1979), 'Wage Formation in Historical Perspective: The French Experience', *Cambridge Journal of Economics*, 3(2): 99–118.

Boyer, Robert (1984) 'Le taylorisme hier: presentation', in Montmollin and Pastré, 1984: 35–49.

Boyer, Robert (1986a) *La théorie de la régulation: une analyse critique*, Paris: La Découverte.

Boyer, Robert (ed.) (1986b) *Capitalismes fin de siècle*, Paris: Presses Universitaires de France.

Boyer, Robert (1988a) 'Technical Change and the Theory of Régulation', in Dosi *et al.* 1988: 67–94.
Boyer, Robert (1988b) 'Formalizing Growth Regimes', in Dosi *et al.* 1988: 608–30.
Boyer, Robert (ed.) (1988c) *The Search for Labour Market Flexibility: The European Economies in Transition*, Oxford: Clarendon.
Boyer, Robert (1988d) 'Alla ricerca di alternative al fordismo: gli anni ottanta', *Stato e mercato*, 24: 387–423.
Boyer, Robert (1989) 'Économie et histoire: vers de nouvelles alliances?', *Annales ESC*, 6: 1397–426.
Boyer, Robert and **Mistral, Jacques** (1978) *Accumulation, inflation crises*, Paris: Presses Universitaires de France.
Braverman, Harry (1974) *Labor and Monopoly Capital*, New York: Monthly Review Press.
Brusco, Sebastiano (1982) 'The Emilian Model: Productive Decentralization and Social Integration', *Cambridge Journal of Economics*, 6(2): 167–84.
Brusco, Sebastiano and **Sabel, Charles** (1981) 'Artisanal Production and Economic Growth', in F. Wilkinson (ed.), *The Dynamics of Labour Market Segmentation*, London: Academic Press, pp. 99–114.
Campbell, Adrian, Currie, Wendy and **Warner, Malcolm** (1989) 'Innovation, Skills and Training: Micro-electronics and Manpower in the United Kingdom and West Germany', in Hirst and Zeitlin 1989a: 133–54.
Christopherson, Susan and **Storper, Michael** (1989) 'The Effects of Flexible Specialization on Industrial Politics and the Labor Market: The Motion Picture Industry', *Industrial and Labor Relations Review*, 42(3): 331–47.
Crosland, C. A. R. (1964) *The Future of Socialism*, 2nd edn, London: Cape.
Cusumano, Michael (1985) *The Japanese Automobile Industry*, Cambridge, Mass.: Harvard University Press.
Cutler, Anthony, Hirst, Paul, Hindess, Barry and **Hussain, Athar** (1977) *Marx's 'Capital' and Capitalism Today*, vol. I, London: Routledge & Kegan Paul.
Dore, Ronald (1986) *Flexible Rigidities: Industrial Policy and structural Adjustment in Japan*, London: Athlone Press.
Dore, Ronald (1987) *Taking Japan Seriously: A Confucian Perspective on Leading Economic Issues*, London: Athlone Press.
Dosi, Giovanni, Freeman, Christopher, Nelson, Richard, Silverberg, Gerald and **Soete, Luc** (eds) (1988) *Technical Change and Economic Theory*, London: Pinter.
Elam, Mark (1990) 'Puzzling Out the Post-Fordist Debate: Technology, Markets and Institutions', forthcoming in *Economic and Industrial Democracy*, 11(1).
Freeman, Christopher (1987) *Technology Policy and Economic Performance: Lessons from Japan*, London: Frances Pinter.
Freeman, Christopher (1988) 'Japan: A New National System of Innovation', in Dosi *et al.* 1988: 330–48.
Freeman, Christopher, Clark, J. and **Soete, Luc** (1982) *Unemployment and Technical Innovation: A Study of Long Waves in Economic Development*, London: Frances Pinter.
Freeman, Christopher and **Perez, Carlota** (1988) 'Structural Crises of Adjustment: Business Cycles and Investment Behaviour', in Dosi *et al.* 1988: 38–66.
Fridenson, Patrick (1987) 'Un tournant taylorien de la société française (1904–1918)', *Annales ESC*, 5: 1031–60.
Friedman, David (1988) *The Misunderstood Miracle: Industrial Development and Political Change in Japan*, Ithaca: Cornell University Press.
Friedman, Milton (1953) 'The Methodology of Positive Economics', in *idem*, *Essays in Positive Economics*, Chicago: Chicago University Press, pp. 3–43.
Gillespie, James and **Whiteside, Noel** (1989) 'Deconstructing Unemployment: British Developments during the Interwar Years', unpublished paper submitted to the *Econmic History Review*.
Giroski, P. A. (1987) 'The choice between Diversity and Scale?' in Centre

for Business Strategy, *1992: Myths and Realities*, London: London Business School.
Goodman, Edward and **Bamford, Julia** with **Saynor, Peter** (eds) (1989) *Small Firms and Industrial Districts in Italy*, London: Routledge.
Gorz, André (1985) *Paths to Paradise: On the Liberation from Work*, London: Pluto Press.
Gramsci, Antonio (1971) 'Americanism and Fordism', in *Selections from the Prison Notebooks* (ed. and trans. Quintin Hoare and Geoffrey Nowell-Smith), London: Lawrence and Wishart, pp. 277–320.
Hall, Peter A. (ed.) (1989) *The Political Power of Economic Ideas: Keynesianism Across Nations*, Princeton: Princeton University Press.
Harvey, David (1989) *The Condition of Postmodernity*, Oxford: Basil Blackwell.
Herrigel, Gary (1989) 'Industrial Order and the Politics of Industrial Change: Mechanical Engineering', in Peter Katzenstein (ed.), *Industry and Political Change in West Germany: Towards the Third Republic*, Ithaca: Cornell University Press.
Hilferding, Rudolf (1981) *Finance Capital*, London: Routledge & Kegan Paul.
Hindess, Barry (1987) *Politics and Class Analysis*, Oxford: Blackwell.
Hirst, Paul (1985) 'Is It Rational to Reject Relativism?', in J. Overing (ed.), *Reason and Morality*, ASA Monographs 24, London: Tavistock, pp. 85–103.
Hirst, Paul (1989) *After Thatcher*, London: Collins.
Hirst, Paul (1990a) 'Democracy: Socialism's Best Reply to the Right?', in Barrry Hindess (ed.), *Reactions to the Right*, London: Routledge.
Hirst, Paul (1990b) 'An Answer to Relativism', *New Formations*, Spring.
Hirst, Paul and **Zeitlin, Jonathan** (eds) (1989a) *Reversing Industrial Decline? Industrial Structure and Policy in Britain and her Competitors*, Oxford: Berg/New York: St Martins.
Hirst, Paul and **Zeitlin, Jonathan** (1989b) 'Flexible Specialization and the Competitive Failure of UK Manufacturing', *Political Quarterly*, 60(3): 164–78.
Hyman, Richard and **Streeck, Wolfgang** (eds) (1988) *New Technology and Industrial Relations*, Oxford: Basil Blackwell.
Jacques, Martin and **Hall, Stuart** (eds) (1989) *New Times*, London: Lawrence & Wishart.
Jessop, Bob (1990) 'Regulation Theories in Retrospect and Prospect', *Economy and Society*, 19(2):
Johnson, Chalmers (1982) *MITI and the Japanese Miracle: The Growth of Industrial Policy, 1925–1975*, Palo Alto: Stanford University Press.
Jones, Bryn (1982) 'Destruction or Redistribution of Engineering Skills? The Case of Numerical Control', in Stephen Wood (ed.), *The Degradation of Work?*, London: Hutchinson, pp. 179–200.
Jones, Bryn (1989) 'Flexible Automation and Factory Politics: Britain in Comparative Perspective', in Hirst and Zeitlin 1989a: 95–121.
Katz, Harry and **Sabel, Charles** (1985) 'Industrial Relations and Industrial Adjustment in the Car Industry', *Industrial Relations*, 24(3): 295–315.
Kenney, Martin and **Florida, Richard** (1988) 'Beyond Mass Production: Production and the Labor Process in Japan', *Politics and Society*, 16(1): 121–58.
Kenney, Martin and **Florida, Richard** (1989) 'Japan's Role in a Post-Fordist Age', *Futures*, 21(2): 136–51.
Kern, Horst and **Sabel, Charles** (1990) 'Trade Unions and Decentralized Production: A Sketch of Strategic Problems in the West German Labor Movement', unpublished paper, January.
Kocka, Jürgen (1974) 'Organisierter Kapitalismus oder Staatsmonopolistischer Kapitalismus? Begriffliche Vorbemerkungen', in H. Winckler (ed.), *Organisierter Kaiptalismus*, Göttingen: Vandenhoeck & Ruprecht.
Kondratiev, Nikolai (1979) 'The Major Economic Cycles', *Review*, 2(4): 519–62.
Lane, Chrystel (1988) 'Industrial

Change in Europe: The Pursuit of Flexible Specialization in Britain and West Germany', *Work, Employment and Society*, 2(2): 141–68.
Lash, Scott and **Urry, John** (1987) *The End of Organized Capitalism*, Cambridge: Polity.
Leborgne, Danielle and **Lipietz, Alain** (1988) 'New Technologies, New Modes of Regulation: Some Spatial Implications', *Society and Space*, 6: 263–80.
Lipietz, Alain (1987) *Mirages and Miracles: The Crises of Global Fordism*, London: New Left Review.
Lipietz, Alain (1989) *Choisir L'Audace*, Paris: La Découverte.
Lorenz, Edward (1989) 'The Search for Flexibility: Subcontracting Networks in French and British Engineering', in Hirst and Zeitlin 1989a: 122–32.
Luria, Dan (1989) 'Automation, Markets and Scale: Can "Flexible Niching" Modernize American Manufacturing?', unpublished paper submitted to the *International Review of Applied Economics*.
Lyberaki, Antigone (1988) 'Small Firms and Flexible Specialisation in Greek Industry', D.Phil. thesis, University of Sussex.
McClelland, Keith and **Reid, Alastair** (1985) 'Wood, Iron and Steel: Technology, Labour and Trade Union Organisation in the Shipbuilding Industry, 1840–1914', in Royden Harrison and Jonathan Zeitlin (eds), *Divisions of Labour*, Brighton: Harvester: 151–84.
Marquand, David (1988) *The Unprincipled Society*, London: Collins.
Mathews, John (1989a) *Tools of Change: New Technology and the Democratization of Work*, Sydney: Pluto Press.
Mathews, John (1989b) *Age of Democracy: The Politics of Post-Fordism*, Melbourne: Oxford University Press Australia.
Maurice, Marc, Eyraud, François, d'Iribarne, Alain and **Rychener, Frédérique** (1986) *Des enterprises en mutation dans la crise: Apprentissage de technologies flexibles et emergence de nouveaux acteurs*, Aix-en-Provence: Laboratoire d'Économie et de Sociologie du Travail.
Mazier, J., Basle, M. and **Vidal, J-F.** (1984) *Quand les crises durent. . .*, Paris: Économica.
Michelsons, Angelo (1987) 'Turin Between Fordism and Flexible Specialization: Industrial Structure and Social Change, 1960–75', Ph.D. thesis, University of Cambridge.
Michelsons, Angelo (1989) 'Local Strategies of Industrial Restructuring and the Changing Relations between Large and Small Firms in Contemporary Italy: The Case of Fiat Auto and Olivetti', in Zeitlin 1989a: 425–47.
Montmollin, Maurice de and **Pastré Olivier** (eds) (1984) *Le taylorisme*, Paris: La Découverte.
Murray, Feargus (1987) 'Flexible Specialisation in the "Third Italy"', *Capital and Class*, 33: 84–95.
Murray, Robin (ed.) (1987) *The Cyprus Industrial Strategy: Report of the UNDP/UNIDO Misson*, 8 vols, Institute of Development Studies, University of Sussex.
Nelson, Daniel (1975) *Managers and Workers: The Origins of the New Factory System in the United States, 1880–1920*, Madison: University of Wisconsin Press.
Nelson, Daniel (1980) *Frederick W. Taylor and the Rise of Scientific Management*, Madison: University of Wisconsin Press.
New, C. and **Myers, A.** (1986) *Managing Manufacturing Operations in the UK, 1975–85*, Brighton: Institute of Manpower Studies.
Okimoto, Daniel (1989) *Between MITI and the Market: Japanese Industrial Policy for High Technology*, Palo Alto: Stanford University Press.
Perez, Carlota (1983) 'Structural Change and the Assimilation of New Technologies in the Economic and Social System', *Futures* 15(4): 357–75.
Perez, Carlota (1985) 'Microelectronics, Long Waves and World Structural Change: New Perspectives for Developing Countries', *World Development*, 13(3): 441–63.
Piore Michael and **Sabel, Charles** (1983) 'Italian Small Business

Development: Lessons for U.S. Industrial Policy', in John Zysman and Laura Tyson (eds), *American Industry in International Competition*, Ithaca: Cornell University Press: 391–421.
Piore, Michael and **Sabel, Charles** (1984) *The Second Industrial Divide: Possibilities for Prosperity*, New York: Basic Books.
Pollert, Anna (1988) 'Dismantling Flexibility', *Capital and Class*, 34: 42–75.
Prais, S. J. (1976) *The Evolution of Giant Firms in Britain*, Cambridge: Cambridge University Press.
Pyke, Frank, Becattini, Giacomo and **Sengenberger, Werner** (eds) (1990) *Industrial Districts and Inter-firm Co-operation in Italy*, Geneva: International Institute for Labour Studies.
Regini, Marino and **Sabel, Charles** (eds) (1989) *Strategie di riaggiustimento industriale*, Bologna: Il Mulino.
Sabel, Charles (1982) *Work and Politics: The Division of Labour in Industry*, Cambridge: Cambridge University Press, 1982.
Sabel, Charles (1984) 'Industrial Reorganization and Social Democracy in Austria', *Industrial Relations*, 23(3): 344–61.
Sabel, Charles (1986) 'Changing Models of Economic Efficiency and Their Implication for Industrialization in the Third World', in Alejandro Foxley, Michael McPherson and Guillermo O'Donnell (eds), *Development, Democracy and the Art of Trespassing: Essays in Honor of Albert O. Hirschman*, Notre Dame: Notre Dame University Press, pp. 27–55.
Sabel, Charles (1989a) 'Flexible Specialisation and the Re-emergence of Regional Economies', in Hirst and Zeitlin 1989a: 17–70.
Sabel, Charles (1989b) 'Equity and Efficiency in the Federal Welfare State', unpublished paper presented to the Nordic Working Group on the New Welfare State, Copenhagen, 8 August.
Sabel, Charles (1990) 'Skills without a Place: The Reorganization of the Corporation and the Experience of Work', unpublished paper presented to the British Sociological Association conference, Guildford, 2–4 April.
Sabel, Charles and **Zeitlin, Jonathan** (1985) 'Historical Alternatives to Mass Production: Politics, Markets and Technology in Nineteenth-Century Industrialization', *Past and Present*, 108: 133–76.
Sabel, Charles and **Zeitlin, Jonathan** (eds) (forthcoming) *Worlds of Possibility: Flexibility and Mass Production in Western Industrialization.*
Sabel, Charles, Herrigel, Gary, Deeg, Richard and **Kazis, Richard** (1989) 'Regional Prosperities Compared: Massachusetts and Baden-Württemberg in the 1980s', in Zeitlin 1989a: 374–404.
Sabel, Charles, Kern, Horst, and **Herrigel, Gary** (1989) 'Collaborative Manufacturing: New Supplier Relations in the Automobile Industry and the Redefinition of the Industrial Corporation', unpublished paper.
Salais, Robert (1989) 'Why Was Unemployment So Low in France during the 1930s?', in Barry Eichengreen and T. J. Hatton (eds), *Interwar Unemployment in International Perspective*, Dordrecht: Kluwer, pp. 247–88.
Salais, Robert and **Storper, Michael** (1990) 'One Industry, Multiple Rationalities: Flexibility and Mass Production in the French Automobile Industry', Working Paper D901, School of Architecture and Urban Planning, University of California Los Angeles.
Salais, Robert, Baverez, Nicholas and **Reynaud, Bénédicte** (1986) *L'invention du chomage*, Paris: Presses Universitaires de France.
Sayer, Andrew (1986) 'New Developments in Manufacturing: The Just-In-Time System', *Capital and Class*, 30: 43–72.
Sayer, Andrew (1989) 'Post-Fordism in Question', *International Journal of Urban and Regional Research*, 13(4): 666–93.
Schmitz, Hubert (1989) 'Flexible Specialisation - A New Paradigm of Small-Scale Industrialisation?', *Institute of Development Studies Discussion Paper*, 261, University of Sussex.
Sengenberger, Werner and **Loveman, Gary** (1988) *Smaller Units of Employment: A Synthesis Report on Industrial*

Reorganization in Industrialised Countries, Geneva: International Institute for Labour Studies.

Smith, Keith (1984) *The British Economic Crisis*, Harmondsworth: Penguin.

Solomou, Solomos (1987) *Phases of Economic Growth, 1850–1973: Kondratieff Waves and Kuznets Swings*, Cambridge: Cambridge University Press.

Sorge, Arndt, Hartmann, G., Warner, Malcolm and **Nicholas, Ian** (1983) *Microelectronics and Manpower in Manufacturing*, Aldershot: Gower.

Sorge, Arndt and **Streeck, Wolfgang** (1988) 'Industrial Relations and Technical Change: The Case for an Extended Perspective', in Hyman and Streeck 1988: 19–47.

Storper, Michael (1989) 'The Transition to Flexible Specialization in the US Film Industry: The Division of Labour, External Economies and the Crossing of Industrial Divides', *Cambridge Journal of Economics*, 13(2): 273–305.

Storper, Michael and **Christopherson, Susan** (1987) 'Flexible Specialization and Regional Industrial Agglomerations: The Case of the US Motion Picture Industry', *Annals of the Association of American Geographers*, 77(1): 104–17.

Storper, Michael and **Harrison, Bennett** (1990) 'Flexibility Hierarchy and Regional Development: The Changing Structure of Industrial Production Systems and their Forms of Governance in the 1990s', Working Paper D903, School of Architecture and Urban Planning, University of California Los Angeles.

Streeck, Wolfgang (1987) 'Industrial Change and Industrial Relations in the Motor Industry: An International View', *Economic and Industrial Democracy*, 8(4): 437–62.

Thompson, Grahame (1987) 'The American Industrial Policy Debate: Any Lessons for Britain?', *Economy and Society*, 16(1): 1–74.

Thompson, Grahame (1991) 'The Role of Economies of Scale in Justifying Free Trade: The Canada–US Free Trade Agreement and Europe 1992 Programme Compared', *International Review of Applied Economics*, 5(1).

Thompson, Paul, Jr. (1989) *The Nature of Work: An Introduction to Debates on the Labour Process*, 2nd edn, London: Macmillan.

Tolliday, Steven and **Zeitlin, Jonathan** (eds) (1987) *The Automobile Industry and Its Workers: Between Fordism and Flexibility*, Cambridge: Polity Press/New York: St Martins, 1986.

Trigilia, Carlo (1986) *Grandi partiti e piccole imprese*, Bologna: Il Mulino.

Tully, James (ed.) (1988) *Meaning and Context: Quentin Skinner and his Critics*, Cambridge: Polity.

Weir, Margaret and **Skocpol, Theda** (1985) 'State Structures and the Possibilities for 'Keynesian' Responses to the Great Depression in Sweden, Britain, and the United States', in Peter Evans, Dietrich Rueschmeyer and Theda Skocpol (eds), *Bringing the State Back In*, Cambridge: Cambridge University Press, pp. 107–68.

Williams, Karel, Williams, John, and **Thomas, Denis** (1983) *Why are the British Bad at Manufacturing?*, London: Routledge & Kegan Paul.

Williams, Karel, Cutler, Tony, Williams, John and **Haslam, Colin** (1987) 'The End of Mass Production?', *Economy and Society*, 16(3): 404–38.

Wood, Stephen (1988) 'Between Fordism and Flexibility?: The Case of the US Car Industry', in Hyman and Streeck 1988: 101–27.

Wood, Stephen (ed.) (1989) *The Transformation of Work*, London: Unwin Hyman.

Zeitlin, Jonathan (ed.) (1989a) 'Local Industrial Strategies', special issue of *Economy and Society*, 18(4).

Zeitlin, Jonathan (1989b) 'Italy's Success Story: Small Firms with Big-Firm Capability', *QED: Quarterly Enterprise Digest*, (October): 5–9.

Zeitlin, Jonathan and **Totterdill, Peter** (1989) 'Markets, Technology and Local Intervention: The Case of Clothing', in Hirst and Zeitlin 1989a: 155–90.

[13]

Review Article by Karel Williams, Tony Cutler, John Williams and Colin Haslam

The End of Mass Production?

Text reviewed:

Michael Piore and Charles Sabel *The Second Industrial Divide: Possibilities for Prosperity* Basic Books, New York, 1984.

As the economic performance of the advanced Western economies has deteriorated since the early 1970s, so there has been growing interest in the nature, origins and outcome of the present difficulties. Piore and Sabel's book *The Second Industrial Divide* is a contribution to this debate. These two American academics present a distinctive account of what has been increasingly perceived as a 'general crisis of the industrial system' (Piore and Sabel p. 165). Their book ranges widely over many themes but the basic thesis is a simple one; Piore and Sabel argue that 'the present deterioration in economic performance results from the limits of the model of industrial development that is founded on mass production (Piore and Sabel p. 4). As for solutions, Piore and Sabel are agnostic about what will or must happen but they present 'flexible specialisation' as an alternative model of industrial development which offers us the possibility of a prosperous future.

In the United States, *The Second Industrial Divide* was well received and 'flexible specialisation' has been taken up as an idea whose time has come. In Britain the reception has been more mixed. Hyman (1986) has produced a trenchant neo-Marxist critique of the 'myth' of flexible specialisation. But others on the left have reacted quite differently. Murray (1985) has taken up the idea of the obsolescence of mass production with enthusiasm in an article which was provocatively titled 'Benetton Britain'. More significantly, Piore and Sabel's concepts are already being used to provide a framework for further research. A forthcoming

Economy and Society Volume 16 Number 3 August 1987

book on the car industry, edited by Tolliday and Zeitlin is boldly titled *Between Fordism and Flexibility*. This reception justifies a review article which summarises and criticises some of Piore and Sabel's main arguments.

Given this objective, our review article is organised in a fairly straight-forward way. It begins by presenting an analytic summary of the *Second Industrial Divide's* main arguments and then moves on to raise a series of critical questions about these arguments. Is it possible to distinguish between mass production and flexible specialisation? Is there a unitary system of mass production which triumphed for reasons which Piore and Sabel identify? Is mass production breaking up? And, finally, how do we regenerate manufacturing and what benefits can be obtained from this regeneration?

Piore and Sabel's argument

The *Second Industrial Divide* is based on a conceptual distinction between two types of industrial production, mass production and flexible specialisation. On the one hand we have 'mass production' which is characterised by 'the use of special purpose (product specific) machines and of semi-skilled workers to product standardized goods (Piore and Sabel p. 4). The more general the goods, the more specialized the machines and the more finely divided the labour that goes into their production (Piore and Sabel, p. 27). On the other hand we have flexible specialisation or craft production which stands in a neat polar opposition to mass production. This type of production is based on skilled workers who produce a variety of customized goods (Piore and Sabel p. 17).

The text builds a large and ambitious superstructure on the basis of this one opposition. The superstructure has three interrelated elements: first, a theory of types of economy, their characteristic problems and how these problems can and have been resolved; second an interpretative meta-history of the development of modern manufacturing since 1800; third, and finally, an analysis of the current crisis of the advanced economies and its possible solutions. Seldom in the history of intellectual endeavour, can so much have been built on the foundation of one opposition. Piore and Sabel's book is best approached by examining the three super structural elements in turn, beginning with the theory of types of economy.

For Piore and Sabel, mass production and flexible specialisation are not only paradigmatic types of production, they can also be historically realised as types of economy where one kind of production dominates over a given geographic area – regionally, nationally or internationally. Thus the United States from the late nineteenth century created a mass production national economy which was successfully imitated by follower countries in the post–1945 period, thereby creating an international 'mass production economy'; the term itself is used in the title of chapter seven. On a regional basis, viable local economies based on flexible specialisation were realised in nineteenth century European industrial districts, from Lyons to Sheffield, which produced textiles and metal goods. (Piore and Sabel p. 28).

If they repeatedly assert and assume that one type of production can dominate a given area, Piore and Sabel never specify criteria which might be used in deciding whether or not one type of production is dominant in a particular case. When it comes to conceptualising mass production, this issue is of some importance because there is no possibility of a real national economy where all production is undertaken on a mass production basis; as Piore and Sabel concede 'some firms in all industries and almost all firms in some industries continued to apply craft principles of production' (Piore and Sabel p. 20). The survival of something other than mass production is necessary when some end user demands are too small or too irregular to justify mass production and the special purpose machinery required for mass production cannot itself be mass produced (Piore and Sabel. p. 27). In a very orthodox way, Piore and Sabel suppose that the industrial *locus classicus* of modern mass production is in the manufacture of consumer durables (especially cars) and in industries linked to consumer durables such as steel, rubber and plate glass (Piore and Sabel p. 77).

One further complication arises because Piore and Sabel argue that follower countries had, and used, discretionary choice about the organisation of work and methods of labour control and therefore about the degree to which they substituted semi-skilled workers for craftsmen as they introduced mass-production (Piore and Sabel, p. 134). The American mass production system of 'shop floor control over the work process' (Piore and Sabel p. 111) involved narrow job definitions and seniority rights (Piore and Sabel p. 173) in an authoritarian system where management directed the semi-skilled. By way of contrast, countries like West Germany and Japan (Piore and Sabel p. 144, 161) retained important elements of an alternative 'craft system of shop floor control' (Piore and Sabel p. 116) in their factories where management co-operated with multi-skilled workers. In such cases, a mass production national economy can include the labour control elements of its craft opposite and the distinctiveness of mass production in these cases rests narrowly on part of the basic opposition, namely the use of specialized machines to make standardised goods.

Although the distinction between types of economy is thus blurred the notion of differences is sustained partly through the argument that the two types of economy have characteristically different secular economic problems. In both mass production and flexible specialisation, economic stagnation always threatens to interrupt economic development and often does so. But the causes of stagnation are different in the two types of economy, and the ways in which development can be, and has been, restored are very distinct.

Mass production is represented as not so much an economic state as a technologically dynamic trajectory. As mass production develops, the supplying enterprise can capture economies of scale and realise ever-lower production costs and selling price through investing in new generations of product-specific equipment which turns out even larger volumes of standardised goods (Piore and Sabel pp. 52–4). But if the market will not absorb the output, then the mass producer suffers the high fixed costs of an inflexible production system. Piore and Sabel argue that we have learnt in the twentieth century 'that the product specific use of resources pays off only when market stability is ensured' (Piore and Sabel p. 163). To resolve this problem, mass production economies require regulatory institutions that secure a 'workable match' between the production and consumption of goods (Piore and Sabel p. 4).

Existing institutional arrangements often fail or are inadequate for the purpose of regulation. Where they do fail the result is a 'regulation crisis' as in America in the 1890s or 1930s (Piore and Sabel p. 5). Such crises can only be solved through institutional re-construction and innovation. Thus, the crisis of the 1890s was ended with the development of the large corporations which at a micro level stabilized their individual markets by such tactics as ensuring that the fluctuating component of demand was supplied by small marginal producers. (Piore and Sabel pp. 55–6). While the crisis of the 1930s was resolved after the second world war at a macro level through an assortment of Keynesian novelties. In the United States these included new state initiatives such as welfare expenditure, high levels of arms expenditure and 'private' arrangements like wage bargaining on the 1946 UAW/GM pattern which ensured expansion of demand through tying wage rises simultaneously to productivity increases and the rate of inflation (Piore and Sabel pp. 79–82).

If the problem of mass production is one of stabilizing the market, the problem of flexible specialisation is one of ensuring that technical dynamism which Piore and Sabel term 'permanent innovation' (Piore and Sabel p. 17). Under flexible specialisation adjustment to the market is not a major problem and macro regulation is not so crucial. This is because flexibly specialised producers employ general purpose equipment (like the Jacquard loom) which enables the enterprise to shift within and between families of products (Piore and Sabel p. 30). But Piore and Sabel argue that systems of flexible production run a high risk of stagnating technologically because variation in product design and process technology can be limited while firms attempt to cut production costs by sweating labour and using inferior materials. (Piore and Sabel p. 263). On this reading of historical experience 'innovation is fostered by removing wages and labour conditions from competition and by establishing an ethos of interdependence among producers in the same market' (Piore and Sabel p. 272). These objectives can be achieved in a variety of ways. In nineteenth century industrial districts, municipalism, paternalism and familialism all provided organising principles for limiting and structuring competition (Piore and Sabel p. 31). But, one way or another, it is presumed that resources can only be mobilised for permanent innovation if the community is involved and there is a fusion of economic activity, or production in the narrow sense, with the larger life of the community.

The theory of types of economy that we have discussed so far is distinct from the meta theory of history which is the second major element in the superstructure that Piore and Sabel erect on the basis of the opposition between flexible specialisation and mass production. It is logically separate because it would be possible to advance a theory of types of economy without developing a meta history. The meta history of manufacturing which Piore and Sabel present is a variant on the stages theories of economic growth and modernisation which were popular in the 1960s. If this kind of meta history is now being revived by Piore and Sabel, it is being revived in a variant form. The notion of unilinear progress to modernity is rejected as is the notion of a single divide which separates the traditional 'before' from the modern 'after'. That much is signalled by Piore and Sabel's title 'the second industrial divide'.

The meta history of Piore and Sabel is built on the assumption that mass production and flexible specialisation are not only concepts but empirical forms which persist and recur throughout the modern period. Although technology changes and techniques of micro and macro regulations develop, the empirical forms retain the same identity in the 1800s or the 1980s. Thus Piore and Sabel can claim that 'throughout the nineteenth century two forms of technological development were in collision' (Piore and Sabel p. 19). Equally, there is nothing new about the kinds of flexible specialisation which are being developed in the 1980s. Piore and Sabel repeatedly claim that the spread of flexible specialisation now amounts to a revival or 'return to craft methods of production regarded since the nineteenth century as marginal' (Piore and Sabel, p. 252; see also pp. 6, 17). On this view, history must be a process which permutates the two empirical forms which are always the same.

As Piore and Sabel set it up there are only rare moments of choice 'when the path of technological development is at issue' (Piore and Sabel, p. 5) and at which societies can choose between a future built on one or other of the two forms. These moments of technological choice are termed 'industrial divides' and Piore and Sabel identify two of them. The first occurred 'in the nineteenth century' when the emergence of mass production technology – initially in Great Britain and then in the United States – limited the growth of less rigid manufacturing technologies which existed primarily in various regions of Western Europe (Piore and Sabel, p. 5). The second industrial divide is contemporary and dates from the stagnation of the international economic system in the 1970s which is still continuing in the 1980s. Although the two 'divides' are separated in time, the choice is necessarily the same in both cases; it can only be between mass production and flexible specialisation.

This schematic meta history is buttressed with arguments about how and why social choice of technological development occurs rarely and with an account of the determinants of that choice. Crises are not unusual in mass production economies. But most of these crises are 'regulation crises' about the institutions which connect production and consumption rather than 'industrial divides' where the technological form of development is at issue. (Piore and Sabel, p. 5). A kind of inertia holds manufacturing economies onto one 'trajectory' after the choice of technology has been made. As Piore and Sabel argue 'technological choices, once made, entail large investments in equipment and know how, whose amortization discourages subsequent different choices' (Piore and Sabel, p. 38). After an industrial divide one of the contending forms of production wins out and 'the tendency towards uniformity is reversed only when some combination of developments in the market and in the capacity to control nature makes it economically feasible to strike out in new directions' (Piore and Sabel, p. 39).

When it comes to conceptualising the determinants of choice at each divide, Piore and Sabel quite reasonably want to deny any iron law of historical necessity and more specifically to avoid the kind of technological or market determinism which the last quotation hints at. Thus, they reject what they call the classical view which attributes the triumph of mass production in the twentieth century to lower production and selling costs. No examples of cost differentials between craft and mass production are presented but throughout the text it is assumed that at the first and second divides flexible specialisation was, and now is, an economically viable and efficient alternative to mass production. On Piore and Sabel's account, the outcome at a divide is settled by the exercise of political power and the commitment of financial resources. 'The technical possibilities that are realised depend on the distribution of power and wealth: those who control the resources and returns from investment choose from among the available technologies the one most favourable to *their* interests' (Piore and Sabel, p. 38, emphasis in original). What follows is that mass production did not succeed because of its superior economic efficiency in prevailing conditions but rather due to the resources thrown behind those engaged in promoting and using mass production techniques.

Piore and Sabel's anti-classical theory of technological choice is garnished with a rhetorical contrast between the reality of openness at each divide and the ideological appearance of closure after the divide. Piore and Sabel imply that the choice could have gone the other way at the first divide and could now go either way at the second divide. We live in 'a world in which technology can develop in various ways; a world that might have turned out differently from the way it did, and thus a world with a history of abandoned but viable alternatives to what exists' (Piore and Sabel, p. 38). They thus offer a 'branching tree view of history' (Piore and Sabel, p. 67) and claim that the limbs of this tree 'thrive or wither according to the outcomes of social struggles, not some natural law of growth' (Piore and Sabel, p. 15). This openness is ideologically obscured after each industrial divide by the triumph of a 'technological paradigm' which presents the newly dominant form of production as the natural and inevitable victor. Piore and Sabel here borrow the concept of paradigm which Kuhn applied to scientific theory and apply it to a 'vision of efficient production'. They claim ' a new technological paradigm . . . creates the conditions for a new orthodoxy . . . at best half aware that their imagination has been circumscribed by convention, technologists push down the new path' (Piore and Sabel, p. 44). When the paradigm operates to confirm certain techniques and excludes others, the triumph of mass production in the twentieth century is a result both of material support and ideological effect.

The meta history outlined above does not completely determine their position on the current 'general crisis of the industrial system' (Piore and Sabel, p. 165). Piore and Sabel's position on these issues can therefore be considered a third element in the superstructure which they build on top of the basic opposition between mass production and flexible specialisation. But the identification of the crisis at a meta historical 'second industrial divide' does influence their treatment of the crisis which is self-consciously 'open' about causes and outcomes. Thus Piore and Sabel present two supposedly 'alternative' accounts of the origins of the crisis which is caused either by external shocks or internal structural problems in the mass production economies. Equally, they maintain the crisis could be resolved with the victory of either flexible specialisation or a revived Keynesianism. Our authors are avowedly neutral; either of the two causal accounts might be correct and both outcomes are possible (Piore and Sabel, pp. 166, 251). But on our reading, this neutrality is a decorous pretence. The two 'alternative' explanations are largely complementary insofar as the external shocks exacerbate the structural difficulties which advanced economies are beset by. While, in terms of outcome, there can be no real alternative to flexible specialisation because their preconditions for a regeneration of Keynesianism cannot be met.

The first account of the causes of the present crisis presents it as an 'accident' caused by external shocks such as the oil price rises of 1973 and 1979 or the breakdown of the post-war regime of fixed exchange rates, (Piore and Sabel, pp. 66–82). These shocks generated uncertainty; the viability of products and processes depended, for example, on the unpredictable future level of oil prices. Such uncertainty inhibits investment and thus has a depressing effect. In the second account what we have is a kind of internal structural crisis of mass production where the problem is the level and composition of demand nationally and internationally (Piore and Sabel, pp. 183–93). On this structural account the problem is 'the saturation of core markets' and 'the break up of mass markets for standardized products'. The structural explanation is complementary because the internal problems inhibit investment just like the external shocks. Thus confusion about the level and composition of demand had the effect of 'reducing the portion of demand that employers saw as sufficiently long term to justify the long-term fixed cost investments of mass production' (Piore and Sabel, p. 83).

The immediate market problem is 'saturation of industrial

markets in the advanced economies' (Piore and Sabel, p. 187). The argument here is focused on the market for long-established consumer durables such as cars and washing machines; 'by the 1960s domestic consumption of the goods that had led the post-war expansion began to reach its limits' (Piore and Sabel, p. 184). This was a problem because 'no new products emerged to stimulate demand for mass produced goods' (Piore and Sabel, p. 189); specifically computers and home entertainment never became mass production industries. The other market problem for the mass producers was the 'break up of mass markets for standardized goods' (Piore and Sabel, p. 183) in a world of increasing product differentiation on the supply side and growing diversity of tastes on the demand side. The effect of mass market saturation and break up in the advanced countries was accelerated by the development strategies of many third world countries. The protectionist Latin American countries closed off their internal markets while the Asian NICs aggravated market congestion in the advanced countries by pursuing strategies of export led growth. (Piore and Sabel, p. 189).

If these are the dominant market trends, they constitute a problem for mass production and an opportunity for flexible specialisation which operates with multi-use low cost capital equipment. Current trends in manufacturing technology, particularly the development of computer controlled equipment, reinforce the advantage of flexible specialisation in meeting such demand. Piore and Sabel maintain that flexible specialisation is dynamic 'independent of any particular state of technology' and recognise that computers can be put to rigid use by mass producing enterprises. But potentially computer control of equipment like machine tools offers major advantages to flexibly specialised firms producing the small batches and short runs which a differentiated market requires. It is not necessary to replace the machines as in mass production or to manually change tools and fixtures as in old fashioned flexible specialisation; with computer technology, the equipment can be put to new uses without physical adjustment 'simply by re-programming' (Piore and Sabel, p. 260). Piore and Sabel's discussion of new technology concludes with a paean of praise for the computer as the contemporary equivalent of the nineteenth century artsian's tool which now has the liberating potential to ease the tyranny of specialized machinery over semi and unskilled workers; 'the advent of the computer restores human control over the production process, machinery is again subordinated to the operator' (Piore and Sabel, p. 261).

Despite all this, Piore and Sabel formally insist that Keynesianism might be revived and markets could be stabilised so that mass production might provide a basis for renewed prosperity on the other side of the present industrial divide. But a new form of 'international Keynesianism' would require large changes which, according to Piore and Sabel, can only be initiated by national governments acting together. If Keynesianism is to make the world safe for mass production, their prerequisite is new global regulatory mechanisms which raise purchasing power in at least some of the less developed countries. Positively, for example, there must be arrangements which ensure that demand expands at a rate equal to the expansion of productive capacity and mechanisms which apportion the expansion of productive capacity between advanced and developing countries (Piore and Sabel, pp. 252–7). No doubt the IMF could act more expansively and perhaps currency exchange rates could be managed. But when the European countries cannot agree on co-ordinated reflation, it is incredible that the United States, Europe and some of the less developed countries could agree on the much more far-reaching changes which Piore and Sabel insist are necessary.

Against this background of a supposed internal blockage of mass production, Piore and Sabel are able to find regional islands of prosperity built on flexible specialisation which has already 'challenged mass production as the paradigm' (Piore and Sabel, p. 207). Industrial districts like Prato or Emilio Romagna in Central and North Western Italy provide a model for our future (Piore and Sabel, p. 206). The challenge now is 'to see how flexibility – until now confined to a relatively small segment within the mass production system – can be extended throughout the economy' (Piore and Sabel, p. 258). These national developments are likely to produce a new international division of labour. As mass production economies of scale become irrelevant 'the more likely each nation would be to produce a wide range of products on its own' (Piore and Sabel, p. 277). Elsewhere Piore and Sabel envisage that mass production will migrate to the LDCs while the advanced countries specialise in high tech, footwear, garments and machine tools.

Making distinctions

It is now time to turn from exposition to criticism and our criticism begins by considering the basic opposition between two types of production; mass production relying on special purpose product-specific equipment and semi-skilled workers to produce standardized goods versus flexible equipment and skilled workers to produce customized goods. Does this opposition provide a secure foundation for a large superstructure of meta history? Our answer is that the opposition is not up to the job. Piore and Sabel never develop criteria for indentifying instances of mass production and flexible specialisation in a way that is intellectually satisfactory. The identifications that Piore and Sabel do make are in our view, arbitrary and unjustified.

This issue has already been raised in our exposition. We noted then that Piore and Sabel fail to state criteria of dominance which would allow us to determine whether and when one form of production comes to dominate a given area thereby creating a distinctive regional or national economy of the mass production or flexible specialisation type. Worse still, our argument below shows that it is very difficult to identify particular enterprises or industries as instances of mass production or flexible specialisation. At a conceptual level, the opposition appears clear cut. When there are national differences about the organisation of the labour process, there are only three invariant dimensions of difference between mass production and flexible specialisation. These differences concern the dedication of equipment, the extent of product differentiation and the length of production runs. They are summarised in our diagram below:

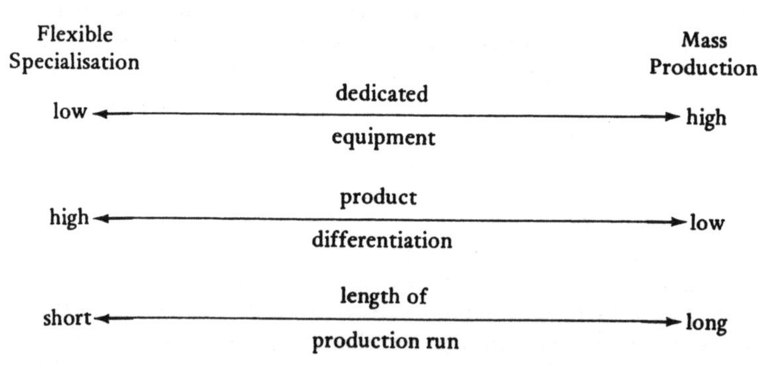

The problem of identification arises because, in most instances, a specific enterprise or industry cannot be neatly situated at one pole of variation (on the far left or right in the diagram) in all three dimensions. It is certain, for example, that many enterprises and industries which Piore and Sabel would classify as mass production do not use completely dedicated equipment to produce a single standardized product. The car industry provides some obvious illustrations of this point.

Many significant items of equipment in car factories are not dedicated. Consider, for example, the large hydraulic presses which have been used for more than fifty years in car factories to produce steel body panels. Such presses have a working life of approximately thirty years and a high rate of throughput (Hartley, 1981, p. 29). A modern car firm would not keep one model in continuous production for the life of the press and, at any one moment in time, the press will be used to produce a variety of panels for one or more models. Different panels are obtained by changing dies and the press is simply equipped with a new set of dies when models are changed over. As so often in the production of consumer goods, the tooling is dedicated and model specific but many items of capital equipment are re-usable. Does that make the car industry less of a mass production industry or not a mass production industry?

As for product differentiation, the output of major firms does not usually consist of a single standardized product which stays in production for decades. The Ford T and the Volkswagen Beetle are exceptions. Other car producers have increasingly imitated the General Motors strategy of providing a range of differently priced models. And, in the European car business, differences in size have always been important. This kind of mass production differentiation is disparaged by Piore and Sabel who observe that GM's models shared many components and were only 'presented as different' (Piore and Sabel, p. 51). But that disparagement is unjustified when Piore and Sabel provide no criteria for discriminating between fundamental difference and trivial styling variation which they admit is commonplace in both flexible specialisation and mass production.

The conclusion must be that dedicated equipment and limited product variety are not unproblematic characteristics which can be used to differentiate the mass production enterprises and industries from the rest. In this case, does length of production run provide an empirical yardstick which can be used to identify mass production and corroborate the meta history? It would, for example, be significant if Piore and Sabel were able to present statistical evidence which showed that, in the consumer durables industries, production runs had grown dramatically shorter over the past fifteen years. But in a three hundred page book, there is no statistical evidence at all on production runs for *any* product of the nineteenth or twentieth centuries. Even if evidence were to be supplied, that would of course raise a whole series of questions about how short a production run has to be before we cross the rubicon from mass production to flexible specialisation? This key question is not posed or answered in *The Second Industrial Divide*, where the deficiencies of the argument are covered up by a regression into circular argument and self reference. If we ask how long is the production run of a mass produced piece of string, Piore and Sabel's answer would be that the production run of craft string is shorter.

If each dimension of difference between mass production and flexible specialisation turns out to be problematic, the difficulties of identification are compounded if we consider the way in which the dimensions of difference are articulated together. As our diagram shows, the simple Piore and Sabel opposition presupposes a particular pattern of joint variation in all three dimensions of difference. A (every?) mass production enterprise will combine limited product differentiation and long run production runs. But it is fairly easy to show that at an enterprise level some 'mass producers' of differentiated goods can sustain long runs while others do not. Japanese majors in consumer electronics can achieve long runs because they have a high share of a protected home market and export successfully to the rest of the world. Thus in the late 1970s, cumulative volume per tv chassis type was 1.2 million in Japan, compared with an average of 400,000 in Europe and just 150,000 in the UK whose national producers did not export and were losing share of their home market (Magaziner and Hout, 1980). In this instance, differences in the size of the available market determined massive differences in production run for enterprises producing similarly differentiated products.

Nor can the existence of a wide range of choice be used to identify an area where flexible specialisation is necessarily making ground. Under free trade conditions, mass production now invariably provides a bewildering choice; British consumers can choose between more than thirty brands and around twenty different makes of washing machines from British and European factories (*Which*, 1986). If that choice does not amount to very much then that is because one product type (the front loader) dominates the European, washing machine market.

The implication of our argument is that mass production and flexible specialisation connot be satisfactorily identified in particular instances, even at the enterprise and industry level. This weakness must undermine much of Piore and Sabel's argument that mass production did displace flexible specialisation in the nineteenth century and that flexible specialisation can now displace mass production. After all if we cannot identify instances of mass production or flexible specialisation how can we determine that one type of production is displacing the other?

Challenging the meta history – the case of Ford

The general problem with meta-history is that it tries to stuff too much into the same bag. This is inevitable when any long and complex historical experience is summarised with the aid of a simple framework which identifies only a couple of stages (or divides). This produces certain characteristic effects, some processes and episodes will be misrepresented because the interpretative presuppositions of the framework have to be satisfied. While other processes and episodes will vanish because they cannot be handled within the framework. When Piore and Sabel are so economical with the concepts and assertive about the connections, it would be surprising if they avoided these problems in their meta-history of manufacturing since 1800. And the issues here are best approached by examining the crucial case of Henry Ford and the Model T. If we except a rather curious attempt to represent the 5$ day as a primitive policy for boosting demand, what Piore and Sabel say about Ford is fairly orthodox. Thus, they note two points which are emphasised in the existing secondary literature: first, Ford's process innovations lowered the price of his product and thus extended the market (Piore and Sabel, p. 51): second, Ford was able to finance his company's expansion without public sales of equity and bank loans (Piore and Sabel, p. 70). What they do not pause to consider is that these points raise fundamental questions about whether the Ford case contradicts their meta history and, more specifically, their grand theory of technical change. In our view it does so and this is the theme which we now wish to develop by counterposing Ford's achievement and Piore and Sabel's meta history.

It is necessary to begin by clarifying the issues. We would not wish to argue, following Chandler (1964), that the triumph of mass production was an inevitable response to the potential mass market which the railways created in the United States. Fords success with the Model T depended on a variety of technical prerequisites including product innovations like the robust and easy to use gear change of the T. Nor would we wish to argue that Ford's process innovations had any 'unique' or 'intrinsic' superiority in all circumstances (see, e.g. Piore and Sabel, p. 40). The Ford T succeeded in particular circumstances when the preconditions for a mass market had been established. This point is proved by the sales success of the T in its early years before the most important process innovations had been introduced. Sales increased ten fold to $42.5 million in the years 1908 to 1912 (Nevins, 1954, p. 645; Hounshell, 1984). The question is why mass production triumphed over craft production in these particular circumstances. Piore and Sabel's general theory of technical change asserts that choice of productive technology is settled by the exercise of political and financial power and that seems to imply that the balance of economic advantage is fairly even.

Ford made radical process innovations in the 1912–3 period by combining technical elements which already existed and applying them to the mass production of a complex consumer good including numerous components and sub-assemblies. There were three key technical elements. The first element was the use of interchangeable parts which had already been pioneered in car manufacture by Cadillac (Nevins, 1954, p. 371). This directly reduced the labour requirement and was a prerequisite for the development of the assembly line. The second key element was the layout of machines in shops according to the sequence of process operations. As a Ford engineer argued, this again reduced labour requirement by eliminating unnecessary internal movement of parts and work in progress (Bornholdt, 1913, p. 277). The third and final element was the introduction of the moving assembly line which had originally been applied to the stripping of carcasses in the meat trade. When applied to the key operation of assembling the chassis of a light touring car, the results were dramatic. In August 1913, 12.5 man hours went into assembling the Model T chassis; by April 1914, after the assembling line had been introduced, the labour requirement was reduced to 1.5 man hours. (Arnold and Faurote, 1972, pp. 136–9).

All this required an increase in capital investment, but the reductions in labour input were so large that Ford dramatically reduced costs of production. The benefit was passed on to the consumer in the form of lower selling prices, as table 1 shows.

The Ford T was a keenly priced, bottom of the market motor car

Table 1 Selling price of the Model T touring car

	$
1908	850
1909	950
1910	780
1911	690
1912	600
1913	550
1914	490
1915	440
1916	360

Source: Nevins (1954)

when it was introduced in 1908. But, by 1916, in real terms allowing for inflation, the price had been quartered. This was not at the expense of Ford Motor Company's profits which increased from $3 million in 1909 to $57 million in 1916 (Nevins, 1954, p. 647).

One further crucial point is that the success of the Ford Motor Company was not politically sponsored or supported by financial institutions. Of the $100,000 initial capital over half was credited to Ford and Malcolmson for machinery and patents, only $28,000 was paid in cash. The eleven original backers who put in money were small businessmen, a brace of lawyers, a carpenter and an office worker (Nevins, 1954, pp. 236–7). These men were not Rockefellers and they did not have a direct line to John Pierpoint Morgan. Equally, none of Ford's backers would have been persona grata at meetings of the executive committee of the bourgeoisie. The inescapable conclusion is that political and financial sponsorship was unnecessary because, in the particular circumstances of the time, Ford's combination of process innovations had an overwhelming economic advantage over the 'craft methods' of production which had been used hitherto to make cars which were toys for rich men.

All the rhetoric about 'branching trees' and openness covers one central deficiency in Piore and Sabel's account of mass production versus flexible specialisation at the first divide; they never present any evidence on the cost of producing cars, or any other complex durable, by alternative methods. The case of the Ford T suggests there never was a choice because there was not a viable craft alternative to mass production of complex consumer durables. That explains why Piore and Sabel are unable to cite any examples of successfully surviving craft production in the key industrial areas where mass production developed. In industries like cars, craft producers only survived precariously by moving up market and meeting the small scale demand for high priced luxury alternatives to mass produced manufacturers.

Challenging the meta history – mass production after Ford

The case of Ford and the T shows how our meta historians can ignore crucial cases which contradict their meta history. We now wish to turn to make the rather different point that Piore and Sabel's account of subsequent mass production fails to register distinctions and differences which are important but do not exist within their framework. In their meta history mass production is always the same and the inevitable outcome, after the triumph of mass production, is a history where nothing really happens except for 'regulation crises'. In an earlier text, Sabel accepted the logic of that position and proposed the concept of 'Fordism' as a kind of shorthand for mass production; 'I will use Fordism as a shorthand term for the organisational and technological principles characteristic of the modern large scale factory' (Sabel, 1983, p. 32). When adequate criteria of instances are never elaborated, it becomes possible to see Fordism everwhere in manufacturing over the past sixty years. Against this we wish to argue that Ford's innovation of the assembly line factory had a limited field of application and Ford did not provide a strategic model which his successors imitated. Ford's production techniques only had an overwhelming cost advantage in the production of complex consumer durables, initially cars and electrical goods, and subsequently in the field of electronics where the products included consumer and producer goods. That gave mass production a substantial field of application; in a recent survey of British manufacturing, 13 per cent of the plants in the sample produced products which contained more than 1,000 components (New and Myers, 1986, p. 6). But for simpler consumer goods, like clothing and furniture, mass production techniques had a limited advantage. Meanwhile the capital intensive process industries, like steel and chemicals, went their own way before and after Ford. It is therefore quite understandable that most plants in the advanced economies do not contain assembly lines; the survey of British manufacturing which we have already cited shows that 31 per cent of plants in the sample used assembly lines and only half of those were mechanically paced (New and Myers, p. 31). Ford's innovations may have been important but they are hardly responsible for the whole trajectory of development in the advanced economies. Rather they created what Mitsui aptly calls assembly industries. Even within this field, Ford did not provide a model which his successors imitated and for that reason alone the concept of 'Fordism' is seriously misleading. As we have already noted, Ford's successors did not generally imitate his product strategy of relying on one long lived model. Most assemblers succeed by making families of inter-related models which are changed over fairly regularly. Equally important, Ford's successors did not aspire to become fully integrated producers who carried out all the operations necessary to production in their own factory. The 'classic' integrated Ford plants of the 1920s and 1930s like Highland Park, Dagenham and Cologne are not typical mass production factories. Ford's successors ran assembly factories whose internal process operations were fed with bought out components. In most European car factories since 1945, bought out components account for half the cost of the finished motor car. These crucial variations on 'Fordism' had important repercussions for the composition of assembly output and for the organisation of production in the assembly industries.

After Ford, assembly was associated with volume and variety at the enterprise level. This was technically possible because assembly lines can be run productively and profitably at relatively low model volume. That point was demonstrated by the way in which Austin and Morris were able to adopt Ford's innovations and achieve significant cost reductions when supplying the relatively small British market for cars in the 1920s and 1930s. Much of the history of mass production in Europe between the wars was one of the adaptation of mass production to lower volume. The next major turning point came after the early 1960s with Japanese innovations in cars and electronics. This showed how the objective of greater variety could be achieved by mass production enterprises which also chased volume increases. Enterprises like Toyota have developed the use of 'mixed lines' where two or more models are assembled on the one line, pressed through inventory reduction through the Kanban system and 'just in time' parts delivery, and also dramatically reduced change over and set up times on equipment like power presses (Shonberger, 1982). The Japanese have demonstrated that productive and marketing advantage can be obtained cheaply because assembly line factories can be used much more flexibly than in most Western countries.

As for the organisation of production, the assembly industries created opportunities for large, medium and small scale enterprises which can be connected in a variety of different ways in input – output terms. Piore and Sabel's treatment of these issues is confused and confusing. They distance themselves from 'the popular view that increasingly associated large plants with mass production' (Piore and Sabel). But they generally treat the survival of something other than large scale production in the mass production industries as evidence of the deficiency of mass production and relate the survival of small firms in that sector simply to the requirements of Fordist firms for specialized equipment to build their standardised products. Against this, we would argue that it is the requirement for components which creates the main market opportunities for small firms. These opportunities are substantial, even in the case of sophisticated products like VCR's where the assembler naturally monpolises the technically difficult high volume process work of head production (Mitsui, 1986). Small or medium firms can also prosper in final assembly if component semi-manufacturers are being produced in volume by large scale enterprises. This is one pattern in areas of electronics where there is large scale production of commodity semi manufacturers like tv tubes, transformers and silicon chips. One of the more eccentric aspects of Piore and Sabel's book is that they effectively identify mass production with the production of final products for the consumer.

The discussion so far has emphasised some of the differences and distinctions which Piore and Sabel neglect. The concept of Fordism should clearly be rejected because it elides too many differences and establishes an uninformative stereotype. Furthermore, any notion of a generic modern system of mass production should be treated with great caution because there are many different ways of organising production, even in the assembly industries. It would be intellectually interesting to analyse these differences. But, until that analysis is provided, it would be foolish to produce substantive work where mass production is a central organising concept.

If the criticism so far has focused on differences and distinctions which Piore and Sabel neglect, it is appropriate finally to examine the one difference which they do recgnise. Culturally and historically determined differences in labour control figure in *The Second Industrial Divide* as the main explanation of variation in the national experience of mass production (Piore and Sabel, p. 162–4). This position rests on a misunderstanding about the general importance of labour in the production process.

As we have argued elsewhere (Williams *et al.*, 1987), labour control is a managerial obsession which attracts attention in a way which is disproportionate to its real significance. If Piore and Sabel accept managerial pre-occupations too readily that is because they believe that 'wages are the major component of costs' (Piore and Sabel, p. 84). But modern manufacture since the industrial revolution has been a system which takes labour out; more specifically, Ford and his successors were successful insofar as they took labour costs out without incurring anything like the same capital costs. If that yields a micro economic advantage, it is also crucial to the macro economic process of economic growth which is all about product output growing faster than production inputs, especially labour. One measure of the proportion of total costs in manufacturing, inside or outside the 'mass production' industries (the New and Myers survey of British manufacturing plants which we have already cited) shows that in their sample, direct labour accounted for an average of just 18 per cent of total production costs (New and Myers, 1986, p. 7). What's done tactically with the labourers who remain is much less important than what is done strategically about taking labour out. Although tactical decisions contribute to, they do not determine the strategic outcome which depends on decision about such matters as investment strategy and marketing. As we have argued in the case of Austin Rover, no amount or variety of labour control can produce viability if management makes major mistakes about investment and marketing, (Williams *et al.*, 1987). Much the same point could be made about the defeat of Ford and the Model T in the American car market of the 1920s. Labour control was irrelevant because Ford could never win with the T when the market increasingly preferred closed cars, and second-hands supplied the basic transport market.

Mass production is presented by Piore and Sabel as a form of production with a technological-cum-market core plus a variable institutional armature. However the main emphasis then falls on labour control which is the one national difference which Piore and Sabel consolidate into their scheme. We have already argued that this variable cannot explain much and we would finally add that other variables can explain more. If the aim is to explain differences in manufacturing performance, it would be more instructive to examine the role of the stock exchange or the lending criteria applied by different national banking systems. We have shown elsewhere (William *et al.*, 1983) how both conditions influence the performance of British manufacturing. If Piore and Sabel represent a variant of institutionalism, this is an impoverished institutionalism with little explanatory power.

Are mass markets breaking up?

Piore and Sabel argue that mass production has reached its limits when the markets for mass produced goods are saturated and breaking up because consumers are now demanding more differentiated goods. If the level and composition of demand necessary for mass production cannot be restored, in Piore and Sabel's framework we are at a 'second industrial divide' and the only way forward is through a revival of flexible specialisation which was marginalised at the first divide. This section of our criticism will examine the *Second Industrial Divide's* interpretations of current market trends. Our conclusion is that Piore and Sabel's account of the market arises out of conceptual confusion and does not rest on any hard evidence.

Piore and Sabel's argument about saturation is focused on the older consumer durables (cars, washing machines, refrigerators) which are mature products with high levels of market penetration. But that does not in every case prevent substantial growth in volume and value of sales. The market for colour tv sets in Britain has more than doubled in size to 3.7 million sets over the past ten years because households now buy small screen tv's as second sets (BREMA *Yearbook*, 1985). Value of sales is buoyant when three quarters of a million teletext sets were sold in 1985 and satellite dishes and high definition tv are just around the corner (BREMA *Yearbook* 1985). TV shows that the most boringly mature product can be re-invented and repackaged to win extra volume and value. In other areas, where the identity of the product is more stable, volume increases are hard to find because market penetration is high. But, for exactly that reason, a huge replacement demand exists. In white goods, for example, 12 million washing machines and 15 million domestic refrigerators are sold in Europe each year. In the new car market, the dominance of replacement demand is associated with cyclical fluctuation as consumers bring forward or postpone their purchase; in most white goods and brown goods the pattern is quite different because replacement demand is extremely stable.

Replacement demand for mature products is not enough for Piore and Sabel who assume that volume increases are necessary for mass producers who seek to move down a long run average cost curve by realizing ever greater economies of scale at higher levels of output. (Piore and Sabel, p. 52). This is formulated by means of a schematic diagram and it is all very hypothetical because a great deal of empirical evidence suggests that the average costs of large firms are often constant over large ranges of output. Even if cost reductions can, in principle, be obtained as output increases, expansion is a risky strategy because it involves fixed investment in capacity increases which will only be profitable if the enterprise and the industry correctly predict the increased size of the future market. There is no good reason why enterprises and industries should not make steady and less risky profits by meeting a large and stable replacement demand which does not tempt producers to invest in over-capacity. If mass production existed, it could be a stable state rather than a trajectory.

In any case if enterprises in consumer goods want increases in volume and value, they can always obtain them by introducing new products, It is salutary here to list some of the new durables which are now being sold in volume in Britain although they did not exist as mass market products ten years ago. In brown and white goods the list would include video cassette recorders, new format cassette players like the 'walkman', compact disc players, micro-wave ovens, dishwashers and food processors. Most of these new products are complementary from the producer's point of view; they can be put together on new lines in existing factories and are sold through existing distribution channels. The development of new products by existing producers is completely ignored by Piore and Sabel. On the issue of new products, Piore and

Sabel's one argument is that computers and home entertainment have failed to become mass production industries because their products are insufficiently universal (Piore and Sabel, pp. 204–5). These industries do not produce something like the T which is 'a machine for everyone and everything' (Piore and Sabel, p. 202), This nonsense is the bizarre result of projecting the shadow of Ford onto the reality of modern industry. VCRs are the products of large scale Japanese assembly factories and are sold in mass markets around the world, even if they are not universal audio-visual pleasure machines like Woody Allen's 'orgasmotron'. Indeed, it is not clear why Piore and Sabel do not follow the logic of their own argument and decide that the T was not a true mass market product because Ford failed to produce a 'travel centre' which was capable of flying and crossing water as well as travelling on and off road.

Even if the mass market is not saturated, it is still conceivable that mass markets are breaking up because consumers demand more differentiated products. With the Piore and Sabel framework, market break up is a much more significant development than market saturation. A problem about a saturated market would in itself only create a 'regulation crisis' of the kind which mass production has solved before through reconstructing the institutions which secure a workable match between supply and demand. But problems about market break up would create a much more fundamental kind of crisis and an industrial divide if markets were breaking up in a way which creates patterns of demand which mass production cannot cope with. The argument of the next couple of paragraphs is that markets may be breaking up, but not in a way which is really threatening.

The orthodox mass producer survives by producing a family of inter-related models. And in the case of the major durables it is unusual for consumers to demand more than a handful of product types. In the case of cars in the European market demand has converged onto four distinct product types. The demand is for small, light, medium and large cars which Ford of Europe meets with the Fiesta, Escort, Sierra and Granada. In every major European national market, except West Germany, 80 per cent plus of sales are taken in the three lower classes where the major manufacturers have similarly packaged look alike models. Volume car firms can no longer survive by making one or two utility models as some did in the 1950s and 1960s, but four basic models is all that is required for the current European car market. Variants like 'hot hatchbacks' or coupes can be easily produced by feeding different components onto the main lines or by setting up lines for 'new models' which simply package components from the enterprise parts bin in a slightly different way. Such competition only threatens those manufacturers who cannot find volume sales and decent runs and thereby cover development costs.

Length of production run will depend on how the enterprise is advantaged or disadvantaged by the parallel process of market fragmentation which has occurred generally in the consumer goods markets of all advanced countries (except Japan) over the past twenty years. With the increasing interchange of manufactures over this period, the area of trade has widened in most product lines and the number of brands and models represented in any one market increases. As the importers move in, they claim volume and the domestic producer (or producers) with market leadership lose market share. In the British case this has been the fate of Austin Rover or of Hoover, Hotpoint and Murphy Richards who had a dominant position in the supply of many kitchen durables in the 1960s. The British problem is that in many product lines, producers have failed to compensate for the inevitable loss of home sales with a sufficiently large expansion of exports. That explains why Austin Rover which is pinned down on its home market is currently making less than 150,000 units each year of the Metro which is its best selling car. Most mainland European producers in cars or white goods also lost out at home, but market fragmentation was not an insuperable problem for them because they won back volume with increased sales to near European markets. Volkswagen for example, prospers with less than 30 per cent of the German market because it takes 5 per cent or more of every other major national market. On this basis, VW makes more than 800,000 units of its best selling model the Golf each year. In any process of market fragmentation there will be trade winners and losers; enterprises or national industries which lose will be marginalised and possibly forced out of business. But that process does not threaten the system of large scale production, any more than the fact of bankruptcy threatens capitalism.

If Piore and Sabel believe mass markets are breaking up that is because they are conceptually confused about what is going on and crucially fail to draw the distinction between simple product differentiation and market fragmentation which has quite different consequences. Equally clearly their position on 'the break up of mass markets for standardized products' (Piore and Sabel, p. 18) does not rest on any sound empirical basis; as we have already noted they provide no statistics on length of production runs and no criteria for discriminating genuinely different products. There is one empirical test of their position on market crisis. It is asserted that difficulties about the market are reflected in a reluctance to undertake fixed investment. In a discussion of the shocks of the 1970s, Piore and Sabel claim that such shocks 'reduced the portion of demand that employers saw as sufficiently long run to justify the long term fixed capital investment of mass production' (Piore and Sabel, p. 183). No statistics are cited to support this claim about inhibited investment. The available evidence on capital formation shows that, if this effect operated in the mass production sector, it was not sufficiently strong to depress the level of gross fixed capital formation in all advanced countries.

As table 2 shows, the real manufacturing investment levels of 1970 were effectively maintained or surpassed in 1983 in four of the six national economies. Two economies, West Germany and Britain, show a sustained fall in real investment levels. If the German record is anomalous, the decline in British manufacturing investment can be fairly easily explained when this country was a trade loser which suffered progressive de-industrialisation. The evidence of healthy investment elsewhere is not explained by, but is broadly consonant with, the evidence which we have presented on market opportunities in the assembly industries of cars and consumer electronics. Piore and

Table 2 Gross fixed capital formation in manufacturing (constant prices)

	1970	1975	1983
USA	100	117,3	134,8
UK	100	84,1	57,7
Japan	100	101,6	181,7
			(1)
W. Germany	100	59,1	80,3
			(2)
France	100	103,7	128,6
			(3)
Italy	100	91,1	96,0

Notes

1 West German figure is for year 1982.
2 All French totals include mining and quarrying as well as electricity and gas.
3 Italian figure is for year 1981. All Italian totals include mining and quarrying.

Source: United Nations Annual Accounts.

Sabel's intimations about market led collapse in these industries appear to be exaggerated melodrama.

The message of new technology

Even if Piore and Sabel are wrong about market demand, they may be correct about the potential of new technology which, on their account, facilitates the resurgence of flexible specialisation. They argue that dedicated equipment and inflexible automation is being superceded by a new generation of micro-electronically controlled machines which allow efficient production in smaller batches. Piore and Sabel take a very definite position on the capability and salience of computer control, but here again they present no solid evidence. Crucially, their book never examines the costs and output potential of specific items of new technology in particular industrial contexts. To redress this absence, we will present some evidence of our own. In modern manufacturing, enterprises are increasingly buying computer integrated manufacturing systems. And we will analyse two cases of such systems; robotised body building lines which are used in the car industry and 'flexible manufacturing systems' (FMS) which are usually used for metal machining. With robots the issue is how new technology changes conditions of production in an industry which has been a central redoubt of volume production in long runs. FMS are more usually installed by capital goods producers and the question here is how new technology changes the economics of batch production.

Commentators like Altshuler (1984, p. 12) and Jones (1985) have recently argued that a new productive flexibility is transforming the economics of the cars business (see also, Piore and Sabel, p. 248). Their position is summarised by Jones who, like Altshuler, relies on asserting the fact of flexibility and provides no evidence about costs.

> . . . the introduction of computer controlled production lines plus the introduction of much more flexible automation involving robots, automated handling, machining cells etc., have changed the economies of scale in production drastically. The use of robots instead of dedicated multi-welders, for instance, gives these plants a greater degree of flexibility to switch models in response to demand and reduces the cost of introducing new models and variants . . . A full range of cars can now be produced in one or two plants at a much lower total volume and at a cost which is competitive with much larger producers. . . .
> (Jones, 1985, pp. 151, 152)

Against this we would argue that the new productive techniques are much less flexible than Jones supposes and the scale economics of the car business have not been fundamentally changed by the new techniques.

There can be no doubt that the process of car body building has been transformed by robots. It is now commonplace for 90 per cent of body welds to be made automatically and, in a modern car factory, most of these welds will be made by robots which are also extensively used in spraying. That explains why in most of the advanced countries, 50 per cent of industrial robots are installed in car factories. These robots are much more flexible than an earlier generation of dedicated jig multiwelders which could only weld one panel of a specific shape. The costs of multi-welders could only be covered on long runs and in Europe only VW (with the Beetle) built whole bodies with multiwelders. Ford and Fiat used them for "structural" components like floor pans which were not changed whenever models were face-lifted. Robots can be set up to perform a variety of spot welding operations on differently shaped panels for a set of models. Every European volume car manufacturer, even little Austin Rover, now operates robot body lines which can produce an output of 200–350,000 shells for two models and several variants on those models if the manufacturer so requires. But a substantial commissioning cost (for software, tools and fixtures) has to be incurred before a robot line is brought into use; one recent estimate suggests that 60 per cent of the cost of a robot system is accounted for by commissioning cost (see Williams *et al.* 1987). Most of this commissioning cost is model generation specific. When a new generation of models is introduced every five years or so the panel geometry will be different and, if the old robots are retained, they will have to be expensively recommissioned. Robots cannot be re-programmed for new models by pressing a few buttons. That is a myth.

It is also certain that the introduction of robots in body building has not transformed the scale economics of the cars business. More flexible automated technology does not allow the small scale producer to become more cost competitive than in the previous era when such firms relied on manual welding and spraying and dedicated automation was the preserve of the Americans and VW. When the commissioning costs of body building equipment is high, the large scale producer with long model runs retains an advantage when it comes to using and re-using body lines. Furthermore body building is only one process in the manufacture of motor cars and throughput requirement remains high in several other processes. Robots are being increasingly used in the final assembly of engines. But engines are still produced on transfer lines dedicated to the production of one engine type for a long period at a rate of 3000,000 units per year or more. For the foreseeable future, small firms in the European volume cars business will buy in engines which they cannot produce at a competitive cost. Even if new technology did change the balance of productive advantage between large and small firms, the large firm would still obtain a general advantage from its ability to spread development costs over longer runs. Because of inability to cover development costs, small car manufacturers always risk being demoted to assembler status.

The case of FMS is inevitably different in detail but it again serves to illustrate the basic discrepancy between reality and the romance of computerised flexibility. A 'flexible manufacturing system' exists where there is computer coordination of two or more manufacturing cells each of which would normally contain several machine tools; the cells will be connected by an automatic transport system which moves pallets, workpieces and tools between machines and to and from workpiece and tool storage; the whole is coordinated by a DNC computer which integrates all operations according to a master programme. (UN, 1986, p. 13). At first sight, the literature on FMS endorses the same doxa as the literature on car body building. The UN survey asserts that 'mass production as a concept is becoming more and more a thing of the past' (UN, 1986, p. 2). While there are familiar claims about added flexibility.

> In mass production environments, computer controlled machines will make it possible to add flexibility to the production system in the sense that the system can be used to manufacture several different product variants with minimal set up times. This opens up important potentials, for dividing large scale production into many smaller batches with the obvious purpose of reducing in-process inventory and achieving a faster adaptation to consumer preference, (UN 1986, p. 2),

But this literature is more sober, because it is not claimed that the availability of FMS changes the economics of the business of metal machining by tilting the balance of advantage against large scale production undertaken by the medium and large sized firm. That claim would be implausible because, as the survey evidence shows, FMS systems are expensive, deliver limited variety and have to be utilised at high volume. On our interpretation, FMS is a way of putting medium volume batch production on a capital intensive basis and therefore represents only a new twist on the old post Fordist story of volume and variety.

FMS systems are not cheap. The most comprehensive available

survey by the UN covers 339 FMS world wide. In this sample, 45 per cent of Japanese systems, 46 per cent European systems and 82 per cent of American systems cost over $3 million. Inevitably, therefore, most FMS are bought by medium and large firms who can afford this kind of expenditure. A large investment in FMS is only sensible if the enterprise has a substantial work load which offers a combination of the variety necessary to use the technical capability of the system and the volume necessary to secure high levels of capacity utilisation. This is another constraint on the adoption of FMS by medium and small sized firms. As the UN survey argued 'for small and medium sized companies it is often difficult to find a large enough product family in terms of the number of parts and the volume per part which over a long period of time will ensure a sufficiently high degree of utilization of FMS' (UN, 1985, p. 55). Smaller firms are also disadvantaged because they will usually lack the necessary in-house expertise to develop and run, a sophisticated custom built system.

The amount of variety delivered is very variable and quite obviously relates to the different requirements of particular enterprises and industries. For obvious reasons, most large firms who buy and use FMS do not want a machine which delivers industrial pump pistons this week and auto clutch components the next. Width of product envelope is built in at the design stage and usually determined by upstream and downstream process requirements. All the surveys show that Japanese FMS deliver substantially more variety than Western FMS. The UN survey showed that 49 per cent of Japanese FMS were able to produce more than 50 'variants' of a product; the comparable European and American figure was 17 and 37 per cent. Jaikumar (1986) has shown that FMS in American factories are used for longer runs than in Japanese factories. The implication is that Western firms are installing expensive over-sophisticated equipment which is not necessary for their business strategies and the end user requirements in the markets defined by those strategies. Variety of output is not an end in itself and many Western users of FMS are incurring hidden costs because they are buying facilities whose full potential will never be exploited.

Whatever variety is planned for and delivered, it is absolutely essential to obtain volume which guarantees high rates of utilisation. In this respect the crucial consideration is not batch size but the cumulative volume of all the batches produced through the year. And this cumulative volume is always high; in Jaikumar's (1986) survey, US FMS delivered an average of 17,000 parts per year while Japanese FMS delivered an average of 24,000 parts per year. When fixed costs are high, an FMS must be driven intensively near to the limit of capacity, if payback in 3 to 5 years is to be achieved. This is always achieved in Japan where two shift working is the norm and where nearly one third of the 60 FMS in Jaikumar's survey were being operated on a continuous three shift basis. As a Western observer reported after a tour of Japanese FMS, 'in all factories, the large product volumes and high throughputs were particularly impressive' (FMS Magazine, April 1986, p. 69). By way of contrast, the British financial experience of operating FMS is disastrous; in the New and Myers survey, more than half of the 64 plants which had installed FMS reported losses or no pay off. British production managers lack the technical ability to use even mundane metal working machinery in an intensive way (Daly and Wagner, 1985). And many British firms have a proven inability to find volume which is as necessary for FMS as for other more traditional forms of capital intensive process equipment which produces less variegated output.

We do not pretend to know the truth about new technology. We doubt whether there is one message in this bottle. If technological change is uneven in its nature and effects, it is unlikely to offer the same benefits to all enterprises and industries. But, in this context, our evidence does cast doubt on the universal validity of Piore and Sabel's over-confident assertions. And, whatever new technology does offer in the two cases which we have considered, it is not going to inaugurate a new era of flexible specialisation which restores those craft methods of production that lost out at the first divide' (Piore and Sabel, p. 5 see also p. 17). New generations of computer controlled equipment may deliver a more varied output but they do not restore an economic system based on redeployable productive resources and low fixed costs. That is a world which we have lost. When robot bodylines or FMS do not change the economics of the business in which they are installed, this new technology is likely to be controlled by medium and large sized firms who will not use it to create a workforce of independent craftsmen.

Piore and Sabel's vision of an artisan future is, like so much else in their book, plausible in some respects. If new technology has one general effect, it is to reduce the relative importance of semi-skilled direct labourers in the manufacturing workforce. In a company like Austin Rover, for example, only half the workforce in 1985 consisted of direct workers of the traditional blue collar kind. But that does not require any return to polyvalent craftsmen who have independent control over entry and the prospect of short term job security through work sharing, (Piore and Sabel, p. 116). These elements of the 'craft model of shop floor control' are not going to be re-created by the modern corporation. The intentions of British managers are disclosed by their interest in, and enthusiasm for, the Atkinson (1984) model of labour control. In this schema, enterprises create an elite of multi-skilled workers whose training and skills are enterprise or plant specific and whose privileges as 'core workers' are granted and can be taken away by the company. The extent and significance of such labour central strategies remains to be investigated. Meanwhile, Piore and Sabel's schema is unhelpful because here again we have an instance where they simply project a stylised shadow of the past onto the present in a way which confuses our understanding of what is going on.

The benefits of flexible specialisation?

We cannot accept Piore and Sabel's diagnosis of what has gone wrong in the advanced economies. The notion of a crisis of mass production must be rejected for two reasons: first, it is based on a concept of mass production which is probably incorrect and certainly elides too many differences; second it is contrary to the available evidence on market trends and new technology. But it must be admitted that the advanced economies are not prospering and it is conceivable that flexible specialisation might provide a basis for industrial regeneration. This last section considers what flexible specialisation can deliver and comes to the conclusion that it is less than Piore and Sabel promise.

When the pre-conditions for a revived Keynesianism cannot, in Piore and Sabel's schema, be met, flexible specialisation is an attractive way forward. Flexible specialisation does not depend on a functioning national or international economic order which delivers a particular level or composition of demand. If flexible specialisation is to maintain technological dynamism, according to Piore and Sable it does require an integration of production into the local community. But this could be managed on a regional or local basis. If neither family nor ethnic identity provide suitable principles of integration in the modern advanced economies, the necessary integration could be provided by municipal or regional government. Within this framework, flexible specialisation provides a simple universal rule for choice of strategy at the enterprise and industry level. On Piore and Sabel's account, if new technology has one message and if there are general trends in the market place then for enterprises and industries there can only be one correct strategy of flexibility. Piore and Sabel do not formally draw this conclusion, but it is the logic of their position. In this context it is significant that in their long book, there is not one case where they commend inflexibility as an appropriate and effective contemporary business strategy. Our question must be what can strategies of flexibility deliver?

Any simple rule which says 'the more flexibility, the better' would not always produce sensible results for the enterprise and the industry. In multi-process activities like steel making or car manufacture, the impact of technological change is almost always uneven and flexible automation is not the invariable correct answer to every question about the design of process technology. Best practice Japanese factories typically use a mix of dedicated 'hard' automation and flexible 'soft' automation. Leading Japanese automation experts like Makino insist that 'hard automation will always be very important' and cost effective because many process functions do not change frequently. In other cases, where flexible technology can be applied across a range of processes, it is impossible to produce the whole of an industry's output using flexible technology. For example, as long as steel mini-mills are scrap-charged, they can do little more than fill a niche in international markets where most of the output is produced in large-scale basic oxygen converters; the price of scrap would go through the roof if all the advanced countries used it as the basic raw materials for steel production.

At an enterprise level, choice of technology depends on the costs of available process technologies, the product strategy of the enterprise and the markets which are available within the limits of investment in distribution. Enterprise calculation is about making an appropriate choice within these parameters. And the choice which is correct for one enterprise will not necessarily be correct for another enterprise. This point is demonstrated by the contrast between VW's success at Wolfsburg with dedicated automation which is used for Golf body building and final assembly, and Austin's failure at Longbridge with the same kind of automation which is used for Metro body building. In both cases, choice of dedicated technology was consistent with the company's product strategy of producing a long life model with few variants. But the market position of the two companies was very different and dedicated equipment which could be intensively used by VW was underutilised by Austin Rover. The strength of VW's European distribution is such that the company can sell 750,000 Golfs each year while Austin Rover, which is pinned down on its home market, has difficulty in selling 150,000 Metros. With this kind of volume, it would have been more sensible for Austin Rover to choose a more flexible line where two or more different models could be built in whatever proportions the company preferred.

The contrast underscores the importance of executing strategy, and making choices work. If choice of technology should be related to the market, when technology has been chosen, execution is about ensuring the right level and composition of demand by managerial (and maybe political) action. This is an area where the British always do badly. We have elsewhere criticised 'giantism' in British nationalised coal, steel and cars where enterprises naively pursued economies of large scale production (Williams *et al.*, 1986). In all three cases massive investment in modernisation only became totally disastrous when projected demand failed to materialise and the nationalised enterprises faced problems about the underutilisation of newly constructed capacity. The problem was not that the strategies were inherently absurd, but that they were executed ineptly because managers and politicians failed to secure the market that was necessary if the new large scale facilities were to run profitably. That adjustment of the market could have been made in at least two of our three cases. Austin Rover's managers could have invested in overseas distribution and the government could have given some preference to British manufactured cars. If that intervention was politically problematic, ministers could have easily prevented the CEGB from forcing down the price of power station coal which is the main product of British Coal.

From our point of view, the weakness of Piore and Sabel's position is that it virtually abolishes the role of enterprise calculation which, in their schema, must become a matter of identifying the particular implications of universal trends in technology and the market. As our examples show, this is an inadequate way of conceptualising the choices and opportunities which face enterprises and industries in the advanced countries. Institutional conditions ensure there are substantial differences in national ability to exploit such opportunities (Williams *et al.*, 1983) and the resulting differences in performance create further difficulties. Trade in manufactures between the advanced countries is a zero sum game, and in an open international economy, successful manufacturing countries like Germany and Japan gain trade share at the expense of unsuccessful manufacturing countries like Britain. The migration of manufacturing output and employment to the more successful advanced countries is a reality which Piore and Sabel avoid by positing the posibility of a future world where all can succeed through flexibility.

If Piore and Sable exaggerate the benefits which can be obtained from flexibility, more fundamentally they exaggerate the benefits which can be obtained from the regeneration of manufacturing. Their book is an argument about the possibility of a prosperous future based on manufacturing. But, any regeneration of manufacturing in the advanced countries is unlikely to solve problems about the distribution of social welfare which are becoming ever more acute. Piore and Sabel cover up these problems in their concluding chapters by putting before us a vision of industrial districts spreading out to create a new republic of craftsmen and smallholders. Against this in our view, institutional forces and free trade are likely to create regional and national islands of prosperity and to sustain substantial wage differentials between manufacturing workers and the rest – even within the islands of prosperity.

The most unsuccessful advanced economies like Britain contain regional islands of prosperity. As Hyman (1986) points out, the Thames valley in Britain provides a model of prosperity built on a diversified base of high tech manufacturing and services. Piore and Sabel attach special significance to those industrial districts like Prato and Emilio Romagna whose prosperity is supposedly built on a foundation of flexible specialisation. If national prosperity built on flexible specialisation is to materialise, then it would be necessary for these islands to expand so that they dominate the whole of the national economy. Before we could determine whether that outcome is plausible, we would need to know the proportion of national employment and output which these districts account for at present. Typically, Piore and Sabel provide no information on these points. In only one case are we given any figures and these relate to performance rather than extent; in Emilio Romagna all we are told is that wage levels and income per capita are rising relative to the rest of Italy and unemployment rates are falling relative to the rest of Italy (Piore and Sabel, p. 227). Such information does not allow us to determine whether industrial districts of this type are filling in occasional niches or can colonise the whole economy. In effect Piore and Sabel only provide homiletic examples of flexible specialisation's supposed success in regional economies. The Italian regions are like Samuel Smiles heroes; they show us all that, with the right kind of effort, it is possible to rise above the disadvantage of humble origins.

Where manufacturing succeeds (with or without flexible specialisation) that success only directly benefits a minority of the population. As Hyman (1986) points out, in all the advanced countries two thirds to three quarters of the workforce is outside the manufacturing section. In most of these countries the proportion employed in manufacturing is in relative decline and in unsuccessful manufacturing countries there are large absolute declines in numbers employed; the numbers employed in British manufacturing have declined by more than two and a half million in the past fifteen years (Cutler, 1986). Steady secular increases in labour productivity have been sustained in manufacturing since the industrial revolution and the expansion of manufacturing output has never required a commensurate increase in manufacturing employment. To make this familiar point in a slightly different way, any increases in manufacturing (net) output are typically appropriated through rises in real wages rather than increases in employment (see Williams et al, 1987). The mechnics of appropriation ensure that the manufacturing sector is always a high wage sector in the advanced countries. Redistribution to the unwaged and to the low paid outside manufacturing will be necessary and such redistribution is most easily carried out by central government which in all the advanced countries has a dominant role in taxation and in secondary income redistribution through social security. That suggests the limits of the Piore and Sabel model of locally guaranteed prosperity. The dynamism of flexible specialisation might be secured by the integration of production into the local community. But, there are clearly limits on what municipal action can achieve. Most municipalities and regions lack the powers of taxation to achieve social objectives like a substantial redistribution of income.

* * *

In conclusion we have demonstrated that Piore and Sabel's basic opposition between mass production and flexible specialisation is unworkable because empirical instances cannot be identified; their meta history is contradicted by the case of Ford and suppresses the subsequent history of the assembly industries; their analysis of market trends and new technology is unconvincing; and their view of where flexible specialisation can take us is incurably romantic. Why then, has this mix of futurology and meta-history been represented as providing a serious and reliable guide to the modern industrial world despite its manifest conceptual and empirical weaknesses? Part of the answer lies in the fact that everywhere it strikes comforting and responsive chords. Thus in Britain Piore and Sabel's work provides a rationale for the local initiatives and plans for socialism in one municipality which have been increasingly popular over the past decade. The reception tells us more about the critical standards of this audience than about the real merits of the text.

Department of Economics
University College of Wales
Aberystwyth and Middlesex polytechnic

References

Altshuler, A., Anderson, M., Jones, D., Roos, D., and Womack, J. (1984), *The Future of the Automobile: the Report of M.I.T's Automobile Program*, London, Allen and Unwin.

Arnold and Faurote (1972), *Ford Methods and the Ford Shops*, New York, Arno Press (reprint, original 1915).

Atkinson, J. (1984), *Flexibility, Uncertainty and Manpower Management*, IMS Report, 89.

Bornholdt (1913), 'Placing machines for sequence of use', *The Iron Age*, Vol. 92.

BREMA *Yearbook* (1985).

Chandler, A. (ed.) (1964), *Giant Enterprise: Ford General Motors and the Automobile Industry*, New York, Harcourt and Brace.

Cutler, T., Williams, K., and Williams, J. (1966), *Keynes, Beveridge and Beyond*, London, Routledge & Kegan Paul.

Daly, A., Hitchens, D., and Wagner, K. (1985), 'Productivity Machinery and Skills in a Sample of British and German Manufacturing Plants'. *National Institute Economic Review*, February.

FMS *Magazine* (1986), April.

Hartley, J. (1981), *The Management of Vehicle Production*, London, Butterworth.

Hounsell, D. (1984), *From the American Systems to Mass Production 1800–1932*. Baltimore, Johns Hopkins, U.P.

Hyman, R. (1986), 'Flexible Specialism: Miracle or Myth'.

Jaikumar, R. (1986), 'Post-Industrial Manufacturing'. *Harvard Business Review*, Nov.-Dec.

Jones, D. (1985), 'Vehicles' in C. Freeman (ed.), *Technical Trends in Employment, 4, Engineering and Vehicles*, Aldershot, Gower.

Magaziner, I. and Hout, T. (1980), *Japan's Industrial Policy*, London, Policy Studies Institute.

Mitsui, I. (1986), 'The Japanese sub-contracting system'. Komazawa Univ., Tokyo.

Murray, J. (1985), 'Benneton Britain'. *Marxism Today*.

Nevins, A. (1954), *Ford: the times, the man and the company*, New York, Charles Scribner and Sons.

New, C. and Myers, A, (1986). *Managing Manufacturing Operations in the UK, 1975–85*, Cranfield.

Sabel, C. (1983), *Work and Politics*, Cambridge, C.U.P.

Schonberger, R. (1982), *Japanese Manufacturing Techniques*, New York, Free Press.

United Nations (1986), *Survey on Flexible Manufacturing Systems*.

Williams, K., Williams, J. and Thomas, D. (1983), *Why are the British Bad at Manufacturing?* London, Routledge & Kegan Paul.

Williams, K. and Haslam, C. (with Williams, J. and Wardlow, A.) (1986), 'Accounting for failure in nationalised industries'. *Economy and Society*, Vol. 15, No. 2.

Williams, K., Williams, J. and Haslam, C. (1987), *The Breakdown of Austin Rover*, Leamington Spa, Berg.

Part II
Lean Production

Work, Employment & Society, Vol. 7, No. 2, pp. 163–188 June 1993

Abstract: One of the most influential books that has been published in recent years is the MIT study *The Machine that Changed the World* by Womack, Jones and Roos (1990). The book combines detailed empirical comparisons with bold and sweeping assertions. The Japanese management system or 'lean production', Womack *et al.* argue, is not only the world's most efficient system for manufacturing cars. It is the one best way of organizing all kinds of industrial production, featuring both dramatic increases in productivity and qualitative improvements in working conditions. According to the MIT team it is predestined to become the standard global production system of the twenty-first century, and they contend: 'Lean production is a superior way for humans to make things. It provides better products in wider variety at lower cost. Equally important, it provides more challenging and fulfilling work for employees at every level, from the factory to the headquarters. It follows that the whole world should adopt lean production, and as quickly as possible' (Womack *et al.*, 1990: 225). The purpose of this article is to challenge this view of lean production as an omnipotent system and unequivocal blessing. Starting with a discussion of the industrial limits of lean production, I then turn to *the* success story of the 1980s – the dramatic expansion of Japanese auto transplants in North America. The social preconditions for this process, largely overlooked in the MIT study, are emphasized, before proceeding to an analysis of the highly ambiguous working conditions at these new plants. The transplants have attracted many American workers, but the relentless production regime has caused growing disillusionment among employees and increasing resistance from union locals, which was demonstrated by the five week strike at CAMI in September–October 1992. Finally, I discuss current developments inside Japan, where automakers face a severe recruitment crisis. In Japan, both unions and environmentalists criticize the JIT-system heralded in the MIT study, and car manufacturers have started to design plants according to new principles. For many reasons lean production will not be the end of history!

LEAN PRODUCTION – THE END OF HISTORY?

Christian Berggren

The Contradictory Views of Japanese Management

In the auto industry, Japanese manufacturing systems have been debated for at least ten years. In this long-standing debate, one aspect is most intriguing, and that is the extremely contradictory assessments of the effects of Japanese production management.

Christian Berggren is an Associate Professor in the Department of Work Science at the Royal Institute of Technology in Stockholm, and author of *Alternatives to Lean Production: Work Organization in the Swedish Auto Industry* (Cornell ILR Press 1992).

The first time I personally encountered this polarized picture was in 1981 at a Swedish so-called Kanban-conference for industrial engineers and production managers. The conference started with a report from a three week study tour organized by a consultant firm, full of enthusiasm for Japanese efficiency and dynamism. Its glowing presentation was followed by a contribution from a general manager of Atlas Copco, an international Swedish maker of air compressors and rock-drilling equipment. For several years he had been responsible for a plant in Japan, and portrayed a grim picture of the conditions of the small supplier firms, the *sh'tauke*, and their workers. The result was widespread confusion and discomfort among the gathered engineers. Here two very contrasting views were presented. Which of them could they believe in? Personally I found the contradiction highly stimulating. Perhaps it was not so much a confusion of minds and information, but contradictions in real life?

A second example of starkly opposing views on Japan and the Toyota Production System was a 1988 seminar for Swedish auto managers in Gothenburg organized by myself and an internal Saab consultant. The Saab consultant had just returned from the start-up of Honda, Alliston in Canada. Previously he had been quite restrained when discussing Japan. In Alliston he had seen the light. Honda's new operation was a 'total new work experience, egalitarian, creative, dynamic, uniquely productive'. At the same seminar the first results of the MIT researcher John Krafcik's international assembly plant study were presented, giving further credit to the assumption of Japanese superiority. Other Saab managers had also been visiting Honda plants in North America. They had seen very different things, however: a frantic work pace, relentless attendance demands, unsafe production equipment, and heavy indoctrination in a quasi-totalitarian culture. Their report seemed to come not only from a different plant but from a different planet!

Finally I would like to contrast the perspective of the MIT book with a quotation from the president of UAW local at a Japanese transplant, whom we interviewed in November 1991,

> Lean production offers a creative tension in which workers have many ways to address challenges. This creative tension involved in solving complex problems is precisely what has separated manual factory work from professional 'think' work in the age of mass production.
> (Womack *et al.* 1990: 102)

> They promised us a rose garden. They gave us a desert.
> (Phil Keeling, UAW President at Mazda's Flat Rock plant, 1990)

One Universal Production System?

The recent international revival of interest in manufacturing, which is very much, in contrast to the post-industrial euphoria of preceding decades, is everywhere accompanied, in Europe and in the US as well as in Australia, by a search for international 'best practice'. That is a central theme in the report of the MIT Commission on Industrial Productivity (Dertouzos, Lester and Solow 1989), and even more so in the report commissioned by the British Department of Trade and Industry (1990), *Manufacturing into the Late 1990s*. In Australia the same perspective is evident in the report on *The Global Challenge* commissioned by the Australian Manufacturing Council (1990). This world-wide convergence is to some extent a result of the internationalization of manufacturing, resulting in a global tendency towards what organizational theorists label isomorphism. It is also a product of the rise of a 'management' industry of business press, management consultants and management schools, all of whom thrive on the identification of alleged best practice and exemplary organizations (Westney 1989: 284).

For nearly all participants in this intellectual crusade, 'best practice' is the same as Japanese practice. The authors of the *Machine that Changed the World* present an impressive documentation of the Japanese superiority – *in car production*. From this study of auto-makers they formulate the 'best practice' perspective in its very extreme. According to Womack *et al.* we have reached a virtual end of history in industrial organization. Lean production is not only the most efficient way of designing and making cars, it will be the *only* way of producing in *every* manufacturing field. Moreover, in this process all the conflicts and contradictions which have plagued industrial history for so long will be solved.[1] Henry Ford's mass production revolution, it is acknowledged, was a doubled-edged sword: it 'made mass consumption possible, while it made factory work barren'. Lean production, in contrast, means benefits for all(Womack *et al.* 1990: 278),

> Lean production will supplant both mass production and the remaining outposts of craft production in all areas of industrial endeavour to become the standard global production system of the twenty-first century. That world will be a very different, and a much better, place.

This fundamentalist perspective, that there is one and only one way of organizing manufacturing, is a kind of global superprojection of Frederic Taylor's famous dictum – the one best way of organizing work. And, as with the approach of Taylor, its gross over-simplification is very American: one best way, the Fordist one, is now superseded by a new best one, Toyota's; one industrial hero, Henry Ford, is followed by a second one, Taichi Ohno, in a neat succession. The far-flung claims of the MIT authors can be contested on several points. I will start with a brief discussion of the overall industrial limits to lean production.

Lean Production – Still a Volume Production

The sweeping assertions of Womack, Jones and Roos are based on their belief that car manufacturing still is the premier industry, just as it was when Sloan wrote his *My Years at General Motors* (1963). Methods that promote productivity in the auto industry will of necessity do the same in other sectors. This view is certainly open to discussion. What, for example, do capital intensive sectors, like the petrochemical or paper-making industries, or research-intensive sectors like pharmaceuticals, have to learn from the almost obsessive focus on hours per unit produced, so evident in the *Machine* book? Toyota's manufacturing system is certainly more flexible than conventional Western ways of car production, but it is still very much a *volume production*. The achievement of Toyota was to introduce small lot manufacturing into this context of volume production. As has been stressed by the Toyota consultant Shigeo Shingo (Shingo 1981: 110), 'The question in this area is not whether or not to mass produce, but whether to produce large or small batches'. The importance of volume is inadvertently acknowledged in the *Machine* book, when it attributes the poor international competitiveness of the Mexican car industry to the small scale of 'only' 25,000 cars annually per model: 'far too low even for today's lean producers to make economically' (Womack *et al.* 1990: 265). Nissan's Australian story confirms the importance of scale in auto production, whether it is lean or not. Nissan was the first Japanese manufacturer to take the Australian market seriously and had already started local assembly in 1966. Mitsubishi and Toyota followed suit. Nissan never reached sufficient volume, however. In 1990, its Australian production totalled 57,000 units; in 1991 the figure decreased to 36,000 (AIA 1992). In February 1992 Nissan's Tokyo headquarters announced that the company would pull out from Australia. The volume was too low to warrant local production. The MIT authors contend that lean production is as flexible as craft production, only much more efficient. A *Business Week* analysis of Nissan's global problem demonstrates that the real world is far more complicated. Flexibility does not come cheap, even in Japan. In order to cut costs and regain profitability Japan's number two maker attempts to reduce the number of variations per model, slow down the model-change cycle and standardize as many parts as possible across models (*Business Week*, 22 April, 1992). And Nissan's problem are not isolated. Every Japanese maker is now moving toward fewer model variations and a common use of parts (*Automotive News*, 30 March, 1992). At the same time they are revamping their production systems to make them more flexible. In short, lean production is still very much a system of large scale manufacturing. *Flexible volume production* and not the antithesis of mass production is an appropriate characterization of the Toyota production system.

The Japanese preference for high volumes and standardization as a basis for a variety of features and options is also stressed by Michael Porter (1990). His massive study *The Competitive Advantage of Nations* contains a much more compelling analysis of Japanese dynamism than is found in the MIT texts. Japanese firms are very competitive in industries such as cars, consumer electronics, semiconductors and standardized machine tools. They have not invented any universal production or management system, however, and Porter finds them much less successful in industries demanding customization and individualized customer relations (Porter 1990: 411): 'Japanese firms do not do well, by and large, in industries or segments involving a high degree of customization to individual buyers, narrow applications, heavy after-sale support, and small lot sizes.'

A core argument in the MIT writings, both in *The Machine* and in the subsequent PhD Thesis by J. P. MacDuffie (1991) is the juxtaposition of lean and mass production. The explicit assumption in this argument is that before the advent of the Toyota system, mass production – plus small vestiges of craft manufacturing – was the all-dominant form of industrial production. This perspective may be approximately correct in car manufacturing, but is utterly misleading as a general frame of analysis, since most industrial activities cannot be placed in either of these categories. For example, an important reason for the strength of most competitive German or Italian industries (many of them medium-sized or grouped as networks of small businesses) is related to the fact that they never adopted the American mass production pattern.

The Success – Japanese Transplants in North America

A most important basis for the universalistic claims of the MIT study is its documentation of the efficiency of the Japanese transplants in North America. These plants have proved that 'lean production' is not confined to the Japanese socio-cultural context, which was an earlier widespread belief. Since they are so important for the overall argument I will discuss these operations at some length. My main source is an extensive field trip to six transplants in Canada and the US in 1990.[2] Fifteen months later, a study group from the Swedish Metalworkers' Union visited four of the same plants plus NUMMI. Their report is another contribution to the 'Scandinavian perspective on lean production'. I will also draw on a number of recent American and Canadian articles and studies of the transplants.

In the 1970s the Japanese were very successful in exporting cars to North America. In the 1980s they were equally successful in exporting their production system. They erected new facilities at a furious tempo, transferring the fierce domestic competition between car producers in Japan to North America. Technically most of the new factories are virtual clones of Japanese sister plants, fully equipped with Japanese machinery, from industrial robotics to transfer presses. More importantly, the Japanese manufacturing culture is also transplanted, as is the supply strategy. In personnel management, however, there have been some important deviations from the Japanese model, which I will come back to below. Honda

Table 1 Auto Transplants in North America 1990

Company	Location	Start year	Prod 1990	Planned prod (thousands)	Planned emp. level
USA					
Honda	Marysville/ Ohio	1982	430	510	8000
Nissan	Smyrna/ Tennessee	1983	240	440	5100
NUMMI (Toyota & GM)	Fremont/ California	1984	200	300	3400
Mazda	Flat Rock/ Michigan	1987	180	240	3400
Diamond Star (Chrysler & Mitsubishi)	Normal/ Illinois	1988	150	240	2900
Toyota	Georgetown/ Kentucky	1988	220	440	5000
Subaru & Isuzu	Lafayette/ Indiana	1989	70	120	1700
Canada					
Honda	Alliston,	1988	100		
Toyota	Cambridge	1989	60		
CAMI Automotive (Suzuki & GM)	Ingersoll, Ontario	1990	50	200	2200

Source: Business Week, August 14, 1989, *Automotive News*, January 7, 1991

and Nissan were the first bold pioneers; now every Japanese car maker has at least one manufacturing facility in North America. The auto firms have been followed by hundreds of Japanese suppliers. Nearly all of these plants are green field sites. Only four are unionized – NUMMI, CAMI, Diamond Star and Mazda – and all of these because of links to The Big Three. Of a total of 350 supplier transplants in USA none is unionized (*Automotive News*, 28 January, 1991).

With the exception of Nissan in Tennessee the facilities are very compactly designed, concentrating press, body, paint and final assembly shops under the same roof. NUMMI, the joint venture between Toyota and GM, has been the most spectacular success. Despite a very conventional technology it quickly reached productivity levels far ahead of all the Big Three plants. By and large all transplants have performed well in terms of productivity, quality and volume increase. In 1990 their output totalled 1.7 million units. That does not mean that they are all profitable, however. In fact, several of them probably never will be. As stated above, lean production is very much a volume production and capacity utilization is the key to profitability. This is threatened by the fierce competition in the American marketplace – in particular the competition between the transplants. In 1992, Japanese makers sold 70 different car models in the United States. As a result, their profit margins are squeezed. The situation is even worse for the supplier transplants. According to a recent analysis in *Fortune* (June 15, 1992), 60 per cent were in the red in 1992, and the situation for Japanese plants in North America will deteriorate further by the end of the 1990s. By then, the magazine predicts, only three Japanese auto transplants will remain. The other manufacturers will leave and never return. It is interesting to compare this forecast with the confident assertion by Womack, Jones and Roos that over-capacity is only a problem for old-style mass producers: 'Today, we hear constantly that the world faces a massive over-capacity crisis . . . This is, in fact, a misnomer. The world has an acute shortage of competitive lean-production capacity and a vast glut of uncompetitive mass-production capacity. The crisis is caused by the former threatening the latter' (Womack *et al.* 1990: 12).

Before analyzing the experience of the auto transplants in more detail, it is important to stress that they are not generally representative of Japanese manufacturing operations in the United States. In a 1989–1990 study of Japanese factories in California, the 'leading magnet of Japanese direct investment in the U.S.', Ruth Milkman surveyed all companies with 100 or more employees in this state. Of a total of 66 firms, she elicited responses from 50. Twenty of these were later visited. More than half of the 50 plants operated in the electronics and electrical equipment industry, many of them turnkey assembly operations. To her surprise Milkman found that these plants differed very much from the NUMMI model. Whereas the Japanese auto transplants have recruited native-born Caucasians, applying a very selective screening process (see below), the Californian plants in Milkman's survey primarily employed low-paid and low-skilled immigrants – Mexican, Salvadorans, Thai, Vietnamese, Filipinos (blacks, however, were conspicuously few). The hiring process was very simple. To quote one interviewed manager (Milkman 1991: 54): 'With what we pay, if they wear shoes, we'll hire 'em.' Basically, these plants had neither adopted lean production principles nor the Japanese form of human resource management. Employee suggestion programs were largely inactive, team structures were few and the celebrated egalitarianism of the auto transplants (no separate parking lots or dining rooms, etc.), was hard to find. Job rotation was rare and 'most of the managers we interviewed laughed outright when we asked about just-in-time delivery and the like' (Milkman 1992: 73). In principle most of them were committed to avoiding lay-offs. Since the labour turnover was very high, a reduction of the workforce was easy to accomplish anyway. These Japanese plants in California had only one important trait in common with auto transplants such as Nissan,

Tennessee or Toyota, Kentucky. That was their strenuous efforts to avoid union representation. From a unionist's perspective, however, this was not very different from the managerial attitudes at American-owned non-union plants.

In spite of their limited value as representatives of Japanese overseas expansion in general, the importance of the auto transplants is difficult to overstate. They have demonstrated that lean production is possible to implement in a large-scale fashion outside Japan. They have become role models, not only for American car plants in North America, but increasingly for the new European operations that are now coming on stream, for example GM's plant in Eisenach in the former East Germany, officially inaugurated in October, 1992. They certainly deserve a closer analysis, both of their productivity performance and working conditions.

Three Basic Preconditions for High Transplant Productivity

The high levels of productivity of Japan's American plants are in most studies attributed to the management system. Production control, plant design, quality assurance and approaches to problem-solving are indeed important. The high transplant efficiency cannot be explained by these factors alone, however. Three basic preconditions play an important role.

- The first is the design of the products. All transplants assemble Japanese products, engineered for easy manufacture. This is an important factor in success, but is much more difficult to measure than assembly hours per plant, and is thus easily missed. The *Machine* book is unfortunately very focused on plant productivity, and reports only one detailed manufacturability analysis, carried out by GM. According to this study the 'design factor' contributed to 41 per cent of the productivity differential between a GM and a Ford plant (Womack *et al.* 1990: 97). In 1988, product engineers at the Swedish Saab company, who selected doors from equivalent Honda and Saab models to perform comparative assembly, using the same methods and technology, found that the Honda doors could be built in a quarter of the time of the Saab doors! Arguably, a Saab 9000 ranks very low on a manufacturability scale, so these figures are hardly representative for all Western manufacturers.
- The second basic precondition is the selection and management of the suppliers. The Japanese performance in the automotive industry is impossible to understand without considering the complex production pyramids of suppliers and subcontractors, which account for 70–75 per cent of the end product value.[3] In a study in 1984 of Toyota suppliers at different hierarchical levels in Aichi prefecture, I found an ambiguous web of dynamic cooperation and outright exploitation (of, for

example, female self-employed home-workers). This was before the Plaza agreement in 1985 which resulted in a very rapid appreciation of the yen. At least at the lower layers of the pyramid, exploitation still seems to be an important aspect of the relationship (Sakai 1990). However, the new 'era of rapid product development', which Japanese auto-makers entered in the second half of the 1980s, triggered a rapid up-grading of many first- and second-tier suppliers, both in terms of manufacturing and Research and Development capacity, making them a very sophisticated part of the Japanese auto industry cluster (Ikeda 1991). Small wonder that the transplants in North America initially chose to import critical and technologically advanced components from Japan. Pressured by the demand for increased local content they have worked hard to select reliable American suppliers. These firms are now facing the most stringent demands they have ever met. Another method to increase 'local' content is to complement the car assembly transplants with component transplants. A number of Japanese suppliers have responded to these calls and erected their own facilities in North America. According to the supplier specialist Shoichiro Sei (1991), this Japanese preference for their own suppliers could not be explained only, or even chiefly, by referring to price, quality and reliable deliveries. Of more fundamental importance is the 'hidden meaning of the social relationship' between car manufacturers and their suppliers. This relationship means that Japanese suppliers incessantly work to improve their performance and offer services which are not spelled out in the contracts. In Japan, Sei emphasizes, contracts mean almost nothing – which is the most fundamental reason why American supplier firms have had such a hard time cooperating with the Japanese.

- A third important precondition for the high transplant productivity is the *extraordinarily careful selection of the personnel.* In Toyota, Kentucky, the screening process consisted of the following steps. First, an IQ test was administered to all applicants. The poorer half was dismissed. Then manual dexterity was tested, and again people with poor scores were dismissed. Tests of ambition, initiative and creativity followed. Role playing to test group orientation and adaptability was another method employed to weed out unsuitable candidates. The process was completed with medical examinations and drug testing. The result was an aggressively achievement-oriented workforce, where workers are competing not only between groups, but also to advance their personal ambitions. The same pattern of rigorous screening was found in all the transplants.

Such a strict selection presupposes an abundance of applicants. That has been the case for all the Japanese plants. When Nissan started in Tennessee there were 100,000 applicants for 3000 jobs. When visiting the plant in 1990 we found the working conditions rather distressing, and in Sweden it would have been difficult to get a stable workforce. In Tennessee however, that was not the problem: quite the opposite. As our taxi-driver told us: 'People could kill to get a job here'. The reason? The best-paying manufacturing jobs in the state, 14–15 dollars per hour, plus health and other benefits, which in the United States means a middle class standard of living. Other unskilled work in the state was paid 5–7 dollars an hour, with few or no benefits. As Bluestone and Harrison have demonstrated, the availability of well-paid jobs for non-professionals in the US has decreased dramatically since the 1970s. Reports in the American business press have supplied ample evidence of the eroded standard of living facing students who finish high school in the early 1990s, compared to the earnings their parents enjoyed.[4]

It is essential to stress these basic preconditions, since they are so often left out of the story. Again, this does not mean that factory management is unimportant. It is difficult to evaluate the relative contribution of the different factors, but there are some indirect measures in John Krafcik's international assembly plant comparisons. In the paint shops, manufacturability and control of complex supplier chains are of less significance than in the welding and final assembly sections. Even so, the 'paint shop productivity' of the American-owned U.S. plants lagged behind Japanese plants by 24 per cent, and the European firms lagged behind Japanese productivity by 100 per cent. Corresponding figures for the final assembly process, where all these factors are important, were considerably higher. Here U.S.-owned plants lagged behind Japanese productivity by 61 and the Europeans lagged behind the Japanese by 161 per cent. (Krafcik 1989).[5]

Job Security and Egalitarian Profile

Perhaps the most contested issue of the transplants is their working conditions. According to the MIT authors, traditional mass production deepened the dualism of industrial work, increased productivity but degraded the work content, and they ask (Womack *et al.* 1990: 100): 'Does lean production restore the satisfaction of work while raising living standards, or is it a sword even more double-edged than Ford's?' Their answer is unequivocal. Lean production makes everyone feel good. Unfortunately they do not substantiate these happy claims with any empirical evidence. My answer is different. Lean production is indeed a double-edged sword and working conditions are deeply contradictory. This is succinctly demonstrated in the experience of the transplants.

On *the positive side* it is possible to list:

(1) Transplants offer more job security to their highly-selected workforce

than American-owned plants do. An important test of this occurred in 1988 when Nissan suffered from poor sales and large stocks of finished cars. In contrast to the customary response among U.S. – and Swedish – companies, the Nissan management did not lay off any workers but retained all its employees until sales improved again.

(2) Their profile is much more egalitarian than traditional American plants where differences between workers and managers, white and blue collar employees, are conspicuous. '*We hated those ties*' a Mazda engineer we met in Flat Rock told us, when describing her feelings towards management at the Ford plant where she formerly worked. Workers have to work hard at the transplants, but that is even more true for managers. Egalitarianism is, however, nc indigenous feature of Japanese companies. In Japan hierarchical status is very important and individualized competitive evaluation of all employees by means of the *satei* system has been a vital aspect of the personnel policy since the late 1960s. In this respect the Japanese plants in North America are not transplants, but have made an important invention of their own. The egalitarian profile of these factories has amazed Japanese researchers. The explanation put forward by Endo (1991a) is the presence of the UAW in the United States, and he argues that UAW has also influenced the policies of non-union plants. However, in Kentucky Toyota plans to gradually introduce a personal assessment system, thus making it a much more orthodox application of Japanese human resource management.

(3) The quality of the products is a central issue and many employees are justly proud of the achievements of their workplace. That was seldom the case when they worked for American companies. This new culture of 'working with pride' seems to encompass also workers who are otherwise critical of many aspects of the Japanese system, such as the high line speed and the unrelenting performance pressure.[6]

(4) Those who pass the exhaustive screening process feel proud of belonging to the selected. The book *Working for the Japanese* (Fucini 1990) reports that according to the employees the high quality of the workforce was the most positive feature of the Mazda plant. The heavy emphasis on team-based problem-solving in the screening process is a social advantage in everyday work. The transplants have also raised the calibre of management considerably. That did not fail to impress the former GM workers at NUMMI and contributed to their acceptance of the new system.

(5) At the transplants, production and problem-solving on the shop floor have top priority. Problems in production involve the entire organization and everybody is supposed to contribute to their solution. At its best, Japanese management has a very methodical and systematic approach to discover defects and problems, to encourage suggestions, to implement

countermeasures and to evaluate the results. This capacity has led Paul Adler to coin the term 'learning bureaucracy' for the Toyota/NUMMI organization and he goes on to emphasize that the 'intensely bureaucratic, formalized and proceduralized work process' of the NUMMI system served 'the purpose of organizational learning. Standardized work, for example, captured learning by codifying best practice and workers were encouraged to constantly improve on this best practice' (Adler 1991: 58–59).

In MacDuffie's work, problem-solving on the shop floor plays a central theoretical role, and he has a lengthy discussion of the integration of conception and execution, of production and innovation, and of the alleged holistic division of labour, which is supposed to characterize the transplants (MacDuffie 1991: 57,61,75). He also conducted a field study of problem-solving activities in relation to recurrent quality issues at a GM, a Ford and a Honda plant. The Ford and GM studies are rigorous and highly critical. The Honda study, alas, is rather elusive and selective. At the American plants the problems in the relation between design department and manufacturing is a central point of analysis. In the Honda study, this relation is not even mentioned. For good reason – the plant had basically no input in the design activities. Nevertheless, Honda as well as several other transplants have put impressive employee suggestion programs in place. A key element is management attention and feedback. According to reports in *Automotive News* (27 April, 1992), Honda's CEO tours the plants every other month to discuss accepted ideas. When feasible, workers are paid overtime to implement the suggestions themselves. At the Kentucky plant, too, management support for employee suggestions is salient. Top managers are reportedly discussing bigger ideas with the workers who suggested the proposals every month. In 1991, this plant implemented more than 20,000 employee suggestions, totalling six ideas per employee. At Ford North America, the figures are reversed (six employees per idea). CAMI, the GM-Suzuki joint-venture is the leader of the pack. In 1990 its 2200 workers submitted 100,000 suggestions, and according to management the plant had started to implement 95,000 of these ideas within half a year.

Management attention and elaborate remuneration systems, that stimulate many small ideas rather than a few big ones, are not the only reasons for these staggering levels. Most of the plants are new, and have many kinks to sort out, and workers are highly selected. Furthermore, in the Toyota system of standardized work, an individual employee is not entitled to change anything by him/herself. In a less strictly controlled system, workers alter the location of materials or re-sequence various operations without informing management. Under the Toyota system they have to submit a formal suggestion before considering any change. That is a significant factor in driving up the numbers.

Lean Production – and Mean

Thus the manufacturing culture of the transplants has several qualities that are attractive to American auto workers. Unfortunately there is another side of the sword too: the unlimited performance demands, the long working hours and requirements to work overtime on short notice, the recurrent health and safety complaints, and the rigorous factory regime that constitutes a new and very strict regime of subordination.

Relentless performance demands

Transplants do not recognize any union regulation of performance demands or other limitations on management's discretion to organize work. Using the *kaizen* technique all slack is eliminated. In the GM car factories that we visited in 1990, the work pace was relatively relaxed. That was the case even at plants, like Buick City, that had achieved high productivity and quality. People had time to talk to visitors and do some reading at their work stations. These things are unthinkable at Japanese transplants. According to the Japanese view, if workers are occasionally able to read a magazine at work, that does not only signify waste (*muda*), but also that workers will lack the motive force to continually make proposals for improvements. The study group from the Swedish Metalworkers summarized their visit to Toyota's Kentucky plant in 1992 by the observation that no one seemed to be idle for one single moment, everybody was working or walking.

Richard Wokutch, author of *Worker Protection, Japanese Style* made a similar reflection when he visited Mazda in Japan: 'Production workers devote their complete attention and tremendous energy to their jobs. There is no horseplay; indeed, workers rarely smile or even talk to one another except about work-related matters. When the line stops for some reason, workers immediately begin cleaning around their work stations, performing maintenance on their tools and equipment or engaging in other work-related activities.' An American trainee sent to the same plant to prepare the start-up of Mazda's operation in Flat Rock observed: 'These people when they go to work they're there to do their job. They don't hardly talk. The little guy that I was working with, I had to almost break his neck to make him say "hi" to me because he's so busy concentrating on his job' (Wokutch 1992: 134,167).

Unlimited Working Hours

In a fundamental sense, lean production is not free of buffers. Long and flexible working hours are the hidden buffer that is utilized if necessary.

The amount of overtime work, often ordered on very short notice, was high in all transplants. The far-reaching management discretion to determine working hours means that, in principle, production quotas will be reached irrespective of what happened during the day or on the shift. This is also an instrument to force the pace of improvements in production. If interruptions require that workers must stay past normal working hours, employees' interest in preventing the recurrence of such interruptions increases. The eight-hour day, a goal for more than a century in the West, is hard to fit into the logic of lean production. 'Toyota expects its employees to be totally committed' observed the Swedish Metalworkers when they visited the Kentucky plant: 'Family, children, etc. should preferably be taken care of by someone else. One example is the way overtime is ordered. Ninety minutes before the end of the shift every employee is told by a big display if s/he has to work overtime and if so, for how long' (Swedish Metalworkers 1992: 60).

In Japan the absolute priority of production is facilitated by the gender division of labour. Regular auto workers are men and their wives take the sole responsibility for the family. Even so the auto workers union, JAW, has become increasingly critical of the long working hours. On average Japanese auto workers work more than 400 hours overtime a year, yielding a typical work-year of 2,300 hours. According to statistics presented by Koshi Endo, the total working time, including 'voluntary' activities such as QC meetings and 'hidden' (non-reported) overtime, is even longer, and often exceeds 2400 hours, fifty per cent more than the normal German work-year (Endo 1991b).

In the US a much larger proportion of women than in Japan are regular workers. At Mazda, for example, female workers made up more than 30 per cent of the workforce. The relentless demand to fulfil production quotas creates even more stress than in Japan. When women work so much overtime, who takes care of the family and the children?

Working with Pain

Japanese plants place considerable emphasis on safety and the avoidance of accidents that can interrupt production. The products are designed for easy manufacture, and great precision characterizes the making of parts. Nevertheless, the sheer repetitiveness of the fragmented jobs, combined with the intense pace and long working hours, lead to significant health risks, above all cumulative trauma disorders (CTD) or repetitive strain injuries (RSI). Incidentally these are not recognized as an occupational injury in Japan. As Wokutch has observed, it is very difficult to discuss the correlation between line speed and health hazards in Japan: 'Most Japanese workers and managers are not willing to talk about the safety and health implications of the work pace for fear of sounding as though they are criticizing the

Japanese work ethic. Complaining would entail a loss of face' (Wokutch 1992: 73).

At Mazda in Flat Rock, there were early reports of an unusually high incidence of the carpal tunnel syndrome, damage to the nerves and tendons in the hands and wrists. In 1989, the total number of work-related injuries was three times higher at Mazda than at comparable American plants (*Detroit Free Press*, 7 July, 1990). In 1990, reports of CTD fell 38 per cent (from 95 to 60 cases), and in 1991 the incidence fell a further 14 per cent. Management attributed the reported reduction to its ergonomics program and improved tools. The union however, claimed that company-induced fear and intimidation were the real reasons for the reported decline (*Automotive News*, 9 December, 1991). At Honda in Ohio, workers were worried about the rapid line pace already in the mid-eighties. In a 1986 interview a worker summed up the situation: 'If it doesn't get you physically, it will mentally sooner or later' (Krause 1986: 44). The company was proud of never having dismissed any workers; employees wondered what job security was worth if they would not be physically able to work past forty or fifty years of age. During our visit to the engine factory in Anna in 1990, managers did not acknowledge that working conditions caused any work injuries at all; instead, they claimed it all depended on the individual: 'There are strong and weak persons. There are right and wrong attitudes.'

Mazda has taken the same approach and systematically disputed workers' compensations claims.[7] The Detroit press has reported a number of cases when injured employees have been dismissed on the ground that there is no suitable work for them: 'Mazda worker Brian Blanton said he is a perfect example of how Mazda treats workers with disputed work-related injuries. Blanton said he injured his back on the job in 1988, had back surgery in 1989 and returned to work with doctor restrictions. "Mazda followed the restrictions for about half a day and then ordered me to do work that involved a lot of bending and lifting, which killed my back", Blanton said. "I told them I couldn't do it so they fired me for disruptive behaviour and work refusal" ' (*Automotive News*, 9 December, 1991). According to critics, Nissan in Tennessee has produced several examples of the same policy: '"As soon as people are injured they have no use for them," says Hardin (a former foreman). "You take the best employee, a hard worker with a good attitude and say an elbow goes out from overwork. They'll say 'Get him the hell outta here'. It is hard for me to believe it, and I have seen it," ' (Junkerman 1987: 18).

The issue of work injuries and trauma disorders is complex. Wokutch carried out an extensive analysis of Mazda (referred to as 'Jidosha'), whereby he compared the Flat Rock plant with Mazda in Japan, with other transplants, and with American plants in the U.S. In the first set of comparisons Wokutch discovered that the illness rate and reported injury rate were dramatically higher in Mazda's U.S. operation than in Japan. One reason for this striking difference was a massive under-reporting of non-fatal injuries in Japan. This under-reporting was indicated by the high proportion of fatalities (which are likely to be reported more accurately) to total lost-work cases in Japan, about 6.5 times higher than in the U.S. for the period 1983–87. 'Soft-tissue disorders', such as CTDs, are typically not regarded as legitimate work injuries in Japan. As an illustration, Wokutch cited an episode from Mazda in Hiroshima. When Mazda's American safety and health director visited this plant he observed a production worker wincing with pain every time he bent over at his work station. He pointed this out to his hosts but their response was only 'Oh, that's nothing' (Wokutch 1992: 192). However, Wokutch also thought that the real level of work injuries was lower in Japan than in America. This he attributed to the much greater worker compliance with rules and specifications in Japan. Worker protection in Japan has a strong behavioural orientation. Because of the obedience and strict discipline of the Japanese workforce this emphasis on correct behaviour is a successful strategy, at least in some areas.

Second, Wokutch examined the experiences of other Japanese transplants in the U.S.. Data was rather fragmentary, but there was no reason to believe that Mazda's plant was substantially out of line with other Japanese transplants (Wokutch 1992: 219). That rendered the Flat Rock operation even more important.

Third, the injuries at Flat Rock in 1988–1989 were analyzed from the perspective of comparable American-owned auto plants. Although Mazda performed relatively well on the dimension of lost workdays, its rates for total injuries and illnesses were far higher than the rates for U.S. auto plants. After a careful sifting of the evidence Wokutch summarized the results in the following way: 'The obvious conclusion is that the stresses of the production system do indeed make soft-tissue disorders more of a problem at Jidosha USA. The problem seems to go to the very core of the production system, which elicits maximum efficiency from both workers and machinery. Although accidents must be avoided at all costs in this production system, slow-developing conditions such as CTDs are evidently viewed as less of a threat' (Wokutch 1992: 195).

A Rigorous Factory Regime

By eliminating buffers, lean production increases management's dependence on employees and their contribution. In the *Machine* book the MIT authors represent 'trust and feelings of reciprocity' as the basis of the system. The elimination of traditional safety nets (buffers, etc) is, however, more than compensated for by the strict personnel selection and the new regime of subordination. Uniforms are compulsory, conduct and discipline codes are spelled out in detail, demands for attendance are absolute, the workplace is minutely regulated and all personal attributes are prohibited. In many respects the transplants involve a militarization of the factory regime. In a society plagued by disorder and delinquency this could be attractive for some employees – but it is far from the democratic quality associated with team-work in Western Europe.

The pressure this regime imposes on workers was demonstrated by events in Flat Rock in 1991. As a part of the new contract (see below) Mazda was forced to relax its perfect attendance policy. Workers were provided with four Paid Absence Allowance (PAA) days, which they could use at their own discretion simply by notifying their supervisor a few hours in advance. Despite the alleged 'trust and feelings of reciprocity', this new right very quickly became a kind of safety valve for many workers. As a result, production came to a stop on Fridays in some departments. To guarantee output without having to add manpower management tried to restrict the use of PAA days, especially on Fridays. In exchange the company offered substantial bonus increases. The workers voted no – the right to decide for themselves on one single issue was obviously too important to be substituted by money. Mazda then introduced the restrictions and the bonus unilaterally, but that is another story, which hardly contributed to feelings of 'trust and reciprocity' at the plant. It might be argued that the Mazda plant represents just bad or insensitive management, and thus cannot be construed as a case against the lean production system. An alternative explanation is that the site of the factory, in the outskirts of Detroit, is particularly unsuitable for the development of a new manufacturing culture in the auto industry. However these counter-arguments, by stressing its cultural and social contingencies, only add to the criticism of the MIT projection of 'lean production' as universally applicable. As a consequence both of Mazda's financial problems and its difficulties in handling the industrial relations, Ford purchased half of the Flat Rock plant in mid 1992. Mazda Motor Manufacturing (U.S.A.) was renamed Autoalliance International Inc., and Ford acquired control of the plant's human resource management.

The contradictory character of work experience at the transplants is borne out by a study of the Canadian CAMI plant, the joint venture between GM and Suzuki. Here a group of researchers and unionists, 'The Canadian Auto Workers Research Group on CAMI', has launched a longitudinal research program, consisting of field studies twice a year during a two year period. The first intervention took place in March 1990, the second in November the same year, when the plant had reached the stage of full production for one of its product lines. These two first field studies were reported to an international colloquium in Quebec in 1991 (Huxley, Wareham *et al*; 1991). On the one hand, the researchers found a consistently high level of participation in suggestion activities (71 per cent of the respondents in the second study) and a majority of workers supporting QC activities. On the other hand there was a deeply ambiguous assessment of the team concept. Workers appreciated its social qualities, but they also thought teams were a way to get people to pressure one another. The proportion of workers viewing teams in this negative way increased from 19 per cent in the first field study to 41 per cent eight months later. Also in the second round of observation, the research team discerned a growing overall disillusionment with CAMI philosophy. Of the interviewed workers 78 per cent argued that CAMI was a factory where management still had all the power. This widespread resentment forms an important backdrop to the strike at CAMI in September–October 1992, which is discussed below.

Unions – Increasing Disillusion

In their study of Nissan's essentially non-union plant in the UK, Garrahan and Stewart point out that the new factory regime is not simply imposed by management but is actively maintained by employees themselves. They pose the question: 'What is the system of legitimacy that allows people not just to be pushed to and beyond the limit, but actually to take responsibility for the failures of the system?' An important part of the answer seems to be Nissan's elaborate 'participative' practices, where 'workers implicate themselves by participation in discipline. In fact, by disciplining others when pointing out faults, they discipline themselves. In such a way, a whole system of self-subordination begins to develop' (Garrahan and Stewart 1992: 116, 138). To prevail, such a system is highly dependent upon the presence of a well-diffused company consciousness and, by the same token, an absence of independent trade unionism that could provide an alternative, collectivist, interpretation of production problems and worker experiences.

These observations make it particularly interesting to study the development at the unionized transplants in America. At the start, all union locals adopted a very cooperative policy. They represented a highly-selected workforce, proud to be there and anxious not to jeopardize their well-paying jobs. The careful screening procedure certainly did not favour union militants, anyway. Moreover, at the national level, UAW viewed far-reaching cooperation with the Japanese as vital to get access to the new breed of plants, and argued very strongly against any revival of adversarial attitudes. The labour-management collaboration and teamwork at NUMMI was heralded by Solidarity House as the future, and an embodiment of central union aspirations. And yet, with increasing experience, resistance and criticism of the new production methods seems to be growing in all union locals.

> Last year, when we were negotiating the first contract, people told us not to be intransigent. The main thing was the jobs and the employment security. They

wanted to believe in the company, just as we did. Today the attitude is altogether different. People do not trust the company any more, even if they tell the truth. (Don Shelby, President of UAW Local 2488, Diamond Star, November 1990)

Many of us came to CAMI naive as to how a plant functions, and had no reason to question the CAMI plant system. We all wanted CAMI to be the employment Utopia described by the employee handbooks ... When I became Vice-President the previous winter I did so because I didn't want a bunch of Union hot-heads running the Local, doing nothing but running management down and bad mouthing everything we had worked so hard to establish as 'The CAMI Way'. In the interim, I experienced and heard about as much reality as I could take, until I realized what a smokescreen it all was. I became exactly what I had hoped to protect this Local and this Company from.
(Rob Pelletier, President of Local 88, CAMI, in *Off the Line*, 5/1990.)

Union Politics at Mazda

The development of industrial relations at Mazda, Flat Rock has been extensively documented in the work of Fucini (1990). The plant started with a very cooperative, not to say acquiescent, union leadership which was put in place by the UAW region. As result of rapidly-growing worker resentment against the new management methods, these UAW-appointees were ousted in the first local elections and replaced by a more militant leadership. At the time of our visit (November 1990) the new president had recently been re-elected and was furiously preparing the negotiations of the new contract. Some time later, 90 per cent of the workers voted to give this leadership the right to call a strike if negotiations stalled. The new Mazda contract was finalized in March 1991 and constitutes the first case where a union, relying on strong membership support, was able to influence and modify the 'lean principles'. As such it was an important reference point in the 1992 contract negotiations at Diamond Star and CAMI, the two other unionized transplants in the North Midwest (CAMI in the Canadian province Ontario is located less than 150 miles from Flat Rock). The most significant novelties of the Mazda contract were:

- More union influence in company decisions about introduction of new technology, outsourcing of work and use of outside contractors;
- Improvements of the union's position regarding health and safety, such as the establishment of a written health and safety grievance procedure, the addition of another full-time health and safety representative and a full-time ergonomics representative, a joint ergonomics training program and union access to information such as symptoms surveys, etc.
- Elimination of the Support Member Pool (temporary employees), and an agreement that temporary employees can only be used if there is mutual agreement between the union and the company about their use

and number and that they will not be used to avoid hiring regular full-time employees.

This is Detroit – but how about NUMMI? Since its inception in the middle of the 1980s, this plant has been operating in a strongly cooperative way, supported by a solid labour-management consensus. Nevertheless, within the union there has been a strong opposition criticizing the high line speed and the constant pressure to work harder and faster, not just smarter. For several years this opposition, 'The People's Caucus', enjoyed a majority in the assembly department. In August 1991 the critics won the majority in local elections. It was no easy task for the new leadership to implement changes in the system, however. When the Swedish Metalworkers visited the plant in 1992 they found the work pace very demanding, the job content extremely restricted and the physical work environment deficient in a number of ways.

CAMI 1992 – The First Transplant on Strike

Of particular importance for the future of industrial relations at the transplants is the development of CAMI in 1992. Following a period of growing tension in conjunction with the new contract negotiations, the CAMI workers walked off the line on 14 September, not to return until 18 October. The Canadian workers wanted an improvement in their compensation to achieve the same wages and benefits as other unionized car workers in Canada. But more important was an increasing general distrust of management. Since early 1991, 500 grievances had been filed at the plant. Workers complained about the inequalities in the treatment of managers and workers and criticized the factory regime, for example the company prohibition against reading newspapers or listening to the radio while on break. A main issue was the vague contract language: '"We're going to have to go back and spell out the rules in a collective agreement," Pellerin [the Canadian Auto Workers' national service representative] said. "No more of this trust and cooperation business"' (*Automotive News*, 21 September, 1992). The contest ended in important union gains, not only in compensation. The new contract included a number of other provisions, such as greater union input in health and safety issues, an expanded training programme and the establishment of a replacement pool for both long and short term absences. When the results of the negotiations were presented for approval or rejection at a union meeting, 1900 members (out of a total of 2300) showed up. At this meeting, the announcement that team leaders now should be elected by the teams, and not appointed by management, was greeted by the biggest cheer. The second most popular announcement was the new language of 'workplace dignity', allowing personal radios during breaks and lunch and emphasizing that personal sweaters and sweatshirts would remain the worker's option, an effective end to management's attempt to enforce uniforms.[8]

Union locals can hardly change the fundamental logic of the new production system. In view of the MIT perspective, that the lean orthodoxy must be accepted piece and parcel, lock, stock and barrel, the modifications they can effect might prove quite significant, however. Most important, the Mazda and CAMI experiences testify that participative management and a highly selective recruitment do not preclude the development of broad union consciousness and mobilization. This will call for a further adaptation of Japanese practices at these plants, and exercise increasing pressure on the nonunion transplants. When UAW in 1989 tried and failed to organize Nissan, Tennessee, management used the NUMMI experience to convince workers that paying union fees would not make any real difference. The new CAMI agreement has enhanced union credibility, and will not fail to impress employees at the so far non-organized Toyota and Honda operations in the Ontario province.

Japan: Automotive Work as 'san-kei': Hard, Dirty and Dangerous

Crucial for the future of auto work are developments in Japan. In the late 1980s the Japanese labour market became very tight, which created increasing difficulties for auto-makers in recruiting young workers. One important reason was that auto work has acquired the reputation of being 3K ('san kei'), that is Kitanai (dirty), Kitsui (hard) and Kiken (dangerous). These negative attitudes are not confined to new entrants to the labour market. According to a large survey carried out by JAW (the national federation of the unions within the auto industry), very few employees would recommend their children to get a job in the automobile industry – of all respondents 43 per cent answered no, 43 per cent found the question difficult to answer, and only 4 per cent gave a positive response. The main reasons among production workers for the reluctance to recommend the auto industry were the following: too low wages compared to the hard work, too much overtime and holiday work, too much shift and night work and too intensive work (JAW 1989: 29). These results were supported by another workplace survey published in 1991. According to this, two thirds or more of the production workers in the auto industry reported too tight manning and too much work, too much overtime and too many difficulties in utilizing paid holidays. Sixty-seven per cent were not satisfied with their working environment, 62 per cent regarded their work as excessively routine and 72 per cent did not find that the company paid enough attention to human resource development (JAW 1991: 50). These results once again demonstrate the biased and one-sided character of MIT's *Machine* book. The book has been sharply criticized by JAW on account of the authors' total neglect of the long working hours Japanese employees are forced to work with no relief in sight. JAW also resents the short product cycles, which are seen as a drain on human and natural resources. The vaunted just-in-time-production system is widely criticized for its detrimental social effects, and has been blamed for urban congestion and pollution. For example, in 1991 a special committee to the Ministry of International Trade and Industry blamed the just-in-time system for overloading the delivery system, creating traffic problems in cities and labour problems nationwide by forcing an insufficient number of drivers to work excessive overtime (*Automotive News*, 2 September, 1991 and 25 March, 1991).

After decades of stunning productivity improvements, Japanese workers demand a better life. JAW urges the automakers to reduce overtime from the industry average of 2,300 hours to 1,800 per year, to lengthen the product cycles and reduce the number of model variations, and argues for less pressure in the workplace and less coercive demands on parts makers, dealers and transport companies (*Automotive News*, 24 February, 1992).

The car makers' recruitment problem is not a transient phenomenon, which will disappear in a time of slower economic growth. According to labour market economist Haruo Shimada, Japan is experiencing a profound demographic change, and there is a long-term shrinkage of the younger end of the labour force. Further, there is a structural change in Japanese values. Affluent and highly educated youngsters increasingly shun dirty and repetitive jobs. As Shimada emphasizes: 'Postwar Japan's rapid economic growth was sustained by the availability of large numbers of young workers who wanted to work and were willing to work for rather low wages' (Shimada 1991: 9). In the 1990s companies will have to accommodate a very different situation. Initially Japanese automakers responded to this pressure in two principal ways – by stepping up their international expansion, and by sharply increasing their investments in automation, seeking a technological solution to the labour problem. The traditional Toyota model of low cost rationalization based on continuous shop floor improvements which is at the heart of the NUMMI success tends to be superseded by a more divided work organization, allowing for a much stronger emphasis on professional specialists, for example permanent and specialist kaizen teams.[9] When capital costs started to soar in the early 1990s this technological strategy became much less feasible. Leading auto executives acknowledged that the development of automation had come to a turning point, and could not be the sole answer to the problems (*Automotive News*, 30 March, 1992).

Already in 1982 the Toyota union had started to demand improvements in the working conditions in the company's assembly plants. They had to wait until 1989 before a process of real change started. Then Toyota began experimenting with so called 'ideal Kumis' (workshops), where the line design was modified in order to achieve a more even workload and new systems of skill formation were launched. In 1991 Toyota opened a new Tahara plant, which is regarded as a pilot factory for the development of a new assembly concept. The motto of the plant is 'factory friendly to the worker'. The degree of automation is high, work shops are spacious and buffers are used to make production more flexible and to improve working conditions. According to Nomura (1992: 12) the heralded 'zero stock philosophy' has been revised, and in the new plant a certain kind of *muda* (waste) is regarded as necessary to alleviate the rigours of assembly work.

Nissan has embarked on the same route. Its new Kyushu plant came on stream in May 1992, and has been presented as a 'dream factory' (*Automotive News*, 13 July, 1992). The conveyor belt is eliminated and each car sits on its own dolly, which can be raised and lowered at every work station to create an optimal working position. The plant is air-conditioned, spacious, and has a low noise level. In many respects is seems to be similar to Volvo's Swedish Kalmar plant – although, as at Kalmar, work is still demanding and bound by the clock. Nevertheless this plant, as well as Toyota's Tahara operation and Mazda's new, highly automated Hofu plant, signify a novel direction in the Japanese automobile industry. All these new facilities address the same two problems – to increase flexibility in order to cope with the increasingly fragmented market, and to create attractive work places, that could safeguard a high worker commitment in the future.

'Lean production' is certainly not the ultimate station of industrial development. Fortunately, history seems both able and keen to provide us with new surprises.

Notes

1. In a study of Nissan's operation in Sunderland, UK, Philip Garrahan and Paul Stewart (1992) have taken on this brave new world theme, so salient in the post-Fordist discourse. By concentrating on the regime of subordination they thoroughly debunk the notion of employee empowerment in lean production and highlight the 'sublimation' of class conflict into peer competition and intergroup rivalry. This aspect of the new management system is of vital importance, but will not be pursued in this article.
2. This study was conducted with two other researchers from the Swedish Royal Institute of Technology, Ernst Hollander and Torsten Björkman. For a full report, see Berggren, Björkman and Hollander (1991).
3. A comprehensive analysis of the various strata of these pyramids is provided by Smitka (1990).
4. See for example 'What Happened to the American Dream?', *Business Week*, 19 August, 1991. This article portrays a couple, Troy and Linda, both working in the service sector. Together they earned 44 per cent less, adjusted for inflation, than Troy's father alone earned as a bluecollar worker of the same age.

5. According to the same study, a Japanese plant spent 1.6 hours/car in the paint shop and 5.0 hours/car in the assembly area. A roughly equivalent German plant spent 9.4 hours/car in the paint shop and 25 + hours/car in the assembly area.
6. Lowell Turner provides some very interesting examples from NUMMI, where members of the oppositional 'People's Caucus' spoke proudly of their high quality performance, and eagerly promoted the products from the factory (Turner 1989: 22–23).
7. Several early cases were reported in *Automotive News*, 13 February, 1989.
8. Dave Robertson, CAW Headquarters Toronto (personal communication, 10 October, 1992).
9. This tendency was reported by several Japanese specialists at the symposium 'Production Strategies and Industrial Relations in the Process of Internationalization', Sendai, Japan, 14–16 October 1991.

References

Adler, P. (1991) '*The 'learning bureaucracy': New United Motor Manufacturing, Inc.*' Draft. School of Business Administration, Univ. of Southern California forthcoming in B.M. Staw and L.L. Cummings (eds.) *Research in Organizational Behavior*, Greenwich, CT: JAI Press.

Automotive Industry Authority (AIA) 1992, *Report on the State of the Automotive Industry 1991*, Canberra: Australian Government Publishing Service.

Automotive News 1989–1992:

13 February, 1989. 'Injury, Training Woes hit new Mazda Plant'.

28 January, 1991. 'UAW loses 3rd attempt to organize'.

25 March, 1991. 'Just-in-time deliveries clog Japanese highways'

2 September, 1991. 'Fast-paced Japanese hit the expansion redline'.

9 December, 1991. 'Mazda, workers disagree on reasons for drop in injuries'.

24 February, 1992. 'Makers vow to cut hours of Japan workers'.

30 March, 1992. 'Debt leaves makers almost broke'.

27 April, 1992. 'Plant floor is fertile soil for ideas'.

13 July, 1992. 'Factory Fantasia. New plants address Japan's reality – few workers many models'.

21 September, 1992. 'CAMI strike wilts transplant rose'.

Australian Manufacturing Council (1990) *The Global Challenge*, Melbourne.

Berggren, C., Björkman, T. and Hollander, E. (1991) *Are they unbeatable? Report from a field trip to study transplants*. Stockholm: Dept. of Work Science, Royal Institute of Technology & Sydney: Centre for Corporate Change, Australian Graduate School of Management.

Bluestone, B. and Harrison, B. (1989) *The Great U-turn*, New York: Basic Books.

British Department of Trade and Industry/PA Consulting Group (1989) *Manufacturing into the Late 1990s*, London.

Business Week 1989–1992:

14 August, 1989. 'Shaking up Detroit'.

19 August, 1991. 'What Happened to the American Dream?'

22 April, 1992. 'Will Nissan get it right this time?'

Dertouzos, M.L., Lester, R.K., Solow, R.M. and the MIT Commission on Industrial Productivity (1989) *Made in America: Regaining the Productive Edge*, New York: Harper Perennial.

Detroit Free Press, 7 July, 1990. 'Danger rises in new auto jobs'.

Endo, K. (1991a). *'Satei (Personal Assessment) and inter-worker competition in Japanese firms'*. Paper presented at the symposium Production Strategies and Industrial Relations in the Process of Internationalization, Sendai 14–16 October 1991.

Endo, K. (1991b) *Working hours in Japan*, Yamagata: Dept. of Economics, Yamagata University. Mimeo.

Fortune, 15 June, 1992. 'How Japan got burned in the U.S.A.'.

Fucini, J. and S. (1990) *Working for the Japanese*, New York: The Free Press.

Garrahan, P. and Stewart, P. (1992) *The Nissan Enigma*, London: Mansell Publishing Ltd.

Huxley, Wareham *et al.* (1991) *Team Concept: A case study of Japanese production management in a unionized Canadian auto plant.* Paper presented at the Université Laval in Quebec, August 1991.

Ikeda, M. (1991) *Development network in the automobile industry – new developments.* Paper presented at the symposium 'Production Strategies and Industrial Relations in the Process of Internationalization', Sendai 14–16 October 1991.

JAW, Confederation of Japanese Automobile Workers' Unions (1989), *Report on consciousness of union members* (in Japanese), Tokyo. Mimeo.

JAW (1991) 'Report on the seventh survey of consciousness of union members' (in Japanese), *Rodo Chosa (Labour Research)*, June, 32–70.

Junkerman, J. (1987) 'Nissan, Tennessee', *The Progressive*, Vol. 51: 6, 16–20.

Krafcik, J. (1989) *Explaining High Performance Manufacturing: The International Automotive Assembly Plant Study*, Houston: Competitive Manufacturing Research. Mimeo.

Krause, K. (1986) 'Americans can build good cars', *Washington Monthly*, July–August, 41–46.

MacDuffie, J. (1991) *Beyond Mass Production. Flexible Production Systems and Manufacturing Performance in the World Auto Industry.* PhD Thesis submitted at MIT/Sloan School of Management, February 1991.

Milkman, R. (1991) *Japan's California Factories – Labor Relations and Economic Globalization*, Los Angeles: Institute of Industrial Relations, University of California.

Nomura, M. (1992) *Farewell to Toyotism'? Recent Trend of a Japanese Automobile Company*, Okayama: Dept of Economics, Okayama University. Mimeo.

Porter, M. (1990) *The Competitive Advantage of Nations*, London: MacMillan.

Smitka, M. (1991) *Competitive Ties: Subcontracting in the Japanese Automobile Industry,* New York: Columbia University Press.

Sakai, K. (1990) 'The feudal world of Japanese Manufacturing', *Harvard Business Review*, Nov–Dec, 38–49.

Sei, S. (1991) *'Is technical innovation all? A hidden meaning of social relationships behind the product development stage in Japanese automotive industry.'* Paper presented at the symposium 'Production Strategies and Industrial Relations in the Process of Internationalization', Sendai 14–16 October 1991.

Shimada, H. (1991) 'Japan's Changing Labor Market', *Journal of Japanese Trade & Industry*, 4, 8–11.

Shingo, S. (1981) *The Toyota Production System*, Tokyo: Japan Management Association.

Sloan, A.P. (1963) *My Years at General Motors*, Garden City, New York: Doubleday.

Swedish Metalworkers' Union (1992) *Japanska produktionskoncept i Nordamerika* (Japanese production concepts in North America), Stockholm.

Turner, L. (1988) *NUMMI in Context*, Berkeley: Department of Political Science, University of California.

Westney, E. (1989) 'Internal and external linkages in the MNC: The case of R&D subsidiaries in Japan', in C.A. Bartlett, Y. Doz and G. Hedlund (eds.) *Managing the Global Firm*, London: Routledge.

Wokutch, R. (1992) *Worker Protection, Japanese Style*, Ithaca, New York: ILR Press.

Womack, J.P., Jones, D.T. and Roos, D. (1990) *The Machine that Changed the World*, New York: Macmillan.

Royal Institute of Technology
Department of Work Science
S100 44 STOCKHOLM
SWEDEN

[15]

Lean Production and the Toyota Production System – Or, the Case of the Forgotten Production Concepts

Ian Hampson
University of New South Wales

Advocates and critics alike have accepted 'lean' images of the Toyota production system. But certain production concepts that are integral to Toyota production system theory and practice actually impede 'leanness'. The most important of these are the concepts of *heijunka*, or levelled ('balanced') production, and *muri*, or waste from overstressing machines and personnel. Actual Toyota production systems exist as a compromise between these concepts and the pursuit of leanness via *kaizen*. The compromise between these contrasting tendencies is influenced by the ability of unions and other aspects of industrial relations regulation to counter practices such as short-notice overtime and 'management by stress'.

Introduction

From the late 1980s, debate around work organization converged on the concept of 'lean production' which was, according to its advocates, a 'post-Fordist' system of work that is at once supremely efficient and yet 'humane', even democratic (Kenney and Florida, 1988: 122; Adler, 1993; Mathews, 1991: 9, 21; 1988: 20, 23). However, critical research found that, rather than being liberating, lean production can actually intensify work to the point where worker stress becomes a serious problem, because it generates constant improvements (*kaizen*) by applying stress and fixing the breakdowns that result. It thus attracted such descriptions as 'management by stress', 'management by blame', 'management by fear' (Parker and Slaughter, 1988; Sewell and Wilkinson, 1992; Dohse et al., 1985).

Economic and Industrial Democracy © 1999 (SAGE, London, Thousand Oaks and New Delhi), Vol. 20: 369–391.
[0143–831X(199908)20:3;369–391;009459]

Ironically, at the same time as lean production was becoming the latest fad in the West, there was an urgent debate in Japan about the quality of work (Berggren, 1995; Benders, 1996). The phenomenon of *karoshi* – death from overwork – was seen as evidence of the way the pursuit of 'leanness' stressed workers, causing highly deleterious social consequences (NDCVK, 1991; Kato, 1994; Nishiyama and Johnson, 1997). The normally pliant Japan Auto Workers' Union produced a report critical of working conditions in the national vehicle assembly industry (Sandberg, 1995: 23). As a result of this and other factors, some key Japanese auto producers retreated from 'leanness', and set up exemplary models of work that were interesting hybrids of the Toyota production system and 'humanized' work principles (Benders, 1996; Shimizu, 1995; Berggren, 1995). Critical researchers have also suggested that the way forward for the organization of work might be to combine elements of lean production with principles of 'humanized work' (Berggren, 1992: 16, 232, Ch. 13; Sandberg, 1995: 2). Such combinations are conceivable, and the Japanese experiments question the assumed identity between 'lean production' and the Toyota production system.

This article argues that the focus on 'leanness' by critics and advocates alike has distracted the gaze of researchers from certain 'anti-lean' concepts contained within Toyota production system theory and practice. The article also suggests that the degree of 'leanness' in particular plants is shaped by surrounding institutional frameworks. Management strategy will seek to move down the lean path ('doing more with less'), while conditions in the labour market, industrial relations and the resources of affected unions may limit leanness. The first section surveys the dominant 'lean' image of the Toyota production system. The second section explicates the Toyota production system's 'forgotten production concepts'. The third section explores the mechanisms by which a particular plant's degree of 'leanness' comes to express the balance of forces in the industrial relations system and its political-institutional surrounds. The section also notes how shifts in this balance have driven a certain 'humanizing' of the Toyota production system in Japan, and canvasses the interesting case of Australia, where characteristics of the industrial relations system deflected the test case Toyota Altona plant from paradigmatic 'leanness'.

The Toyota Production System: 'Lean Production' or 'Management-by-Stress'?

The report of the MIT project into the world car industry attributed Japanese economic success to 'lean production' (Womack et al., 1990). This term quickly shaped images of the Toyota production system, for advocates and critics alike. Resonating with athletic imagery, later purveyors of the concept would emphasize 'agile' production, while critics would emphasize the deleterious effects of 'management by stress' and the dangers of corporate anorexia.

The Toyota production system was developed in the post-war period, and owes much to the production engineer Taiichi Ohno, himself a formidable advocate of 'leanness' (see Ohno, 1988: 44–5). By the late 1980s, owing in part to some deft marketing efforts, 'lean production' became the focal point of the debate about work organization, and was portrayed as nothing less than the future of work.

> Lean production is a superior way for humans to make things. It provides better products in wider variety at lower cost. Equally important, it provides more challenging and fulfilling work for employees at every level. . . . It follows that the whole world should adopt lean production, as quickly as possible. (Womack et al., 1990: 225)

Lean production is lean, its advocates argue, because it uses 'less of everything', even as little as half (Womack et al., 1990: 13). While this is certainly an exaggeration (Williams et al., 1992; Unterweger, 1992: 3), the claim that lean production could combine efficiency with quality of work life quickly became widely accepted. 'Lean production' became seen as 'best practice' (e.g. PCEK/T, 1990: Ch. 4; Dertouzos et al., 1989). The principles were transferred to other countries, as Japanese auto and other producers shifted production facilities overseas, where they met a mixed reception.

Many accounts of the Toyota production system accord centrality to the concept of *kaizen* or 'constant improvement' (Womack et al., 1990: 56; Oliver and Wilkinson, 1992: 35; Fucini and Fucini, 1990: 36, Ch. 3). Improvement means the removal of all activities that do not add value, which are defined as waste, or (in Japanese) *muda*. The concept of *muda* can refer to excessive set-up time, excessive inventory and work in progress, defective materials/products that require rework or repairs, cluttered work areas, overproduction, unnecessary motions, too much quality (overspecification), double handling in conveyance of materials and, above all, idle time (Oliver and Wilkinson, 1992: 26; Monden, 1994: 199–200). Monden (1994: Ch. 13) also refers to such *muda* as *seiri* – or 'dirt', and the removal of *muda* is thus a kind of cleansing (*seiso*). An important catalyst to *kaizen* is 'just-in-time' (JIT) production. Contrasting with allegedly traditional western approaches which accumulate stocks of components, JIT means producing only what is needed, as nearly as possible to when it is needed, and delivering it 'just in time' to be used (Monden, 1994: Ch. 2; Oliver and Wilkinson, 1992: 28).[1] *Kaizen*, 'leanness' and JIT converge on the mythological 'zero-buffer' principle. Buffers permit linked production processes to work at speeds somewhat independent of each other, and therefore enable workers to take short breaks, or to accommodate production irregularities without affecting adjacent production processes. Removing buffers makes visible production imbalances and other problems, prompting operators to fix them (Dohse et al., 1985: 129–30). Thus, the necessary counterpart of JIT production is *heijunka*, or 'levelled production' – a condition in which all parts of the overall production process are synchronized with each other. We presently return to the concept of *heijunka*, which, neglected in the literature, is a focal point of this article.

Kaizen not only seeks to eliminate errors in production, but also to locate their sources (Womack et al., 1990: 56; Ohno, 1988: 17). Workers' 'participation' is crucial, through monitoring and detecting any variations in process or product. Workers also contribute ideas about reorganizing and improving production, and this delivers productivity improvements through incremental innovation (Rosenberg, 1982: 60–6; Sayer, 1986: 53). This provides some basis for the claims that 'lean production' is 'participatory', post-Taylorist and post-Fordist (e.g. Kenney and Florida, 1988: 122; 1989: 137; Womack et al., 1990: 102). But work procedures are closely analysed and written down on standard operating procedure charts, which are displayed in the workplace, and which workers are required to follow closely (Ohno, 1988: 21). Changes to work procedures must be given assent by team leaders, and/or higher management. Standardized work provides the baseline for further improvement, and the charts, which record innovations to the work process, are a mechanism for 'organizational learning' (Adler and Cole, 1993).

The organization 'learns' by appropriating the innovations, sometimes driven by stress, which become standard practice. But the consequence is that workers cannot use their knowledge of the production process to protect themselves against pacing, since the system appropriates such knowledge. The Toyota production system can thus plausibly be portrayed as a solution to the classic problem of management – how to persuade employees to put their knowledge of the production process at the service of management – even where this means increasing their own workload (Dohse et al., 1985: 128). However, critics and advocates alike agree the system has the potential to cause stress.

> Most people . . . will find their jobs more challenging as lean production spreads. And they will certainly be more productive. At the same time they may find their work more stressful, because a key objective of lean production is to push responsibility far down the organizational ladder. Responsibility means freedom to control one's work – a big plus – but also raises anxiety about costly mistakes. (Womack et al., 1990: 14)

Taiichi Ohno, the architect of the Toyota production system, celebrates the role of stress. He once candidly described the thinking at the heart of his system in an interview.[2]

> If I found a job being done efficiently, I'd say try doing it with half the number of men [*sic*], and after a time, when they had done that, I'd say OK, half the number again.

Ohno described his 'philosophy' in these colourful words:

> There is an old Japanese saying 'the last fart of the ferret'. When a ferret is cornered and about to die, it will let out a terrible smell to repel its attacker. Now that's real nous, and it's the same with human beings. When they're under so much pressure that they feel it's a matter of life or death, they will come up with all kinds of ingenuity.

Critics too seek to capture the workings of the Toyota production system in the concept of 'management by stress'. As Slaughter puts it 'the management by stress system stretches the whole production system – workers, the supplier network, managers – like a rubber band to the point of breaking' (Slaughter, 1990: 10). Applying stress causes the system to break down, identifying sites where the production process can be redesigned, and improvements won (also see Dohse et al., 1985: 127–30). Reducing inventories keeps up the pressure for innovation, by denying the buffers that can provide some shelter from the pace of the line.

> *Kaizen* strips away layer after layer of redundant manpower, material and motions until a plant is left with the barest minimum of resources needed to satisfy its production requirements. The system tolerates no waste. It leaves virtually no room for errors. (Fucini and Fucini, 1990: 36)

The more resources are removed from the production system, the more fragile it becomes, making worker cooperation essential. Workers comply because of what Monden calls 'social conventions and institutions [that] can be called the social production system' (Monden, 1994: 336). These permit powerful management techniques. Total quality management (TQM) quickly traces problems to their source, be it mechanical or human. Dohse et al. (1985: 130–1) describe how workers having problems keeping up indicate that by pressing a button that illuminates a display. Management aggregates this information to indicate potential for staffing reductions. Sewell and Wilkinson (1992) note how the systems of quality control actually function as systems of *surveillance* and *discipline* (in a Foucauldian sense), promoting competition, humiliation and peer pressure. Workers are organized into teams, collectively responsible for a production area, and able to cover for one another in times of stress. This presupposes broad job descriptions, multi-skilling and cross-training, which effectively make workers *interchangeable*. Peer group pressure is mobilized against workers who 'let the team down' (Barker, 1993). For instance, when absent workers are not replaced, their colleagues have to pick up the slack, and they therefore police their workmates' sick leave.

In Japan, a complex system of payment and reward, subject to considerable managerial discretion, provides a powerful management control system. Firm-wide pay increases do not filter down to individual workers equally. Workers receive a component related to seniority, and another composed of bonuses related to the team's performance. Another component is allocated according to workers' 'merit'. Thus a wage rise could be from 85 percent to 115 percent of the amount allocated after cross-firm, seniority and team components have been allocated (Dohse et al., 1985: 139; Berggren, 1992: 132). The key figure allocating the 'merit' component is the frontline supervisor, who may also be the workers' union representative, taking a stint on the shopfloor before moving on to a career in management (Moore, 1987: 144).

Historians of the Japanese labour movement note that the Japanese unions were defeated in the postwar period, and structured into pliant enterprise unions, less able to defend their members against work intensification, and allowing the functions of union representation and managerial supervision to blur (Moore, 1987). The renowned practice of 'lifetime employment' and firm-specific training and career paths make for a lack of inter-firm mobility, in turn reinforced by the strong 'core/periphery' division in the labour market. Thus an employee leaving a long-term career in a core firm risks falling into the periphery of insecure and less well paid employment (Dohse et al., 1985: 133–41; Kumazawa and Yamada, 1989). All in all, 'lean production' contains considerable potential for the degradation of work, precisely because it *is* lean.

The Toyota Production System's Forgotten Production Concepts

As argued earlier, popular renditions of the Toyota production system emphasize the elimination of 'waste', with a tendency to focus on 'idle time'. But at least some forms of waste may be eliminated through more efficient production management, as opposed to work intensification (Monden, 1994: 177). Identification of 'waste' in a broader sense may even aid work 'humanization'. Three Japanese words capture a wider range of 'waste', or its sources, than is usual. Fucini and Fucini (1990: 75–6) make reference to the 'three Evil Ms'– *muda, muri and mura*, although no attempt is made to tease out the crucial interrelations between them. *Muri* translates as 'overburden – when workers or machines are pushed beyond their capacity' (Oliver and Wilkinson, 1992: 26), or 'the placing of excessive demands on workers or production equipment' (Fucini and Fucini, 1990: 75–6). This may reduce the production life of both human beings and machines. *Mura* is 'the irregular or inconsistent use of a person or machine' (Fucini and Fucini, 1990: 75), which might result from line imbalance or fluctuations in production pace, and which automatically results in some varieties of *muda*. This is because at least some workers and machines will be working below capacity for some of the time, perhaps while some others at bottlenecks are subjected to excessive stress, while yet others may overproduce. If one part of the production process is working at a low level of capacity utilization, while another is overworked, there is waste from both *mura* and *muri*.

The concept of *heijunka* means 'levelled', 'smoothed' or 'balanced' production, and one of its functions is to counter the kind of imbalance described earlier. Monden (1994: 8) refers to *heijunka* as 'the cornerstone of the Toyota production system'. This article suggests that the concept has been underexplored in the academic literature on the Toyota production system (also see Coleman and Vaghefi, 1994: 31). *Heijunka*, or 'levelled production', is a strategy to meet the demands of the market including fluctuations – while carrying as little work in progress stock as possible. Elimination of work in progress inventory offers savings to the firm in terms of the capital that would have been invested in it, in the space needed to house it, in the workers necessary to count it, in losses due to rust, depreciation and so on (Monden, 1994: 2). The emphasis that Toyota production system literature places on *heijunka* suggests that it may be a more fertile source of productivity gains than simply seeking to eliminate 'idle time'.

Achieving *heijunka* is a difficult task of production management, which poses the problem of balancing losses from down-time against losses from carrying inventory in a situation where multiple products are made on the same line. Systems producing complex manufactures are not infinitely and instantly 'flexible' – that is, able to adjust to changes in demand, or accommodate variations between different models. (The Toyota Corona, for instance, came in S, CS, CSX and Avante models, each with different features, in addition to a choice of sedans or station wagons, with manual or automatic gearboxes.) Production has to be planned in advance – and the difficult task here is to aggregate the atomistic components of demand into a production schedule within the 'flexibility' capacities of existing production technology (especially the ability to quickly change press dies and jigs) while containing work in progress inventory. For instance, a month with high demand at its end but a slack period at the beginning, across a variety of models, could be 'levelled' by allocating an 'averaged' (and projected) demand to each day. The alternative would be to dedicate the line first to one model, then to another. But this would require stockpiling components and finished products of one model or another. Toyota's production engineers developed the 'mixed production system' (also called 'linear', or 'synchronous' production), where various models are produced on the same line on the same day, with quick changeovers. This ensures that all components of the production process are working at a 'synchronized' pace, minimizing buildup of work in progress inventory and other forms of 'waste', and achieving 'uniform plant loading' (Coleman and Vaghefi, 1994: 31; Park, 1993; Monden, 1994: Ch. 4; Shingo, 1989). Production plans are thus the outcome of complex and exacting calculations, that balance losses and economies from a variety of sources, and allow the reduction of work in progress inventory, productive capacity and lead times to the consumer (Coleman and Vaghefi, 1994: 32).

Heijunka also seeks to 'balance' the workload to be performed to the capacity or capability of the process (machines and operators) to complete that work (Shingo, 1989; cited in Coleman and Vaghefi, 1994: 31). It also seeks to balance workload between adjacent components of the production system, including between workers. As indicated above, imbalanced production procedures give rise to waste (*mura*, and possibly *muri*). Thus, and crucially for this article's argument, a strong tension exists between *kaizen* and *heijunka*, which intensifies as the buffers and work in progress inventory is lowered in quest of productivity increases. First, since *heijunka* presupposes 'balancing' the workload to the capacity of the operators and machines, increasing that workload to drive *kaizen* is antithetical to *heijunka*. Second, an important source of waste (*mura*) is unscheduled fluctuations in daily work volume (Monden, 1994: 64), and these often result from *kaizen* activities driven by 'management by stress'. Levelling is a counter-principle to this disruption. There is, therefore, a trade-off between economies attained through *heijunka* (levelled production) and those gained by removing buffers to drive innovation *(kaizen)*. Since real interlinked production processes will never attain perfect balance, a certain amount of buffer stock is necessary to attain continuity of production. Reducing this can induce instability, and the need to rebalance adjacent production processes. Thus the notions of 'balanced', 'levelled' and 'stabilized' production, and *continuous* production are intertwined.

Toyota production systems, in a context of considerable product variation, reach a balance between *kaizen* activities and *heijunka* – a balance which weighs losses caused by carrying 'excessive' resources against down-time resulting from attempts to remove those resources. The balance is shaped by who is to bear the costs and benefits, and their relative power resources. First, the costs of disruption to the smooth flow of production (*mura*), in particular down-time, caused by pursuing *kaizen*, might be externalized through short-notice and/or unpaid overtime. Monden argues that an essential support for the Toyota production system is *shojinka*, or 'the adjustment and rescheduling of human resources', and makes special reference to 'early attendance and overtime' (Monden, 1994: 159, 66). Such practices are in effect a large 'buffer' outside normal working time (Berggren, 1992: 52; 1995: 78). This 'buffer' *externalizes* to workers and communities the costs of excessively enthusiastic *kaizen*, with its attendant disruption of production and *mura*. On the other hand, if communities and unions reject short-notice and/or unpaid overtime, the company would be forced to place a higher value on careful production management to achieve quotas, therefore emphasizing *heijunka* over *kaizen* and the quest for leanness.

Second, in compliant industrial relations systems, the costs of *kaizen*- and stress-driven production strategies may be borne directly by workers. The effects of 'speedup'-induced stress on workers (*muri*) may not show up until after work hours, in the form of fatigue, sleep disturbance, digestive malfunction, headaches, injuries and so on. More immediate problems like occupational overuse syndrome may be 'externalized' by dismissal, and hiring another worker. Many such problems may be paid for by the host country's health system, or by the worker in later life. Such strategies depend on a plentiful supply of willing workers to take such jobs, and 'flexible' industrial relations systems. They also depend on a lax occupational health and safety regime, that either lacks legislation mandating safe work practices, or lacks the means of enforcement. To the extent that particular national social settlements and industrial relations systems permit such strategies, companies can be expected to pursue them.

The Political Shaping of the Toyota Production System

This section argues that the choice of management strategy that emphasizes *heijunka* and production continuity, or leanness and *kaizen*, is shaped by the nature of the surrounding social settlement and industrial relations system. If the latter permits (as the preceding section argued), the costs of pursuing *kaizen* and leanness can be externalized. On the other hand, some industrial relations systems reject extremes of 'leanness'. As Turner (1991: passim, 223–5) has argued, unions' and workers' fortunes in the 'new era' are crucially dependent on their ability to shape the course of industrial restructuring and work reorganization. This ability depends, first, on having an accurate analysis of contending images of work organization and industrial policy and their possible implications for unions and workers, and second, on having the 'power resources' to act out of such an analysis. Such power resources consist of legislation that mandates worker participation in decision-making, and/or 'corporatist' arrangements that enable union influence on public policy – in short, on a favourable position within the industrial relations and political/institutional framework which shapes industrial adjustment. Where both exist, excessively 'lean' versions of the Toyota production system will be rejected, and the converse is true – the absence of these conditions may enable truly 'lean' production. In Japan, through most of the postwar period the balance of social and economic forces has clearly been in favour of the 'lean' version of the Toyota production system. 'Lean' plants are also to be found in the USA and UK, where labour is denied influence. On the other hand, in Japan, the early 1990s saw a certain trend away from 'leanness' and 'management by stress' because of a tightening of the labour market (Benders, 1996; Berggren, 1995). Tight labour markets also drove early experiments with work humanization in Sweden, but lately with rising unemployment and an economic liberal ideological offensive, 'lean' images of work have enjoyed increasing acceptance.

No better example of the systematic externalization of the costs of management by stress exists than the homeland of 'lean production'. Through the mid- to late 1980s, the phenomenon of *karoshi* emerged into Japanese public life. The term was invented in 1982 to refer to the increasing number of deaths, typically from strokes or heart attacks, that were attributed to overwork. The National Defence Council for the Victims of Karoshi (NDCVK), a public advocacy group mainly comprising lawyers seeking redress and compensation for the families of victims, estimated there were 10,000 victims of this condition a year (NDCVK, 1991; see also Nishiyama and Johnson, 1997: 2). While the Japanese government officially denied that *karoshi* existed, even objecting to the use of the term by the ILO, by 1993 nearly half of the Japanese population feared they or an immediate family member might become a victim (Kato, 1994: 2).

The origins of *karoshi* lie in the oil crisis and the 'Nixon shocks' in 1973, which imposed a heavy load on Japanese companies, and caused them to demand greater efforts from their workforces (Kato, 1994: 2). To deal with the oil shocks, Japanese companies emphasized 'stripped down management' (NDCVK, 1991: 98) later to be celebrated in the West as 'lean production'. This contributed to an increasing incidence of *karoshi*, the causes of which 'range from long working hours, a sudden increase in work load and the added mental pressures of expanded responsibilities and production quotas' (NDCVK, 1991: 99). These are precisely the working conditions to be found in excessively 'lean' workplaces which, as we have seen, run on low levels of staffing, with problems that arise during the day fixed by a seemingly endless overtime buffer at day's end. International comparative statistics put the hours worked by Japanese far in excess of other countries (except Korea), at least until the 1990 recession cut them back (NDCVK, 1991; Ross et al., 1998: 347). Furthermore, the widespread practice of unpaid overtime and the aggregation of hours across part-time workers systematically understated hours worked by many Japanese. Of the number of cases reported in a Tokyo Hotline, the majority had been working in excess of 70 hours per week in stressful conditions (NDCVK, 1991: 99). To compound the situation, Japanese workers are allocated far fewer vacation days than their international counterparts, and do not take all of the days owing to them (NDCVK, 1991: v).

The Japanese employer body Keidanren reported in a major survey that 88 percent of employers regularly use overtime (Kato, 1994: 2). This is because of the low overtime premium rates in Japan – 25 percent of the base wage, which itself is only a portion of the total wage (as the account in the second section of this article demonstrated), making the cost of hiring new employees greater than working existing employees on overtime (NDCVK, 1991: 87). Furthermore, the number of hours worked by employees is only weakly regulated in Japan, as state regulations are interpreted flexibly, and the onus for overtime regulation is placed on the company union or a 'collective representative'. Thus, most collective agreements contain a clause effectively giving management the right to demand short-notice overtime in a wide range of circumstances. In 1991 a long-running court battle over the celebrated case of Mr Tanaka – a Hitachi worker sacked for refusing overtime in 1967 – ended, with a determination by the Supreme Court that effectively reinforced managerial prerogatives in this area (see Joint Committee of Trade Unions Supporting Mr Tanaka's Trial, 1989).[3] The bursting of the bubble boom in 1990 caused Japanese companies to cut back on excessive overtime, thus somewhat defusing the issue (Berggren, 1995: 64). However, the case of *karoshi* underscores vividly how a 'flexible' industrial relations system permits the use of overtime to externalize the costs of 'leanness' on to the surrounding society.

The literature on the transplantation of Japanese production techniques contains many examples of pliant industrial relations systems allowing degrees of leanness that impose costs on workers and the surrounding community. As one example, the study by Fucini and Fucini (1990) of the Mazda plant at Flat Rock, Michigan, is replete with instances of leanness run riot. The company chose 3500 applicants from a pool of 96,500, and maintained a steady supply of 'flexible' labour (Fucini and Fucini, 1990: 1). As the plant got underway, and production volumes rose, production strategies increasingly emphasized 'leanness' through understaffing (Fucini and Fucini, 1990: 147). Overtime was compulsory and workers were notified late, and this became a major point of contention (Fucini and Fucini, 1990: 114, 145). Taking vacations was discouraged (Fucini and Fucini, 1990: 155). There was a high and increasing incidence of repetitive strain injury due to the persistent and high production demands. There were few less demanding jobs for older workers or for workers on 'light duties' due to injury. Supervisors pressured workers to return to work before they were ready, in some cases aggravating the original injury, and in others prompting the worker to refuse, leading to dismissal (Fucini and Fucini, 1990: 175–91). Unlike most transplants in the USA, the Mazda Flat Rock plant was unionized (as the result of it being a joint venture with Ford, the domestic operations of which had to accommodate the United Auto Workers) but the union adopted a compliant stance, perhaps showing particular 'flexibility' on the issue of overtime. (The outcome for the union was increasing worker discontent, and, ultimately, the development of a breakaway faction.) The extremes of leanness were permitted by the way the industrial relations system did not support workers' meaningful participation in decisions of work design. Such 'participation' was limited to *kaizen* activities, on terms controlled by management. The union's lack of power resources, and its accommodating stance with management, left workers with no institutional support to resist work intensification and excessive stress-driven *kaizen.*

On the other hand, well-known work organization experiments and traditions in Europe and Scandinavia embodied somewhat opposite calculations and conditions. Jurgens (1991) has identified a 'European model' of work, which contrasts with central principles of the Toyota production system. First, work organization should favour long-cycle jobs with higher degrees of skill, autonomy and discretion over short-cycle assembly line work. Second, notions of professionalism and skill are underpinned by public, not firm-level, skill formation and recognition infrastructure. And third, the 'European model' of work rejects Japanese-style 'team work' in favour of 'group work', where teams have more autonomy. The position of the team leader is more accountable to team members, by rotation or election. Fourth, adequate levels of buffer stocks protect against pacing (Jurgens, 1991: 245; Turner, 1991). While the success in terms of implementation is limited, the struggle over work reorganization in Europe can plausibly be portrayed as a clash between this model and the principles that underlie 'lean production'.

The well-known Swedish experiments in work organization reflected certain aspects of the Swedish social settlement of the 1970s and 1980s – tight labour markets, social democratic incumbency, influential unions and solidaristic wages which prevented employers compensating poor working conditions with extra pay. Although the initial experiments on work humanization were employer initiatives (Cole, 1989) the unions were able to influence them towards a more congenial form in certain areas. The 1970s saw unions' influence via 'corporatist' arrangements bear fruit in the form of 'codetermination' legislation that strengthened the ability of unions to influence work reorganization at shopfloor level (Turner, 1991).

Although the balance of forces in the Swedish social settlement that permitted such experiments is by now a fact of history, their ingredients are worth mentioning as counter-poles to 'lean production'. First, there is the oft-mentioned 'Scandinavian' emphasis on quality of work life, which favours 'buffers'.

> Scandinavian respect for the workers' quality of life requires that the worker have the ability to work quickly for a few minutes in order to take a small personal break without stopping the line. (Klein, 1989: 65)

Attempts to make work life more 'humanized' reached their apotheosis at the Volvo Uddevalla factory, which eschewed the assembly line in favour of dock assembly, in which teams of workers assembled whole cars in very long work cycles (Sandberg, 1995). But as is also well known, those union power resources were considerably wound back in the 1990s. Also, it seems, analysis of the implications of 'lean production' was lacking to the point where 'lean production' would acquire considerable legitimacy, complete with a shift to individualized payment systems, albeit under the guise of 'solidaristic work' (Kjellberg, 1992; Mahon, 1994).

This thesis of how the prominence given to 'leanness' or *heijunka* depends on the balance of forces in the industrial relations/political arena is also supported by relatively recent developments in Japan's automobile assembly industry. From the late 1980s, the Japan Auto Workers' Federation of Unions ran a public campaign against the conditions of work in the auto industry, which it characterized as demanding, dirty and dangerous (Joint Committee of Trade Unions, 1989). It issued a public report criticizing the industry in 1992 (JAW, 1992; Berggren, 1995: 75), and suggested that the working conditions might be improved if managers in the industry gave more attention to the concept of *muri*, and less to the more narrow concept of *muda* (Sandberg, 1995: 23). During the 'bubble boom' period, from 1986 to 1988, tightness in the labour market opened more job opportunities for workers outside the auto industry, leading to labour shortages (Benders, 1996: 14). Part of the employers' response was to undertake work humanization. In one of the ironies of history, this rejection of leaness proceeded just as the West's fascination with the concept grew (Sandberg, 1995; Berggren, 1995: 76; Shimuzu, 1995; Benders, 1996: 11).

The experiments in Japan (specifically Toyota's Tahara and Kyushu plants) moved away from leanness – but within constraints. Most importantly, the practice of short-notice overtime was not available, since work shifts were 'back to back', and this removed the time buffer at the end of the day (Berggren, 1995: 78). Their adherence to JIT was limited by their remote location (Sandberg, 1995: 22; Benders, 1996: 15). They allowed greater emphasis on 'internal' buffers, and the Toyota plant at Kyushu had not one moving assembly line, but a series of 'mini lines' that were linked by buffers (Shimuzu, 1995: 399; Benders, 1996: 18). The lines could be stopped and started independently of each other, thus minimizing losses from down-time, and alleviating the stress that comes from halting the whole plant's production. They made significant changes to the *satei* system, by lowering the proportion of payment that is determined by the individual evaluation, and in some cases removing the productivity-linked component (Shimuzu, 1995: 395–6; Benders, 1996: 21). They made numerous ergonomic improvements, to lessen the risk of injury (Shimuzu, 1995: 397, passim). It remains to be seen if these plants become more typical of auto production in Japan, but they do illustrate that the Toyota production system is capable of considerable social shaping in an 'anti-lean' direction. On the other hand, the experiments also reveal some of the limits of that shaping. Most notably, although some of the experiments shifted away from automation, at least in trim and final assembly, the basic work process remained essentially unchanged. In particular, there was no lengthening of work cycles, and the moving assembly line remained (Sandberg, 1995: 22).

The case of Australia is interesting but more complicated. It is tempting to see the union movement in Australia in the 1980s and early 1990s as having considerable power resources with which to shape the course of work reorganization and industrial adjustment, and this indeed was the interpretation of many commentators (e.g. Kyloh, 1994; Archer, 1992). However, there is considerable evidence of work intensification in a range of surveys, some of them conducted by government departments (DIR, 1995, 1996), some by independent research organizations (ACIRRT, 1998). Union density has fallen, from 51 percent in 1976, to 31.1 percent in 1996 (ABS, 1997), in part because of the failure of unions' 'involvement' in restructuring. While unions did indeed hold some influence over work reorganization in Australia, especially from the late 1980s, this influence acquiesced to work reorganization that amounted to work intensification. In the late 1980s the union movement as a whole did not make an accurate assessment of the dangers posed by 'lean production' (Hampson et al., 1994). The influential doctrine of post-Fordism did not properly distinguish models of work taking shape in Sweden, Germany and Japan, since they were all 'post-Fordist' (e.g. Mathews, 1989: 37; Curtain and Mathews, 1990: 73; Botsman, 1989). Important union strategic documents lacked a critical understanding of the Toyota production system's potential for 'leanness', and what made it different from European models of work (e.g. Anon, 1989a: 11–13; ACTU/TDC, 1987: 135, 155–6). The concept of 'lean production', embedded as it was in notions of 'best practice', gained considerable institutional momentum. A government International Best Practice Demonstration Programme was set up to provide funds to firms and workplace change consultants to implement 'best practice' work organization (Hampson et al., 1994; PCEK/T, 1990).

Even so, Australia's industrial relations system in the auto assembly industry did not prove to be fertile soil for at least the *extremes* of 'leanness', despite the fact that Toyota made union assent to 'lean' working arrangements a condition of the investment at their new plant at Altona, Victoria (*Australian Financial Review*, 12 December 1991). First, the actual layout of the plant had some similarities with the Japanese 'post-lean' experimental plants described earlier. Trim and final assembly consisted of several mini-lines separated by buffers, and this permitted each to start and stop independently of the others. The concept of *heijunka* was prominent at the plant, according to management, and this somewhat tempered the pursuit of 'leanness'.[4] Second, extending the principles of JIT to suppliers risked disruption, as many could not meet strict delivery schedules, so a thorough component inventory was kept on the plant premises. Third, the company encountered considerable 'external' constraints as to selection, training and reward, and thus did not have a free hand to fully implement human resource management strategies supportive of 'leanness'. A pre-existing agreement with the main union, the Victorian branch of the Vehicles Division of the Amalgamated Metals and Engineering Union gave priority to employees of the nearby Dandenong and Port Melbourne plants, which were being phased out of auto assembly, thus limiting managerial prerogatives as to recruitment and selection. Although Toyota itself was a registered provider of training, the firm's autonomy in that respect was somewhat limited by requirements that training be in line with national accreditation standards, in particular the Vehicle Industry Certificate (VIC), with a view to transferability of qualifications and the development of career paths (Anon., 1989b, 1989c, 1995). And Australia's award system, which determines a component of wages and working conditions centrally, prevented the implementation of individualized merit pay.[5] Workplace reform and restructured awards had in any case linked pay increments with competency standards and progression up skills ladders, integrated with the VIC (Anon., 1989b, 1989c, 1995). Interestingly, Toyota put in place suggestion schemes that gave cash rewards for useful suggestions, and these to some extent provided a degree of 'functional equivalence' to the *satei* system (interview with human resource manager, Toyota, 9 December 1994), which could encourage individual participation in *kaizen*. Fourth, the union had in 1991 just changed leadership in favour of the Left, which was far more suspicious of 'lean' ideas and the post-Fordist ideology in which they were set than the Right. Thus the union, while committed to the implementation of 'lean' principles by pre-existing 'structural efficiency' agreements, also sought to impede the full implementation of those principles.

Fifth, the company's ability to schedule short-notice overtime was contested. While the relevant 'Structural Efficiency Agreement' (Anon., 1989b: 18) agreed that 'overtime will be worked on a basis determined by the actual production needs of the enterprise', and the 1995 Workplace Agreement affirmed that 'Toyota reserves the right to assign work in excess of the basic working week' (Anon., 1995: 17), the agreement goes on to state that 'working pattern variations will be discussed with affected employees at least fourteen days prior to the variation being implemented' and limits overtime to 20 hours per calendar month (Anon., 1995: 17, 19). However, the agreement also refers to 'short-notice overtime', which is '*voluntary*'. Even so, 'if employees are *required* to work additional overtime on week days, they will be notified of the actual overtime needed on that day by the beginning of the second relief break. The actual overtime *required* will depend on 'the amount of daily overtime forecast and the production schedule volume which may have been lost due to unforeseen problems' and even 'there may be exceptional circumstances' in which 'shorter notice than that detailed above' is justified (Anon., 1995: 19, emphasis added). Reading between the lines, overtime was a contested issue which was not clearly regulated by the formula of words here. However, and this is the point, nor was it a matter of uncontested managerial prerogative, and thus it could not constitute the endless buffer which could support an emphasis on *kaizen*.

Thus, the Toyota plant at Altona, Australia, hardly conformed to the celebrated 'lean' model, and according to management did not seek 'zero-buffers', but sought to balance the goal of inventory reduction against the advantages to be derived from 'levelled production' (interview with human resource manager, Toyota, 9 December 1994).

Conclusion

Advocates and critics alike of Japanese production methods have neglected important production concepts, the most important of which is *heijunka*. There is a tension between the approach to production emphasized in the *heijunka* concept, and that implicit in approaches driven by 'leanness' and the quest for *kaizen*. An emphasis on *heijunka* values continuity, balance and the avoidance of down-time – *kaizen* accepts disruption in quest of productivity improvements via innovation. This article has argued that the balance struck between the contending principles of *heijunka* and *kaizen* is shaped by the surrounding social settlement within which the industrial relations system and particular work arrangements are set. Strong, strategically adept unions and a supportive industrial relations system that can impede managerial prerogative will be less likely to allow extremes of 'leanness'. On the other hand, industrial relations systems where unions are excluded, lack power resources and/or are ill informed strategically are more congenial to the extremes of 'leanness'. This distinction may offer a rhetorical strategy for progressives to shape the actual outcomes of work reorganization. It seems that Womack et al. have given a less than comprehensive explication of the Toyota production system, and in so doing emphasized 'leanness' at the expense of the 'forgotten production concepts', in particular *heijunka, mura* and *muri. The Machine that Changed the World* is thus set within a long managerialist tradition with particular strengths in the USA (see Hayes and Wheelwright, 1984) that seeks to substitute for management's deficiencies in the organization of manufacturing by intensifying work at the expense of the conditions of workers.

Notes

The research and fieldwork for this article were supported by a grant from the Faculty of Commerce and Economics of the University of New South Wales. A plant visit to the Toyota plant in Kyushu, Japan, was supported by the Japan Institute of Labour. Thanks are extended to both. I am also grateful to Gayle Tierney and Joe Caputo and others of the Victorian branch of the Vehicles Division of the Amalgamated Metal Workers Union, and to Doug Rickarby of the Human Resources Department of Toyota Altona for valuable assistance. Key ideas received a patient and supportive hearing from Åke Sandberg and Christian Berggren while I was on study leave at the National Institute for Working Life in Sweden. Yuki Tagata and Hiromi Nagayoshi of the Japan Institute of Labour provided valuable information on the Japanese concepts. Gustavo Guzman of the Production Engineering Department of the Federal University of Minas Gerais, Brazil, commented valuably on an early draft. Any errors are the author's fault.

1. The idea was derived from American supermarkets, when empty shelf space indicates more stock is needed (Ohno, 1988: 25–6; Shingo, 1989: 90).

2. Interview given to the BBC programme *Nippon* (shown on Australia's Special Broadcasting Service, 22 October 1991).

3. The Supreme Court ruling on the Tanaka case is available at http://www/mol.go.jp/bulletin/year/1992/vol31-05/05.

4. The Toyota Australia Workplace Agreement (Anon., 1995: 6) lists 'balanced and levelled production' among key principles of the Toyota production system.

5. This information and much of the following was gleaned from two plant visits in 1993 and 1994, two semi-structured interviews with management and six with union representatives (8–10 December 1993, 9–11 December 1994).

References

ABS (Australian Bureau of Statistics) (1997) *Trade Union Members*, Cat. 6325. Canberra: ABS.

ACTU/TDC (Australian Council of Trade Unions/Trade Development Council Secretariat) (1987) *Australia Reconstructed.* Canberra: Australian Government Publishing Service.

ACIRRT (Australian Centre for Industrial Relations Research and Training) (1998) *Australia at Work: Just Managing?* Sydney: Prentice-Hall.

Adler, P. (1993) 'Time and Motion Regained', *Harvard Business Review* January–February: 97–108.

Adler, P. and R. Cole (1993) 'Designed for Learning: A Tale of Two Auto Plants', *Sloan Management Review* Spring: 85–94.

Anon. (1989a) *The Australian Vehicle Manufacturing Industry: Award Restructuring. Report of the Tripartite Study Mission to Japan, United States of America, Federal Republic of Germany, and Sweden.* Melbourne: Ramsay Ware.

Anon. (1989b) *Structural Efficiency Agreement between the VBEF and Toyota: Non-Trade Group.* Victoria: Vehicle Builders' Employees Federation.

Anon. (1989c) *Structural Efficiency Agreement between Toyota Motor Corporation, and Vehicle Builders' Employee Federation, Amalgamated Metal Workers Union, Electrical Trade Union, Australian Society of Engineers. Support Mechanism: Training.* Victoria: Vehicle Builders' Employees Federation.

Anon. (1995) 'Toyota Australia Workplace Agreement (Port Melbourne, National Parts Division and Sydney and Regions)'.

Archer, R. (1992) 'The Unexpected Emergence of Australian Corporatism', pp. 377–417 in J. Pekkarinen, M. Pohjola and B. Rowthorn (eds) *Social Corporatism: A Superior Economic System?* Oxford: Clarendon Press.

Barker, J. (1993) 'Tightening the Iron Cage: Concertive Control in Self-Managing Teams', *Administrative Science Quarterly* September: 408–37.

Benders, J. (1996) 'Leaving Lean? Recent Changes in the Production Organization of some Japanese Car Plants', *Economic and Industrial Democracy* 17: 9–38.

Berggren, C. (1992) *Alternatives to Lean Production, Work Organisation in the Swedish Auto Industry.* Ithaca, NY: ILR Press.

Berggren, C. (1995) 'Japan as Number Two: Competitive Problems and the Future of Alliance Capitalism after the Burst of the Bubble Boom', *Work, Employment and Society* 9(1): 53–94.

Botsman, P. (1989) 'Rethinking the Class Struggle: Industrial Democracy and the Politics of Production', *Economic and Industrial Democracy* 10: 123–42.

Cole, R. (1989) *Strategies for Learning: Small Group Activities in American, Japanese and Swedish Industry*. Berkeley: University of California Press.

Coleman, J. and R. Vaghefi (1994) 'Heijunka (?): A Key to the Toyota Production System', *Production and Inventory Management Journal* 4: 31–5.

Curtain, R. and J. Mathews (1990) 'Two Models of Award Restructuring in Australia', *Labour and Industry* 3(1): 58–75.

Dertouzos, M., R. Lester and R. Solow (1989) *Made in America – Regaining the Competitive Edge*. Cambridge, MA: MIT Press.

DIR (Department of Industrial Relations) (1995) *Department of Industrial Relations, Enterprise Bargaining in Australia, Annual Report, 1994*. Canberra: DIR.

DIR (Department of Industrial Relations) (1996) *Department of Industrial Relations, Enterprise Bargaining in Australia, Annual Report, 1995*. Canberra: DIR.

Dohse, K., U. Jurgens and T. Malsh (1985) 'From "Fordism" to Toyotism? The Social Organisation of the Labor Process in the Japanese Automobile Industry', *Politics and Society* 14(2): 115–46.

Fucini, J. and S. Fucini (1990) *Working for the Japanese*. New York: Free Press.

Hampson, I., P. Ewer and M. Smith (1994) 'Post-Fordism and Workplace Change: Towards a Critical Research Agenda', *Journal of Industrial Relations* 36(2): 231–57.

Hayes, R. and S. Wheelwright (1984) *Restoring our Competitive Edge: Competing through Manufacturing*. New York: John Wiley.

JAW (Japan Auto Workers) (1992) *Japanese Automobile Industry in the Future. Towards Coexistence with the World, Consumers and Employees*. Tokyo: JAW.

Joint Committee of Trade Unions Supporting Mr Tanaka's Trial (1989) *Unfair Dismissal in the Hitachi Musashi Plant: Resistance to Zangyo and Karoshi*.

Jurgens, U. (1991) 'Departures from Taylorism and Fordism: New Forms of Work in the Automobile Industry', Ch. 11 in B. Jessop, H. Kastendiek, K. Nielsen and O. Pedersen (eds) *The Politics of Flexibility: Restructuring State and Industry in Britain, Germany and Scandinavia*. Aldershot: Edward Elgar.

Kato, Tetsuro (1994) 'The Political Economy of Japanese *Karoshi*', paper prepared for the XVth Congress of the International Political Science Association, 20–25 August, at http: //www.ff.iij4u.or.jp/~katote/*Karoshi*.html

Kenney, M. and R. Florida (1988) 'Beyond Mass Production: Production and the Labour Process in Japan', *Politics and Society* 16(1): 121–58.

Kenney, M. and R. Florida (1989) 'Japan's Role in a Post-Fordist Age', *Futures* April: 136–51.

Kjellberg, A. (1992) 'Sweden: Can the Model Survive?', Ch. 3 in A. Ferner and R. Hyman (eds) *Industrial Relations in the New Europe*. Oxford: Basil Blackwell.

Klein, J. (1989) 'The Human Costs of Manufacturing Reform', *Harvard Business Review* March–April: 60–6.

Kumazawa, M. and J. Yamada (1989) 'Jobs and Skills Under the Lifetime Nenko Employment System', in S. Wood (ed.) *The Transformation of Work*. London: Unwin Hyman.

Kyloh, R. (1994) 'Restructuring at the National Level: Labour-Led Restructuring and Reform in Australia', Ch. 10 in W. Genberger and D. Campbell (eds) *Creating*

Economic Opportunities: The Role of Labour Standards in Industrial Restructuring. Geneva: International Institute for Labour Studies.

Mahon, R. (1994) 'From Solidaristic Wages to Soldaristic Work: A Post-Fordist Historic Compromise for Sweden?', pp. 285–314 in W. Clement and R. Mahon (eds) *Swedish Social Democracy: A Model in Transition*. Canada: Canadian Scholars Press.

Mathews, J. (1988) *A Culture of Power*. Sydney: Pluto Press/Australian Fabian Society and Socialist Forum.

Mathews, J. (1989) *Tools of Change: New Technology and the Democratisation of Work*. Sydney: Pluto Press.

Mathews, J. (1991) *Ford Australia Plastics Plant: Transition to Teamwork through Quality Enhancement*, University of New South Wales Studies in Organisational Analysis and Innovation No. 3. Sydney: UNSW.

Monden, Y. (1994) *Toyota Production System*, 2nd edn. London: Chapman and Hall.

Moore, J. (1987) 'Japanese Industrial Relations', *Labour and Industry* 1(1): 140–55.

NDCVK (National Defence Counsel for the Victims of Karoshi) (1991) *Karoshi: When the Corporate Warrior Dies*. Tokyo: NDCVK.

Nishiyama, Katsuo and Jeffrey Johnson (1997) '*Karoshi* – Death from Overwork: Occupational Health Consequences of the Japanese Production Management' (sixth draft for International Journal of Health Services, 4 February, http: //bugsy.serve.net/cse/whatsnew/*Karoshi*.htm

Ohno, T. (1988) *Toyota Production System: Beyond Large-Scale Production*. Cambridge, MA: Productivity Press.

Oliver, N. and B. Wilkinson (1992) *The Japanization of British Industry*. Oxford: Blackwell Business Press.

Park, P. (1993) 'Uniform Plant Loading through Level Production', *Production and Inventory Management Journal* 2: 12–17.

Parker, M. and J. Slaughter (1988) 'Management by Stress', *Technology Review* 91(7): 37–44.

PCEK/T (Pappas; Carter; Evans; Koop; Telesis) (1990) *The Global Challenge: Australian Manufacturing in the 1990s*. Melbourne: Australian Manufacturing Council.

Rosenberg, N. (1982) *Inside the Black Box: Technology and Economics*. Cambridge: Cambridge University Press.

Ross, P., G. Bamber and G. Whitehouse (1998) 'Appendix: Employment, Economics and Industrial Relations: Comparative Statistics', Ch. 12 in G. Bamber and R. Lansbury (eds) *International and Comparative Employment Relations*, 3rd edn. Sydney: Allen and Unwin.

Sandberg, Å. (ed.) (1995) *Enriching Production: Perspectives on Volvo's Uddevalla Plant, as an Alternative to Lean Production*. Sydney: Avebury.

Sayer, A. (1986) ' New Developments in Manufacturing: The Just-in-Time System', *Capital and Class* 30: 43–72.

Sewell, G. and B. Wilkinson (1992) 'Someone to Watch Over Me: Surveillance, Discipline and the Just-in-Time Labour Process', *Sociology* 26(2): 271–89.

Shimuzu, K. (1995) 'Humanization of the Production System and Work at Toyota Motor Co and Toyota Motor Kyushu', pp. 383–404 in Å Sandberg (ed.) *Enriching Production: Perspectives on Volvo's Uddevalla Plant, as an Alternative to Lean Production*. Sydney: Avebury.

Shingo, Shigeo (1989) *A Study of the Toyota Production System from an Industrial Engineering Viewpoint*. Cambridge, MA: Productivity Press.

Slaughter, J. (1990) 'Management by Stress', *Multinational Monitor* January/February: 9–12.

Turner, L. (1991) *Democracy at Work: Changing World Markets and the Future of Labor Unions*. Ithaca, NY: Cornell University Press.

Unterweger, P. (1992) 'Lean Production: Myth and Reality', IMF Automotive Department, October.

Williams, K., C. Haslam, J. Williams and T. Cutler, with A. Adcroft and S. Johal (1992) 'Against Lean Production', *Economy and Society* 21(3): 321–54.

Womack, J., D. Jones and D. Roos (1990) *The Machine that Changed the World*. New York: Macmillan.

Ian Hampson

is a lecturer at the School of Industrial Relations and Organisational Behaviour at the University of New South Wales, Sydney, Australia. His research interests include work organization, training and the politics of labour movements. He has published on post-Fordism, corporatism in Australia, and management education policy.

[16]
The Limits of "Lean"

Michael A. Cusumano

SOME OF THE RESULTS OF CONTINUOUS IMPROVEMENT IN JUST-IN-TIME MANUFACTURING AND RAPID PRODUCT DEVELOPMENT HAVE NOT ALWAYS BEEN FAVORABLE. As the author points out, Japan is suffering from increased traffic due to JIT deliveries, a shortage of blue-collar workers, too many product variations, overly stressed suppliers, and a lack of money for new product development. This situation offers an opportunity to companies in the rest of the world to catch up to the Japanese, modify lean production and product development to create a more balanced approach, and seek competitive advantage in new areas, for example, in more flexible automation, new materials and technologies, innovative product features, and expansion into developing markets.

Michael A. Cusumano is associate professor of management at the MIT Sloan School of Management.

Japanese competitiveness in a number of industries is the result of a combination of factors. Among the most important are a series of innovations and practices in manufacturing and product development that have been referred to as "lean": aimed at high productivity as well as high quality in engineering and manufacturing, resulting in high price-performance in the value of products delivered to the customer. This article outlines some of those innovations and practices, particularly those in the Japanese automobile industry. It then addresses two other issues: how transferable these practices are outside Japan, and what limits the Japanese themselves have encountered as they have pursued "continuous improvement" in manufacturing and product development management.

Principles of "Lean" Management

Table 1 lists the principles of "lean" manufacturing and product development. In manufacturing, these practices made it possible for Toyota and other firms that followed its approach to achieve extremely high levels of quality (few defects) and productivity in manufacturing (output per worker that was as much as two or three times higher than U.S. or European plants in the late 1980s). Japanese firms also achieved relatively high levels of flexibility by producing relatively small lots of different models with little or no loss of productivity or quality.[1] Toyota developed this small-lot, just-in-time (JIT) manufacturing approach in response to the needs of the post-World War II Japanese auto market, which was very small, with few exports, but with rapidly growing demand for different types of car and truck models.

During the late 1970s and 1980s, the nine major Japanese automakers gradually took advantage of their manufacturing capabilities to shift the primary competitive domain to product development. Led by Honda and Toyota, this shift resulted in fast development times (estimated at forty-two months compared to sixty-five months or so for the U.S. and European producers[2]), a very aggressive expansion of product lines by all the Japanese automakers, as well as adoption of full model changes every four years (a practice started in the 1950s). This rapid change and expansion of products allowed Japanese automakers to introduce new features and technologies into their vehicles more quickly than U.S. or European automakers, which generally had product replacement cycles of six to eight years or more. As indicated in Table 1, an important part of the Japanese process for product development was the relatively independent project teams led by "heavyweight" project managers who controlled the human and financial resources to determine the product's features and move it quickly through the various phases of design and into manufacturing. This Japanese system contrasted with the use of "functional" departments (such as for engine design,

Table 1 Principles of "Lean" Management

Production (Toyota Model)

JIT "small-lot" production
Minimal in-process inventories
Geographic concentration of assembly and parts production
Manual demand-pull with kanban cards
Production leveling
Rapid setup
Machinery and line rationalization
Work standardization
Foolproof automation devices
Multiskilled workers
High levels of subcontracting
Selective use of automation
Continuous incremental process improvement

Product Development (Honda Model)

Rapid model replacement
Frequent model-line expansion
Overlapping and compressed development phases
High levels of supplier engineering
"Heavyweight" project managers
Design team and manager continuity
Strict engineering schedules and work discipline
Good communication mechanisms and skills
Multiskilled engineers and design teams
Skillful use of computer-aided design tools
Continuous incremental product improvements

body design, or manufacturing preparations), where departments would hand off work slowly to other departments, often in a sequential manner, rather than in overlapping phases guided by a strong project manager.[3]

With this combination of manufacturing and product development skills, the Japanese automobile industry overall rose to exceed the U.S. industry in total production for the first time in 1980, with over 11 million units, and continued to dominate the world industry through the early 1990s. Accordingly, this Japanese style of manufacturing and product development, dubbed the "lean" approach by former MIT student and researcher John Krafcik, has come to be studied and emulated around the world. The best U.S.-owned auto manufacturing plants have now achieved relative parity with all but the most efficient Japanese plants.[4] Some U.S. product development projects are also reported to have been completed as quickly as the average Japanese projects (though they require more people and engineering hours).

But, while U.S. and European automakers continue to study and, at least in part, emulate Japanese manufacturing and engineering practices, it has now become apparent to many Japanese managers, employees, policymakers, and industry observers that the notion of "continuous improvement" — continually pushing for gains in manufacturing and engineering efficiency — has resulted in a new set of problems and some practical limits. The Japanese automakers are now exploring ways to modify or moderate their approaches, even if they become less efficient in manufacturing or less profuse in engineering outputs. Japanese gains in manufacturing productivity and their rapid expansion and replacement of product lines may have indeed reached a limit. If so, given the improvement programs underway at U.S. and European automakers, then the best western and other Asian firms may soon approach parity with the Japanese in basic manufacturing and engineering prowess. This parity will then make it necessary for all firms to seek competitive advantage not simply by following the lean principles that everyone will know and be implementing, but by defining other domains of competition, such as new levels of manufacturing automation, new materials and technologies, innovative product features, or skillful overseas management and expansion into developing markets.

Limitations of "Lean" in Japan

Like U.S., European, and other automakers, not all Japanese companies have been able or willing to follow fully the standards set by Toyota in manufacturing or Honda in product development. Both Toyota and Honda have a unique history and geographic setting that have facilitated practices such as Toyota's famous JIT and kanban systems or Honda's product development system. Other Japanese firms, such as Nissan, encountered some of the problems that JIT and kanban create when they first tried to introduce the techniques into their own organizations during the 1970s. Similarly, no Japanese automaker has matched the product development performance of Honda (at least for models introduced in Japan). There are several reasons why many Japanese, as well as non-Japanese, firms have been unable or unwilling to follow the lean principles to their fullest extent. There are also several solutions or countermeasures to deal with the problems of lean management taken to the extreme (Table 2). The Japanese firms are currently exploring these alternatives, in autos and in other industries.

Urban Congestion and Geographical Distance

During the 1970s, Nissan discovered that the Toyota practice of having suppliers make or deliver components "just in time" to assembly lines several times a day, with deliveries controlled by the physical exchange of production or parts delivery tickets (kanban cards), did not

Table 2 Limitations of "Lean": Japan in the 1990s

	Problems	Solutions
Production	Urban congestion Long geographic distances Overseas locations Stress on suppliers Too much product variety Shortage of blue-collar workers	Less frequent parts deliveries More electronic data transfers More computerized control systems More attention to supplier needs More parts standardization More manufacturable designs More automation More dispersed Japanese production More overseas production
Product Development	High cost of frequent model replacement High cost of frequent model line expansion Environmental and recycling costs Too much product variety	Less frequent model replacement Fewer model lines and variations Less frequent auto purchases by customers More parts and materials recycling More sharing of parts across products Less "heavyweight" project managers

work well in congested urban areas. As more and more Japanese factories in different industries have adopted the Toyota practice, traffic worsened to the point where, in the 1990s, the Japanese government mounted a media campaign encouraging companies to reduce the frequency of their parts deliveries. Traffic congestion pollutes the environment and wastes time while people are stranded in traffic and in manufacturing plants, waiting for components to arrive.

Nissan's plants have always been more dispersed than Toyota's, so Nissan management was convinced that it was indeed more practical and economical to keep a greater amount of inventory on hand than Toyota did. Nissan did this even though it had adopted the practice in the early 1950s, along with Toyota, of reducing unnecessary inventories to save on operating expenses and catch mistakes that might be hidden or take too much time to identify if parts were stored for weeks or months. Ultimately, Nissan reduced average inventories from a month to a day or so, but not to the extreme of a couple of hours as Toyota did. Other Japanese automakers in other parts of Japan encountered similar problems; traffic congestion even in formerly rural areas like Toyoda City and Aichi Prefecture (where most of Toyota's suppliers are located) has forced companies to make JIT a bit less timely.

Similarly, with companies establishing plants in different areas of Japan to escape the congestion and labor shortages in the major urban areas, the once-elegant kanban system, requiring the physical exchange of production or delivery tickets (originally by workers carrying kanban cards on their bicycles from station to station or carrying boxes of components with the kanban cards attached), is no longer practical. Suppliers now need to deliver larger loads, sometimes by ship to different islands in Japan or to North America, Europe, or other parts of Asia. It is not practical to track or control the ordering of components simply by physically exchanging kanban cards or cards attached to boxes, just as it is not practical to make and deliver very small batches of components.

Of course, the Japanese have not reverted completely to the former style of mass production. In the old system, companies made and stored a month or more of components and controlled production by inflexible schedules that "pushed" components into the system, regardless of what was happening at individual production stations, and then tracked the production process through real-time computer systems with inaccurate information. But the days when even Toyota could operate in a highly predictable and geographically small area within Japan are now over. Other companies, especially U.S. firms that made components in one state or country and shipped them thousands of miles, also noticed this limitation of the Toyota practice years ago, even though they benefited considerably, in productivity and quality, by reducing unnecessary levels of inventory and reducing delivery times from suppliers.

Supplier Management

Another obvious limitation of lean manufacturing is the need for cooperative and reliable suppliers, which account for approximately 75 percent of manufacturing work in the automobile industry and approximately half of product development, measured by costs.[5] For the system to work, suppliers must agree to manufacture components in small lots and then deliver frequently to assembly plants — otherwise they will simply hold inventory, raising their own carrying costs and eliminating their ability

As Japanese companies disperse their plants throughout Japan and other parts of the world, they have been able to move only some of their suppliers.

to improve quality and productivity through short production runs and correction of errors or process improvements made with each new setup. As Japanese companies disperse their plants throughout Japan and other parts of the world, however, they have been able to move only some of their suppliers. Non-Japanese suppliers have not complied exactly with Japanese pricing and quality requirements, nor have the Japanese trusted foreign suppliers fully in product development.[6]

Until the recent recession (which is lasting longer than anybody in Japan predicted), Japan had experienced a severe shortage of factory labor domestically. The Japanese government allowed foreign workers from Southeast Asia, the Middle East, and South America to come to Japan and work in Japanese factories, mostly at the smaller suppliers. This practice helped the labor shortage, but it also introduced new problems: the need to train the foreign workers and manage people with little or no literacy in Japanese. Many companies report quality problems and reductions in worker flexibility as a result of using less-skilled foreigners; this has lowered supplier productivity by forcing managers to reduce work schedules and use more inspection and rework to ensure that they still deliver high-quality components to Japanese assembly plants.

The Shortage of Blue-Collar Workers

One of the brilliant contributions of Toyota managers such as Ohno Taiichi, inventor of the kanban system and director of manufacturing operations at Toyota during its system's formative years from the 1950s through the 1970s, was to view automation with skepticism. Automation, unless it was flexible (easily changed or reprogrammed to handle different product models or variations, or volume fluctuations), introduced rigidity into production processes and was not suitable for labor-intensive assembly operations. As a result, Toyota introduced automated transfer machinery cautiously and used robots in modest numbers only in the 1980s, after they had become programmable, reliable, and inexpensive compared to human workers. Instead, Toyota relied mainly on well-trained workers and gave them broad responsibilities, such as doing much of their own inspection, preventive maintenance, and janitorial work. Line "rationalization" efforts started by Ohno after World War II also ruthlessly eliminated "waste" from all assembly and production activities, until Toyota became by far the most efficient automaker in the world, in terms of labor productivity.[7]

The incremental introduction of automated manufacturing systems meant that Toyota and other Japanese automakers that followed its lead had to rely heavily, at least in part, on large numbers of cooperative and skilled human workers. In turn, managers have asked the Japanese workers to work long hours in physically demanding production systems. The Japanese plants have also been relatively flexible, primarily in terms of their ability to produce a large variety of models in relatively small volumes, averaging around 100,000 units or less per year in the early 1990s, compared to around 200,000 units or more per year per model for U.S. and European auto producers.[8]

Today there are usually more factory jobs than there are young Japanese people willing to take these jobs. The result has been intense competition for blue-collar workers, not only by small suppliers but also by the assembly facilities of major companies. In addition, young Japanese workers leave blue-collar jobs and, increasingly, even white-collar jobs, if they feel overworked or unhappy for other reasons. For example, in the early 1990s, Toyota encountered serious difficulties staffing its factories near Toyoda City because of the severe shortage of blue-collar workers (women are still not permitted to work in most Japanese auto assembly factories) and had employee turnover rates in its factories of approximately 30 percent annually, including the seasonal hiring of temporary workers. Although this is not actually a new problem for Toyota, the labor shortage and turnover problem is likely to worsen rather than improve if the Japanese economy recovers. As a result, a necessary change in strategy and tactics will likely reduce the productivity advantage Toyota has enjoyed at home.

Product Variety

The virtual explosion in Japanese product variety during the 1980s and early 1990s, particularly for Japan's domestic market, enabled the most successful companies to expand their market shares and regularly convince customers to buy new versions of automobiles, video recorders, stereos, lap-top computers and word processors, microwave ovens, and dozens, if not hundreds, of other consumer products. Toyota and other companies designed JIT/kanban-like systems to facilitate small-lot production when combined with fast equipment setup or changeover times, synchronized parts production and rapid delivery, and versatile workers who can quickly move to solve problems or shift to parts lines and assembly lines for rapidly selling products.

But large engineering organizations and independent heavyweight project managers, encouraged by marketing organizations, have created too much product variety and offered too many options to customers. The result is that parts makers and assembly plants have to accommodate very small and very rare orders too frequently. This variety requires constant equipment setups and kanban exchanges, as well as many deliveries of small lots of components — just when total sales are stagnant, and workers, suppliers, and traffic systems have reached a sort of practical limit. Not surprisingly, many Japanese firms have concluded that, in the short term, they need better scheduling and control systems to handle so much variety, and, more importantly, they need to treat the root cause of the problem and reduce variety to the 20 percent or so of models and product variations that generate 80 percent of their profits and sales.

Too much product variety has also created environmental concerns.

It has also become impractical to let the manual exchange of kanban cards "pull" new orders of components into the production system and relay all production information. There are now better methods available (such as the use of bar-code readers and other electronic forms of moving information) for plants with very high levels of variety — which covers most Japanese automakers and producers in many other industries.

Too much product variety has also created environmental concerns. Japanese automakers have been introducing replacements of existing models every four years, in addition to continually expanding their product lines, for example, into new luxury segments. Japanese government regulations and mandatory fees or maintenance charges for automobile inspection also encourage consumers to replace their vehicles every four or five years. One outcome is consistently high domestic demand for new Japanese cars and trucks. But another outcome is the need to dispose of all the replaced vehicles. Some become used-car exports to other parts of the world, but Japanese companies now realize they need to think about how to recycle automobile materials more effectively.

But perhaps the most pressing concern for Japanese managers is the cost of new model development and model replacement now that money is expensive in Japan. Bank interest rates have reached international levels, and banks can no longer make large cheap loans because their portfolios of stocks and real estate (needed as a basis for loan limits as a percentage of bank assets) and the portfolios of their customers (normally used as collateral) have declined in value. And companies can no longer raise much capital on the stock market because of the Japanese investors' reluctance to buy securities in a market that has dropped 50 percent in value during the past several years. The only source of truly "free" money — used in the past for product development as well as capital investment — is operating profits. In the current recession, however, operating profits have also declined dramatically for Japanese firms.

Thus, for the intermediate term, Japanese managers have realized that they need to reduce their overall investments in new product development (which also requires major investments in manufacturing preparations) as well as cut the amount of variety they have in components and final products. Companies in the automobile industry, for example, are now reducing unique parts and product varieties by 30 percent to 50 percent or more for new models. Japanese companies have also been reining in the heavyweight project managers, placing some limits on their budgets and discretion by establishing platform managers and chief engineers. These higher-level managers, who are above the project managers, coordinate the development of a group of technically related models, making sure that they share more key components and manufacturing facilities. These reductions in unique parts and greater sharing of components across models should ease problems in assembly plants and at suppliers, as well as save money in engineering and manufacturing-preparation costs. The risk, of course, is that sales will no longer grow as fast as they did when Japanese companies continually introduced streams of new models with lots of new technology and replaced old models quickly. Sales may even decline, although profits may rise as a percentage of sales if the Japanese learn how to generate more profits from each product development effort, rather than simply look for expansion of sales and market share.

Conclusion

In autos and other industries, leading Japanese companies have maddeningly pursued continuous improvement in inventory reduction and just-in-time manufacturing. They have also pursued the continuous expansion of market share through productivity and quality gains and through nonstop investment in new products and upgrading of old products. One result has been great wealth as the Japanese economy has expanded. But another result is that Japan has become a nation in "gridlock": traffic jams are everywhere as factories and retail stores all want just-in-time deliveries. Companies have trouble finding good workers. Banks have trouble making loans. Managers have difficulty finding money for new investment.

In a sense, Japanese companies are now being forced to become more like everybody else in the world: more profit oriented in the short term! How short-term profit oriented the Japanese will become remains to be seen, however. Japanese managers are accustomed to treating themselves and employees as permanent assets (due to lifetime employment and relatively little labor mobility, although these practices are changing). Managers have also tended to evaluate investments for their long-term strategic value. Nonetheless, Japanese companies are facing a host of difficulties that will make them more like U.S. and European companies and less competitive in manufacturing productivity and quality as well as in new product development. This unfortunate situation for the Japanese presents opportunities for companies in the rest of the world — in the United States and Europe, as well as in other parts of Asia, especially Korea and Taiwan. The fate of the Japanese may well depend less on what Japanese companies do than on how well other companies respond to the limitations of "lean" that the Japanese have encountered. ◆

References

1. J.P. Womack, D.T. Jones, and D. Roos, *The Machine That Changed the World* (New York: Rawson/MacMillan, 1990);
Y. Monden, *The Toyota Production System* (Atlanta, Georgia: Industrial Engineering and Management Press, 1983);
M.A. Cusumano, *The Japanese Automobile Industry: Technology and Management at Nissan and Toyota* (Cambridge, Massachusetts: Harvard University Press, 1985);
M.A. Cusumano, "Manufacturing Innovation: Lessons from the Japanese Auto Industry," *Sloan Management Review*, Fall 1988, pp. 29-39;
J.F. Krafcik, "Triumph of the Lean Production System," *Sloan Management Review*, Fall 1988, pp. 41-52;
K.B. Clark and T. Fujimoto, *Product Development Performance: Strategy, Organization, and Performance in the World Auto Industry* (Boston: Harvard Business School Press, 1991); and
M.A. Cusumano and K. Nobeoka, "Strategy, Structure, and Performance in Product Development: Observations from the Auto Industry," *Research Policy* 21 (1992): 265- 293.
2. Clark and Fujimoto (1991).
3. Ibid.
4. Womack et al. (1990).
5. Cusumano (1985, 1988);
Clark and Fujimoto (1991).
6. M.A. Cusumano and A. Takeishi, "Supplier Relations and Management: A Survey of Japanese, Japanese-Transplant, and U.S. Auto Plants," *Strategic Management Journal* 12 (1991): 563-588.
7. Cusumano (1985, 1988);
M.B. Lieberman, L.J. Lau, and M.D. Williams, "Firm-Level Productivity and Management Influence: A Comparison of U.S. and Japanese Automobile Producers," *Management Science* 36 (1990): 1193-1215.
8. Womack et al. (1990).

Reprint 3542

MANAGEMENT-BY-STRESS
The Team Concept in the US Auto Industry

Mike Parker and Jane Slaughter

The team concept has captured the imagination of the American auto companies, and with good reason. It promises to improve efficiency and quality and to banish the union to the sidelines of the shop floor. In many cases it is delivering on that promise, although in widely varying degrees. Already the team concept has been installed at 15 of 36 General Motors assembly plants and at 5 Chrysler plants. This essay describes the chief features of the team concept and its implications for workers' organization.

Official descriptions of the team concept, from both management and union, focus on two aspects:

— the division of the workforce into teams of 5–20 workers who elect a leader, meet to discuss workplace problems and learn each other's jobs, and;

— the new spirit of harmony between union and management as both join in a common effort to make their particular plant competitive. Indeed, 'every worker is a quality inspector', according to Bruce Lee, director of the United Auto Workers (UAW) western region.

Those aspects are only a small part of a package deal called the team concept. Management designed the package and thus far has been successful in imposing most of it on the United Auto Workers. The union at the national level has quietly welcomed the team concept, and local unions have been unable to resist its imposition.

The team concept package includes the following elements:

(1) A union which has few powers and which has adopted company goals.

(2) A pay-for-knowledge system, in which workers are paid more for learning and being willing to do more jobs, or a requirement that all workers know all the jobs in their team and switch jobs if requested.

(3) An emphasis on quality and individual responsibility for quality.

(4) A strong awareness of 'competition' from other

plants within the corporation and where the plant stands in comparative efficiency and quality ratings.

(5) A management attempt to make workers aware of the interrelatedness of the plant's departments and the place of the individual in the whole; an attempt by both union and management to get beyond the 'I just come to work, do my job and mind my own business' mentality.

(6) Less meaning for seniority than in more traditional plants as 'good jobs', usually obtained by seniority, are eliminated.

(7) Worker involvement in continually increasing the workload as efficiency is continually improved.

(8) Increased management flexibility to use workers as they see fit as classifications are eliminated, with a concomitant decrease in worker rights.

(9) Increased Taylorism, as management gain tighter control over how the job is to be done, partly by soliciting workers' help in Taylorizing their own jobs.

(10) An integrated system of control which is often on the borderline of running into trouble and is kept going only by dedicated attention and a rapid pace, i.e. 'management-by-stress'.

The team concept package is not a major reorganization of production technology, as most team concept plants use the traditional moving assembly line. The package is rather a reorganization of labour relations.

General Motors (GM) is by far the most advanced of the US-based automakers in implementing the team concept. Three levels of implementation – with concomitant levels of work intensity and deterioration of worker rights – can be identified, and typical plants at that level described:

(1) New United Motors Manufacturing, Inc. (NUMMI), a joint venture between GM and Toyota and the most efficient plant in the GM system. Sixty per cent of workers produce the same number of cars as when the plant belonged

solely to GM. Union officials are thoroughly identified with the new system.

(2) Shreveport, Louisiana, the pioneer team concept plant in the auto industry, where the team culture is firmly entrenched despite the fact that half the workers are transfers from traditional GM plants. The union has been key to implementing the team concept.

(3) Poletown, Detroit, where GM bulldozed an entire community to build a high-tech plant and now runs it on only one shift. Workers are resentful of broken promises of participation and respect from management; teams function minimally. The union pretends that its functions are the same as in a traditional plant but in fact tells workers, 'We don't do things that way over here.'

Japanization?

A Japanese term associated with the team concept is *kaizen*, or continuous improvement. At NUMMI, for example, management has dealt with reduced market demand by giving (otherwise redundant) workers the task of observing the production process and making suggestions for increased efficiency. Each time a task is made more efficient, the worker is expected to perform even more tasks per unit time. In this very 'flexible working', each worker becomes 'multi-skilled' by learning a series of job-specific tasks that depend mainly on manual dexterity, physical stamina and the willingness to follow instructions precisely. As in Japan, management continually appropriates workers' knowledge for ever-increasing productivity, though *kaizen* does not require the particularities of Japanese culture in order to work.

The debate as to whether US or British industry is being 'Japanized' is in many ways beside the point for purposes of analysing what the new management systems mean for workers. It is, in fact, perhaps preferable not to emphasize the term Japanization because of the potential for putting a racist construction on the characterization of the new management methods. The country of these methods' origin is less relevant than their employer origin, and that they are for the specific purpose of maximizing efficiency at the expense of workers.

That said, it is clear that US companies do look to Japan. In the USA the team concept exists in its purest form at the Japanese-owned or -managed plants, such as Honda in Ohio, Mazda near Detroit, and at NUMMI, which is run by Toyota. The management methods installed were brought from Japan and modified as little as possible to conform to American habits. The summer 1987 joint pilgrimage to Japan by the GM and UAW negotiating committees is only one sign that both management and union see their salvation in Japanese methods.

WHY NUMMI IS THE MODEL

In just two years the Fremont, California, assembly plant of NUMMI went from 'interesting experiment' to *the* success story in US automobile manufacturing. In producing the version of the Toyota Corolla sold as the Chevrolet Nova, NUMMI became both the standard and the model for the US auto industry, in several ways:

(1) There were massive increases in labour productivity. A GM official proudly introduced the plant by pointing out that it only took 14 hours of direct labour to assemble a Nova as against the 22 hours required by General Motors to produce the J car in the same plant (GM, 1985). General Motors circulates weekly comparisons showing the labour efficiency of its various plants. The chart for 11 May 1986 shows NUMMI way below the standard (100.0) on every type of direct labour except inspection: trim – 57.8, paint – 60.8, total – 61.2. For some types of labour which other GM plants use, NUMMI has a score of zero, either because it does not have that category at all (e.g. industrial engineering, vacation replacements) or because it contracts the work out (e.g. plant security). NUMMI was number

one on almost every category of direct and indirect labour.

(2) Despite the fact that it was a new product on a new line with and new management – all conditions which create years of quality problems in US manufacturing experience – quality at NUMMI quickly climbed and stayed at the top.

(3) NUMMI did these things not with a new plant but by taking over a traditional GM assembly plant.

(4) The technology of the plant was not particularly advanced. Indeed, there was less automation than at most US assembly plants.

(5) Most of the workforce had worked in the same plant when it was run by General Motors. At that time it had the reputation in management circles of being militant, causing lost time through wildcat strikes, having major absentee, drug and alcohol problems and a lack of concern about product quality. Most of the old union leaders continued in office. And yet a change in management transformed the plant into one noted for high productivity, co-operation and low absenteeism.

(6) All of this was accomplished without a confrontation with the union. Indeed, local and national union leaders are some of the biggest public relations boosters of NUMMI. Critics of the local leadership refer to President Tony De Jesus as a 'tour guide'.

To management, the lessons of NUMMI were clear cut. You could take an American UAW workforce and an ordinary factory and in two years, by changing the management, cowing the union and altering the collective bargaining agreement, you could achieve productivity and quality competitive with the Japanese.

☐ 'MANAGEMENT-BY-STRESS'

What the changes in management and the union allowed was the implementation of a total management system which we have dubbed 'management-by-stress'. This goes against many traditional management notions and also against common-sense notions of effective management. The goal is to stretch the system like a rubber band on the point of breaking. Breakdowns in the system are thus made inevitable but are in fact welcomed, because they show where the weak points are, weak points that can then be immediately corrected.

Key to understanding the management-by-stress production method is that it is in fact a system. The approach in each part depends heavily on a corresponding and consistent approach in the other parts:

(1) Speedup.
(2) Just-in-time (JIT) inventory control.
(3) Extensive use of outside contracting.
(4) Use of technology.
(5) Design of products to reduce labour hours required.
(6) Methods to reduce scrap and rework.
(7) Flattened management structure.
(8) Absentee control.
(9) Elimination of job classifications for maximum flexibility.

The *andon* board illustrates how management-by-stress

works. *Andon* is a visual display system, usually including a lighted board over the assembly line showing each work station. Most *andon* displays use one or two colours combined with chimes, buzzers or sirens. For illustration purposes, imagine a variation where the status of each station is indicated with one of three lights:

GREEN – production is keeping up and there are no problems;
YELLOW – operator is falling behind or having difficulty keeping up and needs help;
RED – problem requires stopping the line.

A yellow light flashes when an operator pulls a cord. The red light comes on and the line stops if the cord is not pulled again within a fixed interval, which varies from 15 to 60 seconds.

In the traditional US operation, management would want to see nothing but green lights. In fact it would design enough slack into machinery, operations and labour power so that the plant could always run in the green. Individual managers are motivated to have excess stock and excess workers on hand to cover breakdowns or to deal with emergencies. 'CYA' (cover your ass) is the standard operating procedure.

Under management-by-stress, however, 'all green' is not a desirable state. It means that the system is not running as fast or efficiently as it might do. It is far preferable to have yellow lights flashing fairly frequently – indicating that operators, and the system as a whole, are being stretched to their limits. As it becomes obvious wherein problems lie, management can focus on those points and make necessary adjustments.

A NUMMI manager explains the process to new workers:

How many of you worked in this plant before? Ever shut the line down? What happened? Everything broke loose. The plant manager, plant superintendents, assistant superintendents, foremen, general foremen, everybody became unglued. We don't get unglued here. It's a different world. It's okay to shut the line down. It's okay to make a mistake. It's okay to cause a problem because that's an opportunity for us to change and do something just a little better the next time around . . . (GM, 1985)

Once the system has been fine-tuned, it can be further stressed, perhaps by turning up the line speed or by cutting the number of workers. Resources can be taken away from stations which are always green. The ideal state is achieved when the plant is running with all stations just on the line between green and yellow.

The idea that the system 'equilibrates', or drives towards being evenly balanced, has great appeal to the engineer or technocrat, particularly after seeing years of waste and mismanagement. The opportunity for tighter control should be obvious. The problem is that most of this control is being applied to human beings.

☐ Just-in-time

The advantages to management of just-in-time (JIT) are well known: reduction of inventory costs and faster quality control. But there is another attraction: the elimination of cushions and the facilitation of stress.

The traditional system has been dubbed by a British researcher 'just in case'. Banks of parts are maintained, and one section of the production system is thus cushioned from problems in another section. There is time to fix a problem before it affects the next section. JIT, on the other hand, would seem to create the potential for innumerable halts in production.

But this seemingly negative feature of JIT becomes positive under a management-by-stress system. When an operation experiences trouble of any kind, there is no hiding it. The problem becomes immediately apparent to all. Foremen will come running to deal with the immediate problem, and then management at all levels will focus attention on the weak spot. There is no place for cushions in management-by-stress.

Taylorism and speedup

Most of the current industrial relations literature portrays the team concept as a humanistic alternative to the ideas of 'scientific management' originated by Frederick W. Taylor. *Business Week* editorializes: 'Such team-based systems, perfected by Japanese car makers, are alternatives to the "scientific management" system, long used in Detroit, which treats employees as mere hands who must be told every move to make' (31 August 1987).

This is part of the fantasy that is being constructed around these new systems. In fact, the tendency is in the opposite direction – to specify every move a worker is to make in far greater detail than ever before. Far from a retreat from Taylorism, management-by-stress intensifies it, producing a kind of super-Taylorism. The 'takt' time (how many seconds each worker has to work per car) and the 'cycle time' (how long it takes an individual worker to complete her job) are carefully measured. The aim, of course, is to have the cycle time equal the takt time as nearly as possible.

Not only the cycle time for all tasks but every motion within those tasks is carefully measured. At NUMMI a worker is told when to move her left hand towards her gun, how long it must take her to pick it up, how many steps to take back to the car, and so on until up to fifty-nine seconds of work per minute are specified.

Jobs are to be done in precisely the same way every time, by every worker. If the charting calls for holding the part with your right hand and tightening with your left, that is how it must be done. The worker may not change the way he does the job without permission. The company explanation is that this is how quality is maintained. The argument becomes circular when we realize that management's definition of 'quality', now taught to rank-and-file workers in joint management–UAW programmes, is 'conformance to requirements'. The rigidity is illustrated by the comments of one team leader at Mazda (a new plant near Detroit which rivals NUMMI for management-by-stress) who noted with pride what he considered the system's flexibility: 'We make allowances for people who are left-handed.'

While the logic makes sense from an engineering point of view, it is hard on the human elements. There are, of course, problems with falling behind when jobs are so closely charted. Short people might prefer to do a job differently from tall people. There is no chance to change the way of doing a job in the middle of the day, to give some muscles a chance to relax and to use others. Worst of all from the workers' viewpoint, there is no chance to 'work up the line' – work faster for a short while to obtain a breathing spell. Work is continuous and unvarying.

☐ Appropriation of workers' knowledge for job design

One of the hallmarks of Taylorism was the removal of brainwork from the shop floor. Management-by-stress, on the other hand, seeks to utilize workers' sense of observation, recognizing this as a valuable tool which should not go to waste. Management demands that workers make available their thoughts about the production process.

The new Taylorism of management-by-stress recognizes that trying to take all decision-making power off the shop floor to the executive offices has certain limitations. The people who actually do the work understand the process better than distant observers. So the formula is to get management involved on the shop floor:

(1) Lower-level management knows the production process because group leaders (foremen) frequently work on the line itself – something forbidden by traditional UAW contracts.

(2) The team leaders (union members) are effectively

incorporated into management. A key part of their jobs is to document worker knowledge for use by supervisors.

(3) Group leaders and team leaders are essentially the people who have responsibility for designing the jobs and adjusting them. At NUMMI there is no separate industrial engineering or time study department. The group leaders and team leaders have a vested interest in the successful application of the initial job design and charting they did during the 'trial-build' period, and they are in a position to monitor the workers continually.

Proponents of the team concept often say that workers in such plants get to participate in designing their own jobs. There is some truth to this statement, but this prerogative does not work in workers' interests. The plant's basic production processes, with its lines and machinery, are, of course, developed by corporate engineering. In the cases of both NUMMI and Mazda, the processes and the equipment were designed and tested in Japan. The lines are installed and the trial-build period is begun, in a process which can take from weeks to months to build only a few hundred cars. Finally, production is started on one shift but the line runs slowly. Only over a period of months is the line brought up to speed.

During the initial trial-build period, the 'teams' of workers consist of engineers, group leaders, and team leaders, the hourly workers selected by management. These are the people who initially design the jobs. Gradually, regular workers are added. During this slow period they too may have input into job design.

Once production is running, workers are still encouraged to suggest how their jobs could be done 'better', i.e., more efficiently. In fact, they are forced to make such improvements just to keep up. The catch, of course, is that they are not allowed to keep to themselves the tricks they learn about doing the job faster. Such knowledge is appropriated by management – who then uses it to add still more work to the job.

☐ Absentee policy

Another key element in maintaining a taut system is the policy towards absenteeism. At NUMMI, and at Mazda, there are no extra workers hired as absentee replacements. A team consists of 4 to 8 people and a team leader. The team members all have full jobs carefully assigned through the charting system. There is no slack. The team leader has no regular production job but an extensive list of assignments. These include supervisor-type responsibilities such as keeping track of absenteeism and tardiness, distributing tools or gloves, dealing with parts-supply problems, and training team members on new jobs. In addition, the team leader helps out when a team member is having difficulty with a job, fills in to provide bathroom or medical breaks, or when a team member must go to the repair area to correct a defect.

If a member is absent, normally this means the team leader has to do the production job of the missing member. *Then* if team members need relief or help they must depend on the group leader (who supervises 2 to 4 teams), either to fill in directly on the job or to assign someone from another team to help out.

Again, stress built into the system keeps it functioning. All the difficulties of someone being absent fall on those who are in daily contact with the absentee – his co-workers and immediate supervisors. The problems are not shifted upstairs by having the personnel department hire a 'redundant' workforce to cover for absenteeism.

Other team members find it is more difficult to get relief when they need it. Their regular jobs are also harder, because if a replacement team member has difficulty with her job, it can interfere with the pace of the other jobs. As a result workers tend to resent the absentee who, within the assumptions of the system, is the cause of the problem. Several workers interviewed at NUMMI commented that they would like some way to have people who were absent too much removed from their group.

Stopping the line

'Workers can stop the line' is the single feature that has come to symbolize the difference between management-by-stress production systems and 'the old way of doing things'. The companies represent this power as the foundation of their policy of respect for the workers' humanity. 'It is not a conveyor that operates men,' says the Toyota production manual, 'it is men that operate a conveyor' (Monden, 1983). On the line at NUMMI, 'pulling the cord' results in distinctive chimes and flashing lights on the *andon* board. If the cord is not pulled again within a set time to cancel the warning (usually one minute) then the line will stop.

At first hearing there is tremendous appeal to having the cord. In the traditional practice of the Big Three auto firms, workers did not stop the line unless someone was dying. It did not matter that you couldn't keep up or that scrap was going through. You tried to get the foreman's attention and he could decide whether to stop the line (very rarely) or leave the problem to be repaired further on. In most plants, stopping the line meant disciplinary action.

The right to stop the line is supposed to be a substitute for union-negotiated work standards. Why have a cumbersome procedure for deciding whether a job is too hard? Instead we can have a system that trusts the worker. If the worker is making a genuine attempt but cannot keep up, all she has to do is pull the stop cord. There is no penalty.

In traditional plants, in contrast, company industrial engineers set the standards for a job – which operations in what period of time. The union has the right to protest them and even to strike over them during the life of the contract. Once established the standards cannot be changed arbitrarily.

The NUMMI contract specifically allows for easy change of production standards by the group leader. The union is not involved in the initial design of jobs. Production standards are not grievable but use a different procedure which can involve appealing to a joint union/management committee.

During trial-build and training periods, the cord seems to work for everybody. It helps workers get the assistance they need when problems come up; it helps keep up quality throughout the difficult process of establishing a new line; and it helps management identify and quickly respond to the problem.

However, once the job is well defined and most of the bugs are worked out of the production process, the cord can become an instrument of oppression. As the line is speeded up and the whole system is stressed, it becomes harder and harder to keep up all the time. Once the standardized work – so painstakingly charted, refined and recharted – has been in operation for a while, any problem is assumed to be the fault of the worker, who has the burden of proof to show otherwise.

The result is that workers become hesitant to pull the cord. A Mazda worker describes the Catch 22 in which a co-worker found herself.

> She had a hard time one day and pulled the stop cord several times. The next day management literally focussed attention on her. Several management officials observed and they set up a video camera to record her work. She found herself working further and further into the hole. She got too far past her station and fell off the end of the platform and broke her ankle. They told her it was her fault – she didn't pull the stop cord when she fell behind.

To those who have experience with labour relations, it will be obvious that the potential weakness of the management-by-stress system is the union, or union consciousness on the part of the workers.

SHREVEPORT

The Shreveport, Louisiana, GM plant represents an intermediate version of the team concept. The system is not as taut as at NUMMI, although most movement is in that direction. Twenty-five hundred hourly workers on two shifts build 720 pickup trucks and sport utility vehicles per day at Shreveport. They are assisted by only twenty-eight robots.

Shreveport is GM's flagship 'team concept' plant. Team concept was an integral part of the plant's functioning from the time it opened in 1981. Long before NUMMI, the Shreveport plant embodied Japanese-style methods in the way it organized production and in its 'participatory' style. That style was chosen, however, solely by management, which researched other 'employee-centred' companies and culled what it deemed best. At that point GM had not realized the advantages of 'jointness' with the UAW.

Shreveport was originally part of GM's 'Southern strategy', conceived in the 1970s to escape the UAW by building non-union plants in 'right-to-work' states (whose laws banned the closed shop). In 1979, however, the UAW won an agreement from GM to allow Northern union members to transfer to the new Southern plants – and to bring automatic recognition of the union with them. Thus GM lost its opportunity to start from scratch with both a greenfield plant and a brand new, non-union workforce.

GM, however, has done well for itself even with the existing workforce, now half UAW transfers and half local hires. For different reasons, both transfers and locals have embraced or learned to tolerate the team concept. And mutual suspicion between the two groups has also worked to the company's advantage. Local hires, even those new to the auto industry, have been successfully taught that 'traditional' plants are bad and 'modern' ones good; this attitude weakens the position of those workers seeking to defend such union 'traditions' as seniority rights.

Many union officials, almost all of whom are transfers,

say they are not happy with the team concept. Yet the union appears in fact to be quite at home with it. One rank-and-filer says, 'The reason the team concept has worked so well in this plant is because the union worked so hard to make it work.'

Teams are larger at Shreveport than at NUMMI, typically of fifteen workers. They choose their 'team coordinator' by consensus from among themselves (at NUMMI the 'team leader' is appointed by management). They meet with their foreman (redubbed 'general technician') for a half-hour once a week, on overtime. Attendance at team meetings was made voluntary, at the union's insistence, in the 1984 contract. However, attendance is still high because, as one worker put it, 'it's the easiest time-and-a-half you ever made'. (At NUMMI, teams meet only as needed; when they do, attendance is expected.)

UAW officials point with pride to the voluntary clause. They seem to feel that it solves most of the problems union members may have with the team concept: if an individual doesn't like team meetings, he doesn't have to come. There are three problems with this approach. One is that meetings are a very small part of the team concept as a whole; much more important is the way production is organized and the effect on seniority rights. The second is that the structure is still in place, very little disturbed by the absence of the few dissenters who choose not to come. Third, decisions made in team meetings are still binding on those who choose not to attend.

Much is made of the team's decision-making powers. The decisions it can make, however, range from the small to the trivial. Team decisions include (1) how vacant jobs within the team are filled (usually plant seniority or team seniority), (2) how long the team co-ordinator and other positions serve, and (3) when the team meeting is held (before or after work).

Besides the team co-ordinator, each team has a plethora of positions to fill – safety co-ordinator, quality co-ordinator, recording secretary, timekeeper, salvage co-ordinator, planner. Each position is recallable. Sometimes no one volunteers for the jobs and the team co-ordinator must carry out their functions as well.

If it is near model-change time, the team meeting will include a report from the team planner. This worker helps implement job redesign required by changes in the trucks or in the mix of vehicles. Management's general aim, of course, is to reduce the number of jobs, so the planner's job often involves helping to do this.

Larry Spinney, a 'general technician' (foreman) on the Trim line, explained the process:

> In Trim we're becoming more efficient, eliminating operations, and we're not losing quality. The people set up their own jobs. We'll make a presentation to the team: 'This is the efficiency of this team' – that's the minutes per hour on your job description. Our goal is to have every job set up so that they're working 48 to 52 minutes out of every hour. For example, if the line was running at one per minute, and it took you 52 seconds to do your job, that's 52 minutes per hour.
>
> So in the presentation we tell them, if every operation in the team was set at 48–52, it would take x amount of people to do this team's work. That figure could be 12, but there's 15 currently in the team. So they need to eliminate 3.
>
> People will say, 'Well, I can take this on my job,' and someone else will take another part of it. It's worked out by the team planners.

Shreveport operates under the 'pay-for-knowledge' system which is standard in American team concept plants: workers can earn more money by learning more jobs. A worker who knows all the jobs in her or his team earns a base rate of $13.28 plus cost-of-living allowance (which all production workers receive at NUMMI), while a worker who knows the minimum earns only $12.82. Most workers are in the top two pay grades.

One reason workers are willing to learn many jobs is that management does not often take advantage of its right to shift them off their regular jobs onto other ones. This is possible because each team includes one Absentee Replacement Operator, a category which does not exist at NUMMI and a major difference with the Japanese-run plants.

The rotation through different jobs – which is emphasized in the team concept literature and training as one of its most important features – rarely happens. Most workers choose the stability of a regular job, although they may rotate if the foreman does not mind.

Life in the plant is not as regulated as at NUMMI. Workers are permitted to eat, drink or smoke on the line, and even to play radios. Visitors are encouraged, providing a somewhat casual atmosphere very unlike traditional plants, or NUMMI, where workers must sign out and lose pay even to go to the carpark at lunchtime.

☐ Worker involvement in quality

One of the most striking things about the Shreveport plant is the apparent commitment to quality by both workers and management, though this commitment does not include allowing each operator a generous amount of time to do the job. As in other GM team concept plants, management's twice daily audit of vehicles chosen at random is reported on at an open meeting attended by both salaried and hourly employees. The quality co-ordinator from each team is supposed to attend. As with many of the team 'offices', the chance to get off the line for a break is one incentive for taking the job.

At the day shift audit on 14 August 1987, the plant had failed to reach its goal of an average of 3.55 'discrepancies' per vehicle; its score was 4.50. Discrepancies include loose bolts or screws, fenders that don't fit quite right, and 'dings' (tiny dents). The quality co-ordinator from each team where a discrepancy occurred introduced him or herself ('I'm Linda from Team 22') and reported on what had gone wrong.

Usually the cause of the defect was unclear, but most co-ordinators used the same formulation: they blamed operator error while defending the operator in question as a good worker. The quality co-ordinator from team #57 in Trim, for example, said, 'Usually that operator is very competent. We will get back with him.' Not once was it suggested that perhaps the job was designed wrong, although suppliers were occasionally blamed. Although not nearly as efficient (nor flashy) as NUMMI's *andon* system of identifying errors, the public audit and *mea culpa* session also have the effect of pinpointing blame on individual workers.

The meeting, then, was essentially a recitation of operator errors, accompanied by promises and exhortations to 'watch these jobs a lot closer and try to do better'. The hourly workers attending had no control over what went on in other areas of the plant, so it was unclear how they were expected to benefit from knowledge of minor errors made elsewhere. The whole meeting lasted only twelve minutes. The audit session seemed more than anything else to be part of management's general strategy of making workers feel responsible for quality and see the production process as a whole.

Shreveport management realizes that it cannot change attitudes entirely, and the agreement to let team meetings be voluntary was a recognition of this fact. Yet it would like workers to stop feeling 'I just come to work, do my job, and mind my own business.' Most Americans firmly believe they have a right to take this attitude, if they so choose. That one should be one's brother's keeper *on management's behalf*, and as part of the job description, is a new notion for them.

☐ Why workers care about quality and efficiency

At the 14 August audit the Trim Department scored 1.13 against a goal of .70. One of the quality co-ordinators said, 'Trim isn't looking very good this morning, but we're not going to hold our heads down, because tomorrow is another day.' According to other Shreveport workers, this was a sincere expression of sentiment.

The audit ended with the quality analyst warning that 'Moraine's last audit was 2.8; I don't have to tell you people any more than that.' GM's Moraine, Ohio, plant builds the same pickup as Shreveport. Moraine and another GM plant in Michigan are commonly seen as 'the competition'. Fear of the competition creates a deep-seated anxiety about the future of the Shreveport plant.

The plant dates only from 1981 and (according to the union) consistently ranks first or second in efficiency, quality and attendance in the truck division, but Shreveport workers do not feel secure. Engrained in the culture of the plant is the notion that it could be chosen for closure at any time if it does not measure up against its competition. The fear of losing their GM jobs makes Shreveport workers amenable to whatever management suggests is the way to make their plant number one.

This overriding fear exists despite the fact that there has never been a layoff at Shreveport. Any job cuts made possible by improved efficiency have been handled by attrition. Union officials say there are about sixty surplus people currently on the payroll, filling in for absentees or cleaning up.

The fear for job security works for both local hires and transfers. For the locals, the job at GM is almost always the best-paying job they've ever had, and by far the best-paying factory job in the area. And the transfers, generally more union-minded, are if anything even more determined than the locals not to be the victim of (another) plant closure. They have moved once to retain their GM seniority, and they don't want to uproot their families again. Jack Ross, the union's full-time joint programmes staffer, explains: 'They lost their families and homes once and they don't want to do it again. It was a plus for this plant – they got scared, and that's why they're involved more.'

☐ Worker attitudes

Workers at the Shreveport plant display a wide range of attitudes towards the team concept, varying from 'this is great – it's like family' to 'it sucks', with all shades in between. Most union members interviewed shared one complaint, however: that management was not living up to its promises about the team concept and was sliding back into traditional ways.

The team concept philosophy is supposed to include (1) an atmosphere where workers are encouraged to have input into decisions and status differences are broken down, and (2) an atmosphere in which 'we're all in this together' as the plant as a whole strives to beat the competition. In other words, workers are to be treated with dignity and as equals (in some regards) with management, and in return are to adopt management's goals for the plant – even when this means more work for them.

Jim Hodge, vice-president of Local 2166, says that the team concept has benefited the workers because it has saved the corporation money by allowing fewer workers to be employed. He believes, however, that it is 'weaker now than when it first started. The foremen now have a bad attitude, and it's tearing down the team concept.' Fred Smiley, a union representative, believes that the plant 'would be utopia if management had the same commitment to the team concept that the employees have'.

Joe Pietrzyk, who transferred from Buffalo, says:

> In theory the team concept seems like a good idea, but in practice it's the same old garbage, more like a management concept, a one-way street. Any experienced worker will use any theory to his own advantage. If you

> participate, they [management] don't bother you. Opportunities come a little easier if you're a Level VI, like a day off when you want it. If you're a brownie [brown-nose or ass-licker] you'll work out fine.

One measure of the team concept's popularity is two votes held in 1982 and 1984 on whether to try to get rid of the pay-for-knowledge system. Both times the vote was to keep pay-for-knowledge – understood as the team concept package – although the majority was very narrow the second time. Subsequently, whether to keep the team concept was not a question; instead team functioning was codified in the union contract for the first time.

It is difficult to describe an 'average' attitude for a 'typical' Shreveport worker. It does seem clear, however, that there are fewer workers than in most GM plants with either a militant union attitude or an 'I don't care about the job, I'm just here to collect my pay' attitude. This is probably due both to the lack of union experience of half the workforce and to the team concept. In any case, apparently no one feels that it is worthwhile to buck the team concept *itself*. After eight years, it appears to be in Shreveport to stay. Joe Pietrzyk sums up the experience of many transfers: 'Either you accept the team concept or you don't. And sooner or later, you will.'

POLETOWN

At GM's Poletown plant in Detroit, which began production in late 1985, the forms of the team concept exist with little of the content – except for the speedup and the weakening of the union. The familiar structures exist – team co-ordinators, team meetings, pay-for-knowledge (on paper). But workers whose hopes were raised by the team training they received in 1984 and 1985 are no longer even disillusioned. Management continues its autocratic methods. In contrast to the traditional plants from which workers came to Poletown, this plant features speedup, fewer classifications, fewer good jobs and a thoroughly discredited union. Thirty-four hundred workers on one shift build about 300 Cadillacs, Buicks, Oldsmobiles and Allantes.

Ray Church, the first union shop chairman, emphasizes that at Poletown the team concept was imposed by management from the beginning. When he was appointed chairman of the unit and transferred there from the Fleetwood plant in February 1984, team training was already under way: 'They didn't ask us about team concept, or pay-for-knowledge, or combined classifications. We never as a local union signed team concept or pay-for-knowledge, we weren't given the option.'

Teams of 20–30 members hold voluntary meetings once a week for half an hour. Unlike at Shreveport, where workers must demonstrate that they know all the jobs in their team to make the top pay of $13.28, at Poletown $13.28 is the standard wage. In theory, workers are to know the jobs in their team, and management has the flexibility to put them where needed. In practice, however, workers do not necessarily get trained on many jobs and they have regular jobs from which they are seldom moved.

Team meetings accomplish little. One worker, Ron Banks, said that his team had never discussed problems on the job:

> The bathrooms are dirty, there's trash in the parking lot – this is our input. Can we get more tissue paper in the bathroom? Nothing meaningful. Questions about the job never come up. I can't recall a time when it has. Sometimes we'll watch a taped message on our TV from the plant manager, a pep talk on absenteeism or quality.

Most people do attend the team meetings. 'We're usually out within ten minutes,' one worker says. 'Our foreman is good, but some make you stay the whole half-hour.'

Virtually all Poletown workers come from other, traditional Detroit-area plants. Ray Church says that they don't buy or trust the team concept: 'People have seen so many damn programs at GM. They believe it's all a plan to get more work out of you. They saw how QWL [quality of work life] caused favoritism and split the membership at Fleetwood.' Bill McGuire, a rank-and-filer, agrees: 'If QWL was phase one, this is phase 4 or 5 over here.'

Still, many former Fleetwood and Cadillac workers were hopeful that management meant what it said in the 4 to 6 weeks of training that they received before starting work. Training was run jointly by the union and management. Workers were told that they would rotate through different jobs and that these would be ergonomically designed. They would be able to take their problems to the team. The supervisor's decision would still be final, but the need for such an event would be rare.

☐ Work with dignity?

Ron Banks has examples of how he tested management's supposed commitment to treating workers with dignity – before he gave up in disgust:

> When I was on afternoons, our last break was at one a.m., an hour or so before we went home. This suited everybody I knew fine. Then they changed it to midnight. I went to the superintendent and asked wasn't this something we could vote on? He said management wanted it earlier and they had talked to the shop committee and the shop committee said they'd canvassed the membership and the majority wanted it earlier. So then I asked the committeeman about it. *He* said management had just informed them that they were going to change it.
>
> Another time I was working QIS [quality inspection system], punching a VDT [visual display terminal], and I had to stand on concrete all day. My knee hurt, it would throb at night. In the training they had emphasized ergonomics – 'don't work harder, work smarter', 'fit

> the job to the person'. So I thought sure I would get a stool. First I brought it up in the team. The foreman said it couldn't be done. Then I asked the health and safety committeeman, and he practically laughed. He said, 'They'll never give you a stool.' So then I went to the management health and safety guy, while the union health and safety man just sat there and grinned, he didn't say a word for me. The management guy sent me to another management guy, and that one got real belligerent. He acted like I had to be crazy to be asking for a stool.

Rotation is honoured more in the breach than in the observance. When production began and the line was slow, workers learned all the jobs in their team and switched jobs regularly. When the line got up to speed, however, management ordered rotation to stop because of quality problems.

'One of the main things at our team meetings was the company trying to get rotation going again,' says Mary Kowalski. 'They said if you didn't rotate within a year your pay would be cut. But they didn't enforce it. The cars were coming out so bad they couldn't put people on different jobs. You can't go from job to job and do a quality job.'

Kowalski believes that management would prefer to have workers rotate:

> The company wants you to rotate because they don't want you established on the same job so you can take pride in it. You don't have the pride of saying, 'This is *my* job.' A certain job belongs to you, but when other people come on it who don't have to do it all the time, they see ways work can be added onto it.

Bill McGuire believes that GM's initial emphasis on rotation had an additional motive:

> Say they've got 10 people in the team. They've got 5 good jobs and 5 bad jobs, including one or two *real* bad. They want to see who will do the real bad jobs. Everybody will say, 'I can do this for one day, or one week.' You want to keep it up, because you want to be equal with the rest of the team. Before long more of the jobs are bad.

There is no pretence at Poletown that workers help to design their own jobs, not even the Shreveport method of choosing the team planner who is involved in the modifications made at model change time. But management would clearly like to have the teams implicated in speedup decisions. Bill McGuire tells the story of a woman whose foreman (no one says 'team manager') informed her that more work was to be added to her job. When she protested that there was no way she could do more work, he told her, 'Then take it to the team and see how they decide to divide it up.'

The foundering of the team concept at Poletown is due to two factors: old-fashioned management insufficiently committed to a Shreveport way of operating, and the fact that the company mostly has what it wants (although not quality) even with a bare-bones version of the team concept. It has done away with troublesome classifications, it has the right (though seldom exercised) to move workers around at will, and it has a union which refuses to challenge its decisions even at the insistence of members. Workers who want to file grievances on overloaded jobs are told by their union representatives that 'We don't do that over here.' As one Poletown worker has put it, 'The new word for concessions is team concept. It's company, company, company.'

Poletown workers overwhelmingly voted down the shop committee's first attempt at a local agreement in October 1986. According to McGuire and Banks, there was not a great deal in the agreement besides the team concept. Team co-ordinators' pay differential would have been raised from

10c to 50c, which was widely resented. 'There was nothing weird in it,' Banks says. 'They just needed something to latch on to as a reason to vote it down. It was just a vote against Poletown.'

EFFECT ON THE UNION

The UAW International has actively collaborated in the implementation of the team concept in US plants. The International knows full well that the effect of the team concept in an individual plant is speedup and, at least eventually, fewer jobs; it may understand less well that another effect is a dilution of the union's shop-floor power, as shop-floor power is not something union leaders have cared much about for some time now. The other effect of the companies' drive for the team concept has been a lessening of solidarity among the UAW's local unions.

The evolution and hardening of the International's attitude was rather fast. In June 1985 UAW President Owen Bieber insisted, in the face of internal union criticism, that the special agreement the UAW had made with GM to implement teams at its Saturn plant was an isolated case, meant to keep small car production in the USA. 'We're not going to Saturnize the auto industry,' Bieber said.

By early 1986 the International was twisting the arms of local unions at Chrysler to force them to accept 'modern operating agreements'. Chrysler had threatened to close plants or move work if they did not open their local contracts. At the Trenton Engine plant in Michigan, local union members, angered by International representatives' 'take it or quit' attitude, at first voted down the new agreement in spite of the threat of lost jobs. The International scheduled another vote on the same contract, with no changes, inside the plant; it carried easily.

Getting fewer members to do more work is part of the UAW International's overall strategy for saving jobs in the USA – that is, to save *some* jobs by cutting others. Union leaders have taken upon themselves the task of helping their employers to compete with auto companies in Asia and Latin America (which, of course, are often wholly or partially owned by those same US companies). Under this rubric, they are eager to help the employers become more efficient, and the employer-initiated team concept is one fine way to do so. As one Chrysler worker has put it, 'They're cutting out the jobs we used to depend on so we wouldn't work ourselves to death before we retired.'

☐ Whipsawing

General Motors in particular has mastered the art of 'whipsawing' – playing one plant against another with the promise of work and the threat of closure. The company dangled its medium-duty truck production in front of two locals in Wisconsin and Michigan, for example, urging each to adopt the team concept and 'be in the running'. The local which said 'Yes' to the team concept got the truck.

The union cannot stop GM from making such offers; the only effective strategy would be a united union policy of no local deviations from contracts. Far from adopting such a policy, the International has publicly taken a hands-off attitude, leaving each local to decide on its own. At the UAW Convention in April 1987, the International strongly resisted a proposal to include in its bargaining resolution language against whipsawing. In practice, the International has made it clear that it wants the team concept. When GM has installed teams in its newly built plants, such as Poletown, Shreveport and others, the union has accepted the change without a peep.

Clearly the local unions do not have the power on their own to resist whipsawing. Faced with the threat of their plant's closure, most local union leaders and members will choose speedup over loss of their jobs. The result is a sharp decline in solidarity among local unions.

☐ Loss of shop-floor power

Even back in the 1950s, Walter Reuther referred to auto plants as 'gold-plated sweatshops'; workers were well paid for very hard work. Shop-floor organization long ago gave way to a very slow and not very effective grievance procedure. Even so, the traditional contracts gave workers some protection from company whims and speedup. Stewards had some ability to intervene on workers' behalf. Under the team concept, both management and union tell a complaining worker, 'The team agreed on it, there's nothing we can do.' Even the old bureaucratic structures tend to wither. There has not been a quorum at a union meeting at NUMMI for nearly two years.

Only when there are rules can workers mount a 'work to rule', a slowdown. If there are no rules, but only management 'flexibility', one more form of organized rank-and-file level protest is lost. Thus the introduction of the team concept at this juncture means that more and more areas of shop-floor life are removed from the union's jurisdiction, precisely when the union is making a wholehearted commitment to productivity.

So far, the auto companies have not mounted a major offensive against wages. There have been only cost-of-living wage increases since 1984 (although there have been bonuses) but neither have there been cuts. The union seems committed to maintaining its members' pay levels; perhaps it believes that this is what is most important to them and most likely to prevent rank-and-file rejection of the contract now being negotiated.

It is not hard to predict, however, that having once gutted the union on the shop floor and thoroughly inculcated both leaders and members with the competitive imperative, the companies will want whatever competitive necessities are next on their list – a two-tier wage scale, pay cuts, more freedom to outsource. To quote a Poletown worker: 'Once they get your power away they're coming after your money.' The UAW is living a fantasy if it believes that, once weakened, it can give to the companies only the concessions it chooses to give. With its infrastructure in the plants eviscerated and membership respect for the leadership gone, it will have nothing with which to fight back when push comes to shove.

The catch for the US auto industry is that the speed-up and loss of union power involved in the team concept have set off the beginnings of a revolt. In 1989 the UAW hierarchy was shaken by an impressive though unsuccessful challenge from a movement that called itself 'New Directions'. Anger over the team concept, and other forms of company–union collaboration, was the chief fuel of this revolt. Explaining why he campaigned for a dissident candidate for regional director of the union, one Poletown worker said, 'He's the first one that told the truth about the team concept – that it's a company thing.' At GM's Wentzville, Missouri, plant, several hundred workers have held demonstrations during the time allocated for team meetings; management has had to ask the leaders of these protests what the workers want, completely bypassing the elected union officials.

Such revolt is possible on a wider scale. In the 1930s in Flint it was the work pace and company arbitrariness, not low wages, which led workers to occupy their plants and to join the UAW. An auto worker of the 30s whose name is lost to history spoke perhaps for the auto workers of the 90s when he said, 'I ain't got no kick on wages, I just don't like to be drove.'

□ NOTE

This essay is based on the authors' book, *Choosing Sides*, available from Labor Notes, 7435 Michigan Ave, Detroit, MI 48210, USA, price $15 plus $3 shipping (surface) or $11 (airmail). For their research the authors interviewed rank-and-file workers and union officials from NUMMI, Chrysler's SHAP and Trenton Engine plants, General Motors' Shreveport, Poletown, Buick City, Oklahoma City, Oshawa, Lake Orion, Wentzville, Detroit Diesel, Chevy Forge, Fairfax, and Van Nuys plants, Ford's Hermosillo plant, and Mazda's Flat Rock plant. Documents such as union contracts, plant philosophy statements and the like were reviewed as well.

□ REFERENCES

GM (1985) 'This is NUMMI', videotape for GM Managers. Detroit: General Motors Technical Liaison Office.

Monden, Y. (1983) *Toyota Production System: Practical Approach to Production Management*. Industrial Engineering and Management Press.

Parker, M. and Slaughter, J. (1988) *Choosing Sides: Unions and the Team Concept*. Boston: South End Press/Labor Education and Research Project.

This essay is an edited version of a paper originally written in late 1987.

[18]

APPLIED PSYCHOLOGY: AN INTERNATIONAL REVIEW, 1996, *45* (2), 119–152

The Psychology of Lean Production

James P. Womack, Massachusetts Institute of Technology, USA

Commentary on "Compatibility of Human Resource Management, Industrial Relations, and Engineering Under Mass Production and Lean Production: An Exploration" by Koji Taira

As one of the originators of the term "lean production" (although certainly not of the ideas behind the name, which I and my colleagues simply synthesised from the practices of a number of Japanese firms led by Toyota), it is a pleasure to observe the spread of interest in this concept to a range of academic disciplines including applied psychology. Ironically, the main point of my commentary will be that the analysis of lean production in comparison with the previous dominant industrial paradigm of mass production is perhaps weakest in the area of ... applied psychology! However, before pursuing this point let me briefly summarise my view of the current degree of adoption of lean production.

At the time we were conducting our research at MIT in the late 1980s, we could find practically no examples in Western companies in which all of the key elements of lean production—in the primary work team, in overall organisation of physical production, in the product design team, in the method of dealing with suppliers, and in the method of dealing with customers—were in place. Knowing the long history of the diffusion of mass production from the USA to both Japan and Europe and the many wrong steps along the path, our greatest concern was that the most useful aspects of Japanese innovations would be ignored or rejected by managers, union leaders, and employees in Western firms.

In 1995, the situation is much changed and our worst fears have not come to pass. Indeed, in researching a new book, which I and my colleague Professor Daniel Jones are just completing, we found many examples of the introduction of lean production all the way across the firm in the USA, several examples in the UK, and initial success in a few well-known firms in Germany. What is more, we believe that the rate of diffusion is accelerating as firms discover that the substitution of lean production for mass production is one of the few alternatives open to them in the current era of low growth and economic stagnation.

On the other hand, the old mass production paradigm is still the dominant one in the bulk of industrial enterprises, as shown by a key indicator. The aggregate level of inventories, and in particular inventories of partially finished goods within the production process, has not fallen significantly in any Western economy when the level of inventories are adjusted for the business cycle. Inventories are the best output indicator to track because truly lean production, in which there is no rework, little scrap, and a continuous flow of the product from raw material into the hands of the customer, inherently has low inventories.

By contrast, surveys showing significant levels of "teamwork" and "empowerment" on the shop floor are in my experience largely worthless. Managers and union leaders now feel they are not conducting their affairs properly unless something that can be called a work team is in place, but the term is so diffuse that true empowerment is hard to distinguish from a situation where nothing has changed except that the foreman is now called the team leader.

In summary, diffusion is proceeding but not at the pace indicated by some of the input measures commonly used by human resources specialists.

Now to the interesting question: Why is a system that demonstrably produces better results from the standpoint of the customer and for the wellbeing of the firm so hard to introduce rapidly and universally? I believe this is largely a matter of "psychology" in the sense of the psychic needs of the workforce. Let me explain, based on my direct observation of attempts to introduce lean production in Western mass-production firms in the period since the publication of *The machine that changed the world* (Womack, Jones, & Roos) in 1990.

First, let us remember that the basic objective in converting from mass to lean is to reorganise work by transferring indirect tasks (including a substantial portion of what used to be called "management") to the primary work team while linking the efforts of the teams working on a product so that the product moves quickly and without interruption from design to production launch and from raw material into the hands of the customer.

Because lean methods are much more efficient, the first conversion problem emerges immediately: fewer employees are needed to get the same number of products to customers. Management has two fundamental choices at this point: lay off workers or find new work by speeding up product development and finding new markets. The second choice is clearly the correct one because otherwise management is asking employees to cooperate in the task of eliminating their livelihoods. Lean production is by design a system that can only work with the active cooperation of the workforce so it should hardly be surprising when a firm makes a promising start but then cuts jobs and the initial progress melts away as the workforce drags its feet. Yet we see examples every day of firms that refuse to make job guarantees and watch in bewilderment when lean methods are rejected on the shop floor.

If a firm makes appropriate guarantees that the transition to lean production will not cost jobs (although some firms like those in the defence sector may still have to lay off workers if their traditional market shrinks dramatically before they can develop new products for other markets), it has taken a necessary step, but not a sufficient one.

The next hurdle is the problem of careers, because the creation of horizontal work teams threatens traditional career paths (e.g. the product engineer, the quality expert, the logistics expert, the maintenance expert, the skilled machinist or welder) and the move to eliminate re-work and all manner of "fire fighting" by uncovering the root causes of production problems, threatens the self-image of many professions and skilled trades.

In my view, the career imperative is not just driven by the perceived need for a portable skill but by an existential need for self-definition. Most Western workers, when asked to describe "Who are you?" in one sentence will give an answer related to their primary skill ("product engineer", "quality assurance specialist", "welder"). None is likely to say, "A member of Product Team A", and for the very simple reason that the primary work team has a short life, tied to the life of its product.

Japanese employees, when asked the same question will typically say "I'm a Toyota man" or "I'm a Matsushita engineer" which is a valuable advantage for Japanese firms. However, Western firms lack the primary identification to company, and need to give their employees some sense of who they "are" as they attempt a lean transition. In my and Dan Jones's article in the *Harvard Business Review* (March/April 1994) we try to provide this through the concept of the "alternating career" in which firms make clear to employees that their primary skill is vitally important for the success of work teams and that the firm will take on the responsibility of continually upgrading each employee's skills through rotation from team assignments to functional assignments (that is, back to quality control or logistics) and by making every primary production worker a "process expert" through continuous training in lean techniques.

Even in firms guaranteeing jobs and thinking about the psychic need of employees for careers, there is an additional psychic hurdle to overcome centring on the concept of "responsibility". In firms I have directly observed the ability of lean techniques employed by primary work teams to make the whole product development and production process transparent, so that the root cause of every problem can be identified, to raise concerns in the mind of primary workers. What life has taught many employees is that one of the best features of mass production is that problems are always a mystery and therefore no-one's fault. Exposing problems, by contrast, suggests that someone will be assigned the blame and punished.

Punishment is clearly not in management's interest because the inherent advantage of lean systems lies in the ability of work teams to identify problems quickly and fix them permanently. Thus the discovery of problems should be celebrated rather than punished. However, experience indicates that the management of a firm must practise this new theory for an extended period before the bulk of the workforce will begin to believe that the punitive practices of the past have been abandoned. Only then will they participate in problem solving with their full abilities and energy.

In summary, job guarantees, a new concept of careers, and a no-fault approach to a work team's responsibility for the results of its efforts are needed in order to spur the rapid adoption of lean techniques across the industrial landscape.

What is much less clear to me, but which is also a testable hypothesis, is the common perception—particularly in Europe—that workers gain substantial psychic satisfaction from long cycle times in production operations and from the ability to fabricate or assemble an entire product. From my personal observations, I have concluded that workers gain substantial satisfaction from understanding the entire process needed to create a product, from being directly involved in redesigning and improving this process, from the immediate feedback on improvements that *kaizen* can provide, and from job rotation. However, the ability to actually perform each step for each product and to do so in a long cycle is rarely stated to be important by the employees I have observed.

So let me close this commentary with an offer: I would be delighted to talk with any applied psychologists who are interested in survey research on just which features of lean production—including the job, career, and "responsibility" dimensions—are psychically satisfying and which (if any!) are widely found to be undesirable and therefore in need of *kaizen*.

REFERENCES

Womack, J.P. & Jones, D.T. (1994). From lean production to the lean enterprise. *Harvard Business Review, March/April*, 93–103.

Womack, J.P., Jones, D.T., & Roos, D. (1990). *The machine that changed the world.* New York: Macmillan.

[19]

THE INTERNATIONAL MOTOR VEHICLE PROGRAM'S LEAN PRODUCTION BENCHMARK: A CRITIQUE

by JAMES RINEHART

The Lean Production Benchmark

At the time of the first Binghamton conference in 1978 the Big Three automakers were operating with master agreements and pattern bargaining, and there were no Asian auto assembly transplants in North America. Shortly thereafter, increased imports, plant overcapacity, and recession set the stage for a major overhaul of labor-management relations, production processes, and work practices.

The 1979 Chrysler bailout marked the breakdown of pattern bargaining and the beginning of an era of concessions in the industry. Between 1982 and 1990 seven transplants or joint ventures opened in the United States and four in Canada. (One, Hyundai, is now closed.) While concession bargaining emerged in a period of economic crisis, the return of profitability did not deter the automakers' pursuit of concessions. With the arrival of the transplants, concessionary agreements, once restricted to wages, increasingly incorporated human resource policies and work practices modeled after Japanese automakers, especially Toyota. These included team concept; contingent compensation; outsourcing; collapsed job categories and increased flexibility of worker deployment; decentralization of quality responsibilities;

James Rinehart teaches sociology at the University of Western Ontario. He is the author of *The Tyranny of Work: Alienation and the Labour Process* and co-author of *Just Another Car Factory: Lean Production and its Discontents.*

participatory mechanisms; labor-management cooperation; and the elimination of work rules won in more prosperous years. While there are wide operational and management differences across companies and across plants owned by the same company, most Big Three auto plants in North America now can be characterized as lean production-mass production hybrids.

Academics associated with the International Motor Vehicle Program (IMVP), which is funded by every auto company in the world, are on a mission: to convince manufacturers to shift from mass to lean production and to convince everyone else that this is a good idea. In their influential book *The Machine that Changed the World* (1991), the IMVP trio of Womack, Jones, and Roos compared the operational efficiency of over ninety auto assembly plants in seventeen countries. They concluded that Japanese automakers in Japan, especially Toyota, followed by Japanese transplants, were the leanest, most efficient in the world. This research strategy exemplified what recently has become a lucrative business, viz., benchmarking studies. According to the largest management consulting firm in the world, Andersen Consulting, Inc., "benchmarking is the most powerful tool for assessing industrial competitiveness and for triggering the change process in companies by measuring such factors as productivity, work-in-process, defects, space utilization, amount of inventory, set-up times, and so on."

Womack and his MIT colleagues evangelically promoted lean production as not only the one best way to produce vehicles but also as a worker-friendly system. They stressed the system's precision and flexibility, tight inventories, quick die changes, and low per-unit assembly hours. They also made exaggerated, unsubstantiated claims about its benefits for workers: extensive training, multiskilling, challenging work, empowerment, and harmonious labor-management relations. These humane arrangements arise not just from management's employment of sound human resource policies. The fragility of a lean system, especially its just-in-time (JIT) deliveries and production processes and workers' quality responsibilities, allegedly obligates management to use practices that commit workers to the company and its objectives. Like it or not, lean companies have got to be nice to their workers.

IMVP'S "Retreat"

In their recent book *After Lean Production* (1997), Kochan, Lansbury, and MacDuffie retreated from the unsubstantiated argument advanced by their IMVP colleagues, viz., that lean production provides challenging jobs performed by multiskilled workers in a high-trust, participatory environment. Now the IMVP group simply maintains that the optimal efficiency of lean manufacturing techniques *requires* "high involvement" work practices and human resource policies. These are teams; extensive job rotation; problem-solving groups; suggestion programs; inspection by production workers; hiring criteria that stress willingness to learn and get along with others rather than prior manufacturing experience; long hours of training; contingent compensation (merit pay, profit-sharing, bonuses); and few status differentials (no special cafeterias or parking lots, for example). While IMVP's retreat from its earlier position on lean production's impact on workers is welcome, it is open to criticism.

First, many of IMVP's involvement criteria give no indication of involvement. For example, plants get high involvement scores on the team criterion solely on the basis of having a large percentage of their workforce organized in teams. There is no measure of the degree of team autonomy. We concluded from our longitudinal, in-plant study of CAMI, a GM-Suzuki auto assembly plant heralded as a lean production showcase, that there was little scope for team autonomy.[1] Standardized, short-cycled, line-paced work, JIT pull processes and low inventory, as well as several layers of production management precluded any but the most routine kinds of team discretion. CAMI is not exceptional in this regard. Teams in most of the world's auto plants, including the leanest, have only a fraction of the discretion enjoyed by workers in some Scandinavian factories.[2]

The same criticism applies to training. Ratings are based on hours rather than content, and most of the training in lean plants is ideological and pertains to "soft skills." Ditto for job rotation. For IMVP, the more extensive the job rotation the higher the involvement score, but there is no measure of the complexity of the jobs or the skills demanded of the jobs through which workers rotate. Job rotation involves routine variety and workers who are multitasked, not multiskilled.[3]

Nowhere in the literature is there evidence of multiskilled production workers in lean plants.

Second, the existence of IMVP's "high involvement" practices does not preclude a stressful work environment. Independent of the extent to which enlightened human resource practices exist, lean production manufacturing techniques such as JIT, the reduction of buffers, the elimination of off-line sub-assembly stations, highly standardized, short-cycled jobs, and the continuous reduction of non-value-added labor invariably increase the pace and intensity of work. CAMI has a full complement of "high involvement" work practices and human resource policies, but most job cycles range from one and a half to three minutes, and some of the most contentious issues are time study, job standards, and workloads.[4]

IMVP does not measure workloads, work intensity, job complexity, overtime demands, or health and safety problems, especially repetitive strain injuries (RSIs). Recently the Canadian Auto Workers (CAW) conducted its own benchmarking studies. The union's benchmarking criteria, however, reflected workers' rather than management's perspectives.

The first CAW survey was based on the responses of nearly 2,500 workers in GM, Chrysler, Ford, and CAMI auto assembly plants in Canada. It found every plant had become leaner over the past five or six years, but GM had moved the fastest and the farthest, followed by Ford and Chrysler. CAMI was lean to begin with. JIT procedures were implemented, buffers were reduced; work-in-process was maximized; job content was increased; workers' performance was electronically monitored; and there was a continuous effort to reduce waste, i.e., non-value-added labor.[5] Slightly over 70 percent of the respondents said their workload had increased in the last two years. Workers in the leanest plants, GM and CAMI, said they had the heaviest and fastest work, and they reported the highest degree of physical health risks, exhaustion, and stress. Workers at GM, followed by those at CAMI, were substantially more likely than their counterparts at Ford and Chrysler to report low levels of autonomy and control. So much for the empowered workers of lean production.[6]

The CAW also surveyed 1,670 workers employed in the independent auto parts sector in Canada. The companies were placed in one of four categories, ranging from traditional

Fordist to lean. It is worth quoting at length the study conclusions: "[C]ompared with workers in traditional Fordist-style plants, those at lean companies reported their workload was heavier and faster. They reported workloads were increasing and becoming even faster. They did not report it was easier to change things they did not like about their job. They did report that it was becoming more difficult to get time off and were more likely to have to find a replacement worker before they could go to the washroom. They were more likely to report that they would be unlikely to maintain their current work pace until age 60."[7]

Conditions like these have been observed in lean plants throughout the world. There is no more revealing evidence of lean production's negative impact on workers than recent developments at Toyota in Japan. Due to tight labor markets and worker discontent, Toyota modified key elements of lean by relaxing JIT procedures, moving tasks from the main to sub-assembly lines, and by adding buffers between the main lines.

It is difficult to see how even the most enlightened human resource policies might ease workers' daily burdens in lean factories. As we observed in our book, "Wherever it operates, lean production strives to operate with minimal labor. It is a system that aspires to eliminate all buffers save one: an understaffed workforce that is expected to make up for production glitches, line stoppages, unbalanced production scheduling, and injured or absent workers through intensified effort and overtime. The true buffers in this system are workers."

Third, the "high involvement" practices specified by IMVP do not ensure, let alone correlate strongly with, optimal efficiency. At Toyota in Japan, the undisputed champion of lean production, job design is Taylorist, production is planned and managed from the top, and all employees experience "relentless" pressure. Improvement activity (kaizen) is done mainly by production managers, engineers, and team leaders, as workers are too busy with their jobs to be heavily involved.[8] Conversely, efficiency can be achieved without "enlightened" human resource policies. Big Three plants in Canada, for example, rank high on productivity and quality without many of the "high involvement" practices specified by IMVP.[9] IMVP gave high marks on efficiency and quality to the Ford plant in Hermosillo, Mexico. In fact, Womack, Jones, and Roos stated that the plant's quality "was better than that of the best Japanese plants and the best North American transplants." However, there is little evidence of enlightened human resource policies, and labor-management relations are highly conflictual. The plant has been hit by strikes, boycotts, and line stoppages, and labor turnover ranges between 25 and 44 percent.

Evidence from around the world indicates that lean production is most advanced in greenfield sites (where plant location and new workers are carefully screened) and where companies face severe competition, shrinking markets, overcapacity, and falling profits, i.e., where companies' threats of employment cutbacks and/or plant closure are not idle. Relatedly, lean production is least advanced where threats to jobs do not exist and where the union movement or local is strong and militant. Unless we believe these unions are ignorant of the true character of lean production, their scepticism of and opposition to this system are indicative of its detrimental effects on workers.

NUMMI and Saturn as Model Plants

It is surprising that even some left-leaning academics portray NUMMI (jointly owned by GM and Toyota, and run on Toyota principles) as a model plant. For example, in her recent book *Farewell to the Factory: Auto Workers in the Late 20th Century* (1997), Ruth Milkman unfavorably contrasts Big Three plants with NUMMI, attributing NUMMI's efficiency and high-quality products to shopfloor harmony and a participatory environment. There are several problems with this argument.

First, it ignores Toyota's acknowledged superior capacity to produce vehicles that are easy to manufacture. Ease of manufacture means vehicles that have a minimum number of components to be assembled, which translates into low assembly hours required to build a vehicle. Second, Milkman ignores the long periods of joblessness experienced by NUMMI workers, most of whom were thrown out of work for several years by the closure of the GM plant that now houses NUMMI.[10] It is unrealistic to think that workers' "cooperation" is not influenced by this experience and fears of another plant shutdown. Third, Milkman's enthusiasm for participation (even supposing NUMMI's brand is the genuine article) is puzzling in that she cites much evidence showing, at best, a weak relationship between participation and productivity. Fourth, labor-management cooperation is far from perfect. Petitions have protested understaffing and the company's strict absenteeism policy. The rank and file have elected union candidates who ran on slates criticizing NUMMI's heavy workloads, speed-up, a high rate of RSIs, and management favoritism. The plant was hit by a walk-out in 1994. Finally, almost everything written on NUMMI has come from one enthusiastic source, viz., Paul Adler, a professor in the Department of Management Organization at the University of Southern California.

Academics associated with the IMVP portray the Saturn plant in Spring Hill, Tennessee, as a "post-lean" model of efficiency and organizational governance. Saturn is a team concept plant that uses lean manufacturing processes, but its most unusual feature is the extensive web of collaboration between the company and the union. Proponents call it "the boldest experiment in comanagement in the United States today." The company operates with joint union-management decision-making bodies at various organizational levels, and the union is involved in decisions concerning supplier and dealer selection, choice of technology, and product development. At the departmental level, managerial responsibilities are not handled by foremen and general foremen, but by "partnered" union reps and managers. That one half of middle managers are union members "challenges long-held beliefs about the separation of management and labor."

Many of the problems associated with and criticisms leveled against the labor-management cooperative agreements of the 1920s apply to Saturn. How can a union represent and promote the independent interests of workers when it is so heavily involved in managing workers and supporting whatever it takes to increase output and keep labor costs low? To the extent that union representatives are compromised by management and adopt managerial attitudes, workers' interests will be subordinated to the achievement of goals set by the company.

This is not a hypothetical dilemma. Surveys and focus group interviews of workers revealed a build-up of dissatisfaction with the union's capacity to adequately represent its membership. Rank and file dissatisfaction with the union is reflected by the presence of oppositional caucuses and hotly-contested union elections. Shopfloor pressure was successful in changing union crew leader selection from appointment to election and empowering those crew leaders to file members' grievances. Even Solidarity House now questions the union local's ability to effectively address members' problems and is opposed to extending this kind of union-management partnership arrangement to other plants.

With Saturn sales slipping and workers' bonuses falling, the union local held a referendum in March 1998 that could have abolished the collaborative agreement in favor of a traditional UAW-GM contract. Workers voted two-to-one in favor of maintaining the special arrangements. The vote, it appears, was not an endorsement of labor-management cooperation. Rather, it reflected workers' fears about job security. The president of the local warned that changing the contract, which guarantees no lay-offs except for cataclysmic events or severe economic conditions, would result in 2,700 workers being laid off.

Conclusions

Concessions in the U.S. auto industry began as an attack on wages but quickly embraced restructured production processes and work practices. With the arrival of Japanese transplants, concessions-driven bargaining increasingly incorporated elements of lean manufacturing processes, modes of organizing and deploying workers, and outsourcing. Despite the vigorous efforts of IMVP to promote lean production as mutually beneficial to labor and capital, rapidly accumulating evidence from around the world indicates otherwise. Lean production has failed miserably to live up to its promise of humanizing the workplace. This has become so obvious that IMVP has had to retreat from its original claim of an invariable connection between lean manufacturing techniques and enlightened human resource policies and work practices. Now IMVP simply argues that the full advantages of lean manufacturing procedures can only be realized under enlightened management policies, but their definition of enlightened, as we have seen, is highly questionable.

IMVP portrays NUMMI and Saturn as models of efficiency and harmonious labor-management relations. That progressive academics have begun to endorse these two plants says a great deal not only about the influence of the

IMVP but also about the sorry state of work and industrial relations in the U.S. auto industry.

NOTES

1. We found nothing in the lean production work process that necessitated teams. They were not technically required but a product of social engineering. Nearly all operations in the plant could be done without a team. See J. Rinehart, C. Huxley, and D. Robertson, *Just Another Car Factory: Lean Production and its Discontents* (Ithaca: Cornell University Press, 1997).
2. Paul Adler, a proponent of lean production, admits that neither individual workers nor work teams have much autonomy under lean production. However, he argues that worker autonomy is obsolete. Due to *interdependence* of workers arising from the "deepening" social and technical division of labor, the issue now is the "much more modest question of participation." (See Adler's article in *Perspectives on Work*, vol. 1, no. 1, pp. 61-65.) Adler's logic is similar to that of economists who, with each increase in the unemployment rate, raise what they consider to be the natural and, hence, acceptable level of joblessness.
3. Academics associated with IMVP invariably refer to workers in lean plants as multiskilled, as if such repetition will make it so.
4. The existence of such conflicts speaks volumes about the appropriateness of calling lean production post-Fordist.
5. An example of a reduction of non-value-added labor is the elimination of the steps an assembler takes to pick up a bumper that is to be attached to a car. The attachment adds value but the steps do not.
6. See David Roberston, et al., "The CAW Working Conditions Study: Benchmarking Auto Assembly Plants," (CAW-Canada, North York, Ontario, 1996).
7. See Wayne Lewchuk and David Robertson, "Working Conditions Under Lean Production: A Worker-based Benchmarking Study," *Asia Pacific Business Review*, Summer 1996.
8. This description of Toyota is from Mitsuo Ishida's article in *After Lean Production*.
9. That efficiencies can be realized in the absence of even the most genuinely progressive human resource practices is no mystery. Most workers take pride in doing a good job and are concerned about making quality products. Moreover, there are functional alternatives to commitment, alternatives that produce behaviors similar to those exhibited by committed workers. These *compliant* behaviors arise from the more coercive sources of "motivation," such as high unemployment, higher underemployment, job insecurity, etc.
10. NUMMI chronicler and enthusiast Paul Adler refers to NUMMI workers as motivated by "a mature sense of realism" and an understanding that they must earn their money "in the old-fashioned way." This is another way of saying that NUMMI workers are terribly insecure.

[20]

Human Relations, Vol. 50, No. 5, 1997

Lean Production: Denial, Confirmation or Extension of Sociotechnical Systems Design?

Ben Dankbaar[1,2]

This paper makes a comparison between the basic elements of lean production and sociotechnical systems design (STSD) and compares them both with the characteristics of the traditional Fordist system of mass production. It argues that lean production can hardly be considered as an alternative to mass production, as its proponents suggest, but is on the contrary extending the life of mass production methods. However, lean production does appear to contain some building blocks for the innovative production systems that are expected to prevail in the 21st century. STSD, which has always presented itself as an alternative and possible successor to Fordist methods, will need to link its traditional concerns for quality of work and flexibility of work organizations with the new issues of continuous improvement, learning, and innovation.

KEY WORDS: sociotechnical systems design; lean production; automotive industry; Japan; team work; Fordism.

INTRODUCTION

For some time now, the business press is announcing the definitive demise of the traditional system of work organization based on the principles developed by Taylor and Ford. In its well-known study of the automobile industry, a research team of the Massachusetts Institute of Technology (MIT) has presented the Japanese production system, which they call "lean production," as the historical successor to the system of "mass production," that was developed by Henry Ford in the early decades of this century: ". . . we believe, lean production will supplant both mass production and the remaining outposts of craft production in all areas of industrial endeavor to become the standard global production system of the twenty-first century" (Womack et al., 1990, p. 278). The message that

[1]Nijmegen Business School, University of Nijmegen, Nijmegen, The Netherlands.
[2]Requests for reprints should be addressed to Ben Dankbaar, Nijmegen Business School, University of Nijmegen, P.O. Box 9108, NL-6500 HK, Nijmegen, The Netherlands.

Ford's system ("Fordism") is dead is not particularly new. It has been around at least since the mid-1970s (Dankbaar, 1993). In Europe, particularly in Northern and Western Europe, sociotechnical systems design (STSD) had already presented itself as an alternative to the traditional Taylorist design of jobs and organizations since the 1950s. It is not surprising therefore that the understanding and implementation of "lean production" in Europe is often filled with sociotechnical elements (RKW, 1992; Heidenreich & Schmidt, 1993; Schumann, 1993). The MIT study, however, was highly critical of the sociotechnical approach, exemplified by the Volvo factory of Uddevalla. This plant represented in many ways the most radical break with Fordism in car production as it eliminated the moving assembly line completely and had a complete car assembled by small groups of workers (Ellegård et al., 1992; Sandberg, 1995). The MIT researchers, however, described the Uddevalla plant as a step backward in history, back to the traditions of craft production. And indeed, history seemed to agree with the MIT study as the Uddevalla plant was closed in 1993. Since then, other European car assembly plants have also moved away from production systems with stationary workplaces and longer work cycles. Thus, it seems that at the same time that Fordism is being replaced by another system, STSD as the longstanding apparent heir seems to be losing out. This paper discusses several questions in relation to this development. Can lean production (LP) be considered as an alternative approach to replace Fordism? How does it compare to the longstanding alternative STSD? Can STSD be subsumed under the Japanese model? Should STSD be critical of lean production? Or should it try to integrate elements of LP into its own approach? Clearly, the relevance of these questions extends beyond the car industry. Throughout this century the car industry has been a source of inspiration and model for the organization of production in other sectors and the impact of the lean production debate certainly shows that it still plays that role.

CHARACTERISTICS OF THE OLD SYSTEM

Mass production was introduced to the car industry by Henry Ford and involved among other things a continuously moving assembly line and highly mechanized production of standardized parts. The possibility to use expensive dedicated machinery to produce parts with a high degree of precision was central to the system, that relied on large numbers of unskilled or semiskilled workers. In this respect, Ford continued and perfected an American tradition of producing interchangeable parts (Hounshell, 1984). Whereas early car manufacturers had to make use of skilled "fitters" to work on individual parts and fit them together, Ford could use unskilled workers in assembly because the parts were always fitted. As a result, the workforce contained only relatively small numbers of skilled workers and engineers, mostly in supervisory and planning functions. The specialized machines producing parts generated large inventories. In order to keep costs under control, the range of products and therefore the range of choice for the customers was kept limited. This is exemplified by Henry Ford's famous dictum, that the customer could have a car in any color, as long as it was black (Lacey, 1986).

Just 2 years before the introduction of the first assembly line in the auto industry (1913), Frederick W. Taylor had published his "Principles of Scientific Management" (1911). Taylor's star had been rising rapidly in the first decade of the century and even though it lost some of its brilliance in later decades, the basic ideas of "scientific management" remained very much alive throughout the century. Taylorism stands for a new, systematic way of thinking about the organization of production by a new, rapidly growing group of professionals: the engineers (Noble, 1977). At the core of Taylorism, we find, first, a sharp division of labor between management (engineers) and workers, where workers only execute the tasks that have been designed for them by management; and, second, the notion that the best way to execute a job can be determined scientifically, especially by "time and motion" studies. Payment of the worker had to be related to this scientifically determined standard for a "fair day's work." Moreover, workers were to be selected scientifically, meaning that their skill level should be just enough for the job at hand. A whole tradition of testing grew out of this approach and the development of industrial psychology is closely related to the rise of scientific management.

In Ford's production system, elements of Taylorism were clearly present. What Ford added, however, was the importance of mechanization and the use of machinery to pace work. Time and motion studies were employed to determine manning levels for the assembly line. The machine ultimately set the pace of work at Ford. Workers lost the possibility of influencing the standard for a "fair day's work," because the pace of the line determined the volume of production. The strict separation of planning and execution of work resulted in a very high division of labor in production, based on the short cycle times of an assembly line moving at about 60 cars an hour, in combination with an extensive work planning and industrial engineering function. Production work that couldn't be planned in detail, usually involving skilled workers, for instance for quality inspection and maintenance, was organized in special so-called indirect departments, located more or less close to the areas where the "direct" work was carried out. Lower management had the task of translating and explaining the planning of work by the staff departments and controlling the workers.

Although Ford's system of mass production was modified to some extent by the introduction of a larger variety of cars, which led to General Motors overtaking Ford on the American market in the late 1920s, its basic principles were used to organize not only car manufacturing but all mass production of consumer goods all over the world after the Second World War. These basic principles of the "Fordist" production organization are in fact further elaborations and specifications of the old principle of the division of labor, which had been considered as the engine of economic growth since the days of Adam Smith. Smith had already argued that the size of the market was the only limit to the division of labor. The principles can be summarized in the following five points:

- economies of scale based on a high division of labor;
- separation of execution and control;
- short-cycled, machine-paced production work;
- functional specialization of support departments;
- specialized mechanization and/or dedicated automation.

The whole organization was oriented toward mass production, presumably at the lowest cost. As a result, quantity came before quality, in the sense that quality shortcomings were fixed later on in the production process by special repair workers. The ordinary workers on the line had no responsibility for quality and if provided with defective parts could either use them anyway or let the car go through without the part, so that it had to be fixed later on. By the late seventies, the number of repair workers in Western car plants had increased appreciably and they had almost become a kind of "hidden factory" inside the car factory.

WHAT DID THE SOCIOTECHNICAL APPROACH HAVE TO OFFER AS AN ALTERNATIVE?

The sociotechnical systems approach was originally developed in the British mining industry as a reaction to labor unrest and disappointing productivity in relation to mechanization (Emery, 1959; for a historical overview and extensive bibliography, see Van Eijnatten, 1993). The approach advocated a better balance between the social and the technical aspects of production systems and proposed the introduction of semi-autonomous work groups as a key element in high-motivation–high-productivity organizations. Instead of the Taylorist separation of preparation and execution, the sociotechnical approach advocated a unity of preparation, execution, and control at the lowest level possible in an organization. Autonomous work groups were responsible for the results of their work and they had to be given the means to realize that responsibility. The preconditions for such autonomous groups were to be found in a restructuring of production, e.g., by the creation of parallel flows of production, where each flow would be responsible for a complete product or range of products (group technology). This involved the elimination of functionally specialized departments in favor of product-flow-oriented groupings of machinery.

STSD became of practical importance in Europe and North America in the late 1960s and the 1970s. Twenty years after the Second World War, a new generation, better educated and without memories of the Great Depression, entered the factories. This resulted in a long period of labor unrest in many countries (1967–1972), which was frequently analyzed as a logical consequence of the gap between the ambitions and capabilities of the new generation of workers on the one hand, and the monotonous and degraded work in mass manufacturing on the other. This gap became also visible in less activist forms of protest: absenteeism and personnel turnover, as well as in a decreasing quality of products and services. Enterprises, especially the large mass manufacturing ones, were looking for alternatives to the Taylorist work systems. An important source of inspiration was provided by the sociotechnical systems approach, which became particularly successful in Northern Europe. It was supported there by both employers and trade unions as a way to improve motivation and give shape to industrial democracy on the shop floor.

As a result, STSD experiments in job and work re-design were carried out in a large number of enterprises. The STSD tradition became firmly associated with efforts to improve the quality of working life. An important element of this emphasis on the quality of work became the search for alternatives to the short-cycled work that had become characteristic of mass production assembly lines. Some of the most well-known experiments in this respect were undertaken by Volvo in Sweden (Berggren, 1991; Medbo, 1994). After the 1973 oil crisis and the subsequent change in the economic tide, the interest in these experiments began to diminish, and by the end of the 1970s many of them had disappeared again. Nevertheless, many practical experiences had been made, some experiments indeed became regular practice, and the STSD tradition survived in many places as a potential alternative to the traditional work organization. In the Scandinavian countries, Germany, The Netherlands and also Australia, STSD was further developed in various directions, some emphasizing the importance of participation in organization design; others the need to move beyond workplace reform to a more all-encompassing redesign of organizations (Van Eijnatten, 1993). Modern variants of sociotechnical design continue to incorporate the traditional interest in semi-autonomous groups as the basic unit of work organization. This involves a strong interdependence between flexible multiskill tasks within the group, flexible technical equipment, options for coordination, complete internal process control, participation in boundary control, and responsibility for operational and structural improvements and innovations.

STSD has always emphasized organizational solutions rather than technological solutions. Nevertheless, its organizational designs sometimes picked up technologies that appeared to fit them better than traditional technologies. Very early, for instance, STSD was interested in the use of Automated Guided Vehicles, which opened up more possibilities for stationary workplaces with longer work cycles (the Volvo Kalmar plant of 1974). In Volvo's Uddevalla plant too, several new tools were developed together with the new workplace design. Also, in the 1980s, some considered flexible automation an important additional ingredient for any alternative to the traditional Fordist model. This became an important element in the development of ideas about "anthropocentric production concepts," which were encouraged by the European Commission (Brödner, 1985; Wobbe, 1992). Flexible automation, including robotics, appeared to give a new impulse to the use of group technology (the grouping of machines needed to manufacture a specific family of products), which had always been a central element in sociotechnical (re)design of facilities for component manufacturing.

Thus, the sociotechnical alternatives to mass production emphasize the following principles:

- economies flowing from integration of tasks and self-regulation of work groups;
- unity of preparation, execution, and control;
- autonomous groups as the basic unit of organization;
- lengthening of individual work cycles, job enlargement, and job enrichment (self-regulation);
- organization around parallel product flows (group technology);
- flexible automation.

Over the years, STSD has remained an "alternative" approach, often identified with the movement for humanization of working life and therefore with organized labor. It was not always successful in its efforts to prove to management that humanization is possible without a loss in efficiency and indeed can even increase efficiency. In its Scandinavian and also its Australian form, STSD and humanization were closely tied to industrial democracy, which did not help to attract support from managements. In the Dutch STSD-tradition, the emphasis was more on designing organizations for flexibility, where humanization (e.g., the introduction of self-regulating work teams) appeared as a means rather than as an end in itself. The objective has become to find design principles that do not only lead to improvements in the quality of work, but also contribute to an increase in organizational flexibility and to a reduction of bureaucracy. The quality of work and the quality of the organization are seen as two sides of the same coin. The German tradition has given much attention and has limited itself largely to the design of group work (Binkelmann et al., 1993), which has been an important way to accommodate the relatively high level of skills available in the German manufacturing workforce.

CHARACTERISTICS OF LEAN PRODUCTION

In Japan too, Ford's principles were applied with fervor in the postwar construction of the automobile industry. In the course of time, however, and under the influence of circumstances that were unique to Japan, the industry developed its own rules and structures that differed decisively from the American example. As noted in our introduction, the MIT study on the automobile industry is now presenting this production system, developed in its most advanced form at Toyota, as a new, third phase in the history of car production: "lean production" (after craft production in the late nineteenth century and mass production in most of this century) (Womack et al., 1990).

The MIT study argues that lean production combines the advantages of craft and mass production, while avoiding the disadvantages of high cost of craft production and the rigidity of mass production. It makes optimal use of the skills of the workforce, by giving workers more than one task (multiskilling), by integrating direct and indirect work, and by encouraging continuous improvement activities (quality circles). As a result, lean production is able to manufacture a larger variety of products, at lower costs and higher quality, with less of every input, compared to traditional mass production: less human effort, less space, less investment, and less development time.

Ford had introduced the idea of a continuous flow of product in final assembly, but the parts and components that were added to the body in the course of the assembly process were produced in a large series, exploiting all possible economies of scale. As a consequence, the Fordist assembly lines were surrounded by large stockpiles of parts. In his critical search for "unnecessaries," Toyota engineer Taiichi Ohno in the late 1940s identified these inventories as "waste." Not only did the inventories tie up considerable sums of capital, but they also occupied space and, even more important, they encouraged a lax attitude toward quality. Ohno set out to eliminate this waste. The basic idea was that parts and components should only be produced in the quantity needed for current production. Ideally, there would be a single piece flow of parts and components in the direction of the assembly line, where they would arrive just in time to be used. This was of course an old idea, but Ohno went farther than anyone before in systematically eliminating obstacles on the road to single piece flow production (Cusumano, 1985; Ohno, 1978; Monden, 1983).

Of course, many preconditions had to be fulfilled for a "just-in-time" production system to run smoothly: technical, organizational, and social. In order to avoid a sudden surge or fall in demand for any part, special care was taken to smooth out the variety of products in final assembly as much as possible, i.e., a steady mix of vehicles would come off the line. Moreover, the lead times for the production of all parts were shortened tremendously. Producing small lots of components ran counter to the logic of mass production, where economies of scale would be reaped by running expensive (often dedicated) machinery for long periods without interruption. By various technical and organizational devices, however, setup times at Toyota were reduced to just a couple of minutes, which greatly reduced economical lot sizes.

The Fordist approach consisted of having specialized machines in specialized (functional) departments with specialized workers. Batches of workpieces were moved from one specialized department to the next for different operations to be performed. Toyota brought together all machines that were necessary to produce a certain part and placed them in a U-shaped line. Each machine was provided with a device that made it turn off automatically after completing its operation on one part. This was called "autonomation": the machine could operate autonomously, i.e., untended by the worker. A worker could walk from one machine to the next along the U-shaped line, starting an operation on the first machine, carrying the workpiece that he had just taken off that machine to the second operation, etc., so that after completing one round, one unit would be finished. In that way, one worker could operate a large number of machines, maybe as many as 15, simultaneously. With one worker, the cycle time for producing one unit is the time the worker needs to make one round. It is possible, however, to change the cycle time by changing the number of workers. Contrary to Fordist arrangements, workers in this environment had to be able to operate many different machines, i.e., they had to be "multiskilled." That doesn't necessarily mean that they had to be highly skilled. Learning to operate each machine might be a matter of hours. Nevertheless, more skills were required than in the traditional Fordist setup (Monden, 1983). Multiskilling also created possibilities for workers to fill in for each other in case of absenteeism.

A further deviation from the Fordist model, flowing logically from the workings of the just-in-time system, is that responsibility for quality is given to production. In the Fordist system, it had become the rule that separate inspectors and separate quality control departments were responsible for product quality. In the eyes of Taiichi Ohno, these people didn't add any value to the product, so they should wherever possible be eliminated. Workers should be encouraged to take responsibility for the quality of their own work. In the extreme case of one piece flow production, that is almost self-evident. Every time that a worker passes on a defective part, it will usually be noticed at the next station and production will stop. The defective part cannot be used and there are no other parts available. In other words, just-in-time stands for a quick feedback on quality defects that can then be corrected immediately. It is clear that the system can only work with very high levels of quality, because otherwise production would be continuously disrupted.

If workers are responsible for quality, they must have the means to assure that no incomplete or defective component leaves their workplace. A lot of attention was paid at Toyota to the development of techniques that would allow for the detection of defects or abnormalities and a subsequent stoppage of the line or machine. Some of these were technical devices of the type that also stopped a machine automatically after finishing a job. Others were simple statistical methods that allowed workers to check quality through sampling. For workers on a moving assembly line, quality control is difficult because they have only a given cycle time and then the workpiece is leaving their work area again. In traditional Fordist assembly lines, this leads to defective parts being consciously built into the car (to be removed again in later repair areas) or in incomplete cars rolling off the line. Toyota, however, gave its assembly line workers the possibility for stopping the entire line. Although this can be a costly operation, because it stops all work on the (relevant section of) the line, it does focus all attention on quality defects and makes sure that everyone is engaged in finding a quick and definitive solution.

The Toyota Production System did not in any way depart from the short-cycled, machine-paced work on the assembly line that had been introduced by Ford. Nevertheless, there is a different approach to the division of labor. Although functional specialization and division of labor are the main principles of organizational design, the organizational practice emphasizes cooperation between these departments and between individual production workers. This is a logical consequence of the undivided responsibility for quality and the very strong linkages created by just-in-time production logistics.

Lean production is concerned also with other elements of the organization of the overall production process, like the organization of product development and the relations with suppliers of components (see further below). Its principles regarding the organization of production, however, can be summarized in the following way:

- economies of cooperation;
- just-in-time logistics;
- responsibility for quality with production;
- self-inspection and multiskilling;
- group technology in components production;
- autonomation.

IS LEAN PRODUCTION AN ALTERNATIVE TO FORDISM?

If we compare the principles of lean production (LP) with those of STSD there are some obvious similarities. Can we now argue that LP is an alternative to Fordism in the same way that STSD is? This question is surrounded by much controversy. In the following we will argue that the question can be answered by "yes and no." We will start with the no.

Lean Production Is Not an Alternative to Fordism, But Its Most Perfect Form

LP has maintained some elements in the production organization that have been considered as defining elements of Fordism and certainly contributing to the low quality of work in Fordist production lines, first and foremost short-cycled work on the assembly line. Even more fundamental is the absence of autonomous teams in LP, which is a central concept in STSD.

In the STSD tradition, the idea of returning responsibility to production involved a clear reduction in the division of labor and has been expressed by the notion of "(semi)-autonomous" teams. The autonomy of working groups implies that they are responsible for the results of their activities (within the limits that technical and organizational requirements impose). Autonomous (or self-regulating) working groups have a certain measure of independence that is at the same time the source of their responsibility. Sometimes, this autonomy is accentuated by the presence of buffers that separate the group from problems and interruptions in preceding or following production units.

In the Japanese approach, this particular notion of autonomy is not present. There are no autonomous teams, because everything has been done to create an uninterrupted flow of product and all stocks and buffers have been eliminated. There is a strong emphasis on cooperation: cooperation between all workers and all functions. Contrary to both Fordism and STSD, LP sees no contradiction between standardized, short-cycled work and mutual support between production workers. In the same way, the clear distinctions between various functional departments do not appear to preclude close cooperation and interaction wherever this is necessary.

This practice of mutual support and cooperation has been called "teamwork," for instance, by the MIT-study, but it differs considerably from work in a team-based organization in the European and North American sociotechnical traditions. The Japanese notion of "teamwork" refers to a sense of responsibility for the whole enterprise ("Team Toyota"), and to mutual aid and off-line improvement activities (cf. JMA, 1985). It does not refer to working in teams. The limited importance of teams as an organizational device (as in the STSD tradition) is the reason, why in earlier descriptions of the Toyota production system by Monden (1983) and by Taiichi Ohno (1978) himself, the concept of teams is not present. The use of the expression "teamwork" by the MIT authors to denote the organization of work in Japanese car factories has therefore been a source of confusion. In particular, it has created the impression that creating autonomous teams in the STSD tradition was part of the Japanese model. This is more a proof of the impact of STSD on organizational thinking in the U.S. and Europe than a reference to reality in Japanese car firms.

The result of the specific arrangements of lean production is a much more flexible organization than the traditional Fordist organization, but with a quite low quality of work. In fact, it is sometimes argued that the high flexibility of LP systems has even lowered the quality of work compared to traditional short-cycled work on the assembly line. Whereas in the traditional Fordist system, a worker knows when he has finished his job, in the LP system a worker is never finished, because it is part of his job to help others, if he has finished his own task. This has been called "management by stress" (Parker & Slaughter, 1988; cf. also Fucini & Fucini, 1990; JAW, 1992; for an older account, see Kamata, 1973).

There are some good reasons, therefore, to argue that LP is not an alternative for Fordism, but rather a perfection of Fordism. Historically speaking, it can be considered as the Japanese version of Fordism. In fact, much of what is now considered typically Japanese could also be found in Henry Ford's approach of production (Dohse et al., 1985; Cole, 1989). The Japanese took Ford's ideas and applied and developed them in accordance with their needs and circumstances. We don't want to engage here in a scholastic debate about whether or not the Japanese have developed a system that is worthy to be called their own. The point to note is that their system resulted from a long period of intensive experimenting and learning, applying some basic methods and orientations of Taylorism and Fordism in an environment where American levels of mass production were unattainable. Appropriate methods were developed to achieve high levels of productivity and quality for lower levels and smaller series. By the time that Japanese industry reached "American" mass production levels, a system of production was in force that was already effective at much smaller production series and consequently offered enormous competitive advantages.

Recently, Japanese car manufacturers have been introducing various new elements into their production systems, aimed at improving the quality of work (Benders, 1996). More attention has been paid to ergonomics and a new layout of the production-line has introduced small buffers of three to five cars between various segments of the production line, which give the segments a small measure of autonomy. These changes have been made because of labor market shortages, making it difficult to attract and keep new young workers. Work therefore had to be made more bearable for older workers and women. Automation is also investigated again as a possible way out. Interestingly, these are the same reasons why Swedish car manufacturers became interested in STSD and in robotics in the 1970s. In a way, this only shows that LP is simply an advanced form of Fordism, which is running into the same limitations as Fordism did in the Western countries. In the Japanese context, multiskilling led to pressure on workers to take on the work of absentee co-workers, which in turn led to strong peer pressure to reduce absenteeism (Cole, 1989). In the Swedish context, multiskilling was introduced to make work more attractive, which was expected to reduce absenteeism, because people would be happy to go to work. Obviously, the Japanese approach can function only if no clear exit options are available for workers. The Swedish approach, on the other hand, can at least hope to provide an intrinsic motivation for workers to show up at work every day.

LP has managed to maintain some of the main characteristics of traditional mass production, while producing a substantial variety of products. Thus, LP is providing the flexibility that STSD was promising, but in a conventional framework with a lower quality of work. So far, STSD has not been able to convince the car industry that it can offer similarly flexible organizational forms for mass production (in assembly) with a higher quality of work at the same cost. The closure of the Uddevalla plant has not been supportive in that respect. It appears that (some of) the principles of STSD only find support in a mass production environment, if the labor market situation strongly supports them. Apart from the fact that unemployment levels in virtually all industrialized countries have now risen to levels which appear to exclude easy exit options for assembly workers, a new option has become available to enforce traditional production regimes. Over the past decade, an increasing number of sophisticated production facilities for manufacturing of cars and other complex consumer products have been located in poor, high unemployment regions of Europe, North America, and Asia. Whereas in the past, workers from these regions were brought to the traditional production locations in the industrial centers of the advanced countries, now the workers can stay at home and the factory is moved. Improvements in the means of communication and transportation as well as high levels of automation have made this possible. Relocation of the industry to another region or country therefore appears to have become an increasingly more viable alternative to a redesign of the production system.

Lean Production Is Not the Successor of Fordism, But It Does Contain Some Building Blocks for the Organizational Forms of the Twenty-First Century

So far, we have neglected some quite important elements of LP. We focused on some traditional elements of the discussion about production organization: the division of labor, direct and indirect tasks, length of the work cycle, teamwork, etc. We did not discuss the parts of LP that are connected with the concept of continuous improvement. It was possible to do so, because the idea of continuous improvement is quite far removed from traditional Fordism. Henry Ford of course was interested in technological change and constantly searching for new techniques, but his production organization was not focused on constant change, but on the stability of mass production of uniform products. It had come into existence in a Taylorist environment, which emphasized that there is a best way of doing everything. Lean production, on the contrary, emphasizes that there is always a better way.

In the Japanese approach, there is always room for further improvement, not just through new or improved machinery but also through a further refinement of manual motions, a new layout or new economies in the use of materials. The notion of continuous improvement (kaizen) refers in particular to the latter kind of small, incremental improvements. Such improvements are not imposed one-sidedly on production by some far-away engineering department, but they are developed and proposed by individual workers as well as in meetings of small groups of workers (quality control circles). It is quite normal in Japanese factories to find workers doing time-and-motion studies with stop-watch and all. The first-line supervisors play an important role in encouraging workers to come up with improvement ideas and are also themselves actively submitting proposals. Higher level supervisors and engineers are also heavily involved in kaizen activities, which are obviously more encompassing than the improvement ideas of individual workers. A quality circle selects a problem and develops a solution that is submitted for approval and advice. Technical support may be forthcoming from engineering staff. A well-developed structure of evaluation committees ensures that proposals are dealt with quickly and seriously. In this way, a process of continuous improvement is kept in motion, that makes optimal use of the practical insights and ingenuity of the workforce. Workers are not only engaged in the rationalization of their own work process, but they are actually said to be motivated by management's recognition of their capabilities. In fact, it can be argued that the motivational aspect of kaizen at the worker level is more important to the enterprise than the improvements themselves.

These aspects of lean production have brought Adler to characterize lean production as a "democratic version of Taylorism" (Adler, 1992). Moreover, Adler and Cole have argued that the Taylorist paraphernalia of standardized work descriptions and time-and-motion studies are essential in order to support and organize continuous learning. The combination of the requirement that all work is performed according to standards with the requirement that improvements in standards are continuously made, leads to a necessity for continuous communication between work groups, so that improvement ideas spread through the organization and are implemented everywhere. The result is what Cole and Adler call a "learning bureaucracy" (Adler & Cole, 1993). They argue that the design of work organization in plants like Uddevalla may have stimulated learning effects at the individual level, but had no comparable mechanisms for transferring and storing new ideas across the organization. Even if one may doubt the actual learning effects taking place in short-cycled assembly work (as is done by the supporters of the Uddevalla approach, cf. Berggren 1994), the point that modern production systems need to install mechanisms for the transfer and storage of knowledge (knowledge management) is certainly well taken.

Kaizen is only one, important part of the overall organization for continuous change in the LP system. Another part is the organization of new product development (Clark & Fujimoto, 1991). This aspect of LP was discovered relatively late by the outside world. While everyone (including the MIT researchers) was concentrating on the Japanese organization of production work (the core of the Fordist system), the real sources of high productivity and competitiveness were more likely to be found in the organization of cooperation between marketing, design, development, and production departments. Whereas the Fordist emphasis on progress by specialization had resulted in specialized departments that did not communicate, the Japanese had introduced very effective communication lines, based on development teams (teams indeed here!) led by strong project managers. The cooperation between manufacturing and development is especially important in this respect. The notion of simultaneous engineering (having overlapping stages of product and process development) was not invented in Japan (but in the U.S. aerospace industry), but it was certainly perfected there.

A third component of the organization for change is product development in close cooperation with suppliers (Lamming, 1993). This is another aspect of LP that is now attracting much attention. For historical reasons, the Japanese manufacturers have always been less vertically integrated than their Western competitors. The introduction of just-in-time logistics obviously necessitated close cooperation with the suppliers and the exchange of information, for instance on quality. In the course of time, this has led to suppliers (the so-called first-tier main suppliers of complete subsystems of the car) also becoming involved in product development.

Kaizen, quality circles, simultaneous engineering, and intensive cooperation with suppliers (co-makership) together make up a powerful machinery for continuous improvement, continuous innovation, continuous learning, and continuous adaptation to change. From that perspective, lean production does indeed look like the system for the twenty-first century, at least if one is willing to accept the current view of the business environment in the future. Although the new century will naturally also bring still unknown challenges, it is widely recognized that enterprises in the advanced parts of the world will have to engage in innovation-based competition. Being efficient, producing high quality, and showing flexibility is not enough. Innovation, be it product innovation or process innovation, will be the basis for any competitive edge. Mobilization of all the knowledge available in the enterprise will be necessary and lean production clearly has shown some ways and means of achieving that purpose.

CONCLUSIONS

We have argued that lean production is providing car manufacturers with the needed flexibility, without giving up the advantages of the traditional system of mass production, including the undeniable productivity of the assembly line. In that sense, LP is not the end of mass production, but is on the contrary extending the life of mass production methods in more turbulent markets with a larger variety of products. STSD has not been able to make a big impact on the mass production of cars. The case against short-cycled work has apparently only been convincing where it was backed up by shortages on the labor market. As long as the industry remains an (increasingly) global industry, such shortages are increasingly less likely to occur. Besides that, in spite of the greater turbulence and higher product variety, the car industry remains in many ways a stable environment, which will always remain more susceptible to the methods of mass production.

Given that mass production is a diminishing part of economic activities in the advanced countries and employing fewer and fewer people, this is no reason for despair on the part of the proponents of STSD. The fact that lean production appears to prevail in car manufacturing can hardly be accepted as proof that it represents the production system of the twenty-first century. As the most perfect form of the production system of the twentieth century, however, it may well contain some building blocks for the innovative production systems that will prevail in the twenty-first century. Just as the "American System of Manufacturing" contained some major building blocks for Ford's system, so can lean production offer some clues for the "learning organizations" of the next century. STSD would be well advised to pick up these clues, if it still wants to make a claim for the succession of Fordism. STSD has always been concerned with quality of work and flexibility of work organizations. The real challenge now is to link these to the new issues of continuous improvement, learning, and innovation.

STSD may satisfy itself that it can match LP in terms of creating flexible organizations (with a higher quality of work), but it can still learn quite a bit about the design of innovative organizations. If it wants to remain a relevant tradition, it had better learn those lessons. If it does, because of its attention to the quality of work, it may show itself to be far more capable of creating high-motivation–high-innovation organizations than lean production ever will.

REFERENCES

ADLER, P. S. The learning bureaucracy: New United Motors Manufacturing, Inc. In B. M. Staw and L. L. Cummings (Eds.), *Research in organizational behavior.* Greenwich CT: JAI Press, 1992.

ADLER, P. S., & COLE, R. E. Designed for learning: A tale of two auto plants. *Sloan Management Review,* Spring 1993, 85-94.

BENDERS, J. Leaving lean? Recent changes in the production organization of some Japanese car plants. *Economic and Industrial Democracy,* 1996, *17*(1), 9-38.

BERGGREN, C. *Von Ford zu Volvo. Automobilherstellung in Schweden.* Berlin: Springer Verlag, 1991.

BERGGREN, C. Nummi vs. Uddevalla. *Sloan Management Review,* Winter 1994, 37-49.

BINKELMANN, P., BRACZYK, H. J., & SELTZ, R. (Eds.), *Entwicklung der Gruppenarbeit in Deutschland.* Frankfurt/New York: Campus, 1993.

BRÖDNER, P. *Fabrik 2000.* Berlin: Edition Sigma, 1985.

CLARK, K. B., & FUJIMOTO, T. *Product development performance. Strategy, organization and management in the world auto industry.* Boston: Harvard Business School Press, 1991.

COLE, R. E. *Strategies for learning.* Berkeley: University of California Press, 1989.

CUSUMANO, M. A. *The Japanese automobile industry. Technology and management at Nissan and Toyota.* Cambridge: Harvard University Press, 1985.

DANKBAAR, B. *Economic crisis and institutional change.* Maastricht: Universitaire Pers Maastricht, 1993.

DOHSE, K., JÜRGENS, U., & MALSCH, Th. From Fordism to Toyotism? The social organization of the labor process in the Japanese automobile industry. *Politics and Society,* 1985, *14*(2), 114-146.

EIJNATTEN, F. van. *The paradigm that changed the work place.* Assen: Van Gorcum, 1993.

ELLEGÅRD, K., ENGSTRÖM, T., & NILSSON, L. *Reforming industrial work. Principles and realities in the planning of Volvo's car assembly plant in Uddevalla.* Stockholm: Arbetsmiljöfonden, 1992.

EMERY, F. E. *Characteristics of sociotechnical systems.* London: Tavistock, doc. 527, 1959.

FUCINI, Joseph J., & FUCINI, S. *Working for the Japanese. Inside Mazda's American auto plant.* New York: The Free Press, 1990.
HEIDENREICH, M., & SCHMIDT, G. Gruppenarbeit im internationalen Vergleich. In P. Binkelmann, H.-J. Braczyk, and R. Seltz (Eds.), *Entwicklung der Gruppenarbeit in Deutschland.* Frankfurt/New York: Campus, 1993, pp. 105-146.
HOUNSHELL, D. A. *From the American system to mass production 1800–1932.* Baltimore/London: The John Hopkins University Press, 1984.
JAW (Confederation of Japan Automobile Workers Unions). *Japanese automobile industry in the future. Toward coexistence with the world, consumers and employees.* Tokyo: JAW, 1992.
JMA (Japan Management Association). *KANBAN. Just-in-time at Toyota.* Cambridge, MA: Productivity Press, 1989 (Japanese edition: 1985).
KAMATA, S. *Japan aan de lopende band.* Amsterdam: Jan Mets, 1986 (Japanese edition: 1973).
LACEY, R. *Ford. The men and the machine.* New York: Ballantine Books, 1986.
LAMMING, R. *Beyond partnership. Strategies for innovation and lean supply.* New York: Prentice Hall, 1993.
MEDBO, L. *Product and process descriptions supporting assembly in long cycle time assembly.* Göteborg: Chalmers University of Technology, 1994.
MONDEN, Y. *Toyota production system. Practical approach to production management.* Norcross, GA: Institute of Industrial Engineers, 1983.
NOBLE, D. F. *America by design.* Oxford: Oxford University Press, 1977.
OHNO, T. *Das Toyota-Produktionssystem.* Frankfurt & New York: Campus, 1993 (Japanese edition: 1978).
PARKER, M., & SLAUGHTER, J. *Choosing sides: Unions and the team concept.* Boston: South End Press, 1988.
RKW (Rationalisierungskuratorium der deutschen Wirtschaft). *Lean production. Tragweite und Grenzen eines Modells.* Eschborn: RKW, 1992.
SANDBERG, A. (Ed.). *Enriching production. Perspectives on Volvo's Uddevalla plant as an alternative to lean production.* Aldershot: Avebury, 1995.
SCHUMANN, M. Gruppenarbeit und neue Produktionskonzepte. In P. Binkelmann, H.-J. Braczyk, and R. Seltz (Eds.), *Entwicklung der Gruppenarbeit in Deutschland.* Frankfurt/New York: Campus, 1993, pp. 186-203.
WOBBE, W. *What are anthropocentric production systems? Why are they a strategic issue for Europe?* Luxembourg: European Communities (Science and Technology Policy Series), 1992.
WOMACK, J. P., JONES, D. T., & ROOS, D. *The machine that changed the world.* New York: Rawson Associates, 1990.

BIOGRAPHICAL NOTE

BEN DANKBAAR studied economics and social sciences at the University of Amsterdam. He taught economics at the Faculty of Social Sciences at that same University from 1976 to 1982. From 1982 to 1988, he worked on several research projects concerning developments in the automobile industry at the Science Centre Berlin. From 1988 until 1995, he was program leader at the Maastricht Economic Research Institute on Innovation and Technology (MERIT), where he was in charge of research in the field of technology, work, and organization. Research topics included the organizational aspects of electronic data interchange, technology management in enterprises, regional networks of enterprises, training, and organizational change. He is now a professor of business organization at the Nijmegen Business School of the University of Nijmegen. Ben Dankbaar has been a consultant to the European Commission on matters concerning the automotive industry and has led several large international research consortia.

[21]

Work, employment and society

Volume 17(3): 569–573
[0950-0170(200309)17:3;569–573;034899]
SAGE Publications
London, Thousand Oaks,
New Delhi

Review Essay

Workers, unions and the high performance workplace

Andy Danford
University of the West of England, UK

Eileen Appelbaum, Thomas Bailey, Peter Berg and Arne L. Kalleberg
Manufacturing Advantage: Why High Performance Work Systems Pay Off
Ithaca and London: Cornell University Press, 2000, paperback £19.95, 259 pp.

Gregor Murray, Jacques Bélanger, Anthony Giles and Paul-André Lapointe (eds)
Work and Employment Relations in the High-Performance Workplace
London and New York: Continuum, 2002, paperback £19.99, 244 pp.

The high performance workplace (HPW) is the latest derivative of the lean production model that, like such predecessors as the 'high commitment' organization or 'world class manufacturing', is supposed to encapsulate current shifts in the organization of the labour process and employment relations. Whether it is a term that will stay with us or fade away as a result of the ever mutating business and management discourse remains to be seen. For the moment it has become a salient topic for public policy debates on the competitiveness of British industry.

Like human resource management the concept, if not the practice, of the HPW arrived on these shores from the USA and so it is no coincidence that both books in this review have been written by researchers based mostly in North America. Although the different authors have contrasting theoretical positions they share similar definitions of the HPW concept itself. At its core is the notion of a more systematic mobilization of tacit knowledge and worker discretion by managerial practices (high performance work systems) that permit workers to participate in decisions that affect their organizational routines. In the interests of organizational flexibility and efficiency, workers in the HPW should experience more autonomy over job tasks and working methods and enjoy a greater input into managerial decision-making processes through extensive systems of organizational communications. Appelbaum et al. place more emphasis upon the adoption of clusters of practices that provide workers with the opportunity to participate in work decisions, whilst many of the authors in Murray et al.'s edited collection focus on the centrality of social adhesion in fashioning the HPW employment relationship.

Appelbaum et al.'s monograph provides the more substantial empirical base. Relying primarily on quantitative data collected through structured interviews with workers and managers in three sectors in the USA (steel, apparel and medical electronic instrumentation) the authors use multiple regression analysis to investigate the impact of high performance work systems (HPWS) on organizational performance and a variety of worker outcomes. Some of these workers are employed by firms that adopt HPWS practices and others are employed by non-HPWS firms.

Before summarizing the results of this research it may be instructive to review the authors' theoretical framework. Appelbaum et al. use a conceptual framework that posits a series of links between plant performance, discretionary effort, motivation, skills and worker participation. Underpinning this is a theory of 'systems models' or 'horizontal fit' that emphasizes both internal coherence between work organizational and human resource practices and a fit between the use of 'bundles' of practices and exogenous factors such as market and product flexibility. The authors locate their theoretical position along a continuum of historical patterns of management attempts to secure greater worker output and commitment though non-authoritarian means. Thus, HPWS constitute the latest and most sophisticated pattern of work reform that has direct antecedents in the human relations movement associated with the Hawthorne experiments of the late 1920s and the group relations movement associated with the Tavistock Institute and others in the 1950s.

For this reviewer two problems emerge from this theoretical positioning. First, virtually no space is allowed for critiques of the ideas underpinning the authors' framework or, indeed, for the possibility that the management practices associated with HPWS reflect relations of power, exploitation and control at work. Perhaps if the authors had at least considered the work of Braverman (1974: 140, 141) on this subject then we might have been reminded that the recurrent interest in securing worker cooperation through work reform is rarely concerned with the degradation of men and women in capitalist society but instead with the management problems caused by worker reactions to that degradation. Second, and the corollary of this, is the authors' quite naked managerialist assertions of the need for HPWS clusters to enhance organizational efficiency. For example, they argue that the use of incentive systems, production teams and self-hiring practices can reduce 'individual shirking and free-rider problems' (p. 33). An alternative perspective might consider the right of those who labour under HPWS regimes to take a breather now and again. Similarly, they argue for more rigorous selection practices to capture those with the 'appropriate knowledge, skills and abilities to function effectively in an HPWS' (p. 41). Again, an alternative perspective might ponder the nature of those who get excluded from this recruitment strategy, trade unionists for example? This bias towards HPWS also features in some of the statistical analysis and I shall come to that in a moment.

The presentation of the results of the research – and most of the book is devoted to this – is excellent. Three chapters set up sequential 'case studies' of the industrial sectors and each provides contextual material for the quantitative analysis that makes up the remainder of the book. Information on the shifting economic and industrial environments affecting the participating firms is included along with descriptions of the particular HPWS practices utilized. There then follows a concise and very useful chapter that summarizes the main findings from the quantitative research. Some readers might be tempted to stop there but this would be a mistake because these summaries obscure the impressive depth of analysis that follows as well as some of the problems contained in it. The quantitative results are divided into three main chapters that measure the impact of HPWS on organizational performance, worker outcomes and earnings. This review will concentrate on worker outcomes since it is this theme that may be of most interest to readers of *Work, Employment and Society*.

Appelbaum et al. operationalize the core employee dimensions of HPWS by creating a summative scale that they term 'opportunity to participate'. This comprises survey questions on autonomy in decision-making, membership of self-directed teams, membership of off-line continuous improvement teams and extent of communications with peers, managers and technical experts. A series of multiple regressions of trust, intrinsic rewards, commitment, job satisfaction, job stressors and overall job stress are then presented using 'opportunity to participate' and other components of HPWS as independent variables. It must be said that both the description of the statistical method and the systematic presentation of regression models are first class and would be of great interest to students of multivariate analysis as well as researchers in the field. Nevertheless, there are some problems of researcher partiality. One example of this is the impact of the 'opportunity to participate' scale. Most of the regressions show that this scale has a significantly positive impact on worker outcomes such as trust and commitment and this result is pivotal to the authors' core argument. At various points in the book the authors emphasize the centrality to HPWS of two key components of this 'opportunity to participate' scale: participation in self-directed teams and continuous improvement groups. However, the regressions also show – and the authors understate this – that membership of such teams has a much smaller impact than the more nebulous 'autonomy in decision-making' variable and that in some firms and sectors team membership is likely to have a negative impact on worker outcomes. This leaves us with the unremarkable finding that job autonomy, rather than teamworking, can have a positive impact on workers' sense of trust, commitment and satisfaction. Another example is that additional job security variables seem to be far more powerful predictors of positive worker outcomes than 'opportunity to participate'. Yet this is hardly mentioned in the analysis despite emerging research showing that HPWS practices go hand in hand with downsizing and redundancies and thus give rise to job *insecurity* (Biewener, 1997; Osterman, 2000).

Murray et al.'s edited collection provides some substantial theoretical and empirical contributions to the HPW research agenda. It is recommended reading for researchers and students of contemporary work restructuring. The book addresses three themes that are viewed as critical to an understanding of the conditions that support or constrain the emergence of HPW as a new work paradigm: the shape of the new production model, the implications of the model for workplace relations, and the required institutional configurations that might support the new model. Chapters by different researchers based in Canada provide a mix of theoretical and institutional frameworks whilst Eileen Appelbaum (with a summary of the *Manufacturing Advantage* research) and Paul Edwards et al. provide empirical analysis based on workplace surveys. It is not possible here to summarize all of the material so this review will focus on some of the arguments contained in two key theoretical chapters: Jacques Bélanger, Giles and Murray; and Paul-André Bélanger, Lapointe and Lévesque.

Drawing on the traditions of the Regulation School these different authors claim that we can now discern core principles of an emerging model of production, a model that is significantly different from its Fordist predecessor. It is argued that the institutional stability of the Fordist regime of accumulation gave way to instability, multiple tensions and an eventual crisis of efficiency and worker alienation. In these conditions a 'new paradigm', based on managerial initiative, an unbridled neo-liberalism and new forms of work organization characterized by principles of flexibility and worker empowerment, was essential to overcome the contradictions of the Fordist regime. Of course, we have heard all of this paradigmatic talk before through the debates on 'Japanization', TQM, business process re-engineering and the like. Jacques Belanger et al. claim that what makes the high performance workplace really new is its emphasis upon meaningful job participation (as operationalized by Appelbaum et al.) and social adhesion.

Social adhesion refers to the processes by which employees and their trade unions become committed to organizational objectives, a necessary precondition to the mobilization of tacit knowledge and worker discretion. In considering this the authors raise a number of 'tensions' and 'contradictions' in social adhesion at work which for this reviewer are more profound. For example, the implications of the fundamentally unequal capitalist employment relationship; the transfer of substantial risk to employees in the form of task accretion, employment contract flexibility, and job insecurity; the likelihood of 'disappointed outcomes' for employees; and the erosion of the social base of independent union representation.

The role of trade unions in the high performance workplace is tackled directly by Paul-André Bélanger, Lapointe and Lévesque. These authors argue that assuming a workplace union is in a position of organizational strength then its strategic choice can be reduced to two stark options: either to oppose managerial initiatives and focus on distributive bargaining or to adopt an autonomous, proactive position that involves 'exchanging' employee involvement for the acquisition of new partnership rights. In other words, the union abandons its opposition to direct employee involvement in return for greater union input into managerial decision-making. In supporting the latter option Paul-André Bélanger et al. are quite clear as to what this might mean. That is, weakening shopfloor rules that protect workers and constrain managerial prerogatives; shifting the goal of collective bargaining to work organization and the management of the firm; and developing strong cooperation between managers and workplace unions. Without providing much supporting evidence the authors claim that in some areas of Canada, and particularly Quebec, industrial relations are now marked by a sharing of power, unions are more concerned with a firm's economic performance and unions and workers are now performing duties that used to be the prerogative of management. Having stated this, the authors then observe that such strategic engagement creates tensions for unions and their traditional role as the independent representatives and defenders of workers. Again, for this reviewer these 'tensions' amount to something more profound than this. For example, 'how can [unions] take part in strategic management without being ensnared in the logic of management, a logic that subordinates social performances (employment, wages and working conditions), and thus workers' interests, to the demands for return on investment and profitability of capital?' (p. 163). Or, for example, 'can real participation that actually has an influence on decision making be compatible with a weakened union "partner"?' (p. 163). Well quite. The authors address these problems by arguing for the development of an alternative logic of management that subordinates the profitability of capital and shareholders' earnings to social performance and workers' interests, and for national and regional institutional arrangements that encourage long-termism and labour-management cooperation in corporate governance. This is a laudable thesis but once again we have heard much of it before, whether in the form of the Swedish model, Modell Deutschland, the Japanese model, or the flexible specialization of the Third Italy (which it particularly resembles). The problem for advocates of HPWS and the new workplace partnership is that although nobody is denying that anything has changed there is no real evidence of a *paradigmatic* break with the capitalist logic of 'Fordist' or 'Taylorist' work techniques. Neither is there any evidence that the dominance of maximizing profits and shareholder value in Western firms is about to give way to anything more favourable to worker interests.

References

Biewener, J. (1997) 'Downsizing and the New American Workplace: Rethinking the High Performance Paradigm', *Review of Radical Political Economics* 29(4): 1–22.

Braverman, H. (1974) *Labor and Monopoly Capitalism. The Degradation of Work in the Twentieth Century*. New York and London: Monthly Review Press.

Osterman, P. (2000) 'Work Reorganisation in an Era of Restructuring: Trends in Diffusion and Effects on Employee Welfare', *Industrial and Labor Relations Review* 54(2): 179–200.

Part III
McDonaldisation

The McDonaldization Thesis:

Is expansion inevitable?

George Ritzer

University of Maryland

abstract: The issue addressed here is whether the spread of rationalization, or what I have termed 'McDonaldization', is inexorable. That issue is examined spatially and temporally. Spatially, what we are witnessing is the increasingly global dissemination of rationalization, much of it emanating from the United States. The temporal growth is discussed from the point of view of the expansion of McDonaldization to birth and before and death and beyond; rationalization is coming to contain the entire life course (and more). Evidence of expansion does not, of course, necessarily mean that growth is inevitable. There are some hopeful social developments and there are some things that individuals can do, but it is difficult at this point to envision anything powerful enough to stem the tide in the direction of rationalization.

keywords: birth ✦ death ✦ globalization ✦ McDonaldization ✦ rationalization

The 'McDonaldization thesis' (Ritzer, 1983; 1993; 1996) is derived, most directly, from Max Weber's (1968[1921]) theory of the rationalization of the Occident and ultimately the rest of the world (Kalberg, 1980). Weber tended to see this process as inexorable, leading, in the end, to the iron cage of rationalization from which there was less and less possibility of escape. Furthermore, with the corresponding decline in the possibility of individual or revolutionary charisma, Weber believed that there was a decreasing possibility of the emergence of a revolutionary counter-force.

Time has been kind to the Weberian thesis, if not to the social world. Rationalization has progressed dramatically in the century or so since Weber developed his ideas. The social world does seem to be more of an iron cage and, as a result, there does seem to be less possibility of escape.

International Sociology ✦ September 1996 ✦ Vol 11(3): 291–308
SAGE (London, Thousand Oaks, CA and New Delhi)

And, it does appear less likely that any counter-revolution can upset the march toward increasing rationalization.

It is this theory and empirical reality that forms the background for the development of what has been termed the 'McDonaldization thesis'. This thesis accepts the basic premises of rationalization as well as Weber's basic theses about the inexorable character of the process. Its major point of departure from the Weberian theory of rationalization is to argue that the paradigm of the process is no longer, as Weber argued, the bureaucracy, but it is rather the fast-food restaurant. The fast-food restaurant has combined the principles of the bureaucracy with those of other rationalized precursors (for example, the assembly line, scientific management) to create a particularly powerful model of the rationalization process. It is a relatively new paradigm, traceable to the opening of the first of the McDonald's chain in 1955. While there were a number of predecessors to the first McDonald's outlet in the fast-food industry, it is McDonald's that was the truly revolutionary development in not only that industry, but in the history of the rationalization process.

Embodying perfectly the principles of rationalization, McDonald's became the model to be emulated first by other fast-food chains and later by other types of chain stores. It was not long before the success of McDonald's caught the eye of those in other types of businesses, and ultimately in virtually every other sector of society. Today, not only is McDonald's a worldwide success, but it offers an alluring model to those in a wide variety of leadership positions. It is in this role that McDonald's is playing the key role in the still-further expansion of the process of rationalization. Indeed its participation is so central that the contemporary manifestations of this process can be aptly labelled 'McDonaldization'.

Like Weber I have tended to view this process as inexorable in a variety of senses. First, it is seen as migrating from its roots in the fast-food industry in America to other types of businesses and other social institutions. Second, McDonaldization is spreading from its source in the United States to affect more and more societies around the world. Third, McDonaldization is viewed as having first concentrated on the rationalization of processes central to life itself, but more recently it has moved to encompass the birth process (and before) as well as the process of death (and beyond).

To put this expansionism in more contemporary theoretical terms, McDonaldization is expanding in both space and time (Giddens, 1984; Harvey, 1989). Spatially, McDonaldization is encompassing more and more chains, industries, social institutions and geographic areas of the world. Temporally, McDonaldization has moved from the core of life itself both backward to the birth process as well as the steps leading up to it and forward to the process of dying and its aftermath.

The evidence on the spatial and temporal advance of McDonaldization is overwhelming. However, in this essay I want to do more than review this evidence, I want to reexamine the issue of inexorability. Do its past and present successes mean that McDonaldization is truly inexorable? Is there no hope that the process can be slowed down or even stopped? Is it possible to avoid an iron cage of rationalization that encompasses time (from birth and before to death and beyond) and space (geographic areas within the United States and throughout the world)? Before getting to these issues, I need to review the basic parameters of the McDonaldization thesis.

McDonaldization

I begin with a foundational definition:

> *McDonaldization is the process by which the principles of the fast-food restaurant are coming to dominate more and more sectors of American society, as well as of the rest of the world.*

The nature of the McDonaldization process may be delineated by outlining its five basic dimensions: efficiency, calculability, predictability, control through the substitution of non-human for human technology and, paradoxically, the irrationality of rationality.

First, a McDonaldizing society emphasizes *efficiency*, or the effort to discover the best possible means to whatever end is desired. Workers in fast-food restaurants clearly must work efficiently; for example, burgers are assembled, and sometimes even cooked, in an assembly-line fashion. Customers want, and are expected, to acquire and consume their meals efficiently. The drive-through window is a highly efficient means for customers to obtain, and employees to dole out, meals. Overall, a variety of norms, rules, regulations, procedures and structures have been put in place in the fast-food restaurant in order to ensure that *both* employees and customers act in an efficient manner. Furthermore, the efficiency of one party helps to ensure that the other will behave in a similar manner.

Second, there is great importance given to *calculability*, to an emphasis on quantity often to the detriment of quality. Various aspects of the work of employees at fast-food restaurants are timed and this emphasis on speed often serves to adversely affect the quality of the work, from the point of view of the employee, resulting in dissatisfaction, alienation and high turnover rates. Only slightly over half the predominantly part-time, teenage, non-unionized, generally minimum wage work force remains on the job for one year or more (Van Giezen, 1994). Similarly, customers are expected to spend as little time as possible in the fast-food restaurant. In fact, the drive-through window reduces this time to zero, but if the customers desire to eat in the restaurant, the chairs are designed to impel them to leave after about twenty minutes. All of this emphasis on speed clearly has a negative effect on the quality of the 'dining experience' at a fast-food restaurant. Furthermore, the emphasis on how fast the work is to be done means that customers cannot be served high-quality food which, almost by definition, requires a good deal of time to prepare.

Third, McDonaldization involves an emphasis on *predictability*. Employees are expected to perform their work in a predictable manner and, for their part, customers are expected to respond with similarly predictable behavior. Thus, when customers enter, employees will ask, following scripts (Leidner, 1993), what they wish to order. For their part, customers are expected to know what they want, or where to look to find what they want, and they are expected to order, pay and leave quickly. Employees (following another script) are expected to thank them when they do leave. A highly predictable ritual is played out in the fast-food restaurant and it is one that involves highly predictable foods that vary little from one time or place to another.

Fourth, there is great *control* in a McDonaldizing society, and a good deal of that control comes from non-human technologies. While these non-human technologies currently dominate employees, increasingly they will be replacing human technologies. Employees are clearly controlled by such non-human technologies as french-fry machines that ring when the fries are done and even automatically lift the fries out of the hot oil. For their part, customers are controlled both by the employees who are constrained by such technologies as well as more directly by the technologies themselves. Thus, the automatic fry machine makes it impossible for a customer to request well-done, well-browned fries.

Finally, both employees and customers suffer from the various *irrationalities of rationality* that seem inevitably to accompany McDonaldization. Many of these irrationalities involve the opposite of the basic principles of McDonaldization. For example, the efficiency of the fast-food restaurant is often replaced by the inefficiencies associated with long lines of people at the counters or long lines of cars at the drive-through window. While there are many others, the ultimate irrationality of rationality is dehumanization. Employees are forced to work in dehumanizing jobs and customers are forced to eat in dehumanizing settings and circumstances. In Harry Braverman's terms, the fast-food restaurant is a source of degradation for employees and customers, alike (Braverman, 1974).

Expansionism

McDonald's has continually extended its reach, within American society and beyond. As McDonald's Chairman put the company's objective, 'Our goal: to totally dominate the quick service restaurant industry worldwide ... I want McDonald's to be more than a leader. I want McDonald's to dominate' (Papiernik, 1994).

McDonald's began as a suburban and medium-sized-town phenomenon, but in recent years it has moved into big cities and smaller towns (Kleinfeld, 1985; Shapiro, 1990) that supposedly could not support such a restaurant, not only in the United States but also in many other parts of the world. A huge growth area is in small satellite, express or remote outlets opened in areas that are not able to support full-scale fast-food restaurants. These are beginning to appear in small store fronts in large cities, as well as in non-traditional settings like department stores and even schools. These satellites typically offer only limited menus and may rely on larger outlets for food storage and preparation (Rigg, 1994). McDonald's is considering opening express outlets in such locations as museums, office buildings and corporate cafeterias.

Another significant expansion has occurred as fast-food restaurants have moved onto college campuses (the first such facility opened at the University of Cincinnati in 1973), instead of being content merely to dominate the strips that surround many campuses. In conjunction with a variety of 'branded partners' (for example, Pizza Hut and Subway), Marriott now supplies food to almost 500 colleges and universities (Sugarman, 1995).

Another, even more recent, incursion has occurred: we no longer need to leave the highway to dine in our favorite fast-food restaurant. We can obtain fast food quickly and easily at convenient rest stops along the highway and then proceed with our trip. Fast food is also increasingly available *in* service stations (Chan, 1994). Also in the travel realm, fast-food restaurants are more and more apt to be found in hotels (McDowell, 1992), railway stations and airports and their products are even appearing on the trays of in-flight meals. The following newspaper advertisement appeared a few years ago: 'Where else at 35,000 feet can you get a McDonald's meal like this for your kids? Only on United's Orlando flights'. Now, McDonald's so-called 'Friendly Skies Meals' are generally available to children on Delta flights. In addition, in December, 1994, Delta began offering Blimpie sandwiches on its North American flights (*Phoenix Gazette*, 1994). (Subway sandwiches are also now offered on Continental flights.) How much longer before McDonaldized meals will be available on all flights everywhere by every carrier? In fact, on an increasing number of flights, prepackaged 'snacks' have already replaced hot main courses.

In other sectors of society, the influence of fast-food restaurants has been more subtle, but no less profound. While we are now beginning to see the appearance of McDonald's and other fast-food restaurants in high schools and trade schools (Albright, 1995), few lower-grade schools as yet have in-house fast-food restaurants, but many have had to alter school cafeteria menus and procedures so that fast food is readily and continually available to children and teenagers (Berry, 1995). We are even beginning to see efforts by fast-food chains to market their products in these school cafeterias (Farhi, 1990).

The military has been pressed into offering fast-food menus on its bases and on its ships. Despite the criticisms by physicians and nutritionists, fast-food outlets are increasingly turning up *inside* hospitals. No homes have a McDonald's of their own, but dining within the home has been influenced by the fast-food restaurant. Home-cooked meals often resemble those available in fast-food restaurants. Frozen, microwavable, and pre-prepared foods, also bearing a striking resemblance to McDonald's meals and increasingly modeled after them, often find their way to the dinner table. Then there is the home delivery of fast foods, especially pizza, as revolutionized by Domino's.

As powerful as it is, McDonald's has not been alone in pressing the fast-food model on American society and the rest of the world. Other fast-food giants, such as Burger King, Wendy's, Hardee's, Arby's, Big-Boy, Dairy Queen, TCBY, Denny's, Sizzler, Kentucky Fried Chicken, Popeye's, Subway, Taco Bell, Chi Chi's, Pizza Hut, Domino's, Long John Silver, Baskin-Robbins and Dunkin' Donuts, have played a key role, as have the innumerable other businesses built on the principles of the fast-food restaurant.

Even the derivatives of McDonald's and the fast-food industry more generally are, in turn, having their own influence. For example, the success of *USA TODAY* has led to changes in many newspapers across the nation, for example, shorter stories and color weather maps. As one *USA TODAY* editor put it: 'The same newspaper editors who call us McPaper have been stealing our McNuggets' (Prichard, 1987: 232–3).

Sex, like virtually every other sector of society, has undergone a process of McDonaldization. In the movie *Sleeper*, Woody Allen not only created a futuristic world in which McDonald's was an important and highly visible element, but he also envisioned a society in which even sex underwent the process of McDonaldization. The denizens of his future world were able to enter a machine called an 'orgasmatron' that allowed them to experience an orgasm without going through the muss and fuss of sexual intercourse. In fact, we already have things like highly specialized pornographic movies (heterosexual, homosexual, sex with children, sex with animals) that can be seen at urban multiplexes and are available at local video stores for viewing in the comfort of our living rooms. In New York City, an official called a three-story pornographic center 'the McDonald's of sex' because of its 'cookie-cutter cleanliness and compliance with the law' (*New York Times*, 1986: 6). The McDonaldization of sex suggests that no aspect of our lives is immune to its influence.

Is McDonaldization truly inexorable?

Given the preceding description of McDonaldization, the issue to be discussed in this closing section is whether or not the process is truly inexorable? I want to discuss this issue both spatially and temporally. First, there is the spatial issue of whether McDonaldization is destined to spread from its American roots and become a global phenomenon. Second, there is the temporal issue of whether McDonaldization will inevitably spread from its control over the core of life to colonize birth and before as well as death and beyond.

Globalization

We can discuss the first issue under the heading of globalization, or the spread of McDonald's, and more importantly the principles of McDonaldization, around the world. However, in using the term globalization here, it should be pointed out that, as we will see below, there are some differences between its usage here and the way it has been used in the currently voguish globalization theory.

While there are significant differences among globalization theorists, most if not all would accept Robertson's advocacy of the idea that social scientists adopt 'a specifically global point of view', and 'treat the global condition as such' (Robertson, 1992: 61, 64). Elsewhere, Robertson (1990: 18) talks of the 'study of the world as a whole'. More specifically, Robertson argues that we need to concern ourselves with global processes that operate in relative independence of societal sociocultural processes. Thus, Robertson (1992: 60) argues, 'there is a general autonomy and "logic" to the globalization process, which operates in *relative* independence of strictly societal and other conventionally studied sociocultural processes'. Similarly, Featherstone (1990: 1) discusses the interest in processes that 'gain some autonomy on a global level'.

While the reach of McDonaldization is global, it does not quite fit the model proposed by globalization theorists. The differences between them are clear when we outline those things rejected by globalization theorists:

1. A focus on any single nation-state.
2. A focus on the West in general, or the United States in particular.
3. A concern with the impact of the West (westernization) or the United States (Americanization) on the rest of the world.
4. A concern with homogenization (rather than heterogenization).
5. A concern with modernity (as contrasted with postmodernity).
6. An interest in what used to be called modernization theory (Tiryakian, 1991).

The fact is that while McDonaldization *is* a global process, it has all of the characteristics *rejected* by globalization theorists: it does have its source in a single nation-state; it does focus on the West in general and the United States in particular; it is concerned with the impact of westernization and Americanization on the rest of the world; it is attentive to the homogenization of the world's products and services; it is better thought of as a modern than a postmodern phenomenon (because of its rationality, which is a central characteristic of modernity) and it does have some affinity with modernization theory (although it is not presented in the positive light modernization theory tended to cast on all western phenomena). Thus, McDonaldization is a global phenomenon even though it is at odds with many of the basic tenets of globalization theory.

The global character of this American institution is clear in the fact that it is making increasing inroads around the world (McDowell, B. 1994). For example, in 1991, for the first time, McDonald's opened more restaurants abroad than in the United States (Shapiro, 1992). This trend continues and as we move toward the next century, McDonald's expects to build twice as many restaurants each year overseas than it does in the United States. Already, by the end of 1993 over a third of McDonald's restaurants were overseas. As of the beginning of 1995, about half of McDonald's profits came from its overseas operations. As of this writing, one of McDonald's latest advances was the opening of a restaurant in Mecca, Saudi Arabia (*Tampa Tribune*, 1995).

Other nations have developed their own variants of this American institution, as is best exemplified by the now large number of fast-food croissanteries in Paris, a city whose love for fine cuisine might have led one to think that it would prove immune to the fast-food restaurant. India has a chain of fast-food restaurants, Nirula's, which sells mutton burgers (about 80% of Indians are Hindus who eat no beef) as well as local Indian cuisine (Reitman, 1993). Perhaps the most unlikely spot for an indigenous fast-food restaurant was then war-ravaged Beirut, Lebanon; but in 1984 Juicy Burger opened there (with a rainbow instead of golden arches and J.B. the clown replacing Ronald McDonald) with its owners hoping that it would become the 'McDonald's of the Arab world' (Cowan, 1984).

Other countries not only now have their own McDonaldized institutions, but they have also begun to export them to the United States. For example, the Body Shop is an ecologically sensitive British cosmetics chain with, as of early 1993, 893 shops in many countries; 120 of those shops were in the United States, with 40 more scheduled to open that year (Elmer-Dewitt, 1993; Shapiro, 1991). Furthermore, American firms are now opening copies of this British chain, such as the Limited, Inc.'s, Bath and Body Works.

This kind of obvious spread of McDonaldization is only a small part of that process's broader impact around the world. Far more subtle, and important, are the ways in which McDonaldization and its various dimensions have affected the way in which many institutions and systems throughout the world operate. That is, they have come to adopt, and adapt to their needs: efficiency, predictability, calculability and control through the replacement of human by non-human technology (and they have experienced the irrationalities of rationality).

How do we account for the global spread of McDonaldization? The first and most obvious answer is that material interests are impelling the process. That is, there is a great deal of money to be made by McDonaldizing systems and those who stand to profit are the major motor force behind it.

Culture is a second factor in the spread of McDonaldization. There appears to be a growing passion around the world for things American and few things reflect American culture better than McDonald's and its various clones. Thus, when Pizza Hut opened in Moscow in 1990, a Russian student said: 'It's a piece of America' (*Washington Post*, 1990: B10). Reflecting on the growth of Pizza Hut and other fast-food restaurants in Brazil, the president of Pepsico (of which Pizza Hut is part) said of Brazil that his nation 'is experiencing a passion for things American' (Blount, 1994: F1). Many people around the world identify strongly with McDonald's; in fact to some it has become a sacred institution (Kottak, 1983). On the opening of the McDonald's in Moscow, one journalist described it as the 'ultimate icon of Americana', while a worker spoke of it 'as if it were the Cathedral in Chartres ... a place to experience "celestial joy"' (Keller, 1990: 12).

A third explanation of the rush toward McDonaldization is that it meshes well with other changes occurring in American society as well as around the world. Among other things, it fits in well with the increase in dual-career families, mobility, affluence and in a society in which the mass media play an increasingly important role.

A fourth factor in the spread of McDonaldization and other aspects of American culture (the credit card [Ritzer, 1995], for example), is the absence of any viable alternative on the world stage. The path to worldwide McDonaldization has been laid bare, at least in part, because of the death of communism. With the demise of communism the only organized resistance can come from local cultures and communities. While the latter can mobilize significant opposition, it is not likely to be nearly as powerful as one embedded in an alternate worldwide movement.

Given the spread of McDonaldization and the powerful reasons behind it, what can serve to impede this global development? First, there is the fact that many areas of the world offer little in the way of profits to those who push McDonaldization. Many economies are so poor that there is little to be gained by pushing McDonaldized systems on them. Other institutions within such societies may want to McDonaldize their operations, but they are likely to be so overwhelmed by day-to-day concerns that they will have little time and energy to overhaul their systems. Furthermore, they are apt to lack the funds needed for such an overhaul. Thus their very economic weakness serves to protect many areas of the world from McDonaldization.

Second, we cannot overlook the importance and resilience of local cultures. Globalization theorists, in particular, have emphasized the strength of such cultures. While it is true that McDonaldization has the power to sweep away much of local culture, it is not omnipotent. For example, while the eating habits of some will change dramatically, many others will continue to eat much as they always have. Then, even if the eating habits of an entire culture change (a highly unlikely occurrence), other aspects of life may be partly or even wholly unaffected by McDonaldization. It is also likely that too high a degree of McDonaldization will lead to a counter-reaction and a reassertion of local culture. Also worth mentioning are the many ways in which local cultures affect McDonaldizing systems, forcing them to adapt in various ways to local demands and customs (for example, as discussed above, the mutton burgers in India).

The combination of a comparative lack of economic incentive to the forces behind McDonaldization and the opposition of local cultures will serve to impede the global spread of McDonaldization. However, when a given local culture advances economically, those who profit from McDonaldization will begin to move into that domain. In such cases, only local resistance will remain as a barrier to McDonaldization. It seems clear that while some local cultures will successfully resist, most will fail. In the end, and in the main, the only areas of the world that will be free of McDonaldization are those which lack the economic base to make it profitable.

The only hope on the horizon might be international groups like those interested in health and environmental issues. McDonaldized systems do tend to pose health risks for people and do tend to threaten the environment in various ways. There has, in fact, been some organized opposition to McDonaldized systems on health and environmental grounds. One could envision more such opposition, organized on a worldwide basis, in the future. However, it is worth noting that McDonaldized systems have proven to be quite adaptable when faced with opposition on these grounds. That is, they have modified their systems to eliminate the greatest threats to their customers' health and the greatest environmental dangers. Such adaptations have thus far served to keep health and environmental groups at bay.

The Colonization of Life and Death

While spatial expansion is covered in the previous section under the heading of globalization, in this section I deal with temporal expansion.

McDonaldization first focused on a variety of things associated with *life*. That is, it is the day-to-day aspects of living: food, drink, clothing, shelter and so on that were initially McDonaldized. Firmly ensconced in the center of the process of living, McDonaldization has pressed outward in both directions until it has come to encompass as many aspects as possible of both the beginning (birth) and the end of life (death). Indeed, as we will see, the process has not stopped there, but has moved beyond what would, on first glance, appear to be its absolute limits to encompass (again, to the degree that such a thing is possible) 'pre-birth' and 'post-death'. Thus, this section is devoted to what might be termed the 'colonization' (Habermas, 1987) of birth (and its antecedents) and death (and its aftermath) by the forces of McDonaldization.

In recent years a variety of steps have been taken to rationalize the process leading up to birth: burgeoning impotence clinics (chains [Jackson, 1995], or soon-to-be chains); artificial (or, better, 'donor' [Baran and Pannor, 1989]) insemination; in vitro fertilization (DeWitt, 1993); surrogate mothers (Pretorius, 1994); 'granny pregnancies' (*Daily Mail*, 1994); home pregnancy and ovulation-predictor home tests (Cain, 1995); sex-selection clinics (Bennett, 1983); sex-determination tests like amniocentesis (Rapp, 1994); and tests including chorionic villus sampling, maternal serum alpha-feto-protein and ultrasound to determine whether the fetus is carrying such genetic defects as Down's syndrome, hemophilia, Tay-Sachs and sickle-cell disease. All of these techniques are collectively leading to 'high-tech baby making' (Baran and Pannor, 1989) which can be used to produce what have been called 'designer pregnancies' (Kolker and Burke, 1994) and 'designer babies' (Daley, 1994).

The rationalization process is also manifest in the process of giving birth. One measure of this is the decline in the very human and personal practice of midwifery. In 1900 about half of American births were attended by midwives, but by 1986 that had declined to only 4 percent (Mitford, 1993). Then there is the bureaucratization of childbirth. In 1900, less than 5 percent of US births took place in hospitals, by 1940 it was 55 percent, and by 1960 the process was all but complete with nearly 100 percent of births taking place in hospitals (Leavitt, 1986: 190).

Hospitals and the medical profession developed standard, routinized (McDonaldized) procedures for handling childbirth. One of the best known viewed childbirth as a disease (a 'pathologic process') and its procedures were to be followed even in the case of low-risk births (Treichler, 1990). First, the patient was to be placed in the lithotomy position, 'lying supine with legs in air, bent and wide apart, supported by stirrups' (Mitford, 1993: 59). Second, the mother-to-be was to be sedated from the first stage of labor on. Third, an episiotomy[1] was to be performed to enlarge the area through which the baby must pass. Finally, forceps were to be used to make the delivery more efficient. Describing this type of procedure, one woman wrote 'Women are herded like sheep through an obstetrical assembly line [needless to say, one of the precursors of McDonaldization], are drugged and strapped on tables where their babies are forceps delivered' (Mitford, 1993: 61). This procedure had most of the elements of McDonaldization, but it lacked calculability, but that was added in the form of the 'Friedman Curve' created in 1978. This curve envisioned three rigid stages of labor with, for example, the first stage allocated exactly 8.6 hours during which cervical dilation went from 2 to 4 cms (Mitford, 1993: 143).

A variety of non-human technologies (e.g. forceps) have been employed in the delivery of babies. One of the most widespread is the scalpel. Many doctors routinely perform episiotomies during delivery so that the walls of the vagina are not stretched unduly during pregnancy.

The scalpel is also a key tool in caesareans. A perfectly human process has come, in a large number of cases, to be controlled by this non-human technology and those who wield it (Guillemin, 1989). The first modern caesarean took place in 1882, but as late as 1970 only 5 percent of all births involved caesareans. The use skyrocketed in the 1970s and 1980s, reaching 25 percent of all births in 1987 in what has been described as a 'national epidemic' (Silver and Wolfe, 1989). (By 1989 there had been a slight decline to just under 24%.)

Once the baby comes into the world, there is a calculable scoring system, Apgar, used on newborns. The babies are given scores of 0 to 2 on five factors (for example, heart rate, color), with ten being the top (healthiest) score. Most babies have scores between 7 and 9 after a minute of birth; 8 to 10 after five minutes. Babies with scores of 0 to 3 are in distress.

We move now to the other frontier; from the process of being born to that of dying. The McDonaldization of death begins long before a person dies; it commences in the efforts by the medical system to keep the person alive as long as possible: the increasing array of non-human technologies designed to keep people alive; the focus of medicine on maximizing the *quantity* of days, weeks or years a patient remains alive and the lack of emphasis on the *quality* of life during that extra time; computer systems that assess a patient's chances of survival – 90%, 50%, 10%, and so on; and the *rationing* in the treatment of the dying person.

Turning to death itself, it has followed much the same path as birth. That is, it has been moved out of the home and beyond the control of the dying and their family members and into the hands of medical personnel and hospitals. Physicians have played a key role here by gaining a large measure of control of death just as they won control over birth. And death, like birth, is increasingly likely to take place in the hospital. In 1900, only 20 percent of deaths took place in hospitals, in 1949 it was up to 50 percent, by 1958 it was at 61 percent, and by 1977 it had reached 70 percent. By 1993 the number of hospital deaths was down slightly (65%), but to that must be added the increasing number of people who die in nursing homes (11%) and residences such as hospices (22%) (National Center for Health Statistics, 1995). Thus, death has been bureaucratized, which means it has been rationalized, even McDonaldized. The latter is quite explicit in the growth of hospital chains and even chains of hospices, using principles derived from the fast-food restaurant, which are increasingly controlling death. One result of all of this is the dehumanization of the very human process of death as we are increasingly likely to die (as we are likely to be born) impersonally, in the presence of total strangers.

However, even the best efforts of modern, rationalized medicine inevitably fail and patients die. But we are not free of McDonaldization even after we die. For example, we are beginning to witness the development of the changeover from largely family-owned to chains of funeral homes (Corcoran, 1992; Finn, 1991). The chains are leaping into this lucrative and growing market often offering not only funeral services, but cemetery property and merchandise such as caskets and markers.

Perhaps the best example of the rationalization of death is the cremation. It is the parallel to caesareans in the realm of birth. Cremations are clearly more efficient than conventional funerals and burials. Ritual is minimized and cremations have a kind of assembly-line quality; they lead to 'conveyor belt funerals'. Cremations also lend themselves to greater calculability than traditional funerals and burials. For example, instead of lying in state for a day, or more, the city of London crematorium has the following sign: 'Please restrict service to 15 minutes' (Grice, 1992: 10). Then there is the irrationality of the highly rational cremation which tends to eliminate much of the human ceremony associated with a traditional funeral-burial.

The period after one dies has been rationalized in other ways, at least to some degree. There are, for example, the pre-arranged funerals that allow people to manage their affairs even after they are dead. Another example is the harvesting of the organs of the deceased so that others might live. Then there is cryogenics where people are having themselves, or perhaps just their heads, frozen so that they might be brought back to life when anticipated advances in the rationalization of life make such a thing possible.

Given the rationalization of birth and before as well as death and beyond, are there any limits to this expansion? Several are worth mentioning:

- The uniqueness of every death (and birth): 'Every life is different from any that has gone before it, and so is every death. The uniqueness of each of us extends even to the way we die' (Nuland, 1994: 3).

- The often highly non-rational character of the things that cause death (and cause problems at birth):

> Cancer, far from being a clandestine foe, is in fact berserk with the malicious exuberance of killing. The disease pursues a continuous, uninhibited, circumferential, barn-burning expedition of destructiveness, in which it heeds no rules, follows no commands and explodes all resistance in a homicidal riot of devastation. Its cells behave like members of a barbarian horde run amok – leaderless and undirected, but with a single-minded purpose: to plunder everything within reach (Nuland, 1994: 207).

- Similarly, cancer is described as an 'uncontrolled mob of misfits', 'a gang of perpetually wilding adolescents', and the 'juvenile delinquents of cellular society' (Nuland, 1994: 208). If ever there was a daunting non-rational enemy of rationalization, cancer (and the death it often causes), is it.
- Midwifery has enjoyed a slight renaissance *because* of the dehumanization and rationalization of modern childbirth practices. When asked why they have sought out midwives, women complain about things like the 'callous and neglectful treatment by the hospital staff', 'labor unnecessarily induced for the convenience of the doctor', and 'unnecessary caesareans for the same reason' (Mitford, 1993: 13).
- The slight decline in caesareans is reflective of the growing concern over the epidemic of caesareans as well as the fact that the American College of Obstetricians came out for abandoning the time-honored idea, 'once a caesarean, always a caesarean'.
- Advance directives and living wills that, among other things, tell hospitals and medical personnel what they may or may not do during the dying process.
- The growth of suicide societies and books like Derek Humphrey's *Final Exit* that, among other things, give people instructions on how to kill themselves; on how to control their own deaths.
- The growing interest in euthanasia, most notably the work of 'Dr Death', Jack Kevorkian. Kevorkian's goal is to give people back control over their own deaths.

Conclusion

I have sought in this paper to discuss the spatial and temporal expansion of McDonaldization under the headings of globalization and the colonization of birth and death. It is abundantly clear that McDonaldization is expanding dramatically in terms of both time and space. However, there remains the issue of whether or not this growth is inexorable. A number of the barriers to, and limits on, the expansion of McDonaldization have been discussed in this paper. There clearly are such limits and perhaps more importantly, McDonaldization seems to lead to various counter-reactions that serve to limit this spread. The issue, of course, is whether or not these counter-reactions can themselves avoid being McDonaldized.

While there is some hope in all of this, there is not enough to allow us to abandon the Weberian hypothesis about the inexorable march toward the iron cage of, in this case, McDonaldization. In spite of this likely scenario, I think there are several reasons why it is important for people to continue to try to contain this process. First, it will serve to mitigate the worst excesses of McDonaldized systems. Second, it will lead to the discovery, creation and use of niches where people who are so inclined can escape McDonaldization for at least a part of their day or even a larger portion of their lives. Finally, and perhaps most important, the struggle itself is ennobling. As a general rule, such struggles are nonrationalized, individual, and collective activities. It is in such struggles that people can express genuinely human reason in a world that in virtually all other ways has set up rationalized systems to deny people the ability to behave in such human ways; to paraphrase Dylan Thomas, instead of going gently into that next McDonaldized system, rage, rage against the way it's destroying that which makes life worth living.

Note

1. An episiotomy is an incision between the vagina and the anus to enlarge the opening needed for a baby to pass.

References

Albright, M. (1995) 'INSIDE JOB: Fast-food Chains Serve a Captive Audience', *St Petersburg Times* 15 January: 1H.

Baran, A. and Pannor, R. (1989) *Lethal Secrets: The Shocking Consequences and Unresolved Problems of Artificial Insemination*. New York: Warner Books.

Bennett, N., ed. (1983) *Sex Selection of Children*. New York: Academic Press.

Berry, M. (1995) 'Redoing School Cafeterias To Favor Fast-Food Eateries', *The Orlando Sentinel* 12 January: 11.

Blount, J. (1994) 'Frying Down to Rio', *Washington Post-Business* 18 May: F1, F5.

Braverman, H. (1974) *Labor and Monopoly Capital: The Degradation of Work in the Twentieth Century*. New York: Monthly Review Press.

Cain, A. (1995) 'Home-Test Kits Fill an Expanding Health Niche', *The Times Union-Life and Leisure* (Albany, NY) 12 February: 11.

Chan, G. (1994) 'Fast-Food Chains Pump Profits at Gas Stations', *The Fresno Bee* 10 October: F4.

Corcoran, J. (1992) 'Chain Buys Funeral Home in Mt Holly', *Burlington County Times* 26 January.

Cowan, A. (1984) 'Unlikely Spot for Fast Food', *The New York Times* 29 April: 3: 5.
Daily Mail (1994) 'A New Mama, Aged 62', 19 July: 12.
Daley, J. (1994) 'Is Birth Ever Natural?', *The Times* 16 March.
DeWitt, P. (1993) 'In Pursuit of Pregnancy', *American Demographics* May: 48ff.
Elmer-Dewitt, P. (1993) 'Anita the Agitator', *Time* 25 January: 52ff.
Farhi, P. (1990) 'Domino's is Going to School', *Washington Post* 21 September: F3.
Featherstone, M. (1990) 'Global Culture: An Introduction', in M. Featherstone (ed.) *Global Culture: Nationalism, Globalization and Modernity*, pp. 1-14. London:Sage.
Finn, K. (1991) 'Funeral Trends Favor Stewart IPO', *New Orleans City Business* 9 September.
Giddens, A. (1984) *The Constitution of Society: Outline of the Theory of Structuration.* Berkeley: University of California Press.
Grice, E. (1992) 'The Last Show on Earth', *The Times* 11 January: 10.
Guillemin, J. (1989) 'Babies By Caesarean: Who Chooses, Who Controls?' in P. Brown (ed.) *Perspectives in Medical Sociology*, pp.549–58. Prospect Heights, IL: Waveland Press.
Habermas, J. (1987) *The Theory of Communicative Action. Vol. 2, Lifeworld and System: A Critique of Functionalist Reason*. Boston, MA: Beacon Press.
Harvey, D. (1989) *The Condition of Postmodernity: An Inquiry into the Origins of Cultural Change*. Oxford: Blackwell.
Jackson, C. (1995) 'Impotence Clinic Grows Into Chain', *The Tampa Tribune- Business and Finance* 18 February: 1.
Kalberg, S. (1980) 'Max Weber's Types of Rationality: Cornerstones for the Analysis of Rationalization Processes in History', *American Journal of Sociology* 85: 1145–79.
Keller, B. (1990) 'Of Famous Arches, Beeg Meks and Rubles', *The New York Times* 28 January: 1:1,12.
Kleinfeld, N. (1985) 'Fast Food's Changing Landscape', *The New York Times* 14 April: 3:1,6.
Kolker A. and Burke, B. (1994) *Prenatal Testing: A Sociological Perspective*. Westport, CT: Bergin and Garvey.
Kottak, C. (1983) 'Rituals at McDonald's', in M. Fishwick (ed.) *Ronald Revisited: The World of Ronald McDonald*, pp.52–8. Bowling Green, OH: Bowling Green University Press.
Leavitt, J. (1986) *Brought to Bed: Childbearing in America, 1750-1950*. New York: Oxford University Press.
Leidner, R. (1993) *Fast Food, Fast Talk: Service Work and the Routinization of Everyday Life*. Berkeley: University of California Press.
McDowall, B. (1994) 'The Global Market Challenge', *Restaurants & Institutions* 104, 26:52ff.
McDowell, E. (1992) 'Fast Food Fills Menu For Many Hotel Chains', *New York Times* 9 January: D1, D6.
Mitford, J. (1993) *The American Way of Birth*. New York: Plume. *New York Times* (1986) 5 October: 3: 6.
National Center for Health Statistics (1995) *Vital Statistics of the United States, 1992-1993, Volume II – Mortality, Part A*. Hyattsville, MD: Public Health Service.
Nuland, S. (1994) *How We Die: Reflections on Life's Final Chapter*. New York: Knopf.

Ritzer *The McDonaldization Thesis*

Papiernik, R. (1994) 'Mac Attack?', *Financial World* 12 April.

Phoenix Gazette (1994) 'Fast-Food Flights', 25 November: D1.

Pretorius, D. (1994) *Surrogate Motherhood: A Worldwide View of the Issues*. Springfield, IL: Charles C. Thomas.

Prichard, P. (1987) *The Making of McPaper: The Inside Story of USA Today*. Kansas City, MO: Andrews, McMeel and Parker.

Rapp, R. (1994) 'The Power of "Positive" Diagnosis: Medical and Maternal Discourses on Amniocentesis', in D. Bassin, M. Honey and M. Kaplan (eds) *Representations of Motherhood*, pp. 204–19. New Haven, CT: Yale University Press.

Reitman, V. (1993) 'India Anticipates the Arrival of the Beefless Big Mac', *Wall Street Journal* 20 October: B1, B3.

Rigg, C. (1994) 'McDonald's Lean Units Beef Up NY Presence', *Crain's New York Business* 31 October: 1.

Ritzer, G. (1983) 'The McDonaldization of Society', *Journal of American Culture* 6: 100-7.

Ritzer, G. (1993) *The McDonaldization of Society*. Thousand Oaks, CA: Pine Forge Press.

Ritzer, G. (1995) *Expressing America: A Critique of the Global Credit Card Society*. Thousand Oaks, CA: Pine Forge Press.

Ritzer, G. (1996) *The McDonaldization of Society*, rev. edn. Thousand Oaks, CA: Pine Forge Press.

Robertson, R. (1990) 'Mapping the Global Condition: Globalization as the Central Concept', in M. Featherstone (ed.) *Global Culture: Nationalism, Globalization and Modernity*, pp.15–30. London: Sage.

Robertson, R. (1992) *Globalization: Social Theory and Global Culture*. London: Sage.

Shapiro, E. (1991) 'The Sincerest Form of Rivalry', *The New York Times* 19 October: 35, 46.

Shapiro, E. (1992) 'Overseas Sizzle for McDonald's', *The New York Times* April 17: D1, D4.

Shapiro, L. (1990) 'Ready for McCatfish?', *Newsweek* 15 October: 76–7.

Silver L. and Wolfe, S. (1989) *Unnecessary Cesarian Sections: How to Cure a National Epidemic*. Washington, DC: Public Citizen Health Research Group.

Sugarman, C. (1995) 'Dining Out on Campus', *Washington Post/Health* 14 February: 20.

Tampa Tribune (1995) 'Investors with Taste for Growth Looking to Golden Arches', *Business and Finance* 11 January: 7.

Thomas, D. (1952) 'Do Not Go Gentle into That Good Night', in D. Thomas *The Collected Poems of Dylan Thomas*, p.128. New York: New Directions.

Treichler, P. (1990) 'Feminism, Medicine, and the Meaning of Childbirth', in M. Jacobus, E. Keller and S. Shuttleworth (eds) *Body Politics: Women and the Discourses of Science*, pp. 113–38. New York: Routledge.

Van Giezen, R. (1994) 'Occupational Wages in the Fast-Food Industry', *Monthly Labor Review* August: 24–30.

Washington Post (1990) 'Wedge of Americana: In Moscow, Pizza Hut Opens 2 Restaurants', 12 September: B10.

Weber, M. (1968[1921]) *Economy and Society*. Totowa, NJ: Bedminster Press.

Biographical Note: George Ritzer is Professor of Sociology at the University of Maryland. Among his many books are *Sociology: A Multiple Paradigm Science* (1975), *Metatheorizing in Sociology* (1991), *Expressing America: A Critique of the Global Credit Card Society* (1995), as well as the book from which this essay is derived, *The McDonaldization of Society*, revised edition (1996), which has been, or soon will be, translated into 10 languages.

Address: Department of Sociology, University of Maryland, College Park, MD 20742, USA.

Name Index